Lecture Notes in Mathematics

Volume 2276

This series reports on new developments in all areas of mathematics and their applications - quickly, informally and at a high level. Mathematical texts analysing new developments in modelling and numerical simulation are welcome. The type of material considered for publication includes:

1. Research monographs
2. Lectures on a new field or presentations of a new angle in a classical field
3. Summer schools and intensive courses on topics of current research.

Texts which are out of print but still in demand may also be considered if they fall within these categories. The timeliness of a manuscript is sometimes more important than its form, which may be preliminary or tentative.

More information about this series at http://www.springer.com/series/304

Emmanuel Peyre • Gaël Rémond
Editors

Arakelov Geometry and Diophantine Applications

 Springer

Editors
Emmanuel Peyre
Institut Fourier
Université Grenoble Alpes
Grenoble, France

Gaël Rémond
Institut Fourier
Université Grenoble Alpes
Grenoble, France

ISSN 0075-8434   ISSN 1617-9692   (electronic)
Lecture Notes in Mathematics
ISBN 978-3-030-57558-8   ISBN 978-3-030-57559-5   (eBook)
https://doi.org/10.1007/978-3-030-57559-5

Mathematics Subject Classification: Primary: 14G40; Secondary: 11G35, 14G25, 11G50

This Springer imprint is published by the registered company Springer Nature Switzerland AG.
The registered company address is: Gewerbestrasse 11, 6330 Cham, Switzerland

Preface

A summer school about Arakelov geometry and Diophantine applications was held during the summer of 2017 at the Fourier Institute in Grenoble. This 3-week-long summer school intended to present an introduction to various concepts in Arakelov geometry followed by 2 weeks focused on some striking examples of current Diophantine problems to which Arakelov geometry has been or may be applied. During that summer, the main speakers decided to write this volume that covers the topics of the summer school in a quite self-contained form and more depth.

We hope it will be useful for graduate students and researchers interested in the connections between algebraic geometry and number theory.

The editors would like to thank the ANR project Gardio for its support of the summer school in 2017. The authors are also grateful to the anonymous referees for their very careful reading of previous versions of the texts included in this volume.

Saint-Nizier du Moucherotte, France   Emmanuel Peyre
Lans en Vercors, France   Gaël Rémond
Août 2020

Contents

Introduction

Emmanuel Peyre and Gaël Rémond

Arakelov geometry may be seen as a bridge between algebraic geometry and diophantine geometry. It is based upon the classical analogy between number fields, the function fields of curves and Riemannian geometry described as follows by A. Weil in a letter to his sister S. Weil in 1940:

> Mon travail consiste un peu à déchiffrer un texte trilingue ; de chacune des trois colonnes je n'ai que des fragments assez décousus, j'ai quelques notions sur chacune des trois langues ; mais je sais aussi qu'il y a de grandes différences de sens d'une colonne à l'autre, et dont rien ne m'avertit à l'avance. Depuis quelques années que j'y travaille, j'ai des bouts de dictionnaire. Quelquefois c'est sur une colonne que je fais porter mes efforts, quelquefois sur l'autre.

The fundamental tenet of Arakelov geometry is the analogy between the point at infinity on the projective line which defines an absolute value on the function field of the line and the real absolute value on the field of rational numbers. From this analogy, Arakelov geometry produces a whole dictionary relating module structure to hermitian structure and arithmetics to complex analysis.

The founding contribution of S. J. Arakelov [1, 2] gave a theory of intersection for *compactified divisors* for a curve X over a number field K seen as a compact arithmetic surface. Such a divisor was a formal sum including irreducible closed subsets of codimension 1 in a model of the curve over the ring of integers as well as real multiples of the Riemann curves obtained by taking the extension of scalars via an inclusion $K \to \mathbf{C}$. Arakelov also formulated and conjectured an analogue of the Riemann-Roch theorem.

A large part of the development of Arakelov geometry was devoted to the translation of crucial tools of algebraic geometry. An exhaustive description of

E. Peyre (✉) · G. Rémond
Institut Fourier, Université Grenoble Alpes, Grenoble, France
e-mail: emmanuel.peyre@univ-grenoble-alpes.fr ; gael.remond@univ-grenoble-alpes.fr

E. Peyre, G. Rémond (eds.), *Arakelov Geometry and Diophantine Applications*, Lecture Notes in Mathematics 2276, https://doi.org/10.1007/978-3-030-57559-5_1

these translations is beyond the scope of this introduction, but we may mention the intersection theory in higher dimension by H. Gillet and C. Soulé [8] which provides an arithmetic Chow ring and the development of the arithmetic Riemann-Roch theorem by G. Faltings [5] and J.-M. Bismut et al. [3, 9].

Thanks to its position at the interface between algebraic geometry and number theory, Arakelov geometry naturally has many applications to Diophantine problems. For example, the Faltings height used in his proof of Mordell's conjecture [4] is inspired by Arakelov's point of view on heights [7, 10, chapter II,§1]. G. Faltings also used Arakelov geometry in his proof of Mordell-Lang conjecture [6].

The contributions of the different authors have been organised in three parts:

- Part A of the volume is devoted to tools and concepts of Arakelov geometry which are central in the applications to diophantine equations.
- Part B gathers various questions related to rational points on algebraic varieties, like the distribution of points of bounded height, uniform bounds, as well as rational points seen from the point of view of dynamical systems.
- Part C concentrates on applications of Arakelov geometry to Shimura varieties.

Let us now summarize the content of each chapter.

1   Part A: Concepts of Arakelov Geometry

The first chapter of this book, written by C. Soulé as a short introduction to Arakelov geometry, presents the construction of arithmetic intersection.

Arakelov geometry provides a large wealth of invariants. In particular, it extends the notion of heights and, using an analogy to the slopes of vector bundles, the notion of slopes for adelic spaces. Chapter II, by É. Gaudron, presents the Harder-Narasimhan filtration, the slopes and several type of minima associated to rigid adelic spaces over an algebraic extension of $\mathbf{Q}$. This formalism generalises the Minkowski geometry of numbers for ellipsoids, the twisted height theory by D. Roy and J. Thunder as well as the slope theory of Hermitian vector bundles by J.-B. Bost.

As an example of the translations from algebraic geometry to number theory made possible by Arakelov geometry, H. Chen in Chapter III presents arithmetic Hilbert-Samuel theorems. In algebraic geometry, a Hilbert-Samuel theorem describes the asymptotic behaviour of the dimensions of the components of a graded ring. This applies in particular to the asymptotic dimension of the space $H^0(X, L^{\otimes n})$ for a line bundle L. The arithmetic analogue, first considered by Gillet and Soulé, is directly related to the arithmetic Riemann-Roch theorem.

Chapter IV, by J.-B. Bost, constitutes an introduction to the study of Euclidean lattices and of their invariants defined in terms of theta series. A Euclidean lattice is defined as a pair $\overline{E} = (E, \|.\|)$ where E is some free $\mathbf{Z}$-module of finite rank and $\|.\|$ is some Euclidean norm on the real vector space $E_{\mathbf{R}} = E \otimes \mathbf{R}$. The most basic of these invariants is the non-negative real number:

$$h_\theta^0(\overline{E}) = \log \left(\sum_{v \in E} e^{-\pi \|v\|^2} \right).$$

In this chapter J.-B. Bost explains how such invariants naturally arise when one investigates basic questions concerning classical invariants of Euclidean lattices, such as their successive minima, their covering radius, or the number of lattice points in balls of a given radius. He also discusses their significance from the perspective of Arakelov geometry and of the analogy between number fields and function fields, their role (discovered by Banaszczyk) in the derivation of optimal transference estimates, and their interpretation in terms of the formalism of statistical thermodynamics.

Specifically, J.-B. Bost shows that the invariant $h_\theta^0(\overline{E})$ attached to a Euclidean lattice $\overline{E}$ may be understood as Planck's thermodynamic function of a physical system simply associated to $\overline{E}$, while more naive Arakelov invariants of $\overline{E}$, defined in terms of numbers of lattices points in balls inside $\overline{E}$ and its direct sums $\overline{E}^{\oplus n}$, play the role of the entropy of this physical system.

These relations with the thermodynamic formalism are systematically explored in the second half of the chapter, which introduces some original perspectives both on Euclidean lattices and on the mathematical foundations of statistical physics. An important effort has been made to make the discussion self-contained and accessible to a wide circle of mathematicians, and to focus on simple and general mathematical results. Accordingly this chapter contains a significant amount of original material, previously not available in the literature.

2 Part B: Distribution of Rational Points and Dynamics

Chapter V, by E. Peyre, presents a survey on the distribution of rational points of bounded height on algebraic varieties and the insight provided by heights and slopes in the problem of accumulating subsets, which are closed subsets with zero volume in the adelic space but which contain a positive proportion of the points of bounded height.

In Chapter VI, P. Salberger considers uniform bounds for the number of points of bounded size, refining previous works of R. Heath-Brown, N. Broberg, and himself. More precisely, let X be a subvariety of $\mathbf{P}^n$ defined over $\mathbf{Q}$ and $B = (B_0, B_1, \ldots, B_n)$ be an $(n+1)$-tuple of positive numbers. Then let $X(\mathbf{Q}, B)$ be the set of all rational points on X represented by primitive $(n+1)$-tuples of integers $(x_0, x_1, \ldots, x_n)$ with $|x_i| \leq B_i$ for $i \in \{1, \ldots, n\}$. To study the cardinal of $X(\mathbf{Q}, B)$ one first divides this set into congruence classes $X(\mathbf{Q}, B, P)$ consisting of all points in $X(\mathbf{Q}, B)$ that specialise to a given $\mathbf{F}_p$-point P. The main idea is to construct an hypersurface $Y \subset \mathbf{P}^n$, which contains $X(\mathbf{Q}, B, P)$ but not X. This reduces the original counting problem for $X(\mathbf{Q}, B)$ to counting problems for varieties of lower dimension. The primes p will depend on B and be chosen to be sufficiently large such that the degree of Y only depends on the degree and dimension of X and some small number ε. The method is called the p-adic determinant method since the auxiliary hypersurfaces Y are obtained after showing that certain determinants vanish. The refinement presented here is based on Chow forms which may be seen as an alternative to heights of subvarieties.

The interplay between heights of points and heights of subvarieties, as defined by Arakelov geometry, is at the heart of Chapter VII, by A. Chambert-Loir. Following L. Szpiro, E. Ullmo and S.-W. Zhang, the arithmetic Hilbert-Samuel theorem leads to an equidistribution theorem for "small points". From this, E. Ullmo and S. Zhang derived a proof of the Bogomolov conjecture for subvarieties of abelian varieties. This chapter also presents the non-archimedean analogue of the equidistribution theorem, using Berkovich spaces.

Chapter VIII, by P. Autissier, is devoted to other results of equidistribution on arithmetic surfaces. The theorems of M. Fekete and G. Szegő relate the capacity of a compact set in $\mathbf{C}$ to the finiteness of the set of algebraic numbers in this compact set. Using Arakelov geometry, it is possible to extend these results to arithmetic surfaces. The theorems of Y. Bilu and R. Rumely refine these results by describing the asymptotic distribution of the conjugates of algebraic numbers.

In Chapter IX, R. Dujardin explains how arithmetic equidistribution theory can be used in the dynamics of rational maps on $\mathbf{P}^1$. He first introduces the basics of the iteration theory of rational maps on the projective line over $\mathbf{C}$, as well as some elements of iteration theory over an arbitrary complete valued field and the construction of dynamically defined height functions for rational functions defined over $\overline{\mathbf{Q}}$. The equidistribution of small points gives some original information on the distribution of preperiodic orbits, leading to some non-trivial rigidity statements. R. Dujardin then explains some consequences of arithmetic equidistribution to the study of the geometry of parameter spaces of such dynamical systems, notably pertaining to the distribution of special parameters and the classification of special subvarieties.

3   Part C: Shimura Varieties

The generalization of Arakelov theory to higher dimensions by Gillet and Soulé is well suited to study vector bundles with smooth hermitian metrics. Nevertheless when working with non-compact Shimura varieties or with modular varieties, the metrics that appear naturally and that can be studied using group theoretical or modular methods almost never extend to a smooth metric on a compactification. Typically they extend to a log-singular hermitian vector bundle. In Chapter X, J. I. Burgos Gil gives an example of this phenomenon and describes an extension of Arakelov theory that can be used to study log-singular hermitian vector bundles.

The arithmetic Riemann–Roch theorem refines both the algebraic geometric and differential geometric counterparts, and it is stated within the formalism of Arakelov geometry. For some simple Shimura varieties and automorphic vector bundles, the cohomological part of the formula can be understood via the theory of automorphic representations. Functoriality principles from this theory may then be applied to derive relations between arithmetic intersection numbers for different Shimura varieties. In Chapter XI, G. Freixas i Montplet explains this philosophy in the case of modular curves and compact Shimura curves. This indicates that there is some relationship between the arithmetic Riemann–Roch theorem and Selberg's trace formula.

Chapter XII, by F. Andreatta, presents the application of Arakelov geometry to the proof of an averaged version of Colmez's conjecture relating Faltings heights of abelian varieties with complex multiplication to special values of L-functions. A consequence of this version of Colmez's conjecture was the completion of the proof of André-Oort conjecture.

References

1. S.J. Arakelov, Theory of intersections on the arithmetic surface, in *Proceedings of the International Congress of Mathematicians, Vancouver 1974*, vol. 1 (1975), pp. 405–408
2. S.J. Arakelov, Intersection theory of divisors on an arithmetic surface. Math. USSR Izv. **8**, 1167–1180 (1976)
3. J.-M. Bismut, H. Gillet, C. Soulé, Complex immersions and Arakelov geometry, in *The Grothendieck Festschrift, Collect. Artic. in Honor of the 60th Birthday of A. Grothendieck, Vol. I*. Progress in Mathematics, vol. 86 (1990), pp. 249–331
4. G. Faltings, Endlichkeitssätze für abelsche Varietäten über Zahlkörpern. Invent. Math. **73**, 349–366 (1983)
5. G. Faltings, Calculus on arithmetic surfaces. Ann. Math. (2) **119**, 387–424 (1984)
6. G. Faltings, Diophantine approximation on abelian varieties. Ann. Math. (2) **133**, 549–576 (1991)
7. G. Faltings, G. Wüstholz (eds.), *Rational points. Seminar Bonn/Wuppertal 1983/84*. Aspects of Mathematics, 3 enlarged ed., vol. E6 (Friedr. Vieweg & Sohn, Braunschweig, 1992)
8. H. Gillet, C. Soulé, Arithmetic intersection theory. Publ. Math. Inst. Hautes Étud. Sci. **72**, 93–174 (1990)
9. H. Gillet, C. Soulé, An arithmetic Riemann-Roch theorem. Invent. Math. **110**, 473–543 (1992)
10. L. Szpiro, Degrés, intersections, hauteurs, in *Séminaire sur les pinceaux arithmétiques: La conjecture de Mordell, Astérisque*, vol. 127 (Société Mathématique de France, Paris, 1985), pp. 11–28

Part A
Concepts of Arakelov Geometry

Chapter I: Arithmetic Intersection

Christophe Soulé

1 Introduction

Intersection Theory in projective varieties is a topic in algebraic geometry which goes back to the eighteenth century. An example is Bézout's theorem, which says that two projective plane curves C and D, of degrees c and d and which have no components in common, meet in at most cd points. This result can be extended to closed subvarieties Y and Z in the projective space $\mathbf{P}$ of dimension n over k, with $\dim(Y) + \dim(Z) = n$ and for which $Y \cap Z$ consists of a finite number of points: this number is at most $\deg(Y)\deg(Z)$. Here the degree $\deg(Y)$ of $Y \subset \mathbf{P}$ can be defined as follows. Let L be the canonical line bundle on $\mathbf{P}$. The integer $\deg(Y)$ is characterized by the following two properties:

(i) When $Y = y$ is a closed point with residue field $\kappa(y)$

$$\deg(Y) = [\kappa(y) : k].$$

(ii) When s is a non-trivial rational section of L over Y, with divisor $\operatorname{div}(s) = \sum_\alpha n_\alpha Y_\alpha$,

$$\deg(Y) = \sum_\alpha n_\alpha \deg(Y_\alpha).$$

Note that induction on the dimension of Y shows that (i) and (ii) define the degree $\deg(Y)$ uniquely. One proves that it does not depend on the choice of s

C. Soulé (✉)
IHES, Le Bois-Marie, Bures-sur-Yvette, France
e-mail: soule@ihes.fr

E. Peyre, G. Rémond (eds.), *Arakelov Geometry and Diophantine Applications*,
Lecture Notes in Mathematics 2276, https://doi.org/10.1007/978-3-030-57559-5_2

made in (ii). We refer the reader to [7] §2 for a brief introduction to classical intersection theory.

In 1974, Arakelov discovered an intersection theory on arithmetic surfaces [1]. Namely, if C is a smooth projective curve over $\mathbf{Q}$, consider a regular projective scheme X over $\mathbf{Z}$ with generic fiber equal to C. Since the Krull dimension of X is two, one thinks of it as a surface. And since $\mathbf{Z}$ is affine, this surface X is not complete. To complete X, Arakelov adds to it the set of complex points $X(\mathbf{C})$, viewed as the fiber at ∞ of the map $X \rightarrow \mathrm{Spec}(\mathbf{Z})$. An *arithmetic divisor* is a formal sum $D + \lambda$ where D is a classical divisor on X and λ is a real number. Given two arithmetic divisors $D + \lambda$ and $D' + \lambda'$ (such that D and D' have no common components) Arakelov defines their intersection number, which is not an integer but a real number. He proves several properties of these numbers, e.g. an adjunction formula.

It appears that every notion or result in the classical algebraic geometry of varieties over fields has an *arithmetic analog* in the Arakelov geometry of schemes over $\mathbf{Z}$. In the 1980s, the Arakelov intersection theory was extended to higher dimensions by Gillet and myself [9].

In this chapter, we shall discuss the arithmetic analog of the notion of degrees, namely heights of varieties. To be more precise, we fix a regular projective scheme X over $\mathbf{Z}$. As arithmetic analogs of algebraic line bundles we take *hermitian line bundles*, i.e. line bundles L on X equipped with a smooth hermitian metric h on the restriction of L to the set of complex points $X(\mathbf{C})$. If $\bar{L} = (L, h)$ is such an hermitian line bundle on X and if $Y \subset X$ is an integral closed subset of X, the height of Y is a real number $h_{\bar{L}}(Y)$ which can be defined in several ways. It was first introduced by Faltings using arithmetic intersection theory [6] in his work on diophantine approximation on abelian varieties. Alternatively, $h_{\bar{L}}(Y)$ can be defined axiomatically by axioms similar to (i) and (ii) above [2]. This is the point of view we shall take here. When the unicity of $h_{\bar{L}}(Y)$ is easy to deduce from the axioms by induction on the dimension of Y (see Sect. 2), it is more difficult to show that $h_{\bar{L}}(Y)$ is independent on choices. Actually, the existence of $h_{\bar{L}}(Y)$ is the main result of this chapter, as we try to make it self-contained (see Sect. 3). Section 4 is a survey, without proof, of arithmetic intersection theory [9]. We conclude with a third definition of $h_{\bar{L}}(Y)$, as the integral on Y of a suitable power of the first Chern class of $\bar{L}$ (Theorem 4.7).

The interested reader may want to read in [2] several properties of the height, including arithmetic Bézout's theorems.

2 Definition of the Height

Let X be a regular projective flat scheme over $\mathbf{Z}$, and $\bar{L}$ an hermitian line bundle over X. For every integral closed subset $Y \subset X$ we shall define a real number $h_{\bar{L}}(Y)$, called the (Faltings) *height* of Y [6]. For this we need a few preliminaries.

2.1   Algebraic Preliminaries

2.1.1 Length   Let A be a noetherian (commutative and unitary) ring, and M an A-module of finite type.

There exists a filtration

$$M = M_0 \supset M_1 \supset M_2 \supset \ldots \supset M_r = 0 \qquad (1)$$

such that $M_{i+1} \neq M_i$ and $M_i/M_{i+1} = A/\wp_i$, where $\wp_i$ is a prime ideal, $0 \leq i \leq r - 1$ [3, th.1, p. 312].

Definition 2.1   The module M has *finite length* when there exists a filtration as (1) above where, for all i, $\wp_i$ is a maximal ideal in A.

Lemma 2.2 (Jordan-Hölder)   *If M has finite length, r does not depend on the choice of the filtration (1) with $\wp_i$ maximal. We call this number the* length *of M and denote it $\ell(M) \in \mathbf{N}$.*

Lemma 2.3   *Let*

$$0 \to M' \to M \to M'' \to 0$$

be an exact sequence of A-modules of finite length. Then

$$\ell(M) = \ell(M') + \ell(M'') .$$

The proof of Lemma 2.2 (resp. Lemma 2.3) can be found in [4, th. 6, p. 41] (resp. [5, prop. 16, p. 21]).

2.1.2 Order   Let A be as above. The *dimension* of A is

$$\dim(A) = \max\{n | \exists \text{ a chain of prime ideals } \wp_0 \subset \wp_1 \subset \wp_2 \ldots \subset \wp_n \subset A,$$

$$\text{with } \wp_i \neq \wp_{i+1}\} .$$

Let A be an integral ring of dimension 1, and $a \in A$, $a \neq 0$.

Lemma 2.4   *A/aA has finite length.*

Proof   Let

$$\overline{\wp_0} \subset \ldots \subset \overline{\wp_n}$$

be a maximal chain in A/aA, with $\overline{\wp_i} \neq \overline{\wp_{i+1}}$, and $\varphi : A \to A/aA$ the projection. Let $\wp_i = \varphi^{-1}(\overline{\wp_i})$. We get a chain

$$\wp_0 \subset \ldots \subset \wp_n$$

with $\wp_i \neq \wp_{i+1}$. Since A is integral, (0) is a prime ideal. And $\wp_0 \neq (0)$ since $\wp_0$ contains a. We conclude that

$$\dim(A/aA) \leq \dim(A) - 1 \,.$$

Since $\dim(A) = 1$ this implies that every prime ideal of A/aA is maximal. Therefore A/aA has finite length. $\square$

Let A be as in Lemma 2.4 and $K = \mathrm{frac}(A)$ be the field of fractions of A. If $x \in K - \{0\}$ we define, if $x = a/b$,

$$\mathrm{ord}_A(x) = \ell(A/aA) - \ell(A/bA) \in \mathbf{Z} \,.$$

Lemma 2.5

 (i) $\mathrm{ord}_A(x)$ *does not depend on the choice of a and b.*
(ii) $\mathrm{ord}_A(xy) = \mathrm{ord}_A(x) + \mathrm{ord}_A(y)$.

The proof of Lemma 2.5 is left to the reader.

Example 2.6 Assume A is *local* (i.e. A has only one maximal ideal $\mathcal{M}$) and *regular* (i.e. $\dim A = \dim(\mathcal{M}/\mathcal{M}^2)$). When $\dim(A) = 1$, K has a *discrete valuation*

$$v : K \to \mathbf{Z} \cup \{\infty\} \,,$$

$$A = \{x \in K \text{ such that } v(a) \geq 0\}$$

and $\mathrm{ord}_A(x) = n$ if and only if $x \in \mathcal{M}^n$ and $x \notin \mathcal{M}^{n+1}$. Therefore

$$\mathrm{ord}_A(x) = v(x) \,.$$

2.1.3 Divisors Let X be a nœtherian scheme and O_X be the sheaf of regular functions on X.

Definition 2.7 A *line bundle* on X is a locally free O_X-module L of rank one.

In other words L is a sheaf of abelian groups on X with a morphism of sheaves

$$\mu : O_X \times L \to L$$

such that there exists an open cover

$$X = \bigcup_\alpha U_\alpha$$

such that

(i) $L(U_\alpha) \simeq O_X(U_\alpha)$;
(ii) μ on $L(U_\alpha)$ is the multiplication.

Assume now that X is integral (for every open subset $U \subset X$, $O(U)$ is integral). Let $\eta \in X$ be the generic point.

Definition 2.8 A *rational section* of L is an element $s \in L_\eta$.

Let $Z^1(X)$ be the free abelian group spanned by the closed irreducible subsets $Y \subset X$ of codimension one. We call $Z^1(X)$ the group of *divisors of X*.

If $s \in L_\eta$ is a non-trivial rational section, its *divisor* is defined as

$$\mathrm{div}(s) = \sum_Y n_Y[Y] \in Z^1(X),$$

where n_Y is computed as follows. If $Y \subset X$ has codimension 1 and $Y = \overline{\{y\}}$ is integral, the ring $A = O_{X,y}$ is local, integral, of dimension 1. Its fraction field is

$$K = O_{X,\eta}.$$

Choose an isomorphism $L_y \simeq A$, hence $L_\eta \simeq K$. If $s \in L_\eta - \{0\} = K^*$, we let

$$n_Y = \mathrm{ord}_A(s)$$

(we shall also write $n_Y = \mathrm{ord}_Y(s)$).

One can prove that n_Y does not depend on choices, and $n_Y = 0$ for almost all Y.

Example 2.9 Let K be a number field and $X = \mathrm{Spec}\,(O_K)$. Giving L amounts to give

$$\Lambda = L(X),$$

a projective O_K-module of rank one. If $s \in \Lambda$, $s \neq 0$, we have a decomposition

$$\Lambda / O_K s \simeq \prod_{\wp \text{ prime}} (O_K / \wp^{n_\wp})$$

where $n_\wp = \mathrm{ord}_{O_\wp}(s)$, hence

$$\mathrm{div}(s) = \sum_\wp n_\wp [\wp].$$

2.2 Analytic Preliminaries

Let X be an analytic smooth manifold over $\mathbf{C}$, and $O_{X,\mathrm{an}}$ the sheaf of holomorphic functions on X.

Definition 2.10

(a) An *holomorphic line bundle* on X is a locally free $O_{X,\mathrm{an}}$-module of rank one.
(b) A *metric* $\|\cdot\|$ on L consists of maps

$$L(x) \xrightarrow{\;\|\cdot\|\;} \mathbf{R}_+$$

for any x, where $L(x) = L_x/\mathscr{M}_x$ is the fiber at x. We ask that

(i) $\|\lambda s\| = |\lambda|\,\|s\|$ if $\lambda \in \mathbf{C}$;
(ii) $\|s\| = 0$ iff $s = 0$;
(iii) Let $U \subset X$ be an open subset and s a section of L over U vanishing nowhere; then the map

$$x \longmapsto \|s(x)\|^2$$

is C^∞.

We write $\overline{L} = (L, \|\cdot\|)$.

Denote by $A^n(X)$ the $\mathbf{C}$-vector space of C^∞ complex differential forms of degree n on X. Recall that $A^n(X)$ decomposes as

$$A^n(X) = \bigoplus_{p+q=n} A^{p,q}(X),$$

where $A^{p,q}(X)$ consists of those differential forms which can be written locally as a sum of forms of type

$$u\,d\,z_{i_1} \wedge \ldots \wedge d\,z_{i_p} \wedge d\,\overline{z}_{j_1} \wedge \ldots \wedge d\,\overline{z}_{j_q}$$

where u is a C^∞ function, $d\,z_\alpha = d\,x_\alpha + i\,d\,y_\alpha$ and $d\,\overline{z}_\alpha = d\,x_\alpha - i\,d\,y_\alpha$.

The differential

$$d : A^n(X) \to A^{n+1}(X)$$

is a sum $d = \partial + \overline{\partial}$ where

$$\partial : A^{p,q}(X) \to A^{p+1,q}(X)$$

and

$$\overline{\partial} : A^{p,q}(X) \to A^{p,q+1}(X).$$

We have $\partial^2 = \overline{\partial}^2 = d^2 = 0$ and we let

$$d^c = \frac{\partial - \overline{\partial}}{4\pi i},$$

so that

$$dd^c = \frac{\overline{\partial}\,\partial}{2\pi i}.$$

Lemma 2.11 *Let $\overline{L} = (L, \|\cdot\|)$ be an analytic line bundle with metric. There exists a smooth form*

$$c_1(\overline{L}) \in A^{1,1}(X)$$

such that, if $U \subset X$ is an open subset and $s \in \Gamma(U, L)$ is such that $s(x) \neq 0$ for every $x \in U$,

$$c_1(\overline{L})_{|U} = -dd^c \log \|s\|^2.$$

Proof Let $s' \in \Gamma(U, L)$ be another section such that $s(x) \neq 0$ when $x \in U$. We need to show that

$$-dd^c \log \|s'\|^2 = -dd^c \log \|s\|^2 \quad \text{in} \quad A^{1,1}(U). \qquad (2)$$

There exists $f \in \Gamma(U, O_{X_{\mathrm{an}}})$ such that

$$s' = fs.$$

We get

$$-dd^c \log \|s'\|^2 = -dd^c \log \|s\|^2 - dd^c \log |f|^2.$$

But

$$\partial\overline{\partial} \log |f|^2 = \partial \left[\frac{\overline{\partial} f}{f} + \frac{\overline{\partial}\,\overline{f}}{\overline{f}} \right] = -\overline{\partial}\,\frac{\partial\,(\overline{f})}{\overline{f}} = 0,$$

and (2) follows. $\square$

The form $c_1(\overline{L})$ is called the *first Chern form* of $\overline{L}$.

2.3 Heights

Let X be a regular, projective, flat scheme over $\mathbf{Z}$. We denote by $X(\mathbf{C})$ the set of complex points of X; it is an analytic manifold.

Definition 2.12 An *hermitian line bundle* on X is a pair $\overline{L} = (L, \|\cdot\|)$, where L is a line bundle on X and $\|\cdot\|$ is a metric on the holomorphic line bundle

$$L_{\mathbf{C}} = L_{|X(\mathbf{C})}\,.$$

We also assume that $\|\cdot\|$ is invariant by the complex conjugation

$$F_\infty : X(\mathbf{C}) \to X(\mathbf{C})\,.$$

Let $\overline{L}$ be an hermitian line bundle on X. We let

$$c_1(\overline{L}) = c_1(\overline{L}_{\mathbf{C}}) \in A^{1,1}(X(\mathbf{C}))\,.$$

Theorem 2.13 *There is a unique way to associate to every integral closed subset $Y \subset X$ a real number*

$$h_{\overline{L}}(Y) \in \mathbf{R}$$

in such a way that:

(i) *If* $\dim(Y) = 0$, *i.e. when* $Y = \{y\}$ *where* $y \in X$ *is a closed point, we let* $\kappa(y) = O_{X,y}/\mathcal{M}_{X,y}$ *be the residue field. Then* $\kappa(y)$ *is finite and*

$$h_{\overline{L}}(Y) = \log \#(\kappa(y))\,.$$

(ii) *If* $\dim(Y) > 0$, *let* s *be a non-trivial rational section of L over Y. If*

$$\mathrm{div}_Y(s) = \sum_\alpha n_\alpha\, Y_\alpha\,,$$

then

$$h_{\overline{L}}(Y) = \sum_\alpha n_\alpha\, h_{\overline{L}}(Y_\alpha) - \int_{Y(\mathbf{C})} \log \|s\|\, c_1(\overline{L})^{\dim Y(\mathbf{C})}\,.$$

3 Existence of the Height

3.1 Resolutions To prove Theorem 2.13, we first need to make sense of the integral in (ii). For that we use Hironaka's resolution theorem.

Theorem 3.1 (Hironaka) *Let X be a scheme of finite type over* **C***, and $Z \subset X$ a proper closed subset of X such that $X - Z$ is smooth. Then there exists a proper map*

$$\pi : \widetilde{X} \to X$$

such that:

(i) *$\widetilde{X}$ is smooth;*
(ii) *$\widetilde{X} - \pi^{-1}(Z) \xrightarrow{\sim} X - Z$;*
(iii) *$\pi^{-1}(Z)$ is a divisor with normal crossings.*

In the situation of (ii) in Theorem 2.13, we apply Theorem 3.1 to $X = Y(\mathbf{C})$, and to the union $Z = |\mathrm{div}(s)| \cup Y(\mathbf{C})^{\mathrm{sing}}$ of the support of $\mathrm{div}(s)$ and the singular locus of $Y(\mathbf{C})$. Let $\pi : \widetilde{Y} \to Y(\mathbf{C})$ be a resolution of $Y(\mathbf{C})$, d the dimension of $Y(\mathbf{C})$ and $\omega \in A^{dd}(Y(\mathbf{C}) - Z)$ with compact support. Then we define

$$\int_{Y(\mathbf{C})} \log \|s\| \, \omega = \int_{\widetilde{Y}} \log \|\pi^*(s)\| \, \pi^*(\omega) \, .$$

To see that the integral converges choose local coordinates $z_1, \ldots, z_d$ of $\widetilde{Y}$ such that

$$\pi^*(s) = z_1^n u \, ,$$

with u a non-trivial section. Therefore

$$\log \|\pi^*(s)\| = n \log |z_1| + \alpha \, ,$$

with α of class C^∞, and

$$\pi^*(\omega) = \beta \prod_{i-1}^{d} d z_i \, d \bar{z}_i \, ,$$

with β of class $\mathbf{C}^\infty$. Since

$$\int_{|z| \leq \varepsilon} \log |z| \, d z \, d \bar{z} = 2 \int_0^\varepsilon \int_0^{2\pi} \log(r) \, r \, dr \, d\theta < +\infty \, ,$$

the integral converges.

3.2 By induction on $\dim(Y)$, the unicity of $h_{\overline{L}}(Y)$ is clear.

Now we handle the case $X = \mathrm{Spec}\,(O_K)$ for a number field K. If Σ is the set of complex embeddings of K we have

$$X(\mathbf{C}) = \coprod_{\sigma \in \Sigma} \mathrm{Spec}\,(\mathbf{C}).$$

To give $\overline{L} = (L, \|\cdot\|)$ amounts to give a pair $\overline{\Lambda} = (\Lambda, \|\cdot\|_\sigma)$ where $\Lambda = L(X)$ is a projective O_K-module of rank one and, for any $\sigma \in \Sigma$, $\|\cdot\|_\sigma$ is a metric on $\Lambda \underset{\sigma}{\otimes} \mathbf{C} \simeq \mathbf{C}$ such that

$$\|F_\infty(x)\|_{F_\infty \circ \sigma} = \|x\|_\sigma.$$

If $s \in \Lambda$, $s \neq 0$, we have

$$\mathrm{div}(s) = \sum_{\wp} n_\wp\, [\wp]$$

and

$$h_{\overline{L}}(X) = \sum_{\wp} n_\wp \log(N\wp) - \sum_{\sigma \in \Sigma} \log \|\sigma(s)\|_\sigma,$$

where $N\wp = \#(O/\wp)$.

Since

$$\Lambda/O_K s = \prod_{\wp} (O_\wp/\wp^{n_\wp})$$

we get

$$\sum_{\wp} n_\wp \log(N\wp) = \log \#(\Lambda/O_K s).$$

Lemma 3.2 $h_{\overline{L}}(X)$ *does not depend on the choice of s.*

Proof Let

$$d(s) = \log \#(\Lambda/O_K s) - \sum_{\sigma \in \Sigma} \log \|\sigma(s)\|_\sigma.$$

If $s' \in \Lambda$, $s' \neq 0$, we have

$$s' = f s$$

with $f \in K^*$. Therefore

$$d(s') - d(s) = \sum_{\wp} v_{\wp}(f) \log(N\wp) - \sum_{\sigma \in \Sigma} \log \|\sigma(f)\| = 0$$

by the product formula.   □

3.3 Let us prove Theorem 2.13 when Y has dimension one and Y is horizontal, i.e. Y maps surjectively onto Spec $(\mathbf{Z})$. We have then

$$Y = \overline{\{y\}},$$

where y is a closed point in $X \underset{\mathbf{Z}}{\otimes} \mathbf{Q}$. The residue field $K = \kappa(y)$ is a number field and

$$Y = \text{Spec}(R)$$

where R is an integral ring with fraction field K. Denote by $\widetilde{R}$ the integral closure of R in K (i.e. $\widetilde{R} = O_K$) and let

$$\pi : \widetilde{Y} = \text{Spec}(\widetilde{R}) \to Y$$

be the projection. If

$$s \in \Gamma(Y, L) - \{0\},$$

$$\pi^*(s) \in \Gamma(\widetilde{Y}, \pi^*L) - \{0\}.$$

We shall prove that

$$d(s) = d(\pi^*(s)).\tag{3}$$

By 3.2 this will imply that $d(s)$ is independent of the choice of s. To prove (3) we first notice that

$$Y(\mathbf{C}) = \widetilde{Y}(\mathbf{C}) = \coprod_{\sigma \in \Sigma} \text{Spec}(\mathbf{C}),$$

hence

$$\sum_{\sigma \in \Sigma} \log \|s\|_{\sigma} = \sum_{\sigma \in \Sigma} \log \|\pi^*(s)\|_{\sigma}.\tag{4}$$

Next we consider the commutative diagram

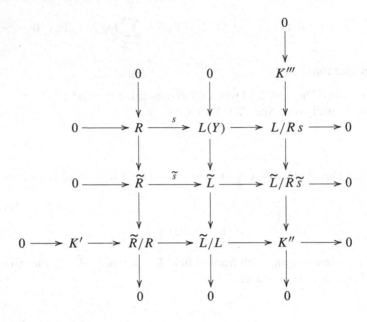

where $\tilde{s} = \pi^*(s) \in \tilde{L} = \pi^*(L)$.

By diagram chase (snake lemma) we get

$$\# K''' = \# K'.$$

On the other hand, for any prime ideal $\wp$ in O_K, we have

$$\# \left(\frac{\tilde{L}}{L} \right)_\wp = \# \left(\frac{\tilde{R}}{R} \right)_\wp$$

since $\tilde{L}_\wp = L_\wp \underset{R_\wp}{\otimes} \tilde{R}_\wp$ and L and $\tilde{L}$ are locally trivial. This implies

$$\# K' = \# K''$$

and $\# K''' = \# K''$. Therefore

$$\# (L/R s) = \# (\tilde{L}/\tilde{R} s). \tag{5}$$

The assertion (3) follows from (4) and (5).

When $\dim(Y) = 1$ and Y is vertical i.e. its image in Spec $(\mathbf{Z})$ is a closed point of finite residue field k, Theorem 2.13 is proved by considering the normalisation

$$\pi : \widetilde{Y} \to Y .$$

The proof is the same as in the case Y is horizontal, the product formula being replaced by the equality

$$\sum_{x \in \widetilde{Y}} v_x(f)\,[k(x) : k] = 0$$

for any $f \in \kappa(y)^*$. Indeed,

$$\log \# k(x) = [k(x) : k] \log \# k .$$

3.4 Assume from now on that $\dim(Y) \geq 2$, with $Y \subset X$ a closed integral subscheme, $Y = \overline{\{y\}}$. If $s \in L_y,\, s \neq 0$,

$$\mathrm{div}(s) = \sum_{\alpha} n_\alpha\, Y_\alpha .$$

Lemma 3.3 *There exists $t \in L_y$ such that, for every α, the restriction of t to Y_α is neither zero nor infinity.*

Proof Let $Y_\alpha = \overline{\{y_\alpha\}}$. The ring

$$R = \varinjlim_{U \text{ s.t.}, \forall \alpha,\ y_\alpha \in U} O(U)$$

is *semi-local*, i.e. it has finitely many maximal ideals $\mathcal{M}_\alpha,\, \alpha \in A$. Let

$$I = \bigcap_{\alpha \in A} \mathcal{M}_\alpha$$

be the radical of R, and

$$\Lambda = \varinjlim_{U \text{ s.t.}, \forall \alpha,\ y_\alpha \in U} L(U) .$$

Note that, for every α,

$$R_{\mathcal{M}_\alpha} = O_{y_\alpha}$$

and, for every pair $\alpha \neq \beta$

$$\mathcal{M}_\alpha + \mathcal{M}_\beta = R .$$

Since L is locally trivial

$$\Lambda \otimes R/I = \prod_\alpha (\Lambda \otimes O_{y_\alpha})/\mathcal{M}_\alpha \simeq \prod_\alpha O_{y_\alpha}/\mathcal{M}_\alpha = R/I \,.$$

Denote by $t \in \Lambda$ an element such that its class in $\Lambda \otimes R/I$ maps to $1 \in R/I$ by the above isomorphism. The module

$$M = \Lambda/Rt$$

is such that $M = IM$. Therefore, by Nakayama's lemma, $M = 0$. Since

$$\Lambda = Rt \,,$$

for every $\alpha \in A$ the restriction of t to Y_α does not vanish.   □

3.5 Given s and t as above we write

$$\mathrm{div}(s) = \sum_\alpha n_\alpha Y_\alpha$$

and

$$\mathrm{div}(t) = \sum_\beta m_\beta Z_\beta \,,$$

with $Z_\beta \neq Y_\alpha$ for all β and α. Consider

$$\mathrm{div}(s) \cdot \mathrm{div}(t) = \sum_\alpha n_\alpha \, \mathrm{div}(t \mid Y_\alpha)$$

and

$$\mathrm{div}(t) \cdot \mathrm{div}(s) = \sum_\beta m_\beta \, \mathrm{div}(s \mid Z_\beta) \,.$$

These are cycles of codimension two in Y.

Proposition 3.4 *We have*

$$\mathrm{div}(s) \cdot \mathrm{div}(t) = \mathrm{div}(t) \cdot \mathrm{div}(s) \,.$$

The proof of Proposition 3.4 will be given later.
Assume $\dim Y(\mathbf{C}) = d$, and define

$$d(s) = h_{\overline{L}}(\mathrm{div}(s)) - \int_{Y(\mathbf{C})} \log \|s\| \, c_1(\overline{L})^d \,.$$

By induction hypothesis we have

$$d(s) = \sum_\alpha n_\alpha \, h_{\overline{L}}(\operatorname{div}(t_{|Y_\alpha}))$$

$$- \sum_\alpha n_\alpha \int_{Y_\alpha(\mathbb{C})} \log \|t\| \, c_1(\overline{L})^{d-1} - \int_{Y(\mathbb{C})} \log \|s\| \, c_1(\overline{L})^d$$

$$= h_{\overline{L}}(\operatorname{div}(s) \cdot \operatorname{div}(t)) - I(s,t)$$

where

$$I(s,t) = \sum_\alpha n_\alpha \int_{Y_\alpha(\mathbb{C})} \log \|t\| \, c_1(\overline{L})^{d-1} + \int_{Y(\mathbb{C})} \log \|s\| \, c_1(\overline{L})^d \, .$$

Proposition 3.5 $I(s,t) = I(t,s)$.

From Propositions 3.4 and 3.5 we deduce that $d(s) = d(t)$ when $\operatorname{div}(s)$ and $\operatorname{div}(t)$ are transverse. When s and s' are two sections of L there exists a section t such that $\operatorname{div}(s)$ and $\operatorname{div}(t)$ (resp. $\operatorname{div}(s')$ and $\operatorname{div}(t)$) are transverse. Therefore

$$d(s) = d(t) = d(s')$$

and Theorem 2.13 follows.

3.6 To prove Proposition 3.4 we write

$$\operatorname{div}(s) \cdot \operatorname{div}(t) = \sum_W n_W \, [W]$$

with $\operatorname{codim}_Y(W) = 2$. Let $W = \overline{\{w\}}$ and

$$R = O_{Y,w} \, .$$

Since $L_w \simeq O_{Y,w}$ one can assume that t (resp. s) corresponds to $a \in R$ (resp. $b \in R$). Since R is integral and $a \neq 0$, we know from the proof of Lemma 2.4 that, if $A = R/aR$,

$$\dim(A) \leq \dim(R) - 1 = 1 \, .$$

Let $\overline{b} \in A$ be the image of b and let $\overline{\wp} \subset A$ be a minimal prime ideal of A. The inverse image $\wp \subset R$ of $\overline{\wp}$ is a minimal nontrivial prime ideal. Since $a \in \wp$, the closed subset defined by $\wp$ in $X = \operatorname{Spec}(R)$ is contained in the image in X of the support of $\operatorname{div}(t)$. Since $\operatorname{div}(t)$ and $\operatorname{div}(s)$ are transverse, b does not belong to $\wp$, hence $\bar{b}$ does not belong to $\bar{\wp}$. According to Theorem 5.15, ii), in [10, chapter 2], it follows that

$$\dim(A/\bar{b}) = \dim(A) - 1 \,.$$

Since $\dim(A) \le 1$ we get $\dim(A) = 1$ and $\dim(A/\bar{b}) = 0$. This implies that $A/\bar{b}$ has finite length. If $\langle a, b \rangle \subset R$ is the ideal spanned by a and b, $A/\bar{b} = R/\langle a, b \rangle$ and we shall prove that

$$n_W = \ell(R/\langle a, b \rangle) \,.$$

3.7 Let A be as above and let M be an A-module of finite type. If $x \in A$ we have an exact sequence

$$0 \to M[x] \to M \xrightarrow{\times x} M \to M/x\,M \to 0 \,. \tag{6}$$

If $M[x]$ and $M/x\,M$ have finite length we define

$$e(x, M) = \ell(M/x\,M) - \ell(M[x]) \in \mathbf{Z} \,.$$

Lemma 3.6

(i) $M[\bar{b}]$ and $M/\bar{b}\,M$ have finite length.
(ii)

$$e(\bar{b}, M) = \sum_{\substack{\wp \subset A \\ \wp \text{ minimal}}} \ell_{A_\wp}(M_\wp)\, e\,(\bar{b}, A/\wp) \,.$$

(iii)

$$e(\bar{b}, A/\wp) = \ell(A/(\wp + b\,A)) \,.$$

Proof of (i) and (ii) Note that both sides in (ii) are additive in M for exact sequences. Therefore we can assume that $M = A/q$ where q is a prime ideal. We distinguish two cases:

(a) If q is maximal, for any minimal prime ideal $\wp$ we have $M_\wp = 0$. Therefore $\ell(M)$ is finite. From Lemma 2.3 and (6) we conclude that

$$e(\bar{b}, M) = 0 \,.$$

(b) Assume $q = \wp$ is minimal. If $\wp' \ne \wp$ is any prime ideal different from $\wp$ we have

$$M_{\wp'} = 0 \,.$$

Therefore the right hand side reduces to one summand and (i) holds. Furthermore

$$\ell_{A_\wp}(M_\wp) = 1$$

and

$$e(\bar{b}, M) = e(\bar{b}, A/\wp)$$

so (ii) is true.

To prove (iii) we let $M = A/\wp$. We saw that $b \notin \wp$ and $A/\wp$ is integral, therefore $M[\bar{b}] = 0$.

On the other hand

$$\dim(A/(\wp + bA)) \leq \dim(A/\wp) - 1 = 0.$$

Therefore

$$e(\bar{b}, A/\wp) = \ell(A/(\wp + bA)). \qquad \square$$

3.8 We shall apply Lemma 3.6 to

$$M = A = R/a\,R.$$

Let $\wp$ be a minimal prime in A and $Y \subset |\mathrm{div}(s)|$ the corresponding component of the support of $\mathrm{div}(s)$. We have

$$\ell(A_\wp) = \mathrm{ord}_Y(t)$$

and

$$\ell(A/(\wp + bA)) - \mathrm{ord}_W(t_{|Y}).$$

Lemma 3.6 (iii) says that

$$e(\bar{b}, A) = n_W.$$

But $\bar{b}$ does not divide zero, so

$$e(\bar{b}, A) = \ell(R/\langle a, b\rangle).$$

Therefore $n_W = \ell(R/\langle a, b\rangle)$. Since $\langle a, b\rangle = \langle b, a\rangle$ we conclude that

$$\mathrm{div}(s) \cdot \mathrm{div}(t) = \sum_W n_W[W] = \mathrm{div}(t) \cdot \mathrm{div}(s).$$

This ends the proof of Proposition 3.4.

3.9 We shall now prove Proposition 3.5. For this we need some more analytic preliminaries. Let X be a smooth complex compact manifold of dimension d.

Definition 3.7 A *current* $T \in D^{p,q}(X)$ is a **C**-linear form

$$T : A^{d-p,d-q}(X) \to \mathbf{C}$$

which is continuous for the Schwartz' topology.

Examples 3.8

(i) If $\eta \in L^1(X) \underset{C^\infty(X)}{\otimes} A^{p,q}(X)$ is an integrable differential, η defines a current by the formula

$$\eta(\omega) = \int_X \eta \wedge \omega .$$

(ii) If $Z = \sum_\alpha n_\alpha Z_\alpha$ is a cycle of codimension p on X, it defines a Dirac current $\delta_Z \in D^{pp}(X)$ by the formula

$$\delta_Z(\omega) = \sum_\alpha n_\alpha \int_{Z_\alpha} \omega ,$$

where the integrals converge by Hironaka's theorem.

We can derivate a current $T \in D^{p,q}(X)$ by the formula

$$\partial T(\omega) = (-1)^{p+q+1} T(\partial \omega)$$

and

$$\bar{\partial} T(\omega) = (-1)^{p+q+1} T(\bar{\partial} \omega) .$$

By the Stokes formula we get a commutative diagram

$$
\begin{array}{ccc}
D^{p,q}(X) & \overset{\partial}{\longrightarrow} & D^{p+1,q}(X) \\
\cup & & \cup \\
A^{p,q}(X) & \overset{\partial}{\longrightarrow} & A^{p+1,q}(X)
\end{array}
$$

and idem for $\bar{\partial}$ and $d = \partial + \bar{\partial}$.

Proposition 3.9 (Poincaré-Lelong) *Let $\overline{L}$ be an hermitian line bundle on X and s a meromorphic section of L. Then we have the following formula in $D^{1,1}(X)$*

$$dd^c(-\log\|s\|^2) + \delta_{\mathrm{div}(s)} = c_1(\overline{L}).\tag{7}$$

3.10 To prove Proposition 3.9 let $Z = |\mathrm{div}(s)|$ be the support of the divisor of s. By Theorem 3.1, there exists a birational resolution

$$\pi : \widetilde{X} \to X$$

where $\pi^{-1}(Z)$ has local equation $z_1 \ldots z_k = 0$. Therefore

$$\pi^*(s) = z_1^{n_1} \ldots z_k^{n_k}$$

locally. If Proposition 3.9 holds for $\pi^*(\overline{L})$ and $\pi^*(s)$, by applying π_* we get (7). So we can assume that $X = \widetilde{X}$. By additivity we can assume that

(a) $\|s\| = |z_1|$

or

(b) $\log\|s\| = \rho \in C^\infty(X)$.

In case (b) $\mathrm{div}(s) = 0$ and (7) is true by definition of $c_1(\overline{L})$ (Lemma 2.11). In case (a) we have to show that, for every differential form ω with compact support in U, and for $\varepsilon > 0$ small enough,

$$-\int_U \log|z_1|^2 \, dd^c(\omega) = \int_{|z_1|=\varepsilon} \omega.$$

But, by Stokes' theorem, we have

$$- \lim_{\varepsilon\to 0} \int_{|z_1|\geq\varepsilon} \log|z_1|^2 \, dd^c(\omega)$$

$$= \lim_{\varepsilon\to 0} \int_{|z_1|=\varepsilon} \log|z_1|^2 \, d^c\,\omega + \lim_{\varepsilon\to 0} \int_{|z_1|\geq\varepsilon} d\log|z_1|^2 \, d^c\,\omega.$$

The first summand vanishes and, applying Stokes' theorem again,

$$\lim_{\varepsilon\to 0} \int_{|z_1|\geq\varepsilon} d\log|z_1|^2 \, d^c\,\omega = -\lim_{\varepsilon\to 0} \int_{|z_1|\geq\varepsilon} d^c\log|z_1|^2 \, d\,\omega$$

$$= \lim_{\varepsilon\to 0} \int_{|z_1|=\varepsilon} d^c\log|z_1|^2 \,\omega - \lim_{\varepsilon\to 0} \int_{|z_1|\geq\varepsilon} dd^c\log|z_1|^2 \,\omega.$$

The second summand vanishes and, taking polar coordinates $z_1 = r_1 e^{i\theta_1}$, we get

$$d^c \log |z_1|^2 = \frac{d\theta_1}{2\pi}$$

and

$$\lim_{\varepsilon \to 0} \int_{|z_1|=\varepsilon} \frac{d\theta_1}{2\pi} \omega = \int_{z_1=0} \omega \,.$$

□

3.11 Coming back to Proposition 3.5 we consider the current

$$T_{s,t} = \delta_{\mathrm{div}(s)} \log \|t\|^2 + \log \|s\|^2 c_1(\overline{L}) \,.$$

Then

$$I(s,t) = T_{s,t}(c_1(\overline{L})^{d-1})/2 \,.$$

Proposition 3.9 implies

$$T_{s,t} = (c_1(\overline{L}) + dd^c \log \|s\|^2) \log \|t\|^2 + \log \|s\|^2 c_1(\overline{L})$$

at least formally: we have to make sense of the product of currents $(dd^c \log \|s\|^2) \log \|t\|^2$. By Stokes' theorem we have (at least formally)

$$dd^c(T_1) \, T_2 = d(d^c(T_1) \, T_2) + d^c(T_1) \, d(T_2)$$
$$= d(d^c(T_1) \, T_2) + d^c(T_1 \, d \, T_2) - T_1 \, d^c d(T_2) \,.$$

Since $d^c d = -dd^c$ and $d(c_1(\overline{L})^{d-1}) = d^c(c_1(\overline{L})^{d-1}) = 0$ we get

$$2I(s,t) = T_{s,t}(c_1(\overline{L})^{d-1}) = T_{t,s}(c_1(\overline{L})^{d-1}) = 2I(t,s) \,.$$

□

3.12 The Height of the Projective Space Let $N \geq 1$ be an integer and $\mathbf{P}^N$ the N-dimensional projective space over $\mathbf{Z}$. The tautological line bundle $O(1)$ on $\mathbf{P}^N$ is a quotient of the trivial vector bundle of rank $N + 1$

$$O_{\mathbf{P}^N}^{N+1} \to O(1) \to 0 \,.$$

We equip $O_{\mathbf{P}^N}^{N+1}$ with the trivial metric and $O(1)$ with the quotient metric.

Proposition 3.10 *The height of* $\mathbf{P}^N$ *is*

$$h_{\overline{O(1)}}(\mathbf{P}^N) = \frac{1}{2} \sum_{k=1}^{N} \sum_{m=1}^{k} \frac{1}{m}.$$

Proof Let s be the section of $O(1)$ defined by the homogeneous coordinate X_0. Then $\operatorname{div}(s) = \mathbf{P}^{N-1}$ and we get, from Theorem 2.13 (ii),

$$h(\mathbf{P}^N) = h(\mathbf{P}^{N-1}) - \int_{\mathbf{P}^N(\mathbf{C})} \log \|s\| \, d\mu$$

where $d\mu$ is the probability measure on $\mathbf{P}^N(\mathbf{C})$ invariant under rotation by $U(N+1)$. If dv is the probability measure on the sphere S^{2N+1} invariant under $U(N+1)$ we have

$$\int_{\mathbf{P}^N(\mathbf{C})} \log \|s\| \, d\mu = \int_{S^{2N+1}} \log |X_0| \, dv$$

and Proposition 3.10 follows from

Lemma 3.11 *The integral on the sphere is given by*

$$\int_{S^{2N+1}} \log |X_0| \, dv = \frac{1}{2} \sum_{m=1}^{N} \frac{1}{m}.$$

4 Arithmetic Chow Groups

4.1 Definition

Let X be a regular projective flat scheme over $\mathbf{Z}$ and $p \geq 0$ an integer. Let $Z^p(X)$ be the group of codimension p cycles on X.

Definition 4.1 A *Green current* for $Z \in Z^p(X)$ is a real current $g \in D^{p-1,p-1}(X(\mathbf{C}))$ such that $F_\infty^*(g) = (-1)^{p-1} g$ and

$$dd^c g + \delta_Z = \omega$$

for a smooth form $\omega \in A^{p,p}(X(\mathbf{C}))$.

We let $\widehat{Z}^p(X)$ be the group generated by pairs (Z, g), $Z \in Z^p(X)$, g Green current for Z, with $(Z_1, g_1) + (Z_2, g_2) = (Z_1 + Z_2, g_1 + g_2)$.

Examples 4.2

(i) Let $Y \subset X$ be a closed irreducible subset with $\mathrm{codim}_X(Y) = p - 1$, and $f \in \kappa(y)$ a rational function on Y. Define $\log |f|^2 \in D^{p-1,p-1}(X(\mathbf{C}))$ by the formula

$$(\log |f|^2)(\omega) = \int_{Y(\mathbf{C})} \log |f|^2 \, \omega$$

(which makes sense by Theorem 3.1). We may think of f as a rational section of the trivial line bundle on Y. Therefore Poincaré-Lelong formula (Proposition 3.9) reads

$$dd^c(-\log |f|^2) + \delta_{\mathrm{div}(f)} = 0.$$

Hence the pair

$$\widehat{\mathrm{div}}(f) = (\mathrm{div}(f), -\log |f|^2)$$

is an element of $\widehat{Z}^p(X)$.

(ii) Given $u \in D^{p-2,p-1}(X(\mathbf{C}))$ and $v \in D^{p-1,p-2}(X(\mathbf{C}))$ we have

$$dd^c(\partial u + \bar{\partial} v) = 0,$$

so $(0, \partial u + \bar{\partial} v) \in \widehat{Z}^p(X)$.

We let $\widehat{R}^p(X) \subset \widehat{Z}^p(X)$ be the subgroup generated by all elements $\widehat{\mathrm{div}}(f)$ and $(0, \partial u + \bar{\partial} v)$.

Definition 4.3 The *arithmetic Chow group* of codimension p of X is the quotient

$$\widehat{\mathrm{CH}}^p(X) = \widehat{Z}^p(X)/\widehat{R}^p(X).$$

4.2 Example

Let $\widehat{\mathrm{Pic}}(X)$ be the group of isometric isomorphism classes of hermitian line bundles on X, equipped with the tensor product. If $\overline{L} = (L, \|\cdot\|) \in \widehat{\mathrm{Pic}}(X)$ and if $s \neq 0$ is a rational section of L we let

$$\widehat{\mathrm{div}}(s) = (\mathrm{div}(s), -\log \|s\|^2) \in \widehat{Z}^1(X)$$

(Proposition 3.9), and we define

$$\widehat{c}_1(\overline{L}) \in \widehat{\mathrm{CH}}^1(X)$$

to be the class of $\widehat{\mathrm{div}}(s)$. It does not depend on the choice of s: if s' is another section of L we have

$$s' = f s$$

with $f \in k(X)$. Therefore

$$\widehat{\mathrm{div}}(s') - \widehat{\mathrm{div}}(s) = \widehat{\mathrm{div}}(f) \in \widehat{R}^1(X).$$

Proposition 4.4 *The map $\widehat{c}_1$ induces a group isomorphism*

$$\widehat{c}_1 : \widehat{\mathrm{Pic}}(X) \to \widehat{\mathrm{CH}}^1(X).$$

Proof To prove Proposition 4.4 we consider the commutative diagram with exact rows

$$
\begin{array}{ccccccccc}
0 & \longrightarrow & C^\infty(X(\mathbf{C})) & \overset{a}{\longrightarrow} & \widehat{\mathrm{Pic}}(X) & \overset{\zeta}{\longrightarrow} & \mathrm{Pic}(X) & \longrightarrow & 0 \\
 & & \| & & \downarrow{\widehat{c}_1} & & \downarrow{c_1} & & \\
0 & \longrightarrow & C^\infty(X(\mathbf{C})) & \overset{a'}{\longrightarrow} & \widehat{\mathrm{CH}}^1(X) & \overset{\zeta'}{\longrightarrow} & \mathrm{CH}^1(X) & \longrightarrow & 0
\end{array}
\qquad (8)
$$

where $a(\varphi)$ is the trivial line bundle on X equipped with the norm such that $\|1\| = \exp(\varphi)$, $\zeta(\bar{L}) = L$, $a'(\varphi) = (0, -\log|\varphi|^2)$ and $\zeta'(Z, g) = Z$. Since c_1 is an isomorphism the same is true for $\widehat{c}_1$. $\qquad\square$

4.3   Products

4.3.1 Denote by $\widehat{\mathrm{CH}}^p(X)_{\mathbf{Q}}$ the tensor product $\widehat{\mathrm{CH}}^p(X) \underset{\mathbf{Z}}{\otimes} \mathbf{Q}$.

Theorem 4.5 *When $p \geq 0$ and $q \geq 0$ there is an intersection pairing*

$$
\begin{array}{ccc}
\widehat{\mathrm{CH}}^p(X) \otimes \widehat{\mathrm{CH}}^q(X) & \longrightarrow & \widehat{\mathrm{CH}}^{p+q}(X)_{\mathbf{Q}} \\
x \otimes y & \longmapsto & x \cdot y
\end{array}
$$

It turns $\underset{p \geq 0}{\oplus} \widehat{\mathrm{CH}}^p(X)_{\mathbf{Q}}$ into a commutative graded $\mathbf{Q}$-algebra.

Let $\zeta : \widehat{\mathrm{CH}}^p(X) \to \mathrm{CH}^p(X)$ be the map sending the class of (Z, g) to the class of Z, and let $\omega : \widehat{\mathrm{CH}}^p(X) \to A^{pp}(X)$ be the map sending (Z, g) to $dd^c g + \delta_Z$. Then

$$\zeta(x \cdot y) = \zeta(x)\,\zeta(y)$$

and

$$\omega(x \cdot y) = \omega(x)\,\omega(y)\,.$$

4.3.2 To sketch a proof of Theorem 4.5, let $y = (Y, g_Y) \in \widehat{Z}^p(X)$ and $z = (Z, g_Z) \in \widehat{Z}^q(X)$.

We first define a cycle $Y \cap Z$. For this we assume that the restrictions $Y_{\mathbf{Q}}$ and $Z_{\mathbf{Q}}$ of Y and Z to the generic fiber $X_{\mathbf{Q}}$ meet properly, i.e. the components of $|Y_{\mathbf{Q}}| \cap |Z_{\mathbf{Q}}|$ have codimension $p + q$ (the moving lemma allows one to make this hypothesis). It follows that there exists a well defined intersection cycle $Y_{\mathbf{Q}} \cdot Z_{\mathbf{Q}} \in Z^{p+q}(X_{\mathbf{Q}})$, supported on the closed set $|Y_{\mathbf{Q}}| \cap |Z_{\mathbf{Q}}|$. Let

$$\mathrm{CH}^p_Y(X) = \ker(\mathrm{CH}^p(X) \to \mathrm{CH}^p(X - Y))$$

be the Chow group with supports in Y, and $\mathrm{CH}^p_{\mathrm{fin}}(X)$ the union of the groups $\mathrm{CH}^p_Y(X)$ when $Y \subset X$ runs over all closed subsets with empty generic fiber. There is a canonical map

$$\mathrm{CH}^p_Y(X) \to \mathrm{CH}^p_{\mathrm{fin}}(X) \oplus Z^p(X_{\mathbf{Q}})\,.$$

One can define an intersection paring

$$\mathrm{CH}^p_Y(X) \otimes \mathrm{CH}^q_Z(X) \to \mathrm{CH}^{p+q}_{Y \cap Z}(X)_{\mathbf{Q}}\,.$$

One method to do so [8, 9, 11] is to interpret $\mathrm{CH}^p_Y(X)_{\mathbf{Q}}$ as the subspace of $K_0^Y(X)_{\mathbf{Q}}$ where the Adams operations ψ^k act by multiplication by k^p $(k \geq 1)$, and to use the tensor product

$$K_0^Y(X) \otimes K_0^Z(X) \to K_0^{Y \cap Z}(X)\,.$$

We let $Y \cap Z \in \mathrm{CH}^{p+q}_{\mathrm{fin}}(X)_{\mathbf{Q}} \oplus Z^{p+q}(X_{\mathbf{Q}})_{\mathbf{Q}}$ be the image of

$$[Y] \otimes [Z] \in \mathrm{CH}^p_Y(X) \otimes \mathrm{CH}^p_Z(X)$$

by the maps

$$\mathrm{CH}^p_Y(X) \otimes \mathrm{CH}^q_Z(X) \to \mathrm{CH}^{p+q}_{Y \cap Z}(X)_{\mathbf{Q}} \to \mathrm{CH}^{p+q}_{\mathrm{fin}}(X)_{\mathbf{Q}} \oplus Z^{p+q}(X_{\mathbf{Q}})_{\mathbf{Q}}\,.$$

Next we define a Green current for $Y \cap Z$. For this we write

$$dd^c\, g_Y + \delta_Y = \omega_Y$$

and

$$dd^c g_Z + \delta_Z = \omega_Z,$$

and we let

$$g_Y * g_Z = \delta_Y g_Z + g_Y \omega_Z.$$

However $g_Y \delta_Z$, being a product of currents, is not well defined a priori. But g_Y is defined up to the addition of a term $\partial(u) + \bar{\partial}(v)$ and one shows that g_Y can be chosen to be an L^1-form on $X(\mathbf{C}) - Y(\mathbf{C})$, with restriction an L^1-form η on $Z(\mathbf{C}) - Z(\mathbf{C}) \cap Y(\mathbf{C})$. We let $g_Y \delta_Z$ be the current defined by η on $Z(\mathbf{C})$ (see above Example 3.8):

$$g_Y \delta_Z(\omega) = \int_{Z(\mathbf{C}) - (Z(\mathbf{C}) \cap Y(\mathbf{C}))} \eta \omega.$$

To see that $g_Y * g_Z$ is a Green current for $Y \cap Z$ we proceed formally:

$$\begin{aligned}
dd^c(g_Y * g_Z) &= dd^c(\delta_Y g_Z) + dd^c(g_Y \omega_Z) \\
&= \delta_Y dd^c(g_Z) + dd^c(g_Y) \omega_Z \\
&= \delta_Y(\omega_Z - \delta_Z) + (\omega_Y - \delta_Y) \omega_Z \\
&= \omega_Y \omega_Z - \delta_Y \delta_Z \\
&= \omega_Y \omega_Z - \delta_{Y \cap Z}.
\end{aligned}$$

We refer to [9] for the justification of this series of equalities.

4.4 Functoriality

Let $f : X \rightarrow Y$ be a morphism.

Theorem 4.6 *For every $p \geq 0$ there is a morphism*

$$f^* : \widehat{\mathrm{CH}}^p(Y) \rightarrow \widehat{\mathrm{CH}}^p(X).$$

If the restriction of f to $X(\mathbf{C})$ is a smooth map of complex manifolds, there are morphisms

$$f_* : \widehat{\mathrm{CH}}^p(X) \rightarrow \widehat{\mathrm{CH}}^{p + \dim(Y) - \dim(X)}(Y).$$

Both f^* and f_* are compatible to ζ and ω. Furthermore

$$f^*(x \cdot y) = f^*(x) \cdot f^*(y)$$

and

$$f_*(x \cdot f^*(y)) = f_*(x) \cdot y.$$

4.5 Heights and Intersection Numbers

4.5.1 Let X be a projective regular flat scheme over $\mathbf{Z}$ and $Y \subset X$ a closed integral subscheme. We assume that X is equidimensional of dimension d and $\mathrm{codim}_X(Y) = p$. One can then define as follows a morphism

$$\int_Y : \widehat{\mathrm{CH}}^{d-p}(X) \to \mathbf{R}.$$

First, assume that $X = Y$ and that $x \in \widehat{\mathrm{CH}}^d(X)$ is the class of (Z, g_Z) where Z is a zero-cycle and $g_Z \in D^{d-1,d-1}(X(\mathbf{C}))$. The cycle Z is then a finite sum

$$Z = \sum_\alpha n_\alpha \, y_\alpha$$

where y_α is a closed point with finite residue field $k(y_\alpha)$, and there exist currents u and v such that $\eta_Z = g_Z + \partial(u) + \bar{\partial}(v)$ is smooth. By definition

$$\int_X x = \sum_\alpha n_\alpha \, \log \# (k(y_\alpha)) - \frac{1}{2} \int_{X(\mathbf{C})} \eta_Z.$$

In general we let g_Y be a Green current for Y in $X(\mathbf{C})$, and $y = (Y, g_Y)$. If $x \in \widehat{\mathrm{CH}}^{d-p}(Y)$ we have $x \cdot y \in \widehat{\mathrm{CH}}^d(X)$ and we define

$$\int_Y x = \int_X x \cdot y - \frac{1}{2} \int_{X(\mathbf{C})} \omega(x) \, g_Y.$$

One checks that this number is independent on the choice of g_Y.

Theorem 4.7 *The height of Y is*

$$h_{\overline{L}}(Y) = \int_Y \widehat{c}_1(\overline{L})^{d-p}.$$

Proof To prove Theorem 4.7 we shall check that the two properties in Theorem 2.13 hold true for the number $\int_Y \widehat{c}_1(\overline{L})^{d-p}$.

When $p = d$, Y is a closed point y and, if x is the class of $(y, 0)$ in $\widehat{CH}^d(X)$, we have

$$\int_X x = \log \#\kappa(y) = h_{\overline{L}}(Y).$$

Assume $\dim(Y) > 0$. Let g_Y be a Green current for Y and $y = (Y, g_Y)$. Choose a rational section s of L on Y, and an extension $\widetilde{s}$ of s to X. Then

$$\widehat{c}_1(\overline{L}) = (\text{div}(\widetilde{s}), -\log \|\widetilde{s}\|^2).$$

If $x = \widehat{c}_1(\overline{L})^{d-p-1}$ we get, from the definition of $\int_Y$,

$$\int_Y x\,\widehat{c}_1(\overline{L}) = \int_X x \cdot \widehat{c}_1(\overline{L}) \cdot y - \frac{1}{2}\int_{X(\mathbb{C})} \omega(x\,\widehat{c}_1(\overline{L}))\,g_Y. \tag{9}$$

But

$$x \cdot \widehat{c}_1(\overline{L}) \cdot y = x \cdot (\text{div}(\widetilde{s} \mid Y), -\log \|\widetilde{s}\|^2 * g_Y)$$
$$= x \cdot (\text{div}(s), -\log \|\widetilde{s}\|^2 \delta_Y + c_1(\overline{L})\,g_Y).$$

If $x =: \widehat{c}_1(\overline{L})^{d-p-1}$ is the class of (Z, g_Z), we get

$$x \cdot \widehat{c}_1(\overline{L}) \cdot y = (Z \cdot \text{div}(s), \omega(x)(-\log \|\widetilde{s}\|^2 \delta_Y + c_1(\overline{L})\,g_Y) + g_Z\,\delta_{\text{div}(s)}). \tag{10}$$

Since

$$\int_X (Z \cdot \text{div}(s), g_Z\,\delta_{\text{div}(s)}) = \int_{\text{div}(s)} x$$

we deduce from (10) that

$$\int_X x \cdot \widehat{c}_1(\overline{L}) \cdot y = \int_{\text{div}(s)} x - \frac{1}{2}\int_{Y(\mathbb{C})} \omega(x)\log \|s\|^2 + \frac{1}{2}\int_{X(\mathbb{C})} \omega(x)\,c_1(\overline{L})\,g_Y. \tag{11}$$

Since $\omega(x\,\widehat{c}_1(\overline{L})) = \omega(x)\,c_1(\overline{L}_{\mathbb{C}})$, (9) and (11) imply that

$$\int_Y \widehat{c}_1(\overline{L})^{d-p} = \int_{\text{div}(s)} \widehat{c}_1(\overline{L})^{d-p-1} - \frac{1}{2}\int_{Y(\mathbb{C})} c_1(\overline{L})^{d-p-1}\log \|s\|. \qquad \square$$

References

1. S.Y. Arakelov, Intersection theory of divisors on an arithmetic surface. Izv. Akad. Nauk SSSR Ser. Mat. **38**, 1179–1192 (1974)
2. J.-B. Bost, H. Gillet, C. Soulé, Heights of projective varieties and positive Green forms. J. Amer. Math. Soc. **7**, 903 (1994)
3. N. Bourbaki, *Algèbre commutative IV, Eléments de Mathématique* (Hermann, Paris)
4. N. Bourbaki, *Algèbre I, Eléments de Mathématique* (Hermann, Paris)
5. N. Bourbaki, *Algèbre II, Eléments de Mathématique* (Hermann, Paris)
6. G. Faltings, Diophantine approximation on abelian varieties. Ann. Math. (2) **133**, 549–576 (1991)
7. H. Gillet, K-theory and intersection theory, in *Handbook of K-theory*, ed. by E.M. Friedlander, D.R. Grayson (Springer, New York, 2005), pp. 235–294
8. H. Gillet, C. Soulé, Intersection theory using Adams operations. Invent. Math. **90**, 243–277 (1987)
9. H. Gillet, C. Soulé, Arithmetic intersection theory. Inst. Hautes études Sci. Publ. Math. **72**, 93–174 (1990)
10. Q. Liu, *Algebraic Geometry and Arithmetic Curves*. Oxford Graduate Texts in Mathematics (Oxford University Press, Oxford, 2002)
11. C. Soulé, D. Abramovich, J.-F. Burnol, J. Kramer, *Lectures on Arakelov Geometry* (Cambridge University Press, Cambridge, 1991)

Chapter II: Minima and Slopes of Rigid Adelic Spaces

Éric Gaudron

1 Introduction

We propose here a lecture on the geometry of numbers for normed (adelic) vector spaces over an algebraic extension of $\mathbb{Q}$. We shall define slopes and several type of minima for these objects and we shall compare them.

First, let us recall some basic notions of the classical geometry of numbers. Let Ω be a free $\mathbb{Z}$-module of rank $n \geq 1$ and let $\| \cdot \|$ be an Euclidean norm on $\Omega \otimes_{\mathbb{Z}} \mathbb{R}$. We shall say that the couple $(\Omega, \| \cdot \|)$ is an Euclidean lattice of rank n. To such a lattice are associated n positive real numbers, called the *successive minima* of $(\Omega, \| \cdot \|)$: for all $i \in \{1, \ldots, n\}$,

$$\lambda_i (\Omega, \| \cdot \|) = \min \{r > 0 ;\ \dim \mathrm{Vect}_{\mathbb{R}} \left(x \in \Omega ;\ \|x\| \leq r \right) \geq i\}.$$

It is also the minimum of the set of max $\{\|x_1\|, \ldots, \|x_i\|\}$ formed with linearly independent vectors $x_1, \ldots, x_i \in \Omega$. We have $0 < \lambda_1(\Omega, \| \cdot \|) \leq \cdots \leq \lambda_n(\Omega, \| \cdot \|)$. Given a $\mathbb{Z}$-basis $e_1, \ldots, e_n$ of Ω, the (co-)volume of Ω is the positive real number

$$\mathrm{vol}(\Omega) = \det \left(\langle e_i, e_j \rangle \right)_{1 \leq i, j \leq n}^{1/2}$$

The author was supported by the ANR Grant Gardio 14-CE25-0015.

É. Gaudron (✉)
Université Clermont Auvergne, CNRS, LMBP, Clermont-Ferrand, France
e-mail: Eric.Gaudron@uca.fr

© The Editor(s) (if applicable) and The Author(s), under exclusive license
to Springer Nature Switzerland AG 2021
E. Peyre, G. Rémond (eds.), *Arakelov Geometry and Diophantine Applications*,
Lecture Notes in Mathematics 2276, https://doi.org/10.1007/978-3-030-57559-5_3

where $\langle \cdot, \cdot \rangle$ denotes the scalar product on $\Omega \otimes_{\mathbb{Z}} \mathbb{R}$ associated to $\| \cdot \|$. Let us define

$$c_{\mathrm{I}}(n, \mathbb{Q}) = \sup \frac{\lambda_1(\Omega, \| \cdot \|)}{\mathrm{vol}(\Omega)^{1/n}}$$

and

$$c_{\mathrm{II}}(n, \mathbb{Q}) = \sup \left(\frac{\lambda_1(\Omega, \| \cdot \|) \cdots \lambda_n(\Omega, \| \cdot \|)}{\mathrm{vol}(\Omega)} \right)^{1/n}$$

where the suprema are taken over Euclidean lattices $(\Omega, \| \cdot \|)$ of rank n. The square $\gamma_n = c_{\mathrm{I}}(n, \mathbb{Q})^2$ is nothing but the famous Hermite constant. Its exact value is only known for $n \leq 8$ and $n = 24$. It can also be characterized as the smallest positive real number c such that, for all $(a_0, \ldots, a_n) \in \mathbb{Z}^{n+1} \setminus \{0\}$, there exists $(x_0, \ldots, x_n) \in \mathbb{Z}^{n+1} \setminus \{0\}$ satisfying $a_0 x_0 + \cdots + a_n x_n = 0$ and

$$\sum_{i=0}^{n} x_i^2 \leq c \left(\sum_{i=0}^{n} a_i^2 \right)^{1/n}.$$

Minkowski proved the following statement (see [19, § 51]):

Theorem (Minkowski) *For every positive integer n, we have $c_{\mathrm{I}}(n, \mathbb{Q}) = c_{\mathrm{II}}(n, \mathbb{Q}) \leq \sqrt{n}$.*

We shall generalize this framework in the following manner:

$$\mathbb{Q} \longrightarrow \text{Algebraic extension } K/\mathbb{Q}$$

$$\text{Euclidean lattice } (\Omega, \| \cdot \|) \longrightarrow \text{Rigid adelic space } E \text{ over } K$$

$$\text{Minimum } \lambda_i(\Omega, \| \cdot \|) \longrightarrow \text{Minimum } \Lambda_i(E)$$

$$\text{Volume } \mathrm{vol}(\Omega) \longrightarrow \text{Height } H(E)$$

$$-\log \mathrm{vol}(\Omega)^{1/n} \longrightarrow \text{Slope } \mu(E).$$

Actually, in the highly flexible world of rigid adelic spaces, there exist numerous types of possible successive minima, having an interest according to the problems addressed. To be over an algebraic extension of $\mathbb{Q}$ which is not necessarily finite brings some new perspectives, issues and results. In particular we shall explain how to compute the Hermite constants of the algebraic closure $\overline{\mathbb{Q}}$ of $\mathbb{Q}$.

2   Rigid Adelic Spaces

Let us begin with a Reader's Digest of [14, § 2].

2.1 Algebraic Extensions of $\mathbb{Q}$

Let $K/\mathbb{Q}$ be an algebraic extension. Let $V(K)$ be the set of places of K (equivalence classes of non trivial absolute values over K). We can write this set as the projective limit $\varprojlim_L V(L)$ over finite subextensions $\mathbb{Q} \subset L \subset K$ of K. The discrete topology on $V(L)$ induces a topology on $V(K)$ by projective limit. It coincides with the topology generated by the compact open subsets $V_v(K) = \{w \in V(K); \ w_{|L} = v\}$ for $v \in V(L)$ and L varies among number fields contained in K. On $V(K)$ can be defined a Borel measure σ characterized by

$$\sigma(V_v(K)) = \frac{[L_v : \mathbb{Q}_v]}{[L : \mathbb{Q}]} \quad \text{for } v \in V(L)$$

($\mathbb{Q}_v = \mathbb{Q}_p$ or $\mathbb{R}$ depending on v, p-adic or archimedean). We have $\sigma(V_p(K)) = 1$ for all $p \in V(\mathbb{Q})$. For $v \in V(K)$ we denote by K_v the topological completion of K at v and $|\cdot|_v$ is the unique absolute value on K_v such that $|p|_v \in \{1, p, p^{-1}\}$ for every prime number p. Then the product formula is written

$$\forall x \in K\backslash\{0\}, \qquad \int_{V(K)} \log |x|_v \, d\sigma(v) = 0.$$

Furthermore, the *adèles* of K is the tensor product $\mathbb{A}_K = K \otimes_\mathbb{Q} \mathbb{A}_\mathbb{Q}$ of K with the adèles of $\mathbb{Q}$:

$$\mathbb{A}_\mathbb{Q} = \left\{ (x_p)_p \in \prod_{p \in V(\mathbb{Q})} \mathbb{Q}_p \, ; \, \{p \text{ prime}; |x_p|_p > 1\} \text{ is finite} \right\}.$$

If K is a number field, $\mathbb{A}_K$ is the usual adèle ring and, for an arbitrary algebraic extension $K/\mathbb{Q}$, one has $\mathbb{A}_K = \bigcup_{L \subset K, [L:\mathbb{Q}]<\infty} \mathbb{A}_L$.

2.2 Rigid Adelic Spaces

From now on, the letter K always denotes an algebraic extension of $\mathbb{Q}$.

Definition 1 *An adelic space E is a K-vector space of finite dimension endowed with norms $\|\cdot\|_{E,v}$ on $E \otimes_K K_v$ for every $v \in V(K)$.*

The (adelic) *standard space* of dimension $n \geq 1$ is the vector space K^n endowed with the following norms:

$$\forall x = (x_1, \ldots, x_n) \in K_v^n, \quad |x|_v = \begin{cases} \left(|x_1|_v^2 + \cdots + |x_n|_v^2\right)^{1/2} & \text{if } v \mid \infty \\ \max\{|x_1|_v, \ldots, |x_n|_v\} & \text{if } v \nmid \infty. \end{cases}$$

Given an adelic space E over K and $v \in V(K)$, a basis $(e_1, \ldots, e_n)$ of $E \otimes_K K_v$ is said to be *orthonormal* if, for all $(x_1, \ldots, x_n) \in K_v^n$, we have $\| \sum_{i=1}^n x_i e_i \|_{E,v} = |(x_1, \ldots, x_n)|_v$.

Definition 2 *A* rigid adelic space *is an adelic space E for which there exist an isomorphism $\varphi \colon E \to K^n$ and an adelic matrix $A = (A_v)_{v \in V(K)} \in \mathrm{GL}_n(\mathbb{A}_K)$ such that*

$$\forall x \in E \otimes_K K_v, \quad \|x\|_{E,v} = |A_v \varphi_v(x)|_v$$

where $\varphi_v = \varphi \otimes \mathrm{id}_{K_v} \colon E \otimes_K K_v \to K_v^n$ is the natural extension of φ to $E \otimes_K K_v$.

In looser terms, a rigid adelic space is a compact deformation of a standard space.

Remarks

(i) Actually, if E is a rigid adelic space over K of dimension n, for every isomorphism $\varphi \colon E \to K^n$, there exists $A \in \mathrm{GL}_n(\mathbb{A}_K)$, upper triangular, such that (φ, A) defines the adelic structure on E.

(ii) If $x \in E \setminus \{0\}$ there exists a number field $K_0 \subset K$ such that $A \in \mathrm{GL}_n(\mathbb{A}_{K_0})$ and $\varphi(x) \in K_0^n$. Thus, outside a compact subset of $V(K)$ (finite union of some $V_v(K)$ with $v \in V(K_0)$), we have $\|x\|_{E,v} = 1$ and A_v is an isometry.

(iii) A rigid adelic space is an adelic space with an orthonormal basis at each $v \in V(K)$ but the converse is not true.

Examples of Rigid Adelic Spaces

(i) K^n (standard space).

(ii) Let $(\Omega, \| \cdot \|)$ be an Euclidean lattice and $(e_1, \ldots, e_n)$ a $\mathbb{Z}$-basis of Ω. We can consider $E_\Omega = \Omega \otimes_{\mathbb{Z}} \mathbb{Q}$ over $K = \mathbb{Q}$, endowed with the norm $\| \cdot \|$ at the archimedean place of $\mathbb{Q}$ and $\| \sum_{i=1}^n x_i e_i \|_{E_\Omega, p} = \max_{1 \le i \le n} \{|x_i|_p\}$ at every prime p ($x_i \in \mathbb{Q}_p$). This definition does not depend on the choice of the $\mathbb{Z}$-basis.

(iii) When K is a number field with ring of integers $\mathcal{O}_K$, we have a one-to-one correspondence between rigid adelic spaces over K and Hermitian vector bundles over $\mathrm{Spec}\, \mathcal{O}_K$. Indeed, let E be a rigid adelic space over K. The projective $\mathcal{O}_K$-module of finite type $\mathcal{E} = \{x \in E ; \ \forall v \in V(K) \setminus V_\infty(K), \ \|x\|_{E,v} \le 1\}$ endowed with the Hermitian norms (invariant by complex conjugation) $\| \cdot \|_\sigma = \| \cdot \|_{E,v}$ at embeddings $\sigma \colon K \hookrightarrow \mathbb{C}$ with associated place $v = \{\sigma, \overline{\sigma}\}$ form a Hermitian vector bundle over $\mathrm{Spec}\, \mathcal{O}_K$. An example is given at the beginning of Sect. 2.2 in Chapter I.

Definition 3 *Given two adelic spaces E, F over K, a linear map $f \colon E \to F$ is an* isometry *if for all $v \in V(K)$ and $x \in E \otimes_K K_v$, we have $\|f_v(x)\|_{F,v} = \|x\|_{E,v}$, where $f_v = f \otimes \mathrm{id}_{K_v}$.*

The adelic spaces E and F will be called *isometric* if there exists an isomorphism $E \to F$ which is an isometry.

Operations on Adelic Spaces Let E, E' be adelic spaces over K and $F \subset E$ a vector subspace. One can consider the following adelic spaces:

Induced Structure F with norms $\| \cdot \|_{E,v}$ restricted to $F \otimes_K K_v$.

Quotient E/F with quotient norms

$$\|\bar{x}\|_{E/F,v} = \inf \left\{ \|z\|_{E,v} ; \ z \in E \otimes_K K_v, \ z = x \mod F \otimes_K K_v \right\}.$$

Dual $E^{\vee} = \operatorname{Hom}_K(E, K)$ (linear forms) with operator norms

$$\forall \ell \in E^{\vee} \otimes_K K_v, \quad \|\ell\|_{E^{\vee},v} = \sup \left\{ \frac{|\ell(z)|_v}{\|z\|_{E,v}} ; \ z \in E \otimes_K K_v \backslash \{0\} \right\}.$$

Given an Euclidean lattice $(\Omega, \| \cdot \|)$, the dual of E_{Ω} corresponds to the dual lattice $\Omega^* = \{\varphi \in (\Omega \otimes_{\mathbb{Z}} \mathbb{R})^{\vee}; \ \varphi(\Omega) \subset \mathbb{Z}\}$ with the gauge[1] of the polar body $C^{\circ} = \{\varphi \in (\Omega \otimes_{\mathbb{Z}} \mathbb{R})^{\vee}; \ \varphi(C) \subset [-1, 1]\}$ of the unit ball $C = \{x \in \Omega \otimes_{\mathbb{Z}} \mathbb{R}; \ \|x\| \le 1\}$.

(Hermitian) Direct Sum $E \oplus E'$ with norm at $v \in V(K)$ given by

$$\|(x, x')\|_{E \oplus E',v} = \begin{cases} \left(\|x\|_{E,v}^2 + \|x'\|_{E',v}^2 \right)^{1/2} & \text{if } v \mid \infty, \\ \max \left\{ \|x\|_{E,v}, \|x'\|_{E',v} \right\} & \text{if } v \nmid \infty, \end{cases}$$

for all $x \in E \otimes_K K_v$ and $x' \in E' \otimes_K K_v$.

Operator Norm $\operatorname{Hom}_K(E, E')$ (linear maps) with

$$\|f\|_v = \sup \left\{ \frac{\|f(x)\|_{E',v}}{\|x\|_{E,v}} ; \ x \in E \otimes_K K_v \backslash \{0\} \right\}$$

for all $f \in \operatorname{Hom}_K(E, E') \otimes_K K_v$. Using the natural isomorphism $E \otimes_K E' \simeq \operatorname{Hom}_K(E^{\vee}, E')$, we get an adelic structure on $E \otimes E'$, denoted $E \otimes_{\varepsilon} E'$ in the sequel (the ε refers to the injective norm for tensor product of Banach spaces).

Tensor Product Assume $E = (\varphi, A)$ and $E' = (\varphi', A')$ are *rigid* adelic spaces. The tensor product $E \otimes_K E'$ is endowed with the (rigid) structure given by $(\varphi \otimes \varphi', A \otimes A')$. It is the same as saying that local orthonormal bases of $E \otimes_K K_v$ and $E' \otimes_K K_v$ give an orthonormal basis by tensor product.

Symmetric Power When $E = (\varphi, A)$ is a rigid adelic space and $i \in \mathbb{N} \backslash \{0\}$, the symmetric power $S^i E$ is endowed with $(S^i(\varphi), S^i(A))$. It corresponds to the quotient structure of the tensor norm by the natural surjection $E^{\otimes i} \to S^i E$. We

---

[1] Recall that the gauge of a set C is the function $j(x) = \inf \{\lambda > 0; \ x/\lambda \in C\}$. When C is a symmetric compact convex set with non-empty interior in a vector space U, then the gauge defines a norm on U.

have $\|x^i\|_{S^i E,v} = \|x\|_{E,v}^i$ for all $x \in E \otimes_K K_v$. If $e_1, \ldots, e_n$ is an orthonormal basis of $E \otimes_K K_v$, then the vectors $e_1^{i_1} \cdots e_n^{i_n}$ with $i_j \in \mathbb{N}$ and $i_1 + \cdots + i_n = i$ form an orthogonal basis of $S^i E$ and

$$\|e_1^{i_1} \cdots e_n^{i_n}\|_{S^i E,v} = \left(\frac{i_1! \cdots i_n!}{i!}\right)^{1/2} \quad \text{if } v \mid \infty \text{ and } 1 \text{ otherwise.}$$

Exterior Power When $E = (\varphi, A)$ is a rigid adelic space with dimension n and $i \in \{1, \ldots, n\}$, the exterior power $\bigwedge^i E$ is endowed with the rigid structure $(\bigwedge^i \varphi, \bigwedge^i A)$. Given $v \in V(K)$, an orthonormal basis $(e_1, \ldots, e_n)$ of $E \otimes_K K_v$ induces an orthonormal basis $(e_{j_1} \wedge \cdots \wedge e_{j_i})_{1 \le j_1 < \cdots < j_i \le n}$ of $\bigwedge^i E \otimes_K K_v$. Note that it differs by a coefficient $\sqrt{i!}$ from the quotient norm $E^{\otimes i} \to \bigwedge^i E$. When $i = \dim E$, the exterior power $\bigwedge^i E$ is called the determinant of E and denoted by $\det E$.

Scalar Extension Let K'/K be an algebraic extension and $E = (\varphi, A)$ be a rigid adelic space. We endow $E \otimes_K K'$ with the rigid adelic structure given by $(\varphi \otimes \mathrm{id}_{K'}, A)$ where $\varphi \otimes \mathrm{id}_{K'} : E \otimes_K K' \to (K')^n$ is induced by φ and A is viewed in $\mathrm{GL}_n(\mathbb{A}_{K'})$ by means of the diagonal embedding $\mathbb{A}_K \hookrightarrow \mathbb{A}_{K'}$. We denote by $E_{K'}$ the adelic space obtained in this way.

These definitions do not depend on the chosen couple (φ, A). Let us mention that every rigid adelic space E over K can be written as the scalar extension $E_0 \otimes_{K_0} K$ of a rigid adelic space E_0 over a number field K_0: choose K_0 such that $A \in \mathrm{GL}_n(\mathbb{A}_{K_0})$ and define $E_0 = \varphi^{-1}(K_0^n)$ with the structure given by $(\varphi_{|E_0}, A)$.

Theorem 4 *If E and E' are rigid adelic spaces, all these adelic structures are rigid except (in general) the one on $\mathrm{Hom}_K(E, E')$ and $E \otimes_\varepsilon E'$ (operator norms). Moreover the canonical isomorphisms $E \simeq (E^\vee)^\vee$ and, for $F \subset E$ a linear subspace, $E/F \simeq (F^\perp)^\vee$ (where $F^\perp$ denotes the annihilator $\{\ell \in E^\vee ; \ \ell(F) = \{0\}\}$) are isometries.*

Proof See [14, Proposition 3.6].   □

We can also prove that, given a rigid adelic space E over K with dimension n and $r \in \{0, \ldots, n\}$, the pairing $\bigwedge^{n-r} E \otimes \bigwedge^r E \to \det E$, $x \otimes y \mapsto x \wedge y$, induces an isometric isomorphism

$$\bigwedge^{n-r} E \simeq (\det E) \otimes \left(\bigwedge^r E\right)^\vee.$$

Moreover the natural map $\bigwedge^r (E^\vee) \to (\bigwedge^r E)^\vee$, $\varphi_1 \wedge \cdots \wedge \varphi_r \mapsto (x_1 \wedge \cdots \wedge x_r \mapsto \varphi_1(x_1) \cdots \varphi_r(x_r))$ is an isomorphism of rigid adelic spaces (whereas it is false if we replace the exterior power by the symmetric power).

Height, Degree and Slope of Rigid Adelic Spaces Let E be a rigid adelic space over K defined by (φ, A).

- The *height of E* is the positive real number

$$H(E) = \exp \int_{V(K)} \log |\det A_v|_v \, d\sigma(v).$$

If $E = \{0\}$ one has $H(E) = 1$. This definition does not depend on the choice of (φ, A) and the integral converges since $|\det A_v|_v = 1$ for v outside a compact subset of $V(K)$.
- The *(Arakelov) degree of E* is

$$\deg E = -\log H(E) = -\int_{V(K)} \log |\det A_v|_v \, d\sigma(v).$$

- The *slope of E* is $\mu(E) = \dfrac{\deg E}{\dim E}$ (only for $E \neq \{0\}$).

In the literature, a rigid adelic space is often denoted with a bar ($\overline{E}$ instead of E) and its degree and slope are accompanied by a hat ($\widehat{\deg}\, \overline{E}$ instead of $\deg E$). Also note that from the definitions, the height and degree of a rigid adelic space are those of its determinant.

Examples

(1) $H(K^n) = 1$, $\deg K^n = \mu(K^n) = 0$ for all $n \in \mathbb{N}\setminus\{0\}$.
(2) If $(\Omega, \|\cdot\|)$ is an Euclidean lattice, then $H(E_\Omega) = \mathrm{vol}(\Omega)$. Indeed, if $(e_1, \ldots, e_n)$ is a $\mathbb{Z}$-basis of Ω, we have $H(E_\Omega) = |\det A|$ where the matrix A characterizes the norm: for every $(x_1, \ldots, x_n) \in \mathbb{R}^n$, $\|x_1 e_1 + \cdots + x_n e_n\| = |(x_1, \ldots, x_n)A|$ that is, $A^t A = (\langle e_i, e_j \rangle)_{1 \le i,j \le n}$.
(3) If K is a number field we have $H(E) = \prod_{v \in V(K)} |\det A_v|_v^{\frac{[K_v : \mathbb{Q}_v]}{[K:\mathbb{Q}]}}$.

Proposition 5 *Let E and E' be rigid adelic spaces over K and $F \subset E$ a linear subspace endowed with its induced adelic structure. Then*

$$H(E/F) = \frac{H(E)}{H(F)} \qquad\qquad (\deg E = \deg F + \deg E/F)$$

$$H(E^\vee) = H(E)^{-1} \qquad\qquad (\deg E^\vee = -\deg E)$$

$$H(E \oplus E') = H(E)H(E') \qquad\qquad (\deg E \oplus E' = \deg E + \deg E')$$

$$H(E \otimes E') = H(E)^{\dim E'} H(E')^{\dim E} \qquad (\mu(E \otimes E') = \mu(E) + \mu(E'))$$

$$H(F^\perp) = \frac{H(F)}{H(E)} \qquad\qquad (\deg F^\perp = \deg F - \deg E).$$

If $n = \dim E$ and $i \in \{1, \ldots, n\}$, we also have $H\left(\bigwedge^i E\right) = H(E)^{\binom{n-1}{i-1}}$, that is, $\mu\left(\bigwedge^i E\right) = i\mu(E)$. Moreover, for all $i \in \mathbb{N}$, we have

$$\mu\left(S^i E\right) = i\mu(E) + \left(2\binom{i+n-1}{n-1}\right)^{-1} \sum_{\substack{(j_1,\ldots,j_n)\in\mathbb{N}^n \\ j_1+\cdots+j_n=i}} \log \frac{i!}{j_1!\cdots j_n!}.$$

Proof See [12, Lemma 7.3], [13, § 2.7], [14, Proposition 3.6]. □

Furthermore, height, degree and slope are invariant by scalar extension: if K'/K is algebraic, then $H(E_{K'}) = H(E)$, $\deg E_{K'} = \deg E$ and $\mu(E_{K'}) = \mu(E)$. Also note that we have an asymptotic estimate

$$\mu\left(S^i E\right) = i\mu(E) + \frac{i}{2}(H_n - 1)(1 + o(1)) \quad \text{when } i \to +\infty$$

in terms of the harmonic number $H_n = \sum_{h=1}^n 1/h$ (see [12, Annex]). It may be viewed as a particular case of the arithmetic Hilbert-Samuel theorem (see Sect. 5 in Chapter III).

The following statement is the key result for the existence of the Harder-Narasimhan filtration of a rigid adelic space which shall be established later (see page 52).

Proposition 6 *Let F and G be linear subspaces of a rigid adelic space over K. Then*

$$H(F + G)H(F \cap G) \leq H(F)H(G)$$

that is, $\deg F + \deg G \leq \deg(F + G) + \deg F \cap G$.

Proof Let $\iota\colon F/F\cap G \to (F+G)/G$ be the natural isomorphism. For all $v \in V(K)$ and $x \in (F/F \cap G) \otimes_K K_v$, we have $\|\iota_v(x)\|_{(F+G)/G,v} \leq \|x\|_{F/F\cap G,v}$ (here $\iota_v = \iota \otimes \mathrm{id}_{K_v}$). In particular, if $e_1, \ldots, e_m$ is an orthonormal basis of $(F/F \cap G) \otimes_K K_v$, then

$$\|(\det \iota_v)(e_1 \wedge \cdots \wedge e_m)\|_{\det(F+G)/G,v}$$
$$= \|\iota_v(e_1) \wedge \cdots \wedge \iota_v(e_m)\|_{\det(F+G)/G,v}$$
$$\underset{\underset{\text{Hadamard inequality}}{\uparrow}}{\leq} \prod_{i=1}^m \|\iota_v(e_i)\|_{(F+G)/G,v} \leq \prod_{i=1}^m \|e_i\|_{F/F\cap G,v} = 1.$$

In other words, the operator norm $\| \det \iota \|_v$ of $\det \iota$ at v is smaller than 1. Thus, using Proposition 5, we get

$$\frac{H(F+G)H(F \cap G)}{H(F)H(G)} = \frac{H((F+G)/G)}{H(F/F \cap G)}$$

$$= H\left((\det F/F \cap G)^\vee \otimes \det ((F+G)/G) \right)$$

$$= \exp \int_{V(K)} \log \| \det \iota \|_v \, d\sigma(v) \leq \exp 0 = 1. \qquad \square$$

A slightly more natural proof can be obtained from Proposition 42.

Heights of Points Let E be an adelic space over K.

Definition 7 *We shall say that the adelic space E is integrable if, for all $x \in E \backslash \{0\}$, the function $V(K) \to \mathbb{R}$, $v \mapsto \log \|x\|_{E,v}$ is σ-integrable.*

A rigid adelic space is integrable as well as ε-tensor products of finitely many rigid adelic spaces. Indeed we have:

Lemma 8 *Let E be rigid adelic space and F be an integrable adelic space over K. Then $E \otimes_\varepsilon F$ is integrable.*

Proof Using the isometric isomorphism $E \otimes_\varepsilon F \simeq \operatorname{Hom}(E^\vee, F)$, it amounts to proving that $\widetilde{f} \colon v \mapsto \log \|f\|_v$ is σ-integrable for every $f \in \operatorname{Hom}(E^\vee, F) \backslash \{0\}$, that is, this function is Borel and its absolute value has finite integral. For the measurability, choose a number field $K_0 \subset K$ such that E, F, f are defined over K_0. Then $\widetilde{f}$ is the composite of $v \mapsto v_{|K_0}$ and $v_0 \in V(K_0) \mapsto \log \|f\|_{v_0}$. This latter function is measurable since every subset of $V(K_0)$ (endowed with its discrete topology) is measurable. As for the restriction map $v \mapsto v_{|K_0}$, it is continuous by definition of the topology put on $V(K)$. Thus $\widetilde{f}$ is Borel and we shall now prove that $\int_{V(K)} |\widetilde{f}| < +\infty$. Let $(e_1, \ldots, e_n)$ be a K-basis of E. Since E is rigid, there exists $a = (a_p)_{p \in V(\mathbb{Q})} \in \mathbb{A}_\mathbb{Q}^\times$ such that, for all $v \in V(K)$ above $p \in V(\mathbb{Q})$ and all $x = \sum_{i=1}^n x_i e_i \in E \otimes_K K_v$, we have

$$|a_p|_v^{-1} \max_{1 \leq i \leq n} \{|x_i|_v\} \leq \|x\|_{E,v} \leq |a_p|_v \max_{1 \leq i \leq n} \{|x_i|_v\}.$$

Since $f(x) = x_1 f(e_1) + \cdots + x_n f(e_n)$, the triangle inequality yields $\|f\|_v \leq |b_p|_v \max_{1 \leq i \leq n} \{\|f(e_i)\|_{F,v}\}$ where $b_\infty = n a_\infty$ and, if p is prime number, $b_p = a_p$. Moreover, since $f \neq 0$, one can choose $m \in \{1, \ldots, n\}$ such that $f(e_m) \neq 0$ and we bound $\|f\|_v^{-1} \leq \|e_m\|_{E,v} / \|f(e_m)\|_{F,v} \leq |b_p|_v / \|f(e_m)\|_{F,v}$. Thus $|\widetilde{f}(v)| = |\log \|f\|_v| = \log \max \left\{ \|f\|_v, \|f\|_v^{-1} \right\}$ is bounded above by

$$\log |b_p|_v + \max_{1 \leq i \leq n} \left\{ \log \|f(e_i)\|_{F,v}, \, -\log \|f(e_m)\|_{F,v} \right\}.$$

In this bound, we can restrict to indices i such that $f(e_i) \neq 0$. Since F is integrable, then each function appearing in the maximum is σ-integrable. We conclude with the fact that the maximum of a finite number of σ-integrable functions is still σ-integrable (since $|\max\{a, b\}| \leq |a| + |b|$). $\square$

The integrability condition is the minimal condition which allows to define the height of a vector of an adelic space.

Definition 9 *Let E be an integrable adelic space over K and $x \in E$. The height $H_E(x)$ is the nonnegative real number:*

$$H_E(0) = 0 \quad \text{and if } x \neq 0, \quad H_E(x) = \exp \int_{V(K)} \log \|x\|_{E,v} \, d\sigma(v).$$

The product formula entails that H_E is a projective height, that is, $H_E(\lambda x) = H_E(x)$ for all $\lambda \in K \backslash \{0\}$.

Examples

1. For all $x = (x_1, \ldots, x_n) \in \mathbb{Z}^n \backslash \{0\}$, one has

$$H_{\mathbb{Q}^n}(x) = \left(x_1^2 + \cdots + x_n^2\right)^{1/2} \gcd(x_1, \ldots, x_n)^{-1}.$$

2. Let $(\Omega, \|\cdot\|)$ be an Euclidean lattice and $x \in E_\Omega$. Then there exists $d_x \in \mathbb{Q} \backslash \{0\}$ such that $H_{E_\Omega}(x) = \|d_x x\|$.
3. When E is a rigid adelic space of dimension 1, one has $H_E(x) = H(E)$ for all $x \in E \backslash \{0\}$.
4. When K is a number field, one has

$$\forall x \in E, \quad H_E(x) = \prod_{v \in V(K)} \|x\|_{E,v}^{[K_v:\mathbb{Q}_v]/[K:\mathbb{Q}]}.$$

5. Let F be the hyperplane $a_1 x_1 + \cdots + a_n x_n = 0$ of K^n (given by $(a_1, \ldots, a_n) \in K^n \backslash \{0\}$). Then $H(F) = H_{K^n}(a_1, \ldots, a_n)$ (since $H(F) = H(F^\perp)$).

Note that when E is a rigid adelic space, the height H_E is invariant by scalar extension: for all $x \in E$, for every algebraic extension K'/K, one has $H_{E \otimes_K K'}(x) = H_E(x)$.

Proposition 10 (Convexity Inequality for Heights) *Let N be a positive integer and $E_1, \ldots, E_N$ be integrable adelic spaces over K. Then the direct sum $E_1 \oplus \cdots \oplus E_N$ is integrable. Moreover, for all $(x_1, \ldots, x_N) \in E_1 \oplus \cdots \oplus E_N$, we have*

$$\left(\sum_{i=1}^N H_{E_i}(x_i)^2\right)^{1/2} \leq H_{E_1 \oplus \cdots \oplus E_N}(x_1, \ldots, x_N).$$

Proof For the integrability, we can restrict to $N = 2$. Observe that for positive real numbers a, b, we have

$$|\log(a + b)| \leq \log 2 + \log \max \left\{ a, \frac{1}{a} \right\} + \log \max \left\{ b, \frac{1}{b} \right\}$$

$$= \log 2 + |\log a| + |\log b|$$

and $|\log \max\{a, b\}| \leq |\log a| + |\log b|$. Applying this to $a = \|x_1\|_{E_1,v}^2$ and $b = \|x_2\|_{E_2,v}^2$ the result comes from the definition of $E_1 \oplus E_2$. As for the height inequality, we proceed as in [13, Lemma 2.2]. Applying Jensen inequality on the probability space $(V_\infty(K), \sigma)$ to the convex function $u \colon \mathbb{R} \to \mathbb{R}$, $u(x) = \log(1 + e^x)$, we get

$$1 + \exp \int_{V_\infty(K)} \log f \leq \exp \int_{V_\infty(K)} \log(1 + f)$$

for every nonnegative function f. By direct induction, we have

$$\sum_{i=1}^{N} \exp \int_{V_\infty(K)} \log f_i \leq \exp \int_{V_\infty(K)} \log(f_1 + \cdots + f_N)$$

for all nonnegative functions $f_1, \ldots, f_N$. Choosing $f_i(v) = \|x_i\|_{E_i,v}^2$ we get the convexity inequality but with only the archimedean part of the heights. To complete with the ultrametric part, we multiply both sides by $\exp \int_{V(K) \setminus V_\infty(K)} \log \max\{f_1, \ldots, f_N\}$ and we bound from below this number by $\exp \int_{V(K) \setminus V_\infty(K)} \log f_i$ for each i. ☐

3 Minima and Slopes

3.1 Successive Minima

Let E be a rigid adelic space with dimension $n \geq 1$. We denote $\Lambda_1(E) = \inf\{H_E(x); \ x \in E \setminus \{0\}\}$. We define three types of successive minima associated to E (still others exist in the literature, see [14]) which have been respectively inspired by the articles [8, 23] and [25]. Let $i \in \{1, \ldots, n\}$.
Bost-Chen minima:

$$\Lambda^{(i)}(E) = \sup\{\Lambda_1(E/F); \ F \subset E \text{ linear subspace, } \dim F \leq i - 1\}$$

Roy-Thunder minima:

$$\Lambda_i(E) = \inf\{\max\{H_E(x_1), \dots, H_E(x_i)\}; \ \dim \mathrm{Vect}_K(x_1, \dots, x_i) = i\}$$

Zhang minima: $Z_i(E) = \inf\left\{\sup_{x \in S} H_E(x); \ S \subset E, \ \dim \mathrm{Zar}(S) \geq i\right\}$

Here $\mathrm{Zar}(S)$ means the Zariski closure of $K.S = \{ax; \ a \in K, \ x \in S\}$ and its dimension is the one of the scheme over $\mathrm{Spec}\,K$ defined by the algebraic set $\mathrm{Zar}(S)$. We have

$$0 < \Lambda^{(1)}(E) \leq \Lambda^{(2)}(E) \leq \cdots \leq \Lambda^{(n)}(E) < \infty$$
$$\| \qquad\qquad |\wedge \qquad\qquad\qquad |\wedge$$
$$\Lambda_1(E) \ \leq \ \Lambda_2(E) \ \leq \cdots \leq \ \Lambda_n(E) \ < \infty$$
$$\| \qquad\qquad |\wedge \qquad\qquad\qquad |\wedge$$
$$Z_1(E) \ \leq \ Z_2(E) \ \leq \cdots \leq \ Z_n(E) \ \leq \infty$$

A field K is a *Northcott field* if, for all positive real number B, the set $\{x \in K; \ H_{K^2}(1, x) \leq B\}$ is finite (for instance, any number field or, according to [5], $\mathbb{Q}(\sqrt{2}, \sqrt{3}, \dots)$ are Northcott fields). It can be proved that, for every integer $n \geq 2$, for every rigid adelic space E with $\dim E = n$, we have $Z_n(E) < \infty$ if and only if K is *not* a Northcott field (see [14, Proposition 4.4]).

Examples

Let n be a positive integer and $i, j \in \{1, \dots, n\}$.

- We have $\Lambda^{(i)}(K^n) = \Lambda_i(K^n) = 1$.
- If K contains infinitely many roots of unity (e.g., $K = \overline{\mathbb{Q}}$), then $Z_i(K^n) = \sqrt{i}$ (consequence of the convexity inequality for heights, see Proposition 10).
- Let $A_n = \{(x_0, \dots, x_n) \in K^{n+1}; \ \sum_{\ell=0}^n x_\ell = 0\} \subset K^{n+1}$. Then $\Lambda_i\left(\bigwedge^j A_n\right) = \sqrt{j+1}$.

In the following, in order to unify notation, we shall sometimes use $\lambda_i^*(E)$ with $* \in \{\mathrm{BC}, \Lambda, Z\}$ to indicate $\lambda_i^{\mathrm{BC}}(E) = \Lambda^{(i)}(E)$, $\lambda_i^\Lambda(E) = \Lambda_i(E)$ or $\lambda_i^Z(E) = Z_i(E)$.

Basic Properties Let E be a rigid adelic space with dimension n and let $i \in \{1, \dots, n\}$.

1. For any non-zero linear subspace $F \subset E$, we have $\lambda_i^*(E) \leq \lambda_i^*(F)$ for all $* \in \{\mathrm{BC}, \Lambda, Z\}$ and $i \leq \dim F$.
2. For every algebraic extension K'/K and every $* \in \{\mathrm{BC}, \Lambda, Z\}$, we have $\lambda_i^*(E_{K'}) \leq \lambda_i^*(E)$.

The latter property is quite easy to prove (see [14, Lemma 4.22]) except, maybe, for $* = \mathrm{BC}$, for which we provide a proof (suggested by G. Rémond): Let us assume that there exists i such that $\lambda_i^{BC}(E_{K'}) > \lambda_i^{BC}(E)$. We choose it as small as possible and we consider a subspace $F \subset E_{K'}$, (necessarily) with dimension $i - 1$, such that $\Lambda_1(E_{K'}/F) > \lambda_i^{BC}(E)$. Let $G \subset E$ be a subspace with maximal dimension satisfying $G \otimes_K K' \subset F$. We have $\dim G \le i - 1$ and so $\lambda_i^{BC}(E) \ge \Lambda_1(E/G)$. Then let us consider $x \in E \backslash G$ such that $\Lambda_1(E_{K'}/F) > H_{E/G}(x \bmod G)$. We have $x \notin F$ otherwise $G \oplus K.x$ has dimension greater than $\dim G$ with $(G \oplus K.x) \otimes_K K' \subset F$. Therefore, we have $H_{E_{K'}/F}(x \bmod F) \ge \Lambda_1(E_{K'}/F) > H_{E/G}(x \bmod G)$, contradicting the fact that the w-norm of $x \bmod F$ is smaller than the w-norm of $x \bmod G$, for all $w \in V(K')$, since $E_{K'}/F$ is a quotient of $(E/G) \otimes_K K'$.

In the following result it is convenient to put $\Lambda_0(E) = 0$ when E is an (integrable) adelic space.

Proposition 11 *Let N be a positive integer and let $E_1, \ldots, E_N$ be integrable adelic spaces over K. Then, for all $i \in \{1, \ldots, \sum_{h=1}^{N} \dim E_h\}$, we have*

$$\Lambda_i(E_1 \oplus \cdots \oplus E_N) = \min \max \left\{ \Lambda_{a_1}(E_1), \ldots, \Lambda_{a_N}(E_N) \right\}$$

where the minimum is taken over all integers $a_h \in [0, \dim E_h]$, $1 \le h \le N$, such that $\sum_{h=1}^{N} a_h = i$.

In particular, $\Lambda_1(E_1 \oplus \cdots \oplus E_N) = \min \{\Lambda_1(E_1), \ldots, \Lambda_1(E_N)\}$.

Proof Fix $(a_1, \ldots, a_N)$ as above. For each $j \in \{1, \ldots, N\}$ such that $a_j \ne 0$, let $x_1^{(j)}, \ldots, x_{a_j}^{(j)}$ be linearly independent vectors of E_j. Then $\{x_h^{(j)}; 1 \le h \le a_j, 1 \le j \le N\}$ forms a free family of i vectors of $E := E_1 \oplus \cdots \oplus E_N$. Thus, by definition of Λ_i, we get

$$\Lambda_i(E) \le \max \left\{ H_E\left(x_h^{(j)}\right); 1 \le h \le a_j, 1 \le j \le N \right\}.$$

The infimum of the right hand side when all $x_h^{(j)}$ vary is precisely $\max \{\Lambda_{a_1}(E_1), \ldots, \Lambda_{a_N}(E_N)\}$ and, then, we can take the infimum over $(a_1, \ldots, a_N)$ to obtain $\Lambda_i(\bigoplus_{j=1}^{N} E_j) \le \min \max_j \{\Lambda_{a_j}(E_j)\}$. For the reverse inequality, consider $x_h^{(j)} \in E_h$ for all $j \in \{1, \ldots, i\}$ and $h \in \{1, \ldots, N\}$ such that the vectors $X_j = (x_1^{(j)}, \ldots, x_N^{(j)})$'s are linearly independent. In particular $X_1 \wedge \cdots \wedge X_i \ne 0$ and, writing this vector as a sum of $x_{\tau(1)}^{(1)} \wedge \cdots \wedge x_{\tau(i)}^{(i)}$ over functions $\tau : \{1, \ldots, i\} \to \{1, \ldots, N\}$, we deduce the existence of τ such that $\{x_{\tau(1)}^{(1)}, \ldots, x_{\tau(i)}^{(i)}\}$ is a free family. For each $h \in \{1, \ldots, N\}$, let n_h be the number of $u \in \{1, \ldots, i\}$ such that $\tau(u) = h$ (the integer n_h may be zero). We have $\sum_h n_h = i$ and the vector space generated by $\{x_h^{(j)}; 1 \le j \le i\}$ has dimension at least n_h. From Proposition 10, we get $H_E(X_j) \ge \max_{1 \le h \le N} H_{E_h}(x_h^{(j)})$ for all $j \in \{1, \ldots, i\}$, so

$$\max_{1 \le j \le i} H_E(X_j) \ge \max_{1 \le j \le i} \max_{1 \le h \le N} H_{E_h}(x_h^{(j)}) \ge \max_{1 \le h \le N} \Lambda_{n_h}(E_h).$$

We then conclude by bounding from below the latter maximum by $\min_{\sum_h a_h = i} \max_h \Lambda_{a_h}(E_h)$. $\qquad\qquad\qquad\qquad\qquad\qquad\qquad\qquad\qquad\qquad\qquad\qquad\quad$ $\square$

3.2 Slopes

In this paragraph, we define the canonical polygon of a rigid adelic space, which gives birth to its successive slopes. These notions have their origin in the works by Stuhler [24] and Grayson [15] (inspired by the article [16] of Harder-Narasimhan). Later on, they have been developed by Bost in two lectures given at the Institut Henri Poincaré in 1997 and 1999 and in his articles [6, 7], then extended in different ways in [12, 1, 10, 8]. Let E be a rigid adelic space over K and $n = \dim E$.

Lemma 12 *There exists a positive constant $c(E)$ such that $H(F) \geq c(E)$ for every linear subspace $F \subset E$.*

Proof Let (φ, A) be a couple defining the adelic structure of E. There exists $a = (a_p)_{p \in V(\mathbb{Q})} \in \mathbb{A}_{\mathbb{Q}}^{\times}$ such that, for all $v \in V(K)$ above $p \in V(\mathbb{Q})$ and for all $x \in E \otimes_K K_v$,

$$|a_p|_v^{-1} |\varphi_v(x)|_v \leq \|x\|_{E,v} \leq |a_p|_v |\varphi_v(x)|_v.$$

Define $|a| = \exp \int_{V(K)} \log |a_p|_v \, d\sigma(v) \geq 1$. For every subspace $F \subset E$ with dimension ℓ, we have $H(F) \geq |a|^{-\ell} H(\varphi(F)) = |a|^{-\ell} H(\det \varphi(F))$. Since $\det \varphi(F)$ is a non-zero vector of $\bigwedge^{\ell} K^n$, which is isometric to $K^{\binom{n}{\ell}}$, we have $H(\det \varphi(F)) \geq 1$ and the conclusion follows with $c(E) = |a|^{-n}$. $\qquad\quad$ $\square$

In other words, the set $\{\deg F \; ; \; F \subset E\}$ is bounded from above. This result allows to define some positive real numbers associated to E: for all $i \in \{0, \dots, n\}$,

$$\sigma_i(E) = \inf\{H(F) \, ; \; F \text{ linear subspace of } E \text{ and } \dim F = i\}.$$

For instance, $\sigma_0(E) = 1$, $\sigma_1(E) = \Lambda_1(E)$ and $\sigma_n(E) = H(E)$. Note that we have $\sigma_{n-1}(E) = \Lambda_1(E^{\vee}) H(E)$ and, more generally, $\sigma_{n-i}(E) = \sigma_i(E^{\vee}) H(E)$ which comes from the isometry $E/F \simeq (F^{\perp})^{\vee}$ (Theorem 4 and Proposition 5). We also have $\sigma_i(E) \geq \Lambda_1(\bigwedge^i E)$. Lemma 12 justifies the following

Definition 13 *Let $P_E \colon [0, n] \to \mathbb{R}$ denote the piecewise linear function delimiting from above the convex hull of the set*

$$\left\{ (\dim F, \deg F) \in \mathbb{R}^2 \, ; \; F \text{ linear subspace of } E \right\}.$$

We shall call P_E the canonical polygon of E.

Naturally, the latter convex hull can be replaced by the one of the (finite) set $\{(i, -\log \sigma_i(E)) \, ; \; i \in \{0, \dots, n\}\}$. By definition, the function P_E is a concave function which satisfies $P_E(0) = 0$ and its slopes

$$\mu_i(E) = P_E(i) - P_E(i-1) \qquad (i \in \{1, \ldots, n\})$$

form a nonincreasing sequence $\mu_1(E) \geq \mu_2(E) \geq \cdots \geq \mu_n(E)$, called the *successive slopes of E*. The greatest slope $\mu_1(E)$ is also denoted $\mu_{\max}(E)$ and the smallest slope $\mu_n(E)$ is $\mu_{\min}(E)$. This terminology is also justified by the following key result.

Lemma 14 *For every rigid adelic space E over K, we have*

$$\mu_{\max}(E) = \max\{\mu(F); \ F \neq \{0\} \text{ linear subspace of } E\}.$$

More precisely, there exists a (single) subspace of E, denoted E_{des}, such that $\mu(E_{\text{des}}) = \mu_{\max}(E)$ and E_{des} contains every linear subspace $F \subset E$ satisfying $\mu(F) = \mu_{\max}(E)$.

The subscript "des" refers to the word *destabilizing*. The proof follows the one of [8, Proposition 2.2].

Proof Let us temporarily denote by c the supremum of slopes $\mu(F)$ when F runs over non-zero linear subspaces of E. This is a real number by Lemma 12. Actually, if $m = \dim F$, we have

$$\mu(F) = \frac{\deg F}{m} \leq \frac{P_E(m)}{m} = \frac{\mu_1(E) + \cdots + \mu_m(E)}{m} \leq \mu_1(E)$$

and so $c \leq \mu_1(E)$. On the other hand, for every linear subspace $F \subset E$, we have $\deg F \leq (\dim F)c$. Since $m \mapsto mc$ is a concave (linear) function we deduce $P_E(m) \leq mc$ for all $m \subset [0, n]$ ($n = \dim E$). Thus $\mu_1(E) = P_E(1) \leq c$ and we get $\mu_1(E) = c = \sup\{\mu(F); \ \{0\} \neq F \subset E\}$. Let us now prove the existence of E_{des}. We proceed by induction on n. The statement is clear for $n = 1$ since $\mu_1(E) = \mu(E)$ in this case. Assume the existence of the destabilizing rigid adelic space when the dimension of the ambient space is at most $n - 1$. Let E be of dimension n. If $\mu_1(E) = \mu(E)$, then $E_{\text{des}} := E$ is the winner. Otherwise the set $\{F \subset E; \ F \neq \{0\} \text{ and } \mu(F) > \mu(E)\}$ is non-empty and we can choose F in it with maximal dimension. By induction hypothesis (and since $\dim F \leq n-1$), there exists F_{des} such that $\mu(F_{\text{des}}) = \mu_{\max}(F)$ and such that, for every linear subspace $G \subset F$ with $\mu(G) = \mu_{\max}(F)$, we have $G \subset F_{\text{des}}$. Let G be a non-zero linear subspace of E. If $G \not\subset F$, then $\dim(F + G) > \dim F$ and, by maximality property of $\dim F$, we have $\mu(F + G) \leq \mu(E)$. Replacing this information in the inequality $\deg F + \deg G \leq \deg(F + G) + \deg F \cap G$ given by Proposition 6, we get

$$(\dim F)\mu(F) + (\dim G)\mu(G) \leq (\dim(F + G))\mu(E) + (\dim F \cap G)\mu_{\max}(F)$$

and so $(\dim G)\mu(G)$ is bounded above by

$$\dim(F + G)\underbrace{(\mu(E) - \mu(F))}_{<0} + (\dim(F + G) - \dim F)\underbrace{\mu(F)}_{\leq \mu_{\max}(F)},$$

$$+ (\dim F \cap G)\,\mu_{\max}(F)$$

which implies $\mu(G) < \mu_{\max}(F)$. If $G \subset F$ we have $\mu(G) \leq \mu_{\max}(F)$. Thus, every non-zero linear subspace of E has its slope at most $\mu_{\max}(F)$ and so $\mu_{\max}(E) = \mu_{\max}(F)$. Then the space $E_{\mathrm{des}} := F_{\mathrm{des}}$ has the required properties.   □

Definition 15 *A rigid adelic space E is* semistable *if $\mu(E) = \mu_{\max}(E)$ (that is, $E_{\mathrm{des}} = E$).*

In this case, the canonical polygon is a straight line. For instance, K^n and A_n (defined on page 48) are semistable (see [13, p. 580] for A_n). Lemma 14 allows to define a unique filtration of E composed of linear subspaces $\{0\} = E_0 \subsetneq E_1 \subsetneq \cdots \subsetneq E_N = E$ such that E_{i+1}/E_i is semistable for every $i \in \{0, 1, \ldots, N-1\}$: the first spaces $E_0, \ldots, E_i$ being chosen, take E_{i+1} satisfying $E_{i+1}/E_i = (E/E_i)_{\mathrm{des}}$. This filtration is called the *Harder-Narasimhan filtration* of E (shortened in *HN-filtration* thereafter). By definition we have $\mu(E_{i+1}/E_{i-1}) < \mu(E_i/E_{i-1})$ and, using $\deg E_{i+1}/E_i = \deg E_{i+1}/E_{i-1} - \deg E_i/E_{i-1}$, we deduce that

$$\mu(E_N/E_{N-1}) < \mu(E_{N-1}/E_{N-2}) < \cdots < \mu(E_1).$$

Theorem 16 *Let $E_0 = \{0\} \subsetneq E_1 \subsetneq \cdots \subsetneq E_N = E$ be the HN-filtration of E. Let $m_i = \dim E_i$. Then $m_1, \ldots, m_{N-1}$ are (exactly) the points at which P_E is not differentiable and $P_E(m_i) = \deg E_i$ for all $i \in \{0, \ldots, N\}$. Moreover, for all $i \in \{1, \ldots, N\}$ and $j \in \{1, \ldots, m_i - m_{i-1}\}$, we have $\mu_{m_{i-1}+j}(E) = \mu(E_i/E_{i-1})$.*

The proof will use the following result (here $n = \dim E$).

Lemma 17 *Let $x \in [0, n]$ such that P_E is not differentiable at x. Then x is an integer and there exists a unique linear subspace $F_x \subset E$ with dimension x such that $P_E(x) = \deg F_x$. Moreover, if P_E is not differentiable at $y \leq x$, then $F_y \subset F_x$.*

Proof By definition of P_E, which is a linear function on each interval $(h, h+1)$ for $h \in \{0, \ldots, n-1\}$, the real number x is necessarily an integer. Since $F_0 = \{0\}$ and $F_n = E$ we may assume $x \in \{1, \ldots, n-1\}$. The construction of P_E and its non differentiability at x entail

$$P_E(x) = \sup\{\deg F \; ; \; F \text{ linear subspace of } E \text{ with dimension } x\}.$$

Then, let us choose some linear subspaces A and B of E, with dimension x, such that $P_E(x) \leq \deg A + \varepsilon$ and $P_E(x) \leq \deg B + \varepsilon$ where

$$\varepsilon = \frac{1}{4}\min\left\{\frac{P_E(\delta) - P_E(i)}{\delta - i} - \frac{P_E(j) - P_E(h)}{j - h}\right\},$$

the minimum being taken over integers i, δ, h, j satisfying $0 \leq i < \delta \leq h < j \leq n$ and $(P_E(\delta) - P_E(i))/(\delta - i) - (P_E(j) - P_E(h))/(j - h) \neq 0$ (in particular $\varepsilon > 0$ by concavity of P_E). Defining $j = \dim(A + B)$ and $i = \dim A \cap B$ and using $\deg A + \deg B \leq \deg(A + B) + \deg A \cap B$ (Proposition 6), we get $2P_E(x) - 2\varepsilon \leq P_E(j) + P_E(i)$ which, if $x \neq i$, implies

$$\frac{P_E(x) - P_E(i)}{x - i} - \frac{P_E(j) - P_E(x)}{j - x} \leq 2\varepsilon \quad \text{since } x - i = j - x.$$

Since P_E is not differentiable at x, the left hand side is positive, contradicting the definition of ε. Thus we have $x - i = j - x = 0$, that is, $A = B$. We proved that there exists $\varepsilon > 0$ such that the set $\{A \subset E; \ \dim A = x \text{ and } P_E(x) \leq \deg A + \varepsilon\}$ is a singleton $\{F_x\}$. The same approach with $A = F_x$ and $B = F_y$ demonstrates $\dim(F_x + F_y) = \dim F_x$ and so $F_y \subset F_x$ when $y \leq x$. $\qquad\square$

Proof of Theorem 16 Let $f_0 = 0 < f_1 < \cdots < f_M = n$ be the abscissae for which P_E is not differentiable and $F_0 = \{0\} \subsetneq F_1 \subsetneq \cdots \subsetneq F_M$ the corresponding subspaces given by Lemma 17. For every linear subspace $F \subset E$ and every $i \in \{1, \ldots, M\}$ such that $F \not\subset F_{i-1}$, the concavity of P_E yields

$$\frac{P_E(\dim(F + F_{i-1})) - P_E(f_{i-1})}{\dim(F + F_{i-1}) - f_{i-1}} \leq \frac{P_E(f_i) - P_E(f_{i-1})}{f_i - f_{i-1}}$$

and this inequality is strict if $\dim(F + F_{i-1}) > f_i$. Bounding from below $P_E(\dim(F + F_{i-1}))$ by $\deg(F + F_{i-1})$ we get $\mu\left((F + F_{i-1})/F_{i-1}\right) \leq \mu\left(F_i/F_{i-1}\right)$ which proves that $\mu\left(F_i/F_{i-1}\right) = \mu_{\max}(E/F_{i-1})$. The equality can occur only if $\dim(F + F_{i-1}) \leq f_i$ and so $F_i/F_{i-1} = (E/F_{i-1})_{\text{des}}$. Thus the sequence $(F_i)_i$ satisfies the same definition as the HN-filtration of E and, by unicity, it is the same: $N = M$ and $F_i = E_i$ and $f_i = m_i$ for all i. As for the equality $\mu_{m_{i-1}+j}(E) = \mu(E_i/E_{i-1})$, it comes from the fact that $\mu_{m_{i-1}+1}(E) = \cdots = \mu_{m_i}(E)$ (since P_E is a line on $[m_{i-1}, m_i]$) and

$$\sum_{j=1}^{m_i - m_{i-1}} \mu_{m_{i-1}+j}(E) = \sum_{j=1}^{m_i \ m_{i-1}} P_E(m_{i-1} + j) - P_E(m_{i-1} + j - 1)$$

$$= P_E(m_i) - P_E(m_{i-1})$$

$$= \deg(E_i/E_{i-1}) = (m_i - m_{i-1})\mu\left(E_i/E_{i-1}\right). \qquad\square$$

From Theorem 16 can be deduced a minimax formula for $\mu_i(E)$.

Proposition 18 *Let E be a rigid adelic space over K and $i \in \{1, \ldots, \dim E\}$. Then*

$$\mu_i(E) = \sup_A \inf_B \mu\left(A/B\right) = \inf_B \sup_A \mu\left(A/B\right)$$

where $B \subset A$ run over linear subspaces of E with $\dim B \leq i - 1 < \dim A$.

Proof Let $E_0 = \{0\} \subsetneq E_1 \subsetneq \cdots \subsetneq E_N = E$ be the HN-filtration of E. Let $h \in \{0, \ldots, N - 1\}$ such that $\dim E_h \leq i - 1 < \dim E_{h+1}$. Let A be a linear subspace of E with dimension $\geq i$ (in particular $A \not\subset E_h$). Using Theorem 16, Lemma 14 and Proposition 6, we get

$$\mu_i(E) = \mu_{\max}(E/E_h) \geq \mu((A + E_h)/E_h) \geq \mu(A/(A \cap E_h))$$

which is greater than $\inf \{\mu(A/B) \, ; \, B \subset A \text{ and } \dim B \leq i - 1\}$. Taking the supremum over A, we obtain $\mu_i(E) \geq \alpha := \sup_A \inf_B \mu(A/B)$. On the other hand, the concavity of P_E implies

$$\mu(E_{h+1}/B) \geq \frac{P_E(\dim E_{h+1}) - P_E(\dim B)}{\dim E_{h+1} - \dim B} \geq \mu(E_{h+1}/E_h) = \mu_i(E)$$

for any linear subspace $B \subset E_{h+1}$ with dimension $< i$. We conclude using $\alpha \geq \inf_B \mu(E_{h+1}/B)$. The same method works with $\inf_B \sup_A \mu(A/B)$. □

In particular we have $\mu_n(E) = \mu_{\min}(E) = \inf \{\mu(E/F) \, ; \, F \subsetneq E\}$ (where $n = \dim E$). Actually the infimum is a minimum as the next proposition and Lemma 14 prove it.

The following statement summarizes several properties of the canonical polygon of a rigid adelic space over an algebraic extension K.

Proposition 19 *Let E be a rigid adelic space over K with dimension n.*

(1) *If L is a rigid adelic space over K with dimension 1, then, for all $x \in [0, n]$, we have $P_{E \otimes L}(x) = P_E(x) + x \deg L$. In particular, for all $i \in \{1, \ldots, n\}$, we have $\mu_i(E \otimes L) = \mu_i(E) + \deg L$.*

(2) *For all $x \in [0, n]$, we have $P_{E^\vee}(x) = P_E(n - x) - \deg E$. In particular, for all $i \in \{1, \ldots, n\}$, we have $\mu_i(E^\vee) = -\mu_{n+1-i}(E)$.*

(3) *Let K'/K be an algebraic extension. Then we have $P_{E_{K'}} = P_E$. In particular, for all $i \in \{1, \ldots, n\}$, we have $\mu_i(E_{K'}) = \mu_i(E)$.*

The last property means that the μ_i's are *absolute minima* (that is, over an algebraic closure of K). The similar feature is not true in general for λ_i^*'s. Moreover this proposition can be restated in terms of the HN-filtration $E_0 = \{0\} \subset E_1 \subset \cdots \subset E_N = E$ of E: The HN-filtrations of $E \otimes_K L$, $E^\vee$, $E_{K'}$ are (respectively) $(E_i \otimes_K L)_i$, $(E_{N-i}^\perp)_i$ and $((E_i)_{K'})_i$.

Proof

(1) Since P_E is a linear function on each interval $[i, i + 1]$, $i \in \{0, \ldots, n - 1\}$, it is enough to prove the equality for $x = m \in \{0, \ldots, n\}$. For every subspace $F \subset E$ with dimension m, we have $\dim F \otimes_K L = m$ and $\deg F + m \deg L = \deg F \otimes_K L \leq P_{E \otimes_K L}(m)$. So $\deg F \leq P_{E \otimes_K L}(m) - m \deg L$ and since the function $m \mapsto P_{E \otimes_K L}(m) - m \deg L$ is concave we deduce $P_E(m) \leq P_{E \otimes_K L}(m) - m \deg L$. The reverse inequality is obtained replacing E by $E \otimes_K L$ and L by $L^\vee$ (using the fact that $L \otimes_K L^\vee$ is isometric to K). The equality for the

i-th slopes arises from this equality relating the canonical polygons and from the definition of μ_i.

(2) As previously, it is enough to prove the equality for $x = m \in \{0, \ldots, n\}$. For a subspace $F \subset E$ with dimension m, the isometric isomorphism $E/F \simeq (F^\perp)^\vee$ yields $\deg F - \deg E = \deg F^\perp$ and $\deg F \le \deg E + P_{E^\vee}(n - m)$. Then we deduce $P_E(m) \le \deg E + P_{E^\vee}(n - m)$ since the right hand side is a concave function of m. For the reverse inequality, replace E by $E^\vee$, m by $n - m$ and use $(E^\vee)^\vee \simeq E$ (Theorem 4).

(3) For every subspace $F \subset E$ with dimension m, we have $\deg F = \deg F \otimes_K K'$ and so $\deg F \le P_{E_{K'}}(m)$ and then $P_E(m) \le P_{E_{K'}}(m)$. It implies that in order to prove the reverse inequality, we may assume that K'/K is Galois. Let $\{0\} = F_0 \subsetneq F_1 \subsetneq \cdots \subsetneq F_N = E \otimes_K K'$ be the HN-filtration of $E_{K'}$ and $d_i = \dim F_i$. Let $e_1, \ldots, e_n$ be a K-basis of E. For every $\tau \in \mathrm{Gal}(K'/K)$, the correspondence $\iota_\tau : E \otimes_K K' \to E \otimes_K K'$ which sends $\sum_{i=1}^n x_i e_i$ ($x_i \in K'$) to $\sum_{i=1}^n \tau(x_i) e_i$ is a bijection that preserves dimension and degree of subspaces of $E_{K'}$. Using Theorem 16 and Lemma 17, we deduce $\iota_\tau(F_i) = F_i$ for all i. Now, let us fix $i \in \{1, \ldots, N\}$. Even if it means permuting the vectors $e_1, \ldots, e_n$, we can find a K'-basis $f_1, \ldots, f_{d_i}$ of F_i and scalars $\alpha_{j,h} \in K'$ for all $1 \le j \le d_i$ and $d_i + 1 \le h \le n$ such that

$$(\star) \qquad \forall j \in \{1, \ldots, d_i\}, \qquad f_j = e_j + \sum_{h=d_i+1}^{n} \alpha_{j,h} e_h$$

(Gaussian elimination). The Galois closure K'_0 of the field generated by K and all algebraic numbers $\alpha_{j,h}$'s is both a subfield of K' and a finite extension of K. So we can consider its normalized trace function $\mathrm{Tr}: K'_0 \to K$ ($\mathrm{Tr}(1) = 1$). Since $\iota_\tau(F_i) = F_i$ for all $\tau \in \mathrm{Gal}(K'_0/K)$, the vector

$$\mathrm{Tr} f_j := \frac{1}{[K'_0 : K]} \sum_{\tau \in \mathrm{Gal}(K'_0/K)} \iota_\tau(f_j)$$

belongs to F_i for all $j \in \{1, \ldots, d_i\}$. With $(\star)$, this element can also be written $e_j + \sum_{h=d_i+1}^{n} \mathrm{Tr}(\alpha_{j,h}) e_h$ which implies, in particular, $\mathrm{Tr} f_j \in E$ and the family $\{\mathrm{Tr} f_1, \ldots, \mathrm{Tr} f_{d_i}\}$ is free. Thus the linear subspace $G_i := \mathrm{Vect}_K (\mathrm{Tr} f_1, \ldots, \mathrm{Tr} f_{d_i})$ of F_i has the same dimension d_i as F_i and so $F_i = G_i \otimes_K K'$. We deduce $P_{E_{K'}}(d_i) = \deg F_i = \deg G_i \le P_E(d_i)$, hence $P_{E_{K'}}(d_i) = P_E(d_i)$. Since $P_{E_{K'}}$ is linear on $[d_i, d_{i+1}]$ and $P_E \le P_{E_{K'}}$ are both concave functions on this interval, we get the equality $P_E = P_{E_{K'}}$ on $[d_i, d_{i+1}]$ and then on $[0, n]$ by varying i. □

In general P_E is difficult to compute, starting with its first value $P_E(1) = \mu_{\max}(E)$. To conclude this paragraph, let us outline an application of the above results to direct sums of rigid adelic spaces, obtaining a counterpart to the formula for $\Lambda_i(E_1 \oplus \cdots \oplus$

E_N) given by Proposition 11. In the following statement, the term $\mu_a(E)$ is $+\infty$ if $a < 1$ and $-\infty$ if $a > \dim E$.

Theorem 20 *Let E, F be some rigid adelic spaces over K and $(E_\ell)_{\ell \in \{0,\ldots,N\}}$ and $(F_h)_{h \in \{0,\ldots,M\}}$ their respective HN-filtrations. Then the HN-filtration of $E \oplus F$ is formed with some subspaces of the shape $E_\ell \oplus F_h$, beginning with $\{0\}$. To go from one notch to the next, the rule is the following:*

$$E_\ell \oplus F_h \begin{cases} \nearrow E_{\ell+1} \oplus F_h & \text{if} \quad \mu\,(E_{\ell+1}/E_\ell) > \mu\,(F_{h+1}/F_h) \\[2mm] \longleftrightarrow E_{\ell+1} \oplus F_{h+1} & \text{if} \quad \mu\,(E_{\ell+1}/E_\ell) = \mu\,(F_{h+1}/F_h) \\[2mm] \searrow E_\ell \oplus F_{h+1} & \text{if} \quad \mu\,(E_{\ell+1}/E_\ell) < \mu\,(F_{h+1}/F_h). \end{cases}$$

In particular we have $\mu_i(E \oplus F) = \max\limits_{\substack{a,b \in \mathbb{N} \\ a+b=i}} \min\{\mu_a(E), \mu_b(F)\}$ for all $i \in \mathbb{N}$.

A straightforward induction yields a formula for the i-th slope of a general direct sum: Let N be a positive integer and $E_1, \ldots, E_N$ be some rigid adelic spaces over K. Then, for all i, we have

$$\mu_i(E_1 \oplus \cdots \oplus E_N) = \max\limits_{\substack{a_1,\ldots,a_N \in \mathbb{N} \\ a_1+\cdots+a_N=i}} \min\{\mu_{a_1}(E_1), \ldots, \mu_{a_N}(E_N)\}.$$

The key statement for proving Theorem 20 is

Lemma 21 *Let A, B be rigid adelic spaces over K. Then we have*

$$(A \oplus B)_{\mathrm{des}} = \begin{cases} A_{\mathrm{des}} & \text{if} \quad \mu_{\max}(A) > \mu_{\max}(B) \\ A_{\mathrm{des}} \oplus B_{\mathrm{des}} & \text{if} \quad \mu_{\max}(A) = \mu_{\max}(B) \\ B_{\mathrm{des}} & \text{if} \quad \mu_{\max}(A) < \mu_{\max}(B). \end{cases}$$

From this lemma we deduce at once the well-known formula

$$\mu_{\max}(A \oplus B) = \max\{\mu_{\max}(A), \mu_{\max}(B)\}.$$

Moreover, it can be checked that the map $A^{\vee} \oplus B^{\vee} \to (A \oplus B)^{\vee}$, $(\varphi, \psi) \mapsto ((a, b) \mapsto \varphi(a) + \psi(b))$ is an (isometric) isomorphism of rigid adelic spaces. In particular their maximal slopes are equal. We then deduce the equality

$$\mu_{\min}(A \oplus B) = \min\{\mu_{\min}(A), \mu_{\min}(B)\}$$

with Proposition 19 and Lemma 21.

Proof of Lemma 21 First note that since A and B are linear subspaces of $A \oplus B$, their maximal slopes are at most the maximal slope of $A \oplus B$ (Lemma 14). Now let us consider $C = (A \oplus B)_{\mathrm{des}}$, $C_A = C \cap A$ and $C^B = \mathrm{Im}(C \to B)$ where $C \to B$ is the restriction to C of the second projection $A \oplus B \to B$. The linear space C^B is isometrically isomorphic to $(A + C)/A$. Then, if C_A and C^B are not reduced to $\{0\}$, Proposition 6 implies

$$\mu(C) \leq \frac{n_A \mu(C_A) + n^B \mu(C^B)}{n_A + n^B}$$

where $n_A = \dim C_A$ and $n^B = \dim C^B$. Since $C_A \subset A$ and $C^B \subset B$, we deduce $\mu_{\max}(A \oplus B) = \mu(C) \leq \max\{\mu_{\max}(A), \mu_{\max}(B)\}$. In view of the lower bound for $\mu(C)$ mentioned at the beginning, it is necessarily an equality and so $\mu(C) = \mu(C_A) = \mu(C^B) = \mu_{\max}(A) = \mu_{\max}(B)$. From Lemma 14 we deduce $C^B \subset B_{\mathrm{des}}$ and the same reasoning with C^A (we still have C^A and C_B non-zero) proves that $C^A \subset A_{\mathrm{des}}$. Finally we get $C \subset A_{\mathrm{des}} \oplus B_{\mathrm{des}}$ and, since this latter space has slope equal to $\mu_{\max}(A \oplus B)$, we have $C = A_{\mathrm{des}} \oplus B_{\mathrm{des}}$. If $C_A = \{0\}$, then $\mu(C) \leq \mu(C^B) \leq \mu_{\max}(B)$; so there is equality and $B_{\mathrm{des}} \subset C \xrightarrow{\sim} C^B \subset B_{\mathrm{des}}$. This proves that $C = B_{\mathrm{des}}$. Besides we have $\mu_{\max}(B) > \mu_{\max}(A)$ since otherwise there would be equality and we should have $A_{\mathrm{des}} \subset C$, then $A_{\mathrm{des}} \subset C_A$, contradicting $C_A = \{0\}$. When $C_A \neq \{0\}$ but $C^B = \{0\}$ we have $C = C \cap A$ and $\mu(C) = \mu_{\max}(A)$, so $A_{\mathrm{des}} = C$. Here again we have $\mu_{\max}(A) > \mu_{\max}(B)$ since otherwise we should have $B_{\mathrm{des}} \subset C$, contradicting $C^B = \{0\}$. $\qquad\square$

To prove Theorem 20 we shall use the fact that if E' is a non-zero linear subspace of a rigid adelic space E, then $\mu_i(E') \leq \mu_i(E)$ for all i (consequence of Proposition 18).

Proof of Theorem 20 We build the HN-filtration of $E \oplus F$ step by step. Let us suppose that we got the notch $E_\ell \oplus F_h$ where the integers ℓ, h may be zero. According to the construction of the HN-filtration (p. 52), the next step in the filtration of $E \oplus F$ is its subspace G, (strictly) containing $E_\ell \oplus F_h$ such that

$$G/E_\ell \oplus F_h = (E \oplus F/E_\ell \oplus F_h)_{\mathrm{des}} = (E/E_\ell \oplus F/F_h)_{\mathrm{des}}.$$

Then we apply the previous lemma to $A = E/E_\ell$ and $B = F/F_h$. It gives the first part of Theorem 20 by observing that $A_{\mathrm{des}} = E_{\ell+1}/E_\ell$ and $B_{\mathrm{des}} = F_{h+1}/F_h$. Let us now establish the formula for the i-th slope of $E \oplus F$. Define $\delta(i) = \max_{a+b=i} \min\{\mu_a(E), \mu_b(F)\}$ and fix $a \in \{1, \ldots, \dim E\}$ and $b \in \{1, \ldots, \dim F\}$ such that $a + b = i$ for some $i \in \{1, \ldots, \dim E + \dim F\}$. There exists $\ell \in \{0, \ldots, N-1\}$ (resp. $h \in \{0, \ldots, M-1\}$) such that $\mu_a(E) = \mu(E_{\ell+1}/E_\ell)$

(resp. $\mu_b(F) = \mu(F_{h+1}/F_h)$). Besides we have $a \in \{\dim E_\ell + 1, \ldots, \dim E_{\ell+1}\}$ and $b \in \{\dim F_h + 1, \ldots, \dim F_{h+1}\}$. Then we have

$$\mu_i(E \oplus F) \geq \mu_{\dim E_{\ell+1} + \dim F_{h+1}}(E \oplus F)$$

$$\geq \mu_{\dim E_{\ell+1} + \dim F_{h+1}}(E_{\ell+1} \oplus F_{h+1}) = \mu_{\min}(E_{\ell+1} \oplus F_{h+1})$$

$$= \min\{\mu_{\min}(E_{\ell+1}), \mu_{\min}(F_{h+1})\} = \min\{\mu_a(E), \mu_b(F)\}.$$

If a or b equals i, this inequality remains true since $\mu_i(E \oplus F) \geq \max\{\mu_i(E), \mu_i(F)\}$. Thus, for all i, we have $\mu_i(E \oplus F) \geq \delta(i)$. Observe now that $i \mapsto \delta(i)$ is a nonincreasing function since, for $a + b = i + 1$, we have

$$\min\{\mu_a(E), \mu_b(F)\} \leq \begin{cases} \min\{\mu_{a-1}(E), \mu_b(F)\} & \text{if } a \geq 1, \\ \min\{\mu_a(E), \mu_{b-1}(F)\} & \text{if } a = 0. \end{cases}$$

Starting from $E_\ell \oplus F_h$, let us call G the next notch in the HN-filtration of $E \oplus F$. We now prove that $\mu_i(E \oplus F) = \delta(i)$ for $i \in \{\dim E_\ell + \dim F_h + 1, \ldots, \dim G\}$. The crucial observation is that either $F_h = \{0\}$ or $F_h \neq \{0\}$ and the appearance of F_h at the notch $E_\ell \oplus F_h$ was caused by the fact that $\ell = 0$ and $\mu(F_h/F_{h-1}) > \mu_{\max}(E)$ or that $\ell \geq 1$ and the slope $\mu(F_h/F_{h-1})$ is greater than $\mu(E_m/E_{m-1}) = \mu_{\dim E_m}(E)$ for some integer $1 \leq m \leq \ell$. In every case, we have $\mu_{\dim F_h}(F) > \mu_{\dim E_{\ell+1}}(E)$. The same reasoning with E_ℓ gives $\mu_{\dim E_\ell}(E) > \mu_{\dim F_{h+1}}(F)$. Now, if $\mu(E_{\ell+1}/E_\ell) > \mu(F_{h+1}/F_h)$, then $G = E_{\ell+1} \oplus F_h$ and

$$\delta(i) \geq \delta(\dim G) \geq \min\left\{\mu_{\dim E_{\ell+1}}(E), \mu_{\dim F_h}(F)\right\} = \mu_{\dim E_{\ell+1}}(E)$$

$$= \mu_i(E \oplus F).$$

The two other possibilities for G are treated in the same way, which allows to conclude. □

4 Comparisons Between Minima and Slopes

4.1 Lower Bounds

The following inequality is as simple as fundamental. It is an extension of the fact that $n \in \mathbb{Z}$ and $n \neq 0$ *implies* $1 \leq |n|$ and can be seen as a variant of the Liouville inequality in transcendence theory. Let E be a rigid adelic space over K with dimension n.

Proposition 22 *We have* $1 \leq \Lambda_1(E) \exp \mu_1(E)$.

Proof Observe that for every $x \in E \setminus \{0\}$, we have $-\log H_E(x) = \deg K.x \leq P_E(1) = \mu_1(E)$. We conclude using the definition of $\Lambda_1(E)$ as the infimum of $H_E(x)$ over $x \in E \setminus \{0\}$. $\qquad\square$

Corollary 23 *For all $i \in \{1, \dots, n\}$, we have $1 \leq \Lambda^{(i)}(E) \exp \mu_i(E)$.*

In particular $1 \leq \lambda_i^*(E) \exp \mu_i(E)$ for all $* \in \{\text{BC}, \Lambda, Z\}$ (since $\Lambda^{(i)}(E) \leq \Lambda_i(E) \leq Z_i(E)$).

Proof Let $i \in \{1, \dots, n\}$ and $E_0 = \{0\} \subsetneq E_1 \subsetneq \cdots \subsetneq E_N = E$ be the HN-filtration of E. Consider the index h such that $\dim E_h \leq i - 1 < \dim E_{h+1}$, so that the maximal slope of E/E_h is equal to $\mu_i(E)$ (Theorem 16). We apply the previous proposition to E/E_h and we conclude bounding from above $\Lambda_1(E/E_h)$ by $\Lambda^{(i)}(E)$. $\qquad\square$

Corollary 24 *We have $H(E) \leq \prod_{i=1}^{n} \Lambda^{(i)}(E)$.*

Proof We multiply the inequalities of the previous corollary and we use $\sum_{i=1}^{n} \mu_i(E) = \deg E = -\log H(E)$. $\qquad\square$

Often, the weaker *Hadamard inequality* $H(E) \leq \Lambda_1(E) \cdots \Lambda_n(E)$ is used.

4.2 Upper Bounds

Let us recall that $\lambda_i^*(E) = \Lambda^{(i)}(E)$, $\Lambda_i(E)$ or $Z_i(E)$ according to $* = \text{BC}, \Lambda$ or Z. Given a positive integer n, let us define several constants:

- $$c_I(n, K) = \sup_{\dim E = n} \Lambda_1(E) H(E)^{-1/n} = \sup_{\dim E = n} \Lambda_1(E) \exp \mu(E)$$

- $$c_{\mathrm{II}}^*(n, K) = \sup_{\dim E = n} \left(\frac{\lambda_1^*(E) \cdots \lambda_n^*(E)}{H(E)} \right)^{1/n}$$

- $$\forall i \in \{1, \dots, n\}, \quad c_i^*(n, K) = \sup_{\dim E = n} \lambda_i^*(E) \exp \mu_i(E).$$

Here the suprema are taken upon all the rigid adelic spaces over K with dimension n. As in [14, § 4.8], we can prove that it is enough to consider hyperplanes of the standard space K^{n+1} (instead of E) to obtain the same numbers. Note that these constants can be infinite (see below).

Some Simple Observations Here n is a positive integer.

1. The number $c_I(n, \mathbb{Q}) = c_{\mathrm{II}}^{\Lambda}(n, \mathbb{Q})$ is the square root of the Hermite constant γ_n mentioned at the beginning of the text.
2. We have $1 \leq c_I(n, K) \leq c_{\mathrm{II}}^{\mathrm{BC}}(n, K) \leq c_{\mathrm{II}}^{\Lambda}(n, K) \leq c_{\mathrm{II}}^{Z}(n, K)$.
3. We have $c_{\mathrm{II}}^*(n, K)^n \leq \prod_{i=1}^{n} c_i^*(n, K)$.
4. For all $i \in \{1, \dots, n\}$, we have $1 \leq c_i^*(n, K) \leq c_{\mathrm{II}}^*(n, K)^n$.
5. The function $n \mapsto c_I(n, K)^n$ is nondecreasing.

To prove this last property, we can take $E \oplus L$ with dim $L = 1$ and $H(L) = \Lambda_1(E)$. To my knowledge it is not known whether $n \mapsto c_I(n, K)$ is nondecreasing, even for $K = \mathbb{Q}$. Nevertheless we shall see it is true when $K = \overline{\mathbb{Q}}$. In the opposite direction, we have the

Theorem (**Mordell Inequality**) *For every integer $n \geq 2$, we have*

$$c_I(n + 1, K) \leq c_I(n, K)^{n/(n-1)}.$$

Proof Let E be a rigid adelic space of dimension $n + 1$. Let ε be a positive real number and $x \in E \backslash \{0\}$ such that $H_E(x) \leq \Lambda_1(E) + \varepsilon$. The hyperplane $F = \{x\}^\perp \subset E^\vee$ satisfies $\Lambda_1(E^\vee) \leq \Lambda_1(F) \leq c_I(n, K)H(F)^{1/n}$. Since $F \simeq (E/K.x)^\vee$, we have $H(F) = H_E(x)/H(E) \leq \Lambda_1(E)/H(E) + \varepsilon/H(E)$. Replacing this bound in the previous inequality and letting $\varepsilon \to 0$ leads to

$$\Lambda_1\left(E^\vee\right) \leq c_I(n, K) \left(\frac{\Lambda_1(E)}{H(E)}\right)^{1/n}.$$

Applying this estimate to $E^\vee$ instead of E and combining both inequalities we obtain $\Lambda_1(E) \leq c_I(n, K)^{n/(n-1)} H(E)^{1/(n+1)}$. $\qquad\qquad\qquad\qquad\qquad\qquad\square$

With a bit more pain, one can also prove that $c_{\amalg}^Z(n, K) \leq c_{\amalg}^Z(2, K)^{2^n}$ (see [14, Proposition 4.14]). Let us also mention the analogue of Minkowski's theorem: *For every positive integer n, we have $c_I(n, K) = c_{\amalg}^\Lambda(n, K)$ (in particular $c_I(n, K) = c_{\amalg}^{BC}(n, K)$).* The proof is based on a deformation metric argument. To a rigid adelic space E over K, we associate another rigid adelic space E' such that $\Lambda_1(E') \geq 1$ and $H(E') = H(E) (\Lambda_1(E) \cdots \Lambda_n(E))^{-1}$ (see [14, Theorem 4.12]).

Definition 25 *An algebraic extension $K/\mathbb{Q}$ is called a* Siegel field *if $c_{\amalg}^\Lambda(n, K) < \infty$ for all $n \geq 1$.*

With the previous observations, K is a Siegel field if and only if $c_I(2, K) < \infty$. In a more elementary approach, K is a Siegel field if and only if it has the following property: There exists a positive real number α such that, for every $(a, b, c) \in K^3 \backslash \{0\}$ there exists $(x, y, z) \in K^3 \backslash \{0\}$ such that $ax + by + cz = 0$ and $H_{K^3}(x, y, z) \leq \alpha \, H_{K^3}(a, b, c)^{1/2}$.

Examples of Siegel Fields

  (i) $\mathbb{Q}$, number fields (Minkowski),
 (ii) $\overline{\mathbb{Q}}$ (Zhang [25] and Roy & Thunder [23]),
(iii) Hilbert class field towers of number fields [14, § 5.5].

Note that a finite extension of a Siegel field is still a Siegel field.
The following result comes from [14, Theorem 1.1 and Corollary 1.2].

Theorem 26

(1) $\forall n \geq 2$, $c_{\amalg}^Z(n, K) < \infty$ if and only if K is a Siegel field of infinite degree.

(2) *A Northcott field is a Siegel field if and only if it is a number field.*

The second claim is a direct consequence of the first one: If K is both a Northcott and a Siegel field, then $Z_i(E) = \infty$ for all $i \in \{2, \ldots, \dim E\}$ and so $c_{\mathbb{I}}^Z(n, K) = \infty$ as soon as $n \geq 2$ and by (1), K is a number field. Besides the implication $\Rightarrow$ in (1) is easy enough: use $c_{\mathbb{I}}^\Lambda(n, K) \leq c_{\mathbb{I}}^Z(n, K)$ and $Z_i(E) = \infty$ for $2 \leq i \leq \dim E$ when K is a number field. So the striking part of Theorem 26 is that it suffices to be a Siegel field of infinite degree to have $c_{\mathbb{I}}^Z(2, K) < \infty$. The proof rests on a deformation metric argument at some ultrametric place, much more subtle than for Minkowski theorem (see [14, § 4.6]). To be a little more precise, let us define the impurity index $u(K)$ of an algebraic extension $K/\mathbb{Q}$. If $v \in V(K)\backslash V_\infty(K)$ we denote by $\sigma(v)$ the measure of the singleton $\{v\}$, by p_v the prime number associated to v, by e_v the ramification index at v and by f_v its residual degree. The *impurity index* of K is

$$u(K) := \sup_{N \geq 1} \inf \left\{ p_v^{\sigma(v)/e_v} \; ; \; v \in V(K)\backslash V_\infty(K), \; p_v^{f_v} \geq N \right\}$$

with the conventions: $p_v^{\sigma(v)/e_v} = 1$ if $e_v = +\infty$ (or $\sigma(v) = 0$) and $p_v^{f_v} \geq N$ is true if $f_v = +\infty$. We can check that $u(K) < +\infty$ if and only if $[K : \mathbb{Q}] = \infty$. Note also that $u(\overline{\mathbb{Q}}) = 1$ and that, for every real number B, there exists an algebraic extension K such that $B < u(K) < +\infty$.

Proposition 27 *Let E be a rigid adelic space over K with dimension n. For each $i \in \{1, \ldots, n\}$, let α_i be a real number such that $0 < \alpha_i < Z_i(E)$. Then there exists a rigid adelic space E' over K with dimension n such that*

$$\frac{\alpha_1 \cdots \alpha_n}{H(E)} \leq \frac{\left(u(K)\Lambda_1(E')\right)^n}{H(E')}.$$

This proposition leads to the bound $c_{\mathbb{I}}^Z(n, K) \leq u(K)c_I(n, K)$ for all $n \geq 1$, thus yielding the first equivalence in Theorem 26.

In short, every constant $c_I(n, K)$, $c_{\mathbb{I}}^*(n, K)$, $c_i^*(n, K)$ is finite when K is a Siegel field of infinite degree and, if $* \neq Z$, this remains true when K is a number field. In everyday life it is useful to have some concrete bounds for $c_{\mathbb{I}}^*(n, K)$. In general it seems to be a difficult problem. Let us mention two cases (see [14, § 5.1 and § 5.2]):

(1) If K is number field with root discriminant $\delta_{K/\mathbb{Q}} = |\Delta_{K/\mathbb{Q}}|^{1/[K:\mathbb{Q}]}$, then

$$c_I(n, K) = c_{\mathbb{I}}^\Lambda(n, K) \leq \left(n\delta_{K/\mathbb{Q}}\right)^{1/2}.$$

(2) If $K = \overline{\mathbb{Q}}$, then

$$c_I(n, \overline{\mathbb{Q}}) = c_{\mathbb{I}}^{BC}(n, \overline{\mathbb{Q}}) = c_{\mathbb{I}}^\Lambda(n, \overline{\mathbb{Q}}) = c_{\mathbb{I}}^Z(n, \overline{\mathbb{Q}}) = \exp\left(\frac{H_n - 1}{2}\right)$$

where $H_n = \sum_{i=1}^{n} 1/i$.

The numbers $c_I(n, \overline{\mathbb{Q}})$ are the only Hermite constants computed for every positive integer n, a situation which contrasts with the classical case $K = \mathbb{Q}$.

Let us now discuss in more details the constants $c_i^*(n, K)$.

Proposition 28 *We have* $c_1^*(n, K) = \max\limits_{1 \le i \le n} c_I(i, K)$ *for every integer* $n \ge 1$.

Proof The constant $c_1^*(n, K)$ does not depend on $*$ since $\lambda_1^*(E) = \Lambda_1(E)$ and so $c_1^*(n, K) = \sup_{\dim E = n} \Lambda_1(E) \exp \mu_1(E)$. Let us consider a rigid adelic space E over K with dimension n and E_{des} its destabilizing space. If $d = \dim E_{\text{des}}$ we have $\Lambda_1(E) \le \Lambda_1(E_{\text{des}})$ and $\mu_1(E) = \mu(E_{\text{des}})$. So we get

$$\Lambda_1(E) \exp \mu_1(E) \le \Lambda_1(E_{\text{des}}) \exp \mu(E_{\text{des}}) \le c_I(d, K) \le \max\limits_{1 \le i \le n} c_I(i, K).$$

Conversely, let F be a rigid adelic space of dimension $i \in \{1, \ldots, n\}$ and $G = L^{\oplus(n-i)}$ where L is a rigid adelic line with $\Lambda_1(L) = \Lambda_1(F)$. Then we have $\dim F \oplus G = n$, $\Lambda_1(F \oplus G) = \min\{\Lambda_1(F), \Lambda_1(G)\} = \Lambda_1(F)$ (Proposition 11) and $\mu(F) \le \mu_1(F \oplus G)$ (since F is a subspace of $F \oplus G$). We get

$$\Lambda_1(F) \exp \mu(F) \le \Lambda_1(F \oplus G) \exp \mu_1(F \oplus G) \le c_1^*(n, K)$$

and so $c_I(i, K) \le c_1^*(n, K)$. $\qquad\qquad\qquad\qquad\qquad\qquad\qquad\qquad\qquad\qquad\qquad$ □

Proposition 29 *For every positive integer* n *and* $* \in \{BC, \Lambda, Z\}$, *we have* $c_1^*(n, K) \le c_2^*(n, K) \le \cdots \le c_n^*(n, K)$.

The proof rests on two auxiliary results.

Lemma 30 *We have* $c_i^*(n, K) \le c_i^*(n + 1, K)$ *for every integer* $i \in \{1, \ldots, n\}$.

Proof Let E be a rigid adelic space of dimension n. For any rigid adelic line L, we have $\min\{\lambda_i^*(E), \Lambda_1(L)\} \le \lambda_i^*(E \oplus L)$. Indeed, for $* = BC$ and $F \subset E$ of dimension $\le i - 1$,

$$\min\{\Lambda_1(E/F), \Lambda_1(L)\} = \Lambda_1((E/F) \oplus L) = \Lambda_1((E \oplus L)/(F \oplus \{0\}))$$

is bounded above by $\lambda_i^{BC}(E \oplus L)$ and we make $\Lambda_1(E/F) \to \lambda_i^{BC}(E)$. For $* = \Lambda$ or Z we use Proposition 10: Consider a subset $S \subset E \oplus L$ such that $\text{Vect}(S)$ or $\text{Zar}(S)$ has dimension $\ge i$. Either $S \subset E$ and we have $\sup_{x \in S} H_{E \oplus L}(x) = \sup_{x \in S} H_E(x) \ge \lambda_i^*(E)$. Or there exists $a \in E$ and $\ell \in L \setminus \{0\}$ such that $(a, \ell) \in S$ and then $\sup_{x \in S} H_{E \oplus L}(x) \ge H_{E \oplus L}(a, \ell) \ge H_L(\ell) = \Lambda_1(L)$. In any case, we get the wanted inequality for $\lambda_i^*(E \oplus L)$. Next, choosing L such that $\Lambda_1(L) = \lambda_i^*(E)$, we get $\lambda_i^*(E) \le \lambda_i^*(E \oplus L)$. Moreover we have $\mu_i(E) \le \mu_i(E \oplus L)$ (see the comment before the proof of Theorem 20, on page 57). So

$$\lambda_i^*(E) \exp \mu_i(E) \le \lambda_i^*(E \oplus L) \exp \mu_i(E \oplus L) \le c_i^*(n + 1, K)$$

and Lemma 30 follows.   □

Lemma 31 *The function* $n \mapsto c_n^*(n, K)$ *is nondecreasing.*

Proof Let L be a rigid adelic line over K such that $\Lambda_1(L) = \exp(-\mu_{\min}(E))$. We have $\lambda_{n+1}^*(E \oplus L) \geq \lambda_n^*(E)$ and, by Theorem 20, we have $\mu_{\min}(E \oplus L) = \min\{\mu_{\min}(E), \mu_{\min}(L)\} = \mu_{\min}(E)$ since $\mu_{\min}(L) = \deg L = \mu_{\min}(E)$. We get

$$\lambda_n^*(E) \exp \mu_{\min}(E) \leq \lambda_{n+1}^*(E \oplus L) \exp \mu_{\min}(E \oplus L) \leq c_{n+1}^*(n+1, K).$$

□

Proof of Proposition 29 We proceed by induction on n assuming $c_i^*(j, K) \leq c_{i+1}^*(j, K)$ is true for all integers $j \leq n-1$ and $i \leq j-1$. Let E be a rigid adelic space of dimension n and an integer $0 \leq i \leq n-1$. If $\mu_i(E) = \mu_{i+1}(E)$, then $\lambda_i^*(E) \exp \mu_i(E) \leq \lambda_{i+1}^*(E) \exp \mu_{i+1}(E) \leq c_{i+1}^*(n, K)$. Otherwise the HN-filtration $\{0\} \subsetneq E_1 \subsetneq \cdots \subsetneq E_N = E$ of E is not trivial (that is, $N \geq 2$).

- If $i + 1 \leq \dim E_{N-1}$, then $\mu_i(E) = \mu_i(E_{N-1})$ (since the restriction of the canonical polygon of E to the interval $[0, \dim E_{N-1}]$ equals $P_{E_{N-1}}$). We deduce

$$\lambda_i^*(E) \exp \mu_i(E) \leq \lambda_i^*(E_{N-1}) \exp \mu_i(E_{N-1})$$

$$\leq c_i^*(\dim E_{N-1}, K)$$

$$\underset{\underset{\text{Induction hypothesis}}{\uparrow}}{\leq} c_{i+1}^*(\dim E_{N-1}, K) \underset{\underset{\text{Lemma 30}}{\uparrow}}{\leq} c_{i+1}^*(n, K).$$

- If $i + 1 > \dim E_{N-1}$, then $\dim E_{N-1} = i$ since $\mu_j(E) = \mu_{\min}(E)$ for $j \geq \dim E_{N-1}+1$ and $\mu_i(E) \neq \mu_{i+1}(E)$ (see Theorem 16). Thus we get $\mu_{i+1}(E) = \mu_{\min}(E)$ and $\mu_i(E_{N-1}) = \mu_{\min}(E_{N-1}) = \mu_i(E)$. We deduce

$$\lambda_i^*(E) \exp \mu_i(E) \leq \lambda_i^*(E_{N-1}) \exp \mu_i(E_{N-1})$$

$$< c_i^*(i, K) \underset{\underset{\text{Lemma 31}}{\uparrow}}{\leq} c_{i+1}^*(i+1, K) \underset{\underset{\text{Lemma 30}}{\uparrow}}{\leq} c_{i+1}^*(n, K).$$

In any case, we have $\lambda_i^*(E) \exp \mu_i(E) \leq c_{i+1}^*(n, K)$, which implies Proposition 29.

□

Actually, for $* = BC$, we have $c_1^{BC}(n, K) = \cdots = c_n^{BC}(n, K)$. Indeed, for every linear subspace $F \subsetneq E$, we have

$$\Lambda_1(E/F) \exp \mu_{\max}(E/F) \leq c_1(n, K) \; (= c_1^{BC}(n, K)).$$

Using Proposition 18, we get $\Lambda_1(E/F) \exp \mu_{\min}(E) \leq c_1(n, K)$ and, taking the supremum over F allows to replace $\Lambda_1(E/F)$ by $\Lambda^{(n)}(E)$. In the end, we obtain $c_n^{BC}(n, K) \leq c_1(n, K)$ and the equality follows from Proposition 29. In summary we have

$$c_I(n, K) = c_{II}^{BC}(n, K) = c_{II}^{\Lambda}(n, K)$$
$$\text{I}\wedge$$
$$c_1^{BC}(n, K) = c_2^{BC}(n, K) = \cdots = c_n^{BC}(n, K)$$
$$\| \qquad \text{I}\wedge \qquad \text{I}\wedge$$
$$c_1^{\Lambda}(n, K) \le c_2^{\Lambda}(n, K) \le \cdots \le c_n^{\Lambda}(n, K)$$
$$\| \qquad \text{I}\wedge \qquad \text{I}\wedge$$
$$c_1^{Z}(n, K) \le c_2^{Z}(n, K) \le \cdots \le c_n^{Z}(n, K)$$
$$\|$$
$$\max_{1 \le i \le n} c_I(i, K)$$

Other Relations Between These Constants

Proposition 32 *Let E be a rigid adelic space of dimension n over K. Then, for every $m \in \{1, \ldots, n\}$ and every $* \in \{BC, \Lambda, Z\}$, we have*

(1) $\lambda_1^*(E) \cdots \lambda_m^*(E) \exp P_E(m) \le c_*^*(n, K)^n$,
(2) $\lambda_1^*(E) \cdots \lambda_m^*(E) \le c_{II}^*(n, K)^m H(E)^{m/n}$.

Question 33 Is the first bound true with $c_{II}^*(n, K)^m$ on the right hand side?

Since $P_E(m) \ge m\mu(E)$, a positive answer to this question would improve both (1) and (2) of the proposition. A weak form of Question 33 might be:

Question 34 Is $c_i^*(n, K) \le c_{II}^*(n, K)^i$ for $1 \le i \le n$?

One can prove that these last two questions have affirmative answers if $n \mapsto c_{II}^*(n, K)$ is a nondecreasing function.

Proof of Proposition 32 For the first inequality, we use the definition of $c_{II}^*(n, K)$ and $\lambda_i^*(E) \exp \mu_i(E) \ge 1$ for every $i \in \{m + 1, \ldots, n\}$ (Corollary 23). We get the result with $\deg E - \sum_{i=m+1}^{n} \mu_i(E) = P_E(m)$. As for the second inequality, we still use the definition of $c_{II}^*(n, K)$ but, i being as above, we bound from below $\lambda_i^*(E)$ by $\left(\lambda_1^*(E) \cdots \lambda_m^*(E)\right)^{1/m}$. $\qquad\square$

Regarding the second point of the proposition, one can prove that if, for $i \in \{1, \ldots, n\}$, we denote by

$$a_i^*(n, K) = \sup_{\dim E = n} \frac{\left(\lambda_1^*(E) \cdots \lambda_i^*(E)\right)^{1/i}}{H(E)^{1/n}}$$

(E varies among rigid adelic spaces over K of dimension n), then $c_I(n, K) = a_1^*(n, K) \le a_2^*(n, K) \le \cdots \le a_n^*(n, K) = c_{II}^*(n, K)$. In particular, when $c_I(n, K) = c_{II}^*(n, K)$, all these constants are equal and $c_{II}^*(n, K)$ is the best constant in Proposition 32-2. There exist other constants in the literature such as the *Rankin* *constant* associated to two integers $1 \leq m \leq n$ and to an algebraic extension K:

$$R(m, n, K) = \sup \left\{ \frac{\sigma_m(E)}{H(E)^{m/n}} ; \ \dim E = n \right\}$$

(recall $\sigma_m(E) = \inf\{H(F); \ F \subset E, \ \dim F = m\}$, see page 50). We leave the following properties as an exercise, whose solution can be found in the book [17, § 2.8] by Martinet. Here $1 \leq i \leq m \leq n$ are integers and $R(m, n, K)$ is shortened in $R(m, n)$ since K is fixed.

(a) $R(1, n) = c_I(n, K)$,
(b) $R(m, n) = R(n - m, n)$,
(c) $R(m, n) \leq c_I(n, K)^m$,
(d) $R(i, n) \leq R(i, m) R(m, n)^{i/m}$ (Generalization of Mordell inequality),
(e) $a_i^*(n, K) \leq a_i^*(m, K) R(m, n)^{1/m}$.

4.3   Transference Theorems

Let E be a rigid adelic space of dimension n over K and let $i \in \{1, \ldots, n\}$. A *transference theorem* gives an upper bound of $\lambda_i^*(E)\lambda_{n-i+1}^*(E^\vee)$ for $* \in \{BC, \Lambda, Z\}$. To establish such a theorem, we are therefore naturally led to introduce the following quantity:

$$t_i^*(n, K) := \sup \left\{ \lambda_i^*(E)\lambda_{n-i+1}^*(E^\vee) ; \ \dim E = n \right\}$$

where E varies among rigid adelic spaces over K with dimension n. From the definition, we get $t_i^*(n, K) = t_{n-i+1}^*(n, K)$ and $t_i^{BC}(n, K) \leq t_i^\Lambda(n, K) \leq t_i^Z(n, K)$. Also note that Proposition 19 implies that the product

$$\lambda_i^*(E)\lambda_{n-i+1}^*(E^\vee) = \left(\lambda_i^*(E) \exp \mu_i(E) \right) \left(\lambda_{n-i+1}^*(E^\vee) \exp \mu_{n-i+1}\left(E^\vee\right) \right)$$

is greater than 1 (Corollary 23) and at most $c_i^*(n, K)c_{n-i+1}^*(n, K)$. Moreover we have

$$\lambda_i^*(E) \exp \mu_i(E) \leq \lambda_i^*(E)\lambda_{n-i+1}^*(E^\vee) \quad \text{and so,} \quad c_i^*(n, K) \leq t_i^*(n, K).$$

That proves that $t_i^*(n, K) < \infty$ is a real number as soon as K is a Siegel field for $* \in \{BC, \Lambda\}$ or a Siegel field of infinite degree for $* = Z$. We do not know if we can expect a polynomial bound in n for $t_i^*(n, K)$ when K is a Siegel field. In particular can we bound $t_i^Z(n, \overline{\mathbb{Q}})$ polynomially in n? In a very optimistic view, we would like to answer positively to the

Question 35 Is $t_i^*(n, K) \leq c_{II}^*(n, K)^2$ true?

The square might be justified by several observations. Firstly, this inequality (and even the equality) is true for $n = 2$ and $* = \Lambda$ (Theorem 37 below). Then, for $K = \overline{\mathbb{Q}}$, we have $c_{\mathbb{I}}^Z(n, \overline{\mathbb{Q}})^2 = \exp(H_n - 1) \simeq n$ whereas $t_i^Z(n, \overline{\mathbb{Q}}) \geq Z_i(\overline{\mathbb{Q}}^n)Z_{n-i+1}(\overline{\mathbb{Q}}^n) = \sqrt{i(n - i + 1)}$ is greater than $n/2$ for $i = \lfloor (n+1)/2 \rfloor$. That proves we cannot replace the square by a lower exponent in Question 35. Moreover, for $* = \mathrm{BC}$, since $c_i^{\mathrm{BC}}(n, K) = c_{n-i+1}^{\mathrm{BC}}(n, K) = c_1(n, K)$ (see page 63), we have $t_i^{\mathrm{BC}}(n, K) \leq c_1(n, K)^2$. However $c_{\mathbb{I}}^{\mathrm{BC}}(n, K) = c_{\mathrm{I}}(n, K) \leq c_1(n, K)$ and so we only got a weak version of Question 35. At last we have the following result valid for any number field and proved by Banaszczyk for $K = \mathbb{Q}$ in his article [2]. The upper bound for $t_i^\Lambda(n, K)$ given hereunder is not too far from $c_{\mathbb{I}}^\Lambda(n, K)^2 = c_{\mathrm{I}}(n, K)^2$ if we take into account the inequalities

$$\frac{n\delta_{K/\mathbb{Q}}^{(1-1/n)}}{25 \max\{1, \log \delta_{K/\mathbb{Q}}\}^{2/n}} \leq c_{\mathbb{I}}^\Lambda(n, K)^2 \leq n\delta_{K/\mathbb{Q}}$$

where the number $\delta_{K/\mathbb{Q}} = |\Delta_{K/\mathbb{Q}}|^{1/[K:\mathbb{Q}]}$ is the root discriminant of K (see [14, Proposition 5.2]).

Theorem 36 *When K is a number field, we have $t_i^\Lambda(n, K) \leq n\delta_{K/\mathbb{Q}}$.*

Let us outline the proof. The problem is to reduce to the case $K = \mathbb{Q}$ and to apply Banaszczyk's theorem. For this, we use the scalar restriction $\mathrm{Res}_{K/\mathbb{Q}} E$ of a rigid adelic space E over a number field K (for a more general finite extension L/K, see [14, Lemma 4.24]). It is the rigid adelic space over $\mathbb{Q}$ built from the space E viewed as a $\mathbb{Q}$-vector space (with dimension $[K : \mathbb{Q}] \dim E$) endowed with

$$\|x\|_{\mathrm{Res}_{K/\mathbb{Q}} E, \infty} = \left(\sum_{v \in V_\infty(K)} [K_v : \mathbb{Q}_v] \|x\|_{E,v}^2 \right)^{1/2}$$

at the archimedean place ∞ of $\mathbb{Q}$ and $\|x\|_{\mathrm{Res}_{K/\mathbb{Q}} E, p} = \max_{v|p} \|x\|_{E,v}^2$ at ultrametric places p of $\mathbb{Q}$. By [14, Lemma 4.29], the height of $\mathrm{Res}_{K/\mathbb{Q}} E$ is

$$H\left(\mathrm{Res}_{K/\mathbb{Q}} E\right) = H(E)^{[K:\mathbb{Q}]} |\Delta_{K/\mathbb{Q}}|^{(\dim E)/2}.$$

Associated to K we also have its differential (rigid) adelic space ω_K over K whose underlying space is $\mathrm{Hom}_{\mathbb{Q}}(K, \mathbb{Q})$ viewed as a K-vector space with the scalar multiplication: $\lambda.\varphi(x) = \varphi(\lambda.x)$ for $\lambda, x \in K$ and $\varphi \in \mathrm{Hom}_{\mathbb{Q}}(K, \mathbb{Q})$. The trace $\mathrm{Tr}_{K/\mathbb{Q}}$ is a basis of ω_K ($\dim \omega_K = 1$) and it allows to define rigid adelic metrics on ω_K by stating $\|\mathrm{Tr}_{K/\mathbb{Q}}\|_v = 1$ if $v \in V_\infty(K)$ and, otherwise, $\|\mathrm{Tr}_{K/\mathbb{Q}}\|_v$ is

$$\inf \left\{ |\lambda|_v \; ; \; \lambda \in K_v \backslash \{0\} \text{ and } \lambda^{-1} \mathrm{Tr}_{K/\mathbb{Q}} \in \mathrm{Hom}_{\mathbb{Z}_p}(\mathcal{O}_{K_v}, \mathbb{Z}_p) \right\}.$$

It is known that $H(\omega_K) = \delta_{K/\mathbb{Q}}^{-1}$ (see [20, p. 219]). Besides, given a rigid adelic space E over K, the $\mathbb{Q}$-linear map $E^\vee \otimes_K \omega_K \to \mathrm{Hom}_\mathbb{Q}(E, \mathbb{Q}), \varphi \otimes \lambda \mapsto \lambda \circ \varphi$ induces an (isometric) isomorphism of rigid adelic spaces over $\mathbb{Q}$:

$$\mathrm{Res}_{K/\mathbb{Q}}\left(E^\vee \otimes_K \omega_K\right) \simeq \left(\mathrm{Res}_{K/\mathbb{Q}} E\right)^\vee$$

(see [9, Proposition 3.2.2]). With these reminders being done, we can now easily prove Theorem 36. Corollary 4.28 of [14] gives $\Lambda_i(E) \leq [K : \mathbb{Q}]^{-1/2}\Lambda_{(i-1)[K:\mathbb{Q}]+1}(\mathrm{Res}_{K/\mathbb{Q}} E)$. Applying this inequality with $E^\vee \otimes_K \omega_K$ and using $\Lambda_{n-i+1}(E^\vee \otimes_K \omega_K) = \Lambda_{n-i+1}(E^\vee)H(\omega_K)$ (since ω_K is a line), we deduce that $\Lambda_i(E)\Lambda_{n-i+1}(E^\vee)$ is bounded above by

$$\frac{\delta_{K/\mathbb{Q}}}{[K : \mathbb{Q}]} \times \Lambda_{i[K:\mathbb{Q}]}\left(\mathrm{Res}_{K/\mathbb{Q}} E\right) \Lambda_{(n-i)[K:\mathbb{Q}]+1}\left(\left(\mathrm{Res}_{K/\mathbb{Q}} E\right)^\vee\right)$$

and the last product of minima is at most $[K : \mathbb{Q}]n$ by [2, Theorem 2.1].

Pekker's Theorem In this paragraph we build on the work of Pekker [21] about $t_i^\Lambda(n, \overline{\mathbb{Q}})$ to generalize it to any Siegel field. It allows to give some general upper bounds for $t_i^\Lambda(n, K)$ exponential in n, as in the following result.

Theorem 37 *Let $1 \leq i \leq n$ be integers. Then, for any algebraic extension $K/\mathbb{Q}$, we have*

(1) $t_1^\Lambda(2, K) = t_2^\Lambda(2, K) = c_1(2, K)^2$
(2) $t_i^\Lambda(n, K) \leq t_1^\Lambda(i, K)t_1^\Lambda(n - i + 1, K)$
(3) *For $n \geq 2$, $t_i^\Lambda(n, K) \leq t_1^\Lambda(n - 1, K)t_1^\Lambda(2, K)$*
(4) $t_i^\Lambda(n, K) \leq t_1^\Lambda(2, K)^{n-1}$

Question 38 Do we have similar results for $t_i^Z(n, K)$?

For the proof of Theorem 37 we need an auxiliary result.

Lemma 39 *Let E be a rigid adelic space of dimension n over K. Let ε be a positive real number. Then there exists a hyperplane $F \subset E$ such that*

$$H(F) \leq (1 + \varepsilon)\sigma_{n-1}(E) \quad and \quad \Lambda_n(E)H(F) \leq (1 + \varepsilon)H(E)t_1^\Lambda(n, K).$$

Proof Let $\varphi \in E^\vee \backslash \{0\}$ such that $H_{E^\vee}(\varphi) \leq (1 + \varepsilon)\Lambda_1(E^\vee)$ and consider $F = \mathrm{Ker}\,\varphi$. The first bound comes from $H(F) = H_{E^\vee}(\varphi)H(E)$ (see Proposition 5) and $\sigma_{n-1}(E) = \Lambda_1(E^\vee)H(E)$ (see page 50). The other one uses in addition the definition of $t_1^\Lambda(n, K)$. $\square$

Proof of Theorem 37

(1) When E is a rigid adelic space with dimension 2, we have $\Lambda_1(E) = \sigma_{2-1}(E) = \Lambda_1(E^\vee)H(E)$ so $\Lambda_1(E)\Lambda_2(E^\vee) = \Lambda_1(E^\vee)\Lambda_2(E^\vee)/H(E^\vee)$. We conclude with Minkowski theorem:

$$t_1^\Lambda(2, K) = \sup_{\dim E = 2} \frac{\Lambda_1(E)\Lambda_2(E)}{H(E)} = c_{\mathbb{I}}^\Lambda(2, K)^2 = c_{\mathbb{I}}(2, K)^2.$$

(2) Let E be a rigid adelic space over K with dimension n and let G be a linear subspace of dimension $i - 1$ of $E^\vee$. We apply Lemma 39 to $G^\perp$ (viewed as a subspace of E): there exists a hyperplane $A \subset G^\perp$ such that $\Lambda_{n-i+1}(G^\perp)H(A) \le (1 + \varepsilon)H(G^\perp)t_1^\Lambda(n - i + 1, K)$. Let us apply again (in the same way) Lemma 39 to $A^\perp \subset E^\vee$: there exists a hyperplane $B \subset A^\perp$ such that $\Lambda_i(A^\perp)H(B) \le (1 + \varepsilon)t_1^\Lambda(i, K)H(A^\perp)$. We have $\dim B = i - 1$ and $H(B) \ge \sigma_{i-1}(E^\vee)$. Moreover $\Lambda_{n-i+1}(G^\perp) \ge \Lambda_{n-i+1}(E)$ and $\Lambda_i(A^\perp) \ge \Lambda_i(E^\vee)$. Multiplying the above inequalities given by Lemma 39 we get

$$\Lambda_{n-i+1}(E)\Lambda_i(E^\vee) \le (1 + \varepsilon)^2 t_1^\Lambda(n - i + 1, K)t_1^\Lambda(i, K) \times \frac{H(A^\perp)H(G^\perp)}{H(B)H(A)}.$$

By Proposition 5 the latter quotient equals $H(G)/H(B)$ and so it is at most $H(G)/\sigma_{i-1}(E^\vee)$. We conclude by letting $H(G) \to \sigma_{i-1}(E^\vee)$ and $\varepsilon \to 0$.

(3) Let E be a rigid adelic space over K with dimension n. Let $\varphi, \psi \in E^\vee$ be linearly independent linear forms. Define $V = \text{Ker}\,\varphi$ and $W = \text{Ker}\,\psi$. By Lemma 39, there exist hyperplanes $V' \subset V$ and $W' \subset W$ such that $\Lambda_{n-1}(V)H(V') \le (1 + \varepsilon)H(V)t_1^\Lambda(n - 1, K)$ and $\Lambda_{n-1}(W)H(W') \le (1 + \varepsilon)H(W)t_1^\Lambda(n-1, K)$. Moreover we have $H(V)/H_{E^\vee}(\varphi) = H(W)/H_{E^\vee}(\psi) = H(E)$. By hypothesis $V \ne W$ and so $V + W = E$ which gives $\Lambda_n(E) \le \max\{\Lambda_{n-1}(V), \Lambda_{n-1}(W)\}$. The latter maximum is attained for one of the two spaces V or W and we denote by G the corresponding hyperplane V' or W'. Hence we get

$$\Lambda_n(E)H(G) \le (1 + \varepsilon)H(E)t_1^\Lambda(n - 1, K)\max\{H_{E^\vee}(\varphi), H_{E^\vee}(\psi)\}.$$

Choosing φ and ψ such that their heights be at most $(1 + \varepsilon)\Lambda_2(E^\vee)$ and using $H(G^\perp) = H(G)/H(E)$, we get $\Lambda_n(E)H(G^\perp) \le (1+\varepsilon)^2 t_1^\Lambda(n-1, K)\Lambda_2(E^\vee)$. Multiplying both sides by $\Lambda_1(E^\vee)$, we find that the quantity $\Lambda_n(E)\Lambda_1(E^\vee)$ is at most

$$(1 + \varepsilon)^2 t_1^\Lambda(n - 1, K) \times \frac{\Lambda_1(E^\vee)\Lambda_2(E^\vee)}{H(G^\perp)}$$

$$\le (1 + \varepsilon)^2 t_1^\Lambda(n - 1, K) \times \frac{\Lambda_1(G^\perp)\Lambda_2(G^\perp)}{H(G^\perp)}$$

and, lastly, smaller than $(1 + \varepsilon)^2 t_1^{\wedge}(n - 1, K) c_{\mathrm{II}}^{\wedge}(2, K)^2$. We conclude with the first statement of Theorem 37 and $\varepsilon \to 0$.

(4) As a direct consequence of (3), we get $t_1^{\wedge}(n, K) \leq t_1^{\wedge}(2, K)^{n-1}$. Then the changeover to $t_i^{\wedge}(n, K)$ arises from point (2) of Theorem 37. $\qquad\square$

The equality $\Lambda_1(E) = \Lambda_1(E^{\vee}) H(E)$ when $\dim E = 2$ allows to prove

$$\sup \{\Lambda_1(E)\Lambda_1(E^{\vee}) ; \ \dim E = 2\} = c_{\mathrm{I}}(2, K)^2.$$

In general, the definition of the number $c_{\mathrm{I}}(n, K)$ provides the upper bound $\sup \{\Lambda_1(E)\Lambda_1(E^{\vee}) ; \ \dim E = n\} \leq c_{\mathrm{I}}(n, K)^2$, but, when $K = \mathbb{Q}$ and $n = 3$, Bergé and Martinet proved that the equality is no longer true [4, Proposition 2.13 (iii)]. Notwithstanding this, when $K = \mathbb{Q}$ and for every $n \geq 1$, a result by Banaszczyk [3, Lemma 5], based on Siegel's mean value theorem, states that

$$\sup \{\Lambda_1(E)\Lambda_1(E^{\vee}) ; \ \dim E = n\} > \frac{n}{2\pi e}$$

(see also Sect. 4.2 in Chapter IV).

Question 40 Do we have a similar lower bound, true for any K and $n \geq 1$, with a function growing to infinity with n, or even linear in n?

5 Heights of Morphisms and Slope-Minima Inequalities

5.1. Until now we have considered only *rigid* adelic spaces. Nevertheless it may be useful (or even crucial) to work with $\mathrm{Hom}_K(E, F)$ endowed with the operator norms, which is not Hermitian in general.

Definition 41 *Let E and F be adelic spaces over K such that $E^{\vee} \otimes_{\varepsilon} F$ is integrable. The height of $\varphi \subset \mathrm{Hom}_K(E, F) \backslash \{0\}$ is*

$$h(\varphi) = h(E, F; \varphi) = \int_{V(K)} \log \|\varphi\|_{E^{\vee} \otimes_{\varepsilon} F, v} \, d\sigma(v).$$

We may also use $H(\varphi) = \exp h(\varphi)$. Here, as defined on page 41,

$$\|\varphi\|_{E^{\vee} \otimes_{\varepsilon} F, v} = \sup \left\{ \frac{\|\varphi(x)\|_{F, v}}{\|x\|_{E, v}} ; \ x \in (E \otimes_K K_v) \backslash \{0\} \right\}$$

is the operator norm. Note that if $E' \subset E$ is a linear subspace, then $h(E', F; \varphi_{|E'}) \leq h(E, F; \varphi)$. When E and F are rigid adelic spaces over K, there is also the Hilbert-Schmidt height for φ built with $\|\varphi\|_{E^{\vee} \otimes F, v}$, which is greater than $h(\varphi)$. In this paragraph, our aim is to compare minima and slopes of two (rigid) adelic spaces connected by a linear map. In the following results, E and F are some rigid adelic spaces over K and $\varphi \colon E \to F$ a linear map.

Proposition 42 *If $\varphi \colon E \to F$ is an isomorphism, then*

(1) $\deg E = \deg F + h(\det E, \det F; \det \varphi)$,
(2) $\mu(E) \leq \mu(F) + h(\varphi)$.

Proof (1) By hypothesis $\det \varphi \colon \det E \to \det F$ is an isomorphism between rigid adelic lines and, for all $v \in V(K)$ and $x \in (\det E) \otimes_K K_v \backslash \{0\}$, we have

$$| \det \varphi |_v = \frac{\| \det \varphi(x) \|_{\det F, v}}{\| x \|_{\det E, v}}.$$

We take logarithms and we integrate over v to conclude. (2) The second statement is a direct consequence of the first one and of Hadamard's inequality $| \det \varphi |_v \leq \| \varphi \|_{E^v \otimes_\varepsilon F, v}^{\dim E}$.   □

Theorem 43 *If $\varphi \colon E \to F$ is injective, then*

$$\mu_{\max}(E) \leq \mu_{\max}(F) + h(\varphi) \quad \text{and} \quad \Lambda_1(F) \leq \Lambda_1(E)H(\varphi).$$

More generally, if $\varphi \neq 0$, then for all $i \in \{1, \ldots, \operatorname{rk} \varphi\}$ and $* \in \{\mathrm{BC}, \Lambda, Z\}$ we have

$$\mu_{i+\dim \operatorname{Ker} \varphi}(E) \leq \mu_i(F) + h(\varphi) \quad \text{and} \quad \lambda_i^*(F) \leq \lambda_{i+\dim \operatorname{Ker} \varphi}^*(E)H(\varphi).$$

Proof First assume that φ is injective. Let $E_0 \subset E$ be a non-zero linear subspace and $F_0 = \varphi(E_0)$. Since φ is injective the induced map $\widetilde{\varphi} \colon E_0 \to F_0$ is an isomorphism and, by Proposition 42,

$$\mu(E_0) \leq \mu(F_0) + h(E_0, F_0; \widetilde{\varphi}) \leq \mu_{\max}(F) + h(\varphi).$$

Taking the supremum over E_0 on the left hand side leads to the first maximal slopes inequality. As for the inequality for the first minima, if $x \in E \backslash \{0\}$, then $\varphi(x) \in F \backslash \{0\}$ so $\Lambda_1(F) \leq H_F(\varphi(x)) \leq H_E(x)H(\varphi)$ and we take the infimum over x to replace $H_E(x)$ by $\Lambda_1(E)$. Now just assume $\varphi \neq 0$. Let $F_0 \subset F$ be a linear subspace with dimension $\leq i - 1$ and $A \subset E$ a linear subspace such that $\dim A \geq i + \dim \ker \varphi$. We have $\dim \varphi^{-1}(F_0) \leq \dim \operatorname{Ker} \varphi + i - 1$ and $\dim \varphi(A) \geq i$. Moreover the induced map $\overline{\varphi} \colon A/A \cap \varphi^{-1}(F_0) \to (\varphi(A) + F_0)/F_0$ is an isomorphism. Using Proposition 42 and $h(\overline{\varphi}) \leq h(\varphi)$, we get $\mu(A/A \cap \varphi^{-1}(F_0)) \leq \mu((\varphi(A)+F_0)/F_0) + h(\varphi)$, from which we deduce

$$\inf \{\mu(A/B) \,; \; B \subset A \text{ and } \dim B \leq i + \dim \ker \varphi - 1\}$$
$$\leq \sup \{\mu(F_1/F_0) \,; \; F_0 \subset F_1 \subset F \text{ and } \dim F_1 \geq i\} + h(\varphi).$$

We conclude with Proposition 18, taking the supremum over A on the left hand side and the infimum over F_0 on the right. As for the analogous inequality for λ_i^*, we distinguish the three cases $* = BC, \Lambda, Z$. For $* = BC$, we proceed as above: $\Lambda_1(F/F_0) \leq \Lambda_1(E/E_0)H(\overline{\varphi})$ (where $E_0 = \varphi^{-1}(F_0)$ and $\overline{\varphi}: E/E_0 \to F/F_0$ is the map induced by φ). Since $\dim E_0 \leq \dim \operatorname{Ker}\varphi + i - 1$ we have $\Lambda_1(E/E_0) \leq \Lambda^{(i+\dim \operatorname{Ker}\varphi)}(E)$. So

$$\Lambda^{(i)}(F) = \sup_{\dim F_0 \leq i-1} \Lambda_1(F/F_0) \leq \Lambda^{(i+\dim \operatorname{Ker}\varphi)}(E)H(\varphi).$$

For $* = \Lambda$ or Z we get an injective map from φ making the quotient by $\operatorname{Ker}\varphi$, which yields $\lambda_i^*(F) \leq \lambda_i^*(E/\operatorname{Ker}\varphi)H(\varphi)$ and we conclude with $\lambda_i^*(E/\operatorname{Ker}\varphi) \leq \lambda_{i+\dim \operatorname{Ker}\varphi}^*(E)$ (for instance, for $* = \Lambda$ it means that if $\{e_1, \ldots, e_{i+\dim \operatorname{Ker}\varphi}\} \subset E$ is a free family, then at least i of the images of the vectors e_j in $E/\operatorname{Ker}\varphi$ are also linearly independent). $\qquad\square$

Corollary 44 *Let $\varphi: E \to F$ be a linear map.*

(1) *If $\varphi \neq 0$, then $\mu_{\min}(E) \leq \mu_{\max}(F) + h(\varphi)$ and $\lambda_1^*(F) \leq \lambda_{\dim E}^*(E)H(\varphi)$.*
(2) *If φ is surjective, then $\mu_{\min}(E) \leq \mu_{\min}(F) + h(\varphi)$ and $\lambda_{\dim F}^*(F) \leq \lambda_{\dim E}^*(E)H(\varphi)$.*
(3) *If φ is surjective, then $\mu_{\max}(F) \leq \deg F - (\dim F - 1)\mu_{\min}(E) + (\dim F - 1)h(\varphi)$.*

Proof (1) Take $i = \operatorname{rk}\varphi$ in Theorem 43, bound from below i by 1 and use $\dim E = \dim \operatorname{Ker}\varphi + \operatorname{rk}\varphi$. (2) Same method but keep $i = \operatorname{rk}\varphi$ which is equal to $\dim F$ since φ is surjective. (3) Observe $\deg F = \mu_{\max}(F) + \sum_{i=2}^{\dim F} \mu_i(F) \geq \mu_{\max}(F) + (\dim F - 1)\mu_{\min}(F)$ and use (2). $\qquad\square$

One can prove that if φ is injective, then for all $i \in \{1, \ldots, \dim E\}$, one has

$$(\star) \qquad P_E(i) \leq P_F(i) + h(\bigwedge^i E, \bigwedge^i F; \bigwedge^i \varphi).$$

For this, observe that, for all $v \in V(K)$, the function

$$i \mapsto \|\bigwedge^i \varphi\|_v / \|\bigwedge^{i-1} \varphi\|_v$$

is a nonincreasing function, so $\left(h(\bigwedge^i \varphi) - h(\bigwedge^{i-1} \varphi)\right)_{1 \leq i \leq \operatorname{rk}\varphi}$ is a decreasing sequence and $i \mapsto P_F(i) + h\left(\bigwedge^i E, \bigwedge^i F; \bigwedge^i \varphi\right)$ is a concave function. The case $i = 1$ in $(\star)$ corresponds to the first statement of Theorem 43.

To conclude this part, let us prove a variant of the first bound in Corollary 44 where we replace the height built with the operator norms by the Hilbert-Schmidt height $h_{\mathrm{HS}}(\varphi) := \log H_{E^\vee \otimes F}(\varphi)$ of φ.

Proposition 45 *Let $\varphi \colon E \to F$ be a linear map. Then*

$$\mu_{\min}(E) + \frac{1}{2}\log \mathrm{rk}\,\varphi \leq \mu_{\max}(F) + h_{\mathrm{HS}}(\varphi).$$

This statement derives almost immediately from the following result.

Lemma 46 *If $\varphi \colon E \to F$ is an isomorphism, then*

$$\mu(E) + \frac{1}{2}\log \dim E \leq \mu(F) + h_{\mathrm{HS}}(\varphi).$$

Proof With Proposition 42, it amounts to proving

$$h(\det E, \det F; \det \varphi) \leq n\left(h_{\mathrm{HS}}(\varphi) - \frac{1}{2}\log n\right)$$

where n denotes the dimension of E. If $v \in V(K)$ is an ultrametric place, we simply bound $|\det \varphi|_v \leq \|\varphi\|_v^n = \|\varphi\|_{E^{\vee} \otimes F, v}^n$. If v is archimedean, we use the Hermitian adjoint φ_v^* of $\varphi_v \colon E \otimes_K K_v \to F \otimes_K K_v$ to write $|\det \varphi|_v = \det\left(\varphi_v^* \varphi_v\right)^{1/2}$. We conclude with the inequality of arithmetic and geometric means applied to the eigenvalues of the positive operator $\varphi_v^* \varphi_v$:

$$n\left(\det \varphi_v^* \varphi_v\right)^{1/n} \leq \mathrm{Tr}\left(\varphi_v^* \varphi_v\right) = \|\varphi\|_{E^{\vee} \otimes F, v}^2. \qquad \square$$

We use this lemma with the isomorphism $\overline{\varphi} \colon E/\mathrm{Ker}\,\varphi \to \mathrm{Im}\,\varphi$ and the bounds $\mu_{\min}(E) \leq \mu(E/\mathrm{Ker}\,\varphi)$ and $\mu(\mathrm{Im}\,\varphi) \leq \mu_{\max}(F)$ (Proposition 18) as well as $h_{\mathrm{HS}}(\overline{\varphi}) = h_{\mathrm{HS}}(\varphi)$ to get Proposition 45.

When $K = \mathbb{Q}$ or $K = \overline{\mathbb{Q}}$, we have $\log \Lambda_1(F) \leq -\mu(F) + \frac{1}{2}\log \dim F$ since $c_{\mathrm{I}}(n, K) \leq \sqrt{n}$ (see page 61). In particular, for every non-zero $\varphi \colon E \to F$, the same technique gives $\log \Lambda_1(\mathrm{Im}\,\varphi) + \mu_{\min}(E) \leq h_{\mathrm{HS}}(\varphi)$ and so $\log \Lambda_1(F) + \mu_{\min}(E) \leq \log \Lambda_1(E^{\vee} \otimes F)$ for all rigid adelic spaces E, F over $\overline{\mathbb{Q}}$ (see [13, Theorem 1.3] for $K = \overline{\mathbb{Q}}$). This leads us to the last part of our course.

5.2 Tensor Product

We will conclude this lecture by raising the problem of the behaviour of minima and slopes (only the first ones) with respect to tensor product. It has been seen that $\mu(E \otimes F) = \mu(E) + \mu(F)$ for every rigid adelic spaces E and F over K. What happens for $\Lambda_1(E \otimes F)$ and $\mu_1(E \otimes F)$? Let us start with two inequalities, always true:

$$\Lambda_1(E \otimes F) \leq \Lambda_1(E)\Lambda_1(F) \quad \text{and} \quad \mu_{\max}(E) + \mu_{\max}(F) \leq \mu_{\max}(E \otimes F).$$

To prove the first one, we can observe that, for all $v \in V(K)$, $x \in E \otimes_K K_v$, $y \in F \otimes_K K_v$, we have $\|x \otimes y\|_{E \otimes F, v} \le \|x\|_{E,v} \|y\|_{F,v}$. That gives $H_{E \otimes F}(e \otimes f) \le H_E(e)H_F(f)$ for $e \in E$ and $f \in F$. When e and f are not zero, $e \otimes f$ is not zero either and $H_{E \otimes F}(e \otimes f) \ge \Lambda_1(E \otimes F)$, leading to the first bound. As for the second one, we can note that

$$\mu_{\max}(E) + \mu_{\max}(F) = \mu(E_{\mathrm{des}}) + \mu(F_{\mathrm{des}}) = \mu(E_{\mathrm{des}} \otimes F_{\mathrm{des}}) \le \mu_{\max}(E \otimes F).$$

So the problem is whether these inequalities are equalities. For Λ_1 the answer is *no, in general*. Actually it has been proved by Steinberg that, for any integer $n \ge 292$, there exists a rigid adelic space E over $\mathbb{Q}$ with dimension n such that $\Lambda_1(E \otimes E) \ne \Lambda_1(E)^2$ [18, p. 47]. Coulangeon obtained similar results for some imaginary quadratic fields K [11]. The author and Rémond proved that for every integers n, m both ≥ 2, there exist some rigid adelic spaces E and F over $\overline{\mathbb{Q}}$ with $\dim E = n$ and $\dim F = m$ such that $\Lambda_1(E \otimes F) \ne \Lambda_1(E)\Lambda_1(F)$ [13, Theorem 1.5]. Actually all the difficulty of the proof lies in the case $n = m = 2$. Here we shall give a different proof, due to Gaël Rémond, not yet published.

Proposition 47 *There exists a rigid adelic plane E over $\overline{\mathbb{Q}}$ such that $\Lambda_1(E \otimes E^\vee) \ne \Lambda_1(E)\Lambda_1(E^\vee)$.*

Proof Let us recall that $c_{\mathrm{I}}(2, \overline{\mathbb{Q}}) = \exp \frac{H_2 - 1}{2} = \exp \frac{1}{4}$ (see page 61). Let us choose $0 < \varepsilon < 1$ such that $e(1 - \varepsilon)^4 > 2$. Let E be a rigid adelic plane over $\overline{\mathbb{Q}}$ such that $\Lambda_1(E) \ge (1 - \varepsilon)H(E)^{1/2}c_{\mathrm{I}}(2, \overline{\mathbb{Q}})$ (definition of $c_{\mathrm{I}}(2, \overline{\mathbb{Q}})$). Since $\dim E = 2$ we have $\Lambda_1(E^\vee)/H(E^\vee)^{1/2} = \Lambda_1(E)/H(E)^{1/2}$. From this equality, we deduce

$$\Lambda_1(E)\Lambda_1(E^\vee) = \left(\frac{\Lambda_1(E)}{H(E)^{1/2}}\right)^2 \ge (1 - \varepsilon)^2 \exp \frac{1}{2}.$$

Furthermore, considering a basis $\{e_1, e_2\}$ of E and the identity map $x = e_1 \otimes e_1^\vee + e_2 \otimes e_2^\vee$ we have $\Lambda_1(E \otimes E^\vee) \le H_{E \otimes E^\vee}(x) = \sqrt{2}$. The choice of ε makes it possible to conclude $\Lambda_1(E \otimes E^\vee) < \Lambda_1(E)\Lambda_1(E^\vee)$. $\square$

The heart of the proof is a lower bound for $\Lambda_1(E)\Lambda_1(E^\vee)$ which is compared to $\Lambda_1(E \otimes E^\vee) \le \sqrt{n}$. Using Banaszczyk's lower bound given before Question 40, we can easily prove that $\Lambda_1(E)\Lambda_1(E^\vee) \ne \Lambda_1(E \otimes E^\vee)$ when $K = \mathbb{Q}$ and $\sqrt{n} < n/(2\pi e)$, that is for $n \ge 292$ (the integer 292 is the upper part of $(2\pi e)^2$). In this way we obtain a variant of Steinberg's result.

It may be the phenomenon enlighted by Proposition 47 does not occur for the maximal slope.

Bost's Conjecture *For all rigid adelic spaces E and F over K, we have $\mu_{\max}(E \otimes F) = \mu_{\max}(E) + \mu_{\max}(F)$.*

The field K is not important here since we can freely replace it by (one of) its algebraic closure, due to the invariance by scalar extension of the maximal slope (Proposition 19). This conjecture is known to be true when $\dim E \times \dim F \le$

9 (see [8]) or, as recently proved by Rémond, when there is a group acting isometrically on $E \otimes_K \overline{K}$ such that this vector space is a direct sum of some non isomorphic irreducible vector subspaces [22]. Here we shall prove a weaker result, also due to Bost and Chen (ibid.). We recall that $H_n = 1 + 1/2 + \cdots + 1/n$ is the harmonic number.

Theorem 48 *Let E and F be some rigid adelic spaces over K and $n = \dim E$. Then we have*

$$\mu_{\max}(E \otimes F) \leq \mu_{\max}(E) + \mu_{\max}(F) + \frac{H_n - 1}{2}.$$

Lemma 49 *For any rigid adelic space E and integrable adelic space F over K, we have*

$$\Lambda_1(F) \leq \Lambda^{(\dim E)}(E)\Lambda_1(E^{\vee} \otimes_{\varepsilon} F).$$

Proof As for the first statement of Corollary 44 with $* = BC$, extended to an integrable space (same proof), we have $\Lambda_1(F) \leq \Lambda^{(\dim E)}(E)H(\varphi)$ for every non-zero $\varphi \in E^{\vee} \otimes_{\varepsilon} F$. We conclude making $H(\varphi)$ tend to $\Lambda_1(E^{\vee} \otimes_{\varepsilon} F)$. □

Lemma 50 *For all rigid adelic spaces A and B, we have*

$$\exp\{-\mu_{\max}(A) - \mu_{\max}(B)\} \leq \Lambda_1(A \otimes_{\varepsilon} B).$$

Proof If $\varphi \in A \otimes_{\varepsilon} B = \mathrm{Hom}(A^{\vee}, B)$ and $\varphi \neq 0$, we saw $\mu_{\min}(A^{\vee}) \leq \mu_{\max}(B) + h(\varphi)$ (Corollary 44). The left hand side equals to $-\mu_{\max}(A)$ (Proposition 19). We conclude with making $h(\varphi)$ tend to $\log \Lambda_1(A \otimes_{\varepsilon} B)$. □

In particular, for every rigid adelic space E, we have $\Lambda_1(E \otimes_{\varepsilon} E_{\mathrm{des}}^{\vee}) = 1$. Indeed, by this lemma, the first minimum is greater than 1 but it is also at most 1 since the injection map $E_{\mathrm{des}} \hookrightarrow E$ has height at most 1.

Lemma 51 *For all rigid adelic spaces A, B, E over K, we have*

$$1 \leq \Lambda_1(E^{\vee} \otimes_{\varepsilon} A \otimes_{\varepsilon} B)\Lambda^{(\dim E)}(E) \exp\{\mu_{\max}(A) + \mu_{\max}(B)\}.$$

Proof Replace F by $A \otimes_{\varepsilon} B$ in Lemma 49 and apply Lemma 50. □

Lemma 52 *For all rigid adelic spaces E and F over K, we have*

$$\mu_{\max}(E \otimes F) \leq \mu_{\max}(F) + \log \Lambda^{(\dim E)}(E^{\vee}).$$

Proof Let us replace E by its dual $E^{\vee}$ and take $A = F$ and $B = (E \otimes F)^{\vee}_{\text{des}}$ in Lemma 51. We get

$$1 \leq \Lambda_1(E \otimes_{\varepsilon} F \otimes_{\varepsilon} (E \otimes F)^{\vee}_{\text{des}}) \Lambda^{(\dim E)}(E^{\vee})$$
$$\times \exp\{\mu_{\max}(F) - \mu_{\max}(E \otimes F)\}.$$

Then, since the operator norm is smaller than the Hilbert-Schmidt norm, we can bound from above the first minimum on the right by $\Lambda_1(E \otimes F \otimes_{\varepsilon} (E \otimes F)^{\vee}_{\text{des}}) = 1$.
□

Proof of Theorem 48 Let us apply Lemma 52 and use the inequality

$$\Lambda^{(\dim E)}(E^{\vee}) \exp \mu_{\min}(E^{\vee}) \leq c_n^{\text{BC}}(n, \overline{\mathbb{Q}})$$

(definition of the constant on the right). By Proposition 19, we have $\mu_{\min}(E^{\vee}) = -\mu_{\max}(E)$ and we saw that $c_n^{\text{BC}}(n, \overline{\mathbb{Q}})$ is equal to the number $\exp((H_n - 1)/2)$ (see pages 61 and 63).
□

Acknowledgements I thank Pascal Autissier, Gaël Rémond and the anonymous referees for their remarks on a previous version of this course. I also thank the organizers of the Summer School, Huayi Chen, Emmanuel Peyre and Gaël Rémond for their invitation.

References

1. Y. André, Slope filtrations. Confluentes Math. **1**, 1–85 (2009)
2. W. Banaszczyk, New bounds in some transference theorems in the geometry of numbers. Math. Ann. **296**, 625–635 (1993)
3. W. Banaszczyk, Inequalities for convex bodies and polar reciproqual lattices in $\mathbb{R}^n$ II: application of K-convexity. Discrete Comput. Geom. **16**, 305–311 (1996)
4. A.-M. Bergé, J. Martinet, Sur un problème de dualité lié aux sphères en géométrie des nombres. J. Number Theory **32**, 14–42 (1989)
5. E. Bombieri, U. Zannier, A note on heights in certain infinite extensions of $\mathbb{Q}$. Rend. Mat. Acc. Lincei **12**, 5–14 (2001)
6. J.-B. Bost, Périodes et isogénies des variétés abéliennes sur les corps de nombres (d'après D. Masser et G. Wüstholz), in *Séminaire Bourbaki*, vol. 237. Astérisque (Société Mathématique de, France, 1996), pp. 115–161
7. J.-B. Bost, Algebraic leaves of algebraic foliations over number fields. Publ. Math. Inst. Hautes Études Sci. **93**, 161–221 (2001)
8. J.-B. Bost, H. Chen, Concerning the semistability of tensor products in Arakelov geometry. J. Math. Pures Appl. (9) **99**, 436–488 (2013)
9. J.-B. Bost, K. Künnemann, Hermitian vector bundles and extension groups on arithmetic schemes. I. Geometry of numbers. Adv. Math. **223**, 987–1106 (2010)
10. H. Chen, Harder-Narasimhan categories. J. Pure Appl. Algebra **214**, 187–200 (2010)
11. R. Coulangeon, Tensor products of Hermitian lattices. Acta Arith. **92**, 115–130 (2000)
12. É. Gaudron, Pentes des fibrés vectoriels adéliques sur un corps global. Rend. Semin. Mat. Univ. Padova **119**, 21–95 (2008)

13. É. Gaudron, G. Rémond, Minima, pentes et algèbre tensorielle. Israel J. Math. **195**, 565–591 (2013)
14. É. Gaudron, G. Rémond, Corps de Siegel. J. Reine Angew. Math. **726**, 187–247 (2017)
15. D. Grayson, Reduction theory using semistability. Comment. Math. Helv. **59**, 600–634 (1984)
16. G. Harder, M.S. Narasimhan, On the cohomology groups of moduli spaces of vector bundles on curves. Math. Ann. **212**, 215–248 (1975)
17. J. Martinet, *Perfect Lattices in Euclidean Spaces*. Grundlehren der Mathematischen Wissenschaften 327 (Springer, Berlin, 2003)
18. J. Milnor, D. Husemoller, *Symmetric Bilinear Forms*. Ergebnisse der Mathematik und ihrer Grenzgebiete, Band 73 (Springer, Berlin, 1973)
19. H. Minkowski, *Geometrie der Zahlen*. (Teubner, Stuttgart, 1910). http://gallica.bnf.fr/ark:/12148/bpt6k99643x
20. J. Neukirch, *Algebraic Number Theory*. Grundlehren der Mathematischen Wissenschaften 322 (Springer, Berlin, 1999)
21. A. Pekker, On successive minima and the absolute Siegel's lemma. J. Number Theory **128**, 564–575 (2008)
22. G. Rémond, Action de groupe et semi-stabilité du produit tensoriel. Confluentes Math. **11**, 53–57 (2019)
23. D. Roy, J. Thunder, An absolute Siegel's lemma. J. Reine Angew. Math. **476**, 1–26 (1996). Addendum and erratum. ibid. **508**, 47–51 (1999)
24. U. Stuhler, Eine Bemerkung zur Reduktiontheorie quadratischen Formen. Arch. Math. (Basel) **27**, 604–610 (1976)
25. S. Zhang, Positive line bundles on arithmetic varieties. J. Amer. Math. Soc. **8**, 187–221 (1995)

Chapter III : Introduction aux théorèmes de Hilbert-Samuel arithmétiques

Huayi Chen

1 Introduction

En géométrie algébrique, un théorème de Hilbert-Samuel au sens général consiste à décrire le comportement asymptotique d'un anneau gradué par $\mathbb{N}$. Par exemple, étant donnée une algèbre graduée $R_\bullet = \bigoplus_{n \in \mathbb{N}} R_n$ sur un corps de base k, telle que chaque composante homogène R_n soit de rang fini sur k, la *fonction de Hilbert* de $R_\bullet$ est définie comme l'application de $\mathbb{N}$ dans $\mathbb{N}$ qui envoie $n \in \mathbb{N}$ sur le rang de R_n sur k. Le théorème de Hilbert-Samuel dans ce cadre-là affirme que, si $R_\bullet$ est engendrée comme k-algèbre par R_1, alors il existe un polynôme $P \in \mathbb{Q}[T]$ tel que $P(n) = \mathrm{rg}_k(R_n)$ pour $n \in \mathbb{N}$ suffisamment grand. Ce résultat peut être obtenu en utilisant la méthode classique de dévissage.

Le théorème de Hilbert-Samuel est aussi valable pour un anneau gradué (non nécessairement une algèbre sur un corps) $R_\bullet$ engendré comme R_0-algèbre par R_1, tel que chaque composante homogène R_n soit un R_0-module de longueur finie (voir §2.1 du chapitre I *infra*), où on considère la fonction qui envoie $n \in \mathbb{N}$ sur la longueur de R_n (appelée *fonction de Hilbert-Samuel de $R_\bullet$*). Par ailleurs, ce résultat se généralise naturellement aux modules gradués de type fini sur un tel anneau, voir [7, VIII §4] pour les détails.

Dans beaucoup de situations pratiques, l'anneau gradué en question s'écrit comme un système linéaire gradué d'un faisceau inversible. Soient X un schéma projectif sur un corps de base k et L un $\mathcal{O}_X$-module inversible. La somme directe $\bigoplus_{n \in \mathbb{N}} H^0(X, L^{\otimes n})$ forme une k-algèbre graduée. Cette algèbre est de type fini notamment lorsque L est ample. Dans ce cas-là on peut déduire du théorème de Riemann-Roch-Hirzebruch et du théorème d'annulation de Serre le comportement

H. Chen (✉)

Institut de Mathématiques de Jussieu - Paris Rive Gauche, Université Paris Diderot, Paris, France

e-mail: huayi.chen@imj-prg.fr ; webusers.imj-prg.fr/~huayi.chen/

© The Editor(s) (if applicable) and The Author(s), under exclusive license to Springer Nature Switzerland AG 2021

E. Peyre, G. Rémond (eds.), *Arakelov Geometry and Diophantine Applications*, Lecture Notes in Mathematics 2276, https://doi.org/10.1007/978-3-030-57559-5_4

asymptotique de la fonction de Hilbert de cette algèbre graduée. Il est aussi possible d'adapter la méthode de dévissage dans le cadre géométrique pour relier le coefficient du terme principal de la fonction de Hilbert au nombre d'intersection de L, toujours sous une hypothèse de positivité convenable du fibré inversible.

Dans le cas où le fibré inversible n'est pas ample, l'algèbre graduée des sections globales n'est pas nécessairement de type fini et le théorème d'annulation de Serre ne s'applique plus. La méthode de dévissage n'est donc pas adéquate pour étudier la fonction de Hilbert de ce genre de fibrés inversibles. En revanche, les résultats de Fujita [23] et Takagi [41] (approximation de Fujita) montrent que, au moins dans le cas où X est un schéma *intègre* et projectif sur un corps et L est un $\mathcal{O}_X$-module inversible gros (c'est-à-dire qu'une puissance tensorielle de L admet un sous-faisceau inversible ample), la fonction de Hilbert de $\bigoplus_{n \in \mathbb{N}} H^0(X, L^{\otimes n})$ est équivalente, lorsque $n \to +\infty$, à un polynôme dont le degré est égal à la dimension de Krull de X. Cependant, contrairement au cas où L est ample, le coefficient du terme dominant pourrait être un nombre irrationnel, comme l'a montré Cutkosky [20, exemple 1.6]. En utilisant la méthode combinatoire d'Okounkov [37], le théorème d'approximation de Fujita a été généralisé dans [31, 34] au cas d'un système linéaire gradué qui contient un diviseur ample.

L'analogue arithmétique du théorème de Hilbert-Samuel avait d'abord été démontré par Gillet et Soulé [28, §5] en utilisant leur théorème de Riemann-Roch arithmétique (voir aussi [27] pour la théorie de l'intersection arithmétique, voir aussi §4 du chapitre I *infra*). Ensuite ce résultat a été revisité et généralisé dans divers contextes par différents auteurs.

On considère un morphisme projectif et plat $\pi : \mathscr{X} \to \operatorname{Spec} \mathbb{Z}$ et un fibré inversible hermitien $\overline{\mathscr{L}}$ sur $\mathscr{X}$. En géométrie d'Arakelov, un théorème de Hilbert-Samuel arithmétique cherche à décrire le comportement asymptotique de la fonction de Hilbert-Samuel arithmétique du système de fibrés vectoriels normés $\pi_*(\overline{\mathscr{L}^{\otimes n}})$ sur $\operatorname{Spec} \mathbb{Z}$, où $n \in \mathbb{N}$. Il y a cependant des variations dans cette formulation. D'abord il y a plusieurs choix naturels pour la structure du fibré vectoriel normé $\pi_*(\overline{\mathscr{L}^{\otimes n}})$, comme par exemple la norme L^2 et la norme sup. Si le théorème de Gromov (voir [40] VIII.2.5 Lemma 2) permet de comparer la norme L^2 à la norme sup, et a été utilisé dans des travaux comme ceux d'Abbes-Bouche [1] et Randriambololona [38], l'article [3] de Berman et Freixas i Montplet considère des métriques hermitiennes singulières, où la norme sup sur l'image directe du fibré peut ne pas avoir de sens (voir aussi [22] pour le cas de l'espace de modules des courbes pointées). Un autre point est le choix de l'invariant arithmétique dans la définition de la fonction de Hilbert-Samuel arithmétique. Dans la littérature, il y a au moins deux possibilités : le degré d'Arakelov (ou la caractéristique d'Euler-Poincaré) et le logarithme du nombre de petites sections. On peut comparer par exemple [46] à [36]. Même si le faisceau inversible $\mathscr{L}$ est ample et sa métrique est semi-positive, en général ces deux constructions donnent des fonctions de Hilbert-Samuel arithmétiques non équivalentes.

Similairement à la situation géométrique, la méthode du théorème de Riemann-Roch arithmétique et la combinaison du dévissage avec des méthodes analytiques s'appliquent difficilement au cas où la fibre générique de $\mathscr{L}$ n'est pas ample. Inspiré par des travaux sur la méthode combinatoire dans le cadre de la géométrie algébrique, deux approches différentes ont été proposées pour étudier le comportement asymptotique des fonctions de Hilbert-Samuel arithmétiques. Celle de [47, 48] adapte la méthode de Lazarsfeld-Mustață dans le cadre arithmétique en considérant la fibre du $\mathbb{Z}$-schéma au-dessus d'une place non archimédienne, tandis que celle de [13, 6] repose sur le résultat même du théorème d'approximation de Fujita des systèmes linéaires gradués, ainsi que la méthode de $\mathbb{R}$-filtration développée dans [14]. On renvoie les lecteurs à [9, 10] pour le cas torique, où la méthode est aussi d'une nature combinatoire.

Je ne cherche pas à donner un aperçu panoramique des résultats du type théorème de Hilbert-Samuel, mais plutôt me limite au cas d'une algèbre graduée sur un corps de base, ainsi que quelques analogues arithmétiques, en mettant une attention particulière sur la similitude et la différence des méthodes utilisées. Le cours est organisé comme suit. Dans le deuxième paragraphe, on traite le cas d'une algèbre de semi-groupe et la méthode combinatoire ; dans le troisième paragraphe, une interprétation géométrique du problème sera présentée en faisant le lien avec la théorie de l'intersection ; dans le quatrième paragraphe, on discute l'analogue métrique du problème ; enfin, dans le cinquième paragraphe, on traite des analogues arithmétiques du théorème de Hilbert-Samuel.

Remerciement

Je suis reconnaissant au collègue anonyme pour sa soigneuse lecture du texte et pour ses commentaires.

2   Méthode combinatoire

2.1   Algèbre de polynômes

Le prototype du théorème de Hilbert-Samuel est le cas des algèbres de polynômes. Soient k un corps commutatif et $R_\bullet = k[T_0, \ldots, T_d]$ l'algèbre des polynômes à $d + 1$ variables (où $d \in \mathbb{N}$), graduée par le degré des polynômes (autrement dit, R_n est l'espace vectoriel sur k engendré par les monômes de degré n). La fonction de Hilbert de $R_\bullet$ est

$$(n \in \mathbb{N}) \longmapsto \binom{d + n}{d}.$$

C'est un polynôme de degré d en n, dont le coefficient du terme dominant est $1/d!$:
on a

$$\binom{d+n}{d} = \frac{(n+d)\cdots(n+1)}{d!} = \frac{1}{d!}n^d + O(n^{d-1}), \quad n \to +\infty. \tag{1}$$

2.2 Algèbre de semi-groupes

L'algèbre de polynômes est un cas particulier d'algèbres de semi-groupes. Soient
$d \in \mathbb{N}$ et Γ un sous-semi-groupe de $\mathbb{N} \times \mathbb{Z}^d$ (muni de la loi d'addition). On désigne
par $k[\Gamma]$ l'algèbre du semi-groupe Γ. Par définition $k[\Gamma]$ est la somme directe d'une
famille de k-modules libres de rang 1 paramétrée par Γ. Pour tout $\gamma \in \Gamma$, on désigne
par e^γ le générateur canonique du k-module libre de rang 1 indexé par γ.[1] La loi
de multiplication sur $k[\Gamma]$ est choisie de sorte que $e^{\gamma+\gamma'} = e^\gamma e^{\gamma'}$ pour tout couple
$(\gamma, \gamma') \in \Gamma^2$. L'algèbre $k[\Gamma]$ est canoniquement munie d'une $\mathbb{N}$-graduation :

$$k[\Gamma] = \bigoplus_{n\in\mathbb{N}} \bigoplus_{\substack{\alpha\in\mathbb{Z}^d \\ (n,\alpha)\in\Gamma}} ke^{(n,\alpha)}.$$

2.3 Semi-groupe d'un corps convexe

Un exemple typique de semi-groupe dans $\mathbb{N} \times \mathbb{Z}^d$ est celui qui provient d'un *corps
convexe* dans $\mathbb{R}^d$, par lequel on entend une partie convexe et compacte de $\mathbb{R}^d$, dont
l'intérieur est non vide. Étant donné un corps convexe $\Delta \subset \mathbb{R}^d$, on construit un
sous-ensemble $\Gamma(\Delta)$ de $\mathbb{N} \times \mathbb{Z}^d$ comme suit :

$$\Gamma(\Delta) := \{(n, \alpha) \in \mathbb{N} \times \mathbb{Z}^d \mid n \geqslant 1, \tfrac{1}{n}\alpha \in \Delta\} \cup \{(0, 0)\}.$$

Par la convexité de Δ, on obtient que $\Gamma(\Delta)$ est un sous-semi-groupe de $\mathbb{N} \times \mathbb{Z}^d$. En
outre, la fonction de Hilbert de l'algèbre $k[\Gamma(\Delta)]$ est donnée par la formule

$$(n \in \mathbb{N}) \longmapsto \#\{\alpha \in \mathbb{Z}^d \mid \tfrac{1}{n}\alpha \in \Delta\}.$$

Par la convergence des sommes de Riemann vers l'intégrale de Lebesgue, on obtient

$$\#\{\alpha \in \mathbb{Z}^d \mid \tfrac{1}{n}\alpha \in \Delta\} = \mathrm{vol}(\Delta)n^d + o(n^d), \quad n \to +\infty,$$

---

[1]L'expression e^γ est purement formelle, qui ne sert qu'à écrire la loi de composition de Γ d'une
manière multiplicative.

où vol est la mesure de Lebesgue sur $\mathbb{R}^d$. Dans le cas où Δ est le simplexe

$$\{(x_1, \ldots, x_d) \in \mathbb{R}_+^d \mid x_1 + \cdots + x_d \leqslant 1\},$$

dont le volume est $1/d!$, on retrouve le résultat dans le cas des algèbres de polynômes.

2.4 Fonction de Hilbert d'une algèbre de semi-groupe

Khovanskiĭ [32] a montré que la méthode du corps convexe peut être appliquée à étudier un sous-semi-groupe général de $\mathbb{N} \times \mathbb{Z}^d$. Soit Γ un sous-semi-groupe de $\mathbb{N} \times \mathbb{Z}^d$. On suppose que $\Gamma \cap (\{0\} \times \mathbb{Z}^d) = \{(0, 0)\}$ pour des raisons de simplicité.[2] Soient $\Gamma_{\mathbb{Z}}$ et $\Gamma_{\mathbb{R}}$ le sous-groupe de $\mathbb{Z} \times \mathbb{Z}^d$ et le sous-espace vectoriel de $\mathbb{R} \times \mathbb{R}^d$ engendrés par Γ, respectivement. Pour tout $n \in \mathbb{N}$, soit

$$\Gamma_n := \{\alpha \in \mathbb{Z}^d \mid (n, \alpha) \in \Gamma\}.$$

Similairement, pour tout $n \in \mathbb{Z}$, soit

$$\Gamma_{\mathbb{Z},n} := \{\alpha \in \mathbb{Z}^d \mid (n, \alpha) \in \Gamma_{\mathbb{Z}}\}.$$

Il s'avère que

$$\mathbb{N}(\Gamma) := \{n \in \mathbb{N} \mid \Gamma_n \neq \varnothing\}$$

est un sous-semi-groupe de $\mathbb{N}$. En outre, le sous-groupe de $\mathbb{Z}$ engendré par $\mathbb{N}(\Gamma)$ s'identifie à

$$\mathbb{Z}(\Gamma) := \{n \in \mathbb{Z} \mid \Gamma_{\mathbb{Z},n} \neq \varnothing\}.$$

On suppose que le groupe $\mathbb{Z}(\Gamma)$ ne se réduit pas à $\{0\}$. Alors $\Gamma_{\mathbb{R}} \cap (\{1\} \times \mathbb{R}^d)$ est un espace affine dans $\mathbb{R} \times \mathbb{R}^d$, dont le sous-espace vectoriel sous-jacent est $\Gamma_{\mathbb{R}} \cap (\{0\} \times \mathbb{R}^d)$. En outre, l'ensemble $\Gamma_{\mathbb{Z}} \cap (\{0\} \times \mathbb{Z}^d)$ est un réseau[3] dans $\Gamma_{\mathbb{R}} \cap (\{0\} \times \mathbb{R}^d)$. On munit ce dernier de l'unique mesure de Haar telle que le réseau $\Gamma_{\mathbb{Z}} \cap (\{0\} \times \mathbb{Z}^d)$ soit de covolume 1.

Soit $A(\Gamma)$ la projection canonique de $\Gamma_{\mathbb{R}} \cap (\{1\} \times \mathbb{R}^d)$ dans $\mathbb{R}^d$. C'est un espace affine dans $\mathbb{R}^d$. La mesure de Haar sur $\Gamma_{\mathbb{R}} \cap (\{0\} \times \mathbb{R}^d)$ induit (par translation puis

---

[2]Cette hypothèse supplémentaire est anodine car $(\Gamma \cap (\mathbb{N}_{\geqslant 1} \times \mathbb{Z}^d)) \cup \{(0, 0)\}$ est aussi un sous-semi-groupe de $\mathbb{N} \times \mathbb{Z}^d$.

[3]c'est-à-dire un sous-$\mathbb{Z}$-module discret de rang maximal.

par projection) une mesure borélienne sur $A(\Gamma)$ que l'on note vol$(\cdot)$. En outre, on désigne par $\Delta(\Gamma)$ l'adhérence de l'ensemble

$$\bigcup_{n\in\mathbb{N},\, n\geqslant 1} \{n^{-1}\alpha \mid \alpha \in \Gamma_n\}.$$

C'est une partie fermée et convexe dans $A(\Gamma)$ qui engendre $A(\Gamma)$ comme espace affine. En particulier, l'intérieur relatif de $\Delta(\Gamma)$ dans $A(\Gamma)$ est non vide.

En utilisant le théorème de Bézout, par un argument élémentaire on obtient que l'ensemble $(\mathbb{N} \cap \mathbb{Z}(\Gamma))\backslash\mathbb{N}(\Gamma)$ est fini. La méthode de Khovanskiĭ est une généralisation aux dimensions supérieures de ce résultat (on renvoie à [32, Proposition 1] pour une démonstration, voir aussi [31, §1] et [5, §1]).

Lemme 2.1 *Pour tout sous-ensemble convexe et compact E de $A(\Gamma)$ qui est contenu dans l'intérieur relatif de $\Delta(\Gamma)$, on a*

$$E \cap \{\tfrac{1}{n}\alpha \mid \alpha \in \Gamma_n\} = E \cap \{\tfrac{1}{n}\alpha \mid \alpha \in \Gamma_{\mathbb{Z},n}\}$$

pour tout entier $n \in \mathbb{N}$ suffisamment grand.

Encore par la convergence des sommes de Riemann vers l'intégrale de Lebesgue on déduit du lemme précédent le résultat suivant. Remarquons qu'aucune condition de finitude n'est exigée pour le sous-semi-groupe Γ.

Théorème 2.2 *Soient Γ un sous-semi-groupe de $\mathbb{N} \times \mathbb{Z}^d$ et $R_\bullet := k[\Gamma]$. Alors*

$$\lim_{n\in\mathbb{N}(\Gamma),\, n\to+\infty} \frac{\mathrm{rg}_k(R_n)}{n^{\dim A(\Gamma)}} = \mathrm{vol}(\Delta(\Gamma)).$$

On renvoie les lecteurs à [31, Theorem 1.15] pour les détails.

2.5   Cas d'une algèbre graduée générale

Okounkov [37] a généralisé cette méthode combinatoire à l'étude de la log-concavité de la fonction de multiplicité. Son approche a été développée par Lazarsfeld et Mustață [34], et Kaveh et Khovanskiĭ [31] respectivement pour étudier le comportement asymptotique de la fonction de Hilbert d'une algèbre graduée intègre (non nécessairement de type fini). L'idée remonte à la méthode très classique d'associer à une algèbre $\mathbb{N}$-filtrée son algèbre graduée correspondante. Cette procédure, qui change la structure d'algèbre en général, permet de garder toutes les informations de la fonction de Hilbert. On introduit une relation d'ordre total $\leqslant$ sur $\mathbb{Z}^d$, supposée être «additive», autrement dit, $\alpha \leqslant \beta$ entraîne $\alpha + \gamma \leqslant \beta + \gamma$ pour tout $(\alpha, \beta, \gamma) \in (\mathbb{Z}^d)^3$. Cette propriété est par exemple satisfaite par la relation d'ordre lexicographique.

Soit $R_\bullet$ une k-algèbre graduée telle que $R_0 = k$. On entend par *valuation homogène à valeurs dans* $(\mathbb{Z}^d, \leqslant)$ sur $R_\bullet$ toute application[4]

$$v : \coprod_{n \in \mathbb{N}} R_n \longrightarrow \mathbb{Z}^d \cup \{+\infty\}$$

qui vérifie les propriétés suivantes :

(a) pour tout élément homogène f de $R_\bullet$, $v(f) \in \mathbb{Z}^d$ si et seulement si f est non nul, et $v(f) = 0$ si $f \in R_0 \backslash \{0\}$.
(b) si f et g sont deux éléments homogènes de $R_\bullet$, on a $v(fg) = v(f) + v(g)$;
(c) si $n \in \mathbb{N}$ et si x et y sont deux éléments de R_n, on a $v(x+y) \geqslant \min(v(x), v(y))$;
(d) pour tout $n \in \mathbb{N}$ et tout $\alpha \in \mathbb{Z}^d$, l'espace vectoriel quotient

$$\mathrm{gr}^\alpha(R_n) := \{x \in R_n \mid v(x) \geqslant \alpha\}/\{x \in R_n \mid v(x) > \alpha\}$$

a pour dimension 0 ou 1 sur k.

Sous ces conditions la somme directe

$$\bigoplus_{n \in \mathbb{N}} \bigoplus_{\alpha \in \mathbb{Z}^d} \mathrm{gr}^\alpha(R_n)$$

est munie d'une structure de k-algèbre graduée. En outre, elle est isomorphe à l'algèbre du semi-groupe

$$\Gamma(R_\bullet) := \{(n, \alpha) \in \mathbb{N} \times \mathbb{Z}^d \mid \mathrm{gr}^\alpha(R_n) \neq \{0\}\}.$$

Il s'avère que les algèbres graduées $R_\bullet$ et $k[\Gamma(R_\bullet)]$ ont la même fonction de Hilbert. Par le théorème 2.2 on peut utiliser la partie convexe $\Delta(\Gamma(R_\bullet))$ pour décrire le comportement asymptotique de la fonction de Hilbert de $R_\bullet$. Il faut cependant remarquer que l'existence d'une valuation homogène sur $R_\bullet$ n'est pas immédiate. En tout cas elle implique que l'anneau $R_\bullet$ est intègre. Dans le cas où $R_\bullet$ est un système linéaire gradué d'un diviseur de Cartier sur un k-schéma projectif et intègre possédant un point rationnel régulier, une telle valuation homogène peut être construite en choisissant un système de paramètres de l'anneau local en ce point rationnel régulier. Cependant l'existence d'un point rationnel régulier entraîne que le K-schéma est géométriquement intègre. On renvoie les lecteurs à [15, 16] pour une approche alternative basée sur la géométrie d'Arakelov relativement à un corps de fonctions où cette hypothèse n'est pas nécessaire.

---

[4]On convient que $\alpha \lneqq +\infty$ et $\alpha + (+\infty) = +\infty$ pour tout $\alpha \in \mathbb{Z}^d$ et que $(+\infty) + (+\infty) = +\infty$.

3 Approche géométrique

3.1 Interprétation géométrique des algèbres graduées

Soit $R_\bullet$ une k-algèbre graduée de type fini, qui est engendrée comme k-algèbre par R_1. Le spectre projectif de $R_\bullet$ est un sous-schéma fermé de $\mathbb{P}(R_1)$ et l'homomorphisme canonique

$$\alpha_n : R_n \longrightarrow H^0(\mathrm{Proj}(R_\bullet), \mathscr{O}_{\mathrm{Proj}(R_\bullet)}(n))$$

est un isomorphisme lorsque $n \in \mathbb{N}$ est assez grand (voir [29] III.2.3.1). On peut donc ramener l'étude asymptotique de la fonction de Hilbert au cas d'un système linéaire gradué.

Soient X un schéma projectif sur $\mathrm{Spec}\, k$ et L un $\mathscr{O}_X$-module inversible ample. On considère la k-algèbre graduée

$$R(L)_\bullet := \bigoplus_{n \in \mathbb{N}} H^0(X, L^{\otimes n})$$

et on cherche à comprendre le comportement asymptotique de la fonction de Hilbert de L définie par

$$(n \in \mathbb{N}) \longmapsto F_L(n) := \mathrm{rg}_k(H^0(X, L^{\otimes n})).$$

Comme L est ample, il existe un entier $n_0 \geqslant 1$ tel que $H^0(X, L^{\otimes n_0})$ soit non nul. Pour simplifier on suppose qu'il existe une section non nulle $s \in H^0(X, L)$ telle que l'homomorphisme de $\mathscr{O}_X$-modules $f_s : L^\vee \to \mathscr{O}_X$ défini par s soit injectif. Soient Y le sous-schéma fermé de X défini par l'annulation de s et $i : Y \to X$ le morphisme d'inclusion. Pour tout entier $n \in \mathbb{N}$, $n \geqslant 1$, on a une suite exacte d'homomorphismes de $\mathscr{O}_X$-modules

$$0 \longrightarrow L^{\otimes(n-1)} \xrightarrow{\mathrm{Id} \otimes f_s} L^{\otimes n} \longrightarrow L^{\otimes n} \otimes i_*(\mathscr{O}_Y) \longrightarrow 0 ,$$

qui induit une suite exacte de groupes de cohomologie :

$$0 \to H^0(X, L^{\otimes(n-1)}) \xrightarrow{\cdot s} H^0(X, L^{\otimes n}) \to H^0(Y, i^*(L^{\otimes n})) \to H^1(X, L^{\otimes(n-1)}),$$

où on a remplacé $H^0(X, L^{\otimes n} \otimes i_*(\mathscr{O}_Y))$ par $H^0(Y, i^*(L^{\otimes n}))$ via les isomorphismes suivants (en utilisant l'adjonction entre i_* et i^*) :

$$H^0(X, L^{\otimes n} \otimes i_*(\mathscr{O}_Y)) \cong \mathrm{Hom}_{\mathscr{O}_X}(\mathscr{O}_X, L^{\otimes n} \otimes i_*(\mathscr{O}_Y))$$

$$\cong \mathrm{Hom}_{\mathscr{O}_X}(L^{\vee \otimes n}, i_*(\mathscr{O}_Y)) \cong \mathrm{Hom}_{\mathscr{O}_Y}(i^*(L)^{\vee \otimes n}, \mathscr{O}_Y)$$

$$\cong \mathrm{Hom}_{\mathscr{O}_Y}(\mathscr{O}_Y, i^*(L)^{\otimes n}) \cong H^0(Y, i^*(L)^{\otimes n}).$$

Comme L est ample, le théorème d'annulation de Serre montre que $H^1(X, L^{\otimes n}) = \{0\}$ et donc

$$F_L(n) - F_L(n-1) = F_{i^*(L)}(n) \tag{2}$$

pour n assez grand, où F_L et $F_{i^*(L)}$ sont respectivement les fonctions de Hilbert de L et $i^*(L)$. Ainsi par récurrence on montre qu'il existe un polynôme P_L de degré $\dim(X)$ tel que $F_L(n) = P_L(n)$ pour $n \in \mathbb{N}$ assez grand.

3.2 Nombre d'intersection et théorème de Hilbert-Samuel

Pour tout entier naturel i, on désigne par $Z_i(X)$ le groupe abélien libre engendré par les sous-schémas fermés intègres de dimension i de X, dont les éléments sont appelés *cycles de dimension i dans X*. Si f est une fonction rationnelle non nulle sur un sous-schéma fermé intègre W de dimension $i + 1$, on désigne par (f) l'élément

$$\sum_Y \operatorname{ord}_Y(f)Y \in Z_i(X),$$

où Y parcourt l'ensemble des sous-schémas fermés intègres de dimension i de W, et $\operatorname{ord}_Y(f)$ est défini dans §2.1.3 du chapitre I *infra*. L'élément (f) est appelé *cycle associé à la fonction rationnelle f*. On dit qu'un cycle de dimension i est *rationnellement équivalent à zéro* s'il appartient au sous-groupe $R^i(X)$ de $Z^i(X)$ engendré par les cycles associés aux fonctions rationnelles non nulles sur les sous-schémas fermés intègres de dimension $i + 1$ de X. On désigne par $\mathrm{CH}_i(X)$ le groupe quotient $Z_i(X)/R_i(X)$, appelé *groupe de Chow de dimension i* de X. Si α est un élément de $\mathrm{CH}_0(X)$, qui est représenté par le cycle $\sum_{j=1}^n \lambda_j x_j$, où $\lambda_1, \ldots, \lambda_n$ sont des entiers, et $x_1, \ldots, x_n$ sont des points fermés de X, on désigne par $\deg(\alpha)$ l'entier

$$\sum_{j=1}^n \lambda_j [\kappa(x_j) : k],$$

où $\kappa(x_j)$ désigne le corps résiduel du point x_j.

Soient $i \in \mathbb{N}_{\geq 1}$ et Z un sous-schéma fermé intègre de dimension i de X. Soient s une section rationnelle non nulle de $L|_Z$ et (s) le cycle

$$\sum_Y \operatorname{ord}_Y(s)Y,$$

où Y parcourt l'ensemble des sous-schémas fermé intègre de dimension $i - 1$ de Z. On désigne par $c_1(L) \cap [Z]$ l'élément de $CH_{i-1}(X)$ représenté par ce cycle (il s'avère que cette classe d'équivalence rationnelle ne dépend pas du choix de la section rationnelle s). Plus généralement, si α est un élément de $CH_i(X)$, qui est représenté par le cycle $\sum_{j=1}^m a_j Z_j$, où $a_1, \ldots, a_m$ sont des entiers, et $Z_1, \ldots, Z_m$ sont des sous-schémas intègres de dimension i de X, on désigne par $c_1(L) \cap \alpha$ la classe de cycles $\sum_{j=1}^m a_j c_1(L) \cap [Z_j]$ dans $CH_{i-1}(X)$.

La formule de récurrence (2) montre que le coefficient du terme dominant du polynôme P_L est

$$\frac{\deg(c_1(L)^{\dim(X)} \cap [X])}{\dim(X)!}.$$

En conclusion, on obtient le résultat suivant (voir [33, théorème 1.1.24]).

Théorème 3.1 *Soient X un schéma projectif de dimension d sur* $\operatorname{Spec} k$ *et L un* $\mathscr{O}_X$*-module inversible ample, alors on a*

$$\lim_{n \to +\infty} \frac{\operatorname{rg}_k(H^0(X, L^{\otimes n}))}{n^d/d!} = \deg(c_1(L)^d \cap [X]).$$

Une démonstration alternative (dans le cas où X est un schéma régulier) consiste à appliquer le théorème de Riemann-Roch-Hirzebruch, qui affirme que (voir [33, exemple 1.1.27])

$$\chi(X, L^{\otimes n}) := \sum_{i=0}^d \operatorname{rg}_k(H^i(X, L^{\otimes n})) = \deg(\operatorname{ch}(L^{\otimes d})\operatorname{Td}(X))$$

$$= \deg(c_1(L)^d \cap [X])\frac{n^d}{d!} + O(n^{d-1}).$$

On utilise aussi le théorème d'annulation de Serre pour identifier $\chi(X, L^{\otimes n})$ à $H^0(X, L^{\otimes n})$ quand n est assez grand.

3.3 Diviseurs de Cartier et systèmes linéaires

Soient X un schéma projectif et intègre sur $\operatorname{Spec} K$ et L un $\mathscr{O}_X$-module inversible. On définit le *volume* de L comme

$$\operatorname{vol}(L) := \varlimsup_{n \to +\infty} \frac{\operatorname{rg}_k(H^0(X, L^{\otimes n}))}{n^d/d!}, \tag{3}$$

où d est la dimension de X. On a vu plus haut que, si L est un $\mathcal{O}_X$-module ample, alors $\mathrm{vol}(L) = \deg(c_1(L)^d \cap [X])$. En outre, la limite supérieure dans la formule précédente est en fait une limite.

Dans le cas où X est un schéma intègre, il est plus commode d'utiliser le langage des diviseurs de Cartier. Soit $\mathcal{O}_X^{\times}$ le faisceau en groupes abéliens défini comme suit : pour tout sous-ensemble ouvert U de X, $\mathcal{O}_X^{\times}(U)$ est l'ensemble des $a \in \mathcal{O}_X(U)$ tels que l'homothétie $a : \mathcal{O}_U \to \mathcal{O}_U$ soit un isomorphisme. Similairement, soient $\mathcal{M}_X$ le faisceau des fonctions rationnelles sur X et $\mathcal{M}_X^{\times}$ le faisceau en groupes abéliens qui envoie U sur l'ensemble des $a \in \mathcal{M}_X(U)$ tels que l'homothétie $a : \mathcal{M}_U \to \mathcal{M}_U$ soit un isomorphisme. On voit aussitôt que $\mathcal{O}_X^{\times}$ est un sous-faisceau en groupes abéliens de $\mathcal{M}_X^{\times}$. On entend par *diviseur de Cartier* toute section sur X du faisceau quotient $\mathcal{M}_X^{\times}/\mathcal{O}_X^{\times}$. On dit qu'un diviseur de Cartier D est *effectif* et on note $D \geqslant 0$ s'il appartient à $\Gamma(X, (\mathcal{M}_X^{\times} \cap \mathcal{O}_X)/\mathcal{O}_X^{\times})$. On désigne par $\mathrm{Div}(X)$ le groupe des diviseurs de Cartier sur X et par $\mathrm{Div}^+(X)$ le sous-semi-groupe de $\mathrm{Div}(X)$ des diviseurs effectifs.

Les diviseurs de Cartier sont étroitement liés aux faisceaux inversibles. Soit L un $\mathcal{O}_X$-module inversible. A toute section globale non nulle s de $L \otimes_{\mathcal{O}_X} \mathcal{M}_X$, on peut associer un diviseur de Cartier $\mathrm{div}(s)$ comme suit. Pour tout sous-ensemble ouvert U de X sur lequel le $\mathcal{O}_X$-module L se trivialise par une section $s_0 \in \Gamma(U, L)$, il existe une unique fonction rationnelle non nulle f telle que $s = f s_0$. La classe de f dans $\Gamma(U, \mathcal{M}_X^{\times}/\mathcal{O}_X^{\times})$ ne dépend pas du choix de la trivialisation locale s_0. Donc le recollement des classes de fonctions rationnelles quand U varie donne un diviseur de Cartier $\mathrm{div}(s)$.

Réciproquement, si D est un diviseur de Cartier sur X, on désigne par $\mathcal{O}_X(D)$ le sous-$\mathcal{O}_X$-module de $\mathcal{M}_X$ engendré par $-D$. C'est un $\mathcal{O}_X$-module inversible. L'application θ de $\mathrm{Div}(X)$ dans $\mathrm{Pic}(X)$ qui envoie tout diviseur de Cartier D sur la classe d'isomorphisme du $\mathcal{O}_X$-module inversible $\mathcal{O}_X(D)$ est un morphisme surjectif de groupes (voir [29] IV.21.3.4–5). On dit que le diviseur D est *ample* si le faisceau inversible $\mathcal{O}_X(D)$ est ample.

Les constructions ci-dessus peuvent aussi être obtenues par la suite exacte de cohomologie associée à la suite exacte de faisceaux abéliens

$$0 \longrightarrow L^{\otimes(n-1)} \xrightarrow{\mathrm{Id}\otimes f_s} L^{\otimes n} \longrightarrow L^{\otimes n} \otimes i_*(\mathcal{O}_Y) \longrightarrow 0 \, ,$$

qui s'écrit comme

$$1 \longrightarrow H^0(X, \mathcal{O}_X^{\times}) \longrightarrow H^0(X, \mathcal{M}_X^{\times}) \xrightarrow{\mathrm{div}} \mathrm{Div}(X) \xrightarrow{\theta} H^1(X, \mathcal{O}_X^{\times}).$$

On désigne par $\mathrm{Div}(X)_{\mathbb{R}}$ le produit tensoriel $\mathrm{Div}(X) \otimes_{\mathbb{Z}} \mathbb{R}$. Les éléments dans $\mathrm{Div}(X)_{\mathbb{R}}$ sont appelés $\mathbb{R}$-diviseurs (de Cartier) sur X. Un $\mathbb{R}$-diviseur D sur X est dit *effectif* (et on note $D \geqslant_{\mathbb{R}} 0$) s'il peut s'écrire comme une combinaison linéaire à coefficients positifs de diviseurs de Cartier effectifs. Il est clair que, si D est un diviseur de Cartier effectif, alors son image dans $\mathrm{Div}(X)_{\mathbb{R}}$ est un $\mathbb{R}$-diviseur effectif. La réciproque est vraie lorsque X est un schéma normal.

Soit $K = k(X)$ le corps des fonctions rationnelles sur X. On appelle *système linéaire* sur X tout sous-k-espace vectoriel de K qui est de rang fini sur k. Si V est un système linéaire sur X, on désigne par $k(V)$ la sous-extension de K/k engendrée par les éléments de la forme a/b, où a et b appartiennent à $V \setminus \{0\}$.

Exemple 3.2 Pour tout diviseur de Cartier D sur X, on définit

$$H^0(D) := \{f \in K^\times \mid D + \mathrm{div}(f) \geqslant 0\} \cup \{0\}.$$

C'est un système linéaire sur X. En outre, on a un isomorphisme canonique entre $H^0(D)$ et $H^0(X, \mathscr{O}_X(D))$. Similairement, pour tout $\mathbb{R}$-diviseur de Cartier E, on définit

$$H^0_\mathbb{R}(E) := \{f \in K^\times \mid E + \mathrm{div}(f) \geqslant_\mathbb{R} 0\} \cup \{0\}.$$

C'est aussi un système linéaire sur X. Rappelons que, si D est un diviseur de Cartier sur X, alors on a $H^0(D) \subset H^0_\mathbb{R}(D \otimes 1)$. Ces deux espaces sont identiques quand X est normal, mais ils sont en général différents (voir [19, exemple 2.4.15] pour un contre-exemple).

3.4 Cas torique

Dans le cas d'une variété torique, les systèmes linéaires et la géométrie des corps convexes sont étroitement liés. Le livre de Fulton [24] donne une vue panoramique sur ce sujet. On se contente ici d'une description succincte. Rappelons qu'une *variété torique de dimension d sur k* est par définition un schéma intègre et normal X sur $\mathrm{Spec}\, k$ muni d'une immersion ouverte du tore $\mathbb{T} \cong \mathbb{G}^d_{\mathrm{m},k}$ dans X ainsi qu'une action $\mu : \mathbb{T} \times X \to X$ qui prolonge l'action canonique de $\mathbb{T}$ sur lui-même, où $\mathbb{G}_{\mathrm{m},k}$ désigne le groupe multiplicatif sur k.

Les variétés toriques peuvent être décrites de façon combinatoire. On désigne par $N = \mathrm{Hom}(\mathbb{G}_{\mathrm{m},k}, \mathbb{T}) \cong \mathbb{Z}^d$ le réseau des sous-groupes à un paramètre de $\mathbb{T}$. Soit $M = \mathrm{Hom}(N, \mathbb{Z})$ son réseau dual, qui s'identifie au groupe des caractères de $\mathbb{T}$. On appelle *cône polyédral rationnel* dans $N_\mathbb{R}$ tout sous-ensemble σ de $N_\mathbb{R}$ de la forme

$$\sigma = \{\lambda_1 v_1 + \cdots + \lambda_r v_r \mid (\lambda_1, \ldots, \lambda_r) \in \mathbb{R}^r_+\},$$

où $\{v_1, \ldots, v_r\}$ est une famille finie d'éléments de N (on dit que σ est *engendré* par la famille $\{v_1, \ldots, v_r\}$). Étant donné un cône polyédral rationnel σ dans $N_\mathbb{R}$, on désigne par $\sigma^\vee$ le sous-ensemble de $M_\mathbb{R}$ des vecteurs α tels que $\alpha(x) \geqslant 0$ pour tout $x \in \sigma$ (on identifie $M_\mathbb{R}$ à l'espace des formes linéaires sur $N_\mathbb{R}$). Il s'avère que $\sigma^\vee$ est un cône polyédral rationnel dans $M_\mathbb{R}$, appelé *cône dual* de σ. En outre, on a $\sigma^{\vee\vee} = \sigma$ si on identifie N au bidual de lui-même. Rappelons que le lemme de Gordan montre que $M_\sigma := M \cap \sigma^\vee$ est un sous-semi-groupe de type fini de M.

Soit σ un cône dans $N_{\mathbb{R}}$. On appelle *face* de σ tout sous-ensemble de la forme $\sigma \cap \mathrm{Ker}(\varphi)$, où φ est un élément de $\sigma^{\vee}$. Toute face de σ est un cône polyédral rationnel. On dit qu'un cône polyédral rationnel σ est *strictement convexe* si $\{0\}$ est une face de σ, ou de façon équivalente, $\sigma^{\vee}$ engendre $M_{\mathbb{R}}$ comme espace vectoriel sur $\mathbb{R}$.

Pour tout cône polyédral rationnel σ qui est strictement convexe, on désigne par X_{σ} le schéma affine $\mathrm{Spec}\, k[M_{\sigma}]$. Si τ est une face de σ, alors $k[M_{\tau}]$ est une localisation de $k[M_{\sigma}]$ par un élément et donc X_{τ} s'identifie à un sous-schéma ouvert de X_{σ}. En particulier, le schéma $X_{\{0\}}$ correspondant à la face $\{0\}$ s'identifie au tore $\mathbb{G}_{\mathrm{m},k}^{d} = \mathrm{Spec}\, k[M]$. On a une action de $\mathbb{G}_{\mathrm{m},k}^{d}$ sur X_{σ}, qui correspond à l'homomorphisme canonique

$$k[M_{\sigma}] \longrightarrow k[M_{\sigma}] \otimes_{k} k[M]$$

qui envoie e^{α} sur $e^{\alpha} \otimes e^{\alpha}$ ($\alpha \in M_{\sigma}$). Cette action étend l'action canonique de $\mathbb{G}_{\mathrm{m},k}^{n}$ sur lui-même. Donc X_{σ} est une variété torique affine. On peut montrer que le schéma X_{σ} est toujours intègre et normal ; il est régulier si et seulement si σ est engendré par une base de N sur $\mathbb{Z}$ (cf. [24, §2.1]).

On appelle *éventail* toute famille finie Σ de cônes polyédraux rationnels strictement convexes dans $N_{\mathbb{R}}$ telle que, pour tout $(\sigma, \sigma') \in \Sigma^2$, l'intersection $\sigma \cap \sigma'$ est une face commune de σ et σ'. Étant donné un éventail Σ, le recollement des schémas affines X_{σ}, $\sigma \in \Sigma$, forme un k-schéma que l'on note X_{Σ}. Les actions du tore $\mathbb{T}$ sur X_{σ}, $\sigma \in \Sigma$, se recollent en une action de $\mathbb{T}$ sur X_{Σ} qui prolonge l'action canonique de $\mathbb{T}$ sur lui-même. Ainsi X_{Σ} est une variété torique, appelé variété torique *correspondant à* Σ. Le k-schéma X_{Σ} est propre si et seulement si $|\Sigma| := \bigcup_{\sigma \in \Sigma} \sigma = N_{\mathbb{R}}$ (cf. [24, §2.4]).

Soient Σ un éventail et X_{Σ} la variété torique correspondant à Σ. On appelle $\mathbb{T}$-*diviseur de Cartier* sur X_{Σ} tout diviseur de Cartier invariant par $\mathbb{T}$. Soit D un $\mathbb{T}$-diviseur de Cartier sur X_{Σ}. Localement sur un ouvert affine X_{σ}, $\sigma \in \Sigma$, le $\mathbb{T}$-diviseur de Cartier D est représenté par une fonction rationnelle sur X_{σ} qui est invariante par l'action de $\mathbb{T}$, qui est donc de la forme $e^{-\alpha_{\sigma}}$, où α_{σ} est un élément de M. La condition de recollement s'interprète comme suit : pour tout couple $(\sigma, \sigma') \in \Sigma^2$, la différence $\alpha_{\sigma} - \alpha_{\sigma'}$ appartient à $M_{\sigma} \cap M_{\sigma'}$. On peut ainsi faire correspondre biunivoquement l'ensemble des $\mathbb{T}$-diviseurs de Cartier à celui des *fonctions de support virtuelles* sur $|\Sigma|$, à savoir les fonctions $\psi : |\Sigma| \to \mathbb{R}$ telles que, pour tout $\sigma \in \Sigma$, la restriction de ψ à σ s'identifie à la restriction d'une forme linéaire dans M. Si $\psi : |\Sigma| \to \mathbb{R}$ est une fonction de support virtuelle, on désigne par D_{ψ} le $\mathbb{T}$-diviseur de Cartier correspondant. Le diviseur D_{ψ} est effectif si et seulement si la fonction ψ est négative. En outre, deux $\mathbb{T}$-diviseurs de Cartier D_{ψ} et D_{φ} sont rationnellement équivalents si et seulement si la différence $\psi - \varphi$ s'identifie à la restriction d'une forme linéaire dans M.

Soient Σ un éventail et $\psi : |\Sigma| \to \mathbb{R}$ une fonction de support virtuelle. L'espace $H^0(D_\psi)$ est canoniquement isomorphe à l'espace vectoriel libre engendré par (voir [24, §3.5] pour les détails)

$$\{e^\alpha \mid \alpha \in M, \ \psi(x) \leqslant \alpha(x) \text{ pour tout } x \in |\Sigma|\}.$$

En particulier, si on désigne par Δ_ψ l'ensemble $\bigcap_{\sigma \in \Sigma} \Delta_\psi(\sigma)$, où

$$\Delta_\psi(\sigma) := \{\alpha \in M_\mathbb{R} \mid \forall x \in \sigma, \ \psi(x) \leqslant \alpha(x)\},$$

alors on a

$$H^0(D_\psi) \cong \bigoplus_{\alpha \in \Delta_\psi \cap M} k e^\alpha.$$

Ainsi l'algèbre graduée $\bigoplus_{n \in \mathbb{N}} H^0(nD_\psi)$ s'identifie à l'algèbre du semi-groupe

$$\Gamma(\Delta_\psi) = \{(n, \alpha) \in \mathbb{N} \times M \mid n \geqslant 1, \ \tfrac{1}{n}\alpha \in \Delta_\psi\} \cup \{(0, 0)\}$$

dans $\mathbb{N} \times M \cong \mathbb{N} \times \mathbb{Z}^d$, et la méthode combinatoire décrite au §2 peut être utilisée pour étudier la fonction de Hilbert-Samuel d'un diviseur invariant par le tore dans une variété torique.

3.5 Corps convexe de Newton-Okounkov

La méthode du corps de Newton-Okounkov permet de relier l'étude des systèmes linéaires gradués et la géométrie des corps convexes dans le cas général. Soient X un k-schéma intègre (non nécessairement torique) et K le corps des fonctions rationnelles sur X. On entend par *système linéaire gradué* sur X toute sous-k-algèbre graduée

$$V_\bullet = \bigoplus_{n \in \mathbb{N}} V_n T^n$$

de l'anneau des polynômes

$$K[T] = \bigoplus_{n \in \mathbb{N}} K T^n$$

telle que chaque V_n soit un système linéaire.

Supposons que le k-schéma X possède un point rationnel régulier x. L'anneau local $\mathscr{O}_{X,x}$ est régulier et son corps des fractions s'identifie à K. Son complété est isomorphe à l'algèbre des séries formelles $k[\![T_1, \ldots, T_d]\!]$ (cf. [21, Proposition 10.16]). On munit le groupe $\mathbb{Z}^d$ de la relation d'ordre lexicographique et on considère l'application $k[\![T_1, \ldots, T_d]\!] \to \mathbb{Z}^d \cup \{+\infty\}$ qui envoie toute série formelle

$$\sum_{\alpha=(\alpha_1,\ldots,\alpha_d)\in\mathbb{N}^d} \lambda_\alpha T_1^{\alpha_1} \cdots T_d^{\alpha_d},$$

sur (avec la convention $\inf \varnothing = +\infty$)

$$\inf\{\alpha \in \mathbb{N}^d \mid \lambda_\alpha \neq 0\}.$$

La composition de cette application avec l'homomorphisme d'inclusion $\mathscr{O}_{X,x} \to k[\![T_1, \ldots, T_d]\!]$ donne une application $v : \mathscr{O}_{X,x} \to \mathbb{N}^d \cup \{+\infty\}$ qui s'étend en une application $v : K \to \mathbb{Z}^d \cup \{+\infty\}$ en prenant

$$v(f/g) = v(f) - v(g) \text{ pour } (f, g) \in K^2, g \neq 0.$$

Il s'avère que, pour tout système linéaire gradué $V_\bullet$, l'application v induit une valuation homogène sur $V_\bullet$ à valeurs dans $\mathbb{Z}^d$. On peut donc appliquer la méthode combinatoire du paragraphe précédent pour obtenir le résultat suivant (voir [31, théorème 4] et [34, théorème 2.13]).

Théorème 3.3 *Soit $V_\bullet$ un système linéaire gradué sur X. On suppose que $V_\bullet$ est birationnel, c'est-à-dire $K = k(V_n)$ pour un certain $n \in \mathbb{N}_{\geqslant 1}$. Soit*

$$\mathbb{N}(V_\bullet) := \{n \in \mathbb{N} \mid V_n \neq \{0\}\}.$$

Alors la limite

$$\operatorname{vol}(V_\bullet) := \lim_{n\in\mathbb{N}(V_\bullet),\, n\to+\infty} \frac{\operatorname{rg}_k(V_n)}{n^d/d!}$$

existe dans $]0, +\infty]$. En outre, on a

$$\frac{\operatorname{vol}(V_\bullet)}{d!} = \operatorname{vol}(\Delta(\Gamma(V_\bullet))).$$

Si de plus $V_\bullet$ est contenu dans un système linéaire gradué de type fini, alors $\Delta(\Gamma(V_\bullet))$ est compact et $\operatorname{vol}(V_\bullet)$ est fini.

Dans le cadre géométrique, la méthode combinatoire permet de tirer plus d'information géométrique que l'interprétation de la fonction volume par la mesure de la partie convexe $\Delta(V_\bullet) := \Delta(\Gamma(V_\bullet))$ (appelée *corps de Newton-Okounkov* dans la littérature). Par exemple, en utilisant le fait que la valuation homogène à valeurs dans $\mathbb{Z}^d$ provient d'une valuation sur K, on déduit de l'inégalité de Brunn-Minkowski le résultat suivant (cf. [6, Proposition 2.10]) pour un variant de cet énoncé pour les systèmes linéaires gradués filtrés. Voir [39] pour un survol sur l'inégalité de Brunn-Minkowski dans le cadre de la géométrie convexe, voir aussi [42] pour des liens avec l'inégalité d'indice de Hodge.

Théorème 3.4 *Soient $U_\bullet$, $V_\bullet$ et $W_\bullet$ trois systèmes linéaires gradués sur X qui sont birationnels. On suppose que, pour tout $n \in \mathbb{N}$ assez grand, on a*

$$U_n \cdot V_n := \{fg \mid f \in U_n,\ g \in V_n\} \subset W_n.$$

Alors on a

$$\Delta(U_\bullet) + \Delta(V_\bullet) := \{x + y \mid x \in \Delta(U_\bullet),\ y \in \Delta(V_\bullet)\} \subset \Delta(W_\bullet).$$

En particulier,

$$\mathrm{vol}(U_\bullet)^{\frac{1}{d}} + \mathrm{vol}(V_\bullet)^{\frac{1}{d}} \leqslant \mathrm{vol}(W_\bullet)^{\frac{1}{d}}.$$

4 Version métrique

Dans ce paragraphe, on suppose que le corps k est muni d'une valeur absolue $|\cdot|$ telle que la topologie sur k induite par $|\cdot|$ est complète.

4.1 Norme déterminant

Soit $(V, \|\cdot\|)$ un espace vectoriel normé de rang fini sur k. On munit $\det(V)$ du *déterminant* $\|\cdot\|_{\det}$ de la norme $\|\cdot\|$, défini comme

$$\forall \eta \in \det(V), \qquad \|\eta\|_{\det} := \inf\{\|x_1\| \cdots \|x_r\| \mid \eta = x_1 \wedge \cdots \wedge x_r\}.$$

Soit $R_\bullet = \bigoplus_{n \in \mathbb{N}} R_n$ une k-algèbre graduée telle que chaque R_n soit un espace vectoriel de rang fini sur k. Pour tout entier $n \in \mathbb{N}$, on fixe une norme $\|\cdot\|_n$ sur l'espace vectoriel R_n. La version métrique du problème de Hilbert-Samuel cherche à déterminer le comportement asymptotique de la suite $\{(\det(R_n), \|\cdot\|_{n,\det})\}$ d'espaces vectoriels normés de rang 1 sur k.

4.2 Espace de Berkovich

Dans le cadre géométrique où $R_\bullet$ est un système linéaire gradué, la suite de normes $(\|\cdot\|_n)_{n \in \mathbb{N}}$ provient souvent d'une métrique, où la théorie de Berkovich est un cadre adéquate pour sa construction, surtout quand la valeur absolue $|\cdot|$ est non archimédienne. Soit X un schéma sur $\operatorname{Spec} k$. Ensemblistement l'*espace de Berkovich* associé à X est par définition l'ensemble des couples $x = (j(x), |\cdot|_x)$, où $j(x)$ est un point schématique de X, et $|\cdot|_x$ est une valeur absolue sur le corps résiduel $\kappa(j(x))$, qui prolonge la valeur absolue $|\cdot|$ sur k. On désigne par $\widehat{\kappa}(x)$ le séparé complété du corps $\kappa(j(x))$ par rapport à la valeur absolue $|\cdot|_x$. Si s est une fonction régulière sur un voisinage ouvert de $j(x)$, on désigne par $s(x)$ l'image de la classe résiduel de s en $j(x)$ par l'application d'inclusion $\kappa(j(x)) \to \widehat{\kappa}(x)$.

L'ensemble X^{an} est naturellement muni d'une topologie de Zariski, celle la moins fine qui rend l'application j continue. Berkovich a défini une autre topologie sur X^{an}, qui est plus fine que la topologie de Zariski (voir [2]). Soit U un sous-schéma ouvert de X et f une fonction régulière sur U. Pour tout point $x \in U^{\mathrm{an}}$, on désigne par $|f|(x)$ la valeur absolue de $f(x)$ par rapport à $|\cdot|_x$. On obtient ainsi une fonction $|f| : U^{\mathrm{an}} \to \mathbb{R}_+$. La *topologie de Berkovich* sur X^{an} est définie comme la topologie la moins fine qui rend continue l'application j et toutes les fonctions de la forme $|f|$, où f parcourt l'ensemble des fonctions régulières sur les sous-schémas ouverts de X.

La construction de l'espace de Berkovich est fonctorielle. Si $\phi : Y \to X$ est un morphisme de k-schémas, on désigne par $\phi^{\mathrm{an}} : Y^{\mathrm{an}} \to X^{\mathrm{an}}$ l'application qui envoie $y = (j(y), |\cdot|_y) \in Y^{\mathrm{an}}$ sur $(\phi(j(y)), |\cdot|_y \circ \phi^{\sharp}_{j(y)})$, où $\phi^{\sharp}_{j(y)} : \kappa(\phi(j(y))) \to \kappa(j(y))$ est le morphisme de corps résiduel défini dans la structure du morphisme de schémas $\phi : Y \to X$. Cette application est continue si on munit Y^{an} et X^{an} de la topologie de Berkovich.

4.3 Faisceaux inversibles métrisés

Soit L un $\mathscr{O}_X$-module inversible. On appelle *métrique* sur L toute famille $\varphi := (|\cdot|_\varphi(x))_{x \in X^{\mathrm{an}}}$, où chaque $|\cdot|_\varphi(x)$ est une norme sur $L(x) := L \otimes_{\mathscr{O}_X} \widehat{\kappa}(x)$. On dit que la métrique φ est *continue* si, pour tout sous-schéma ouvert U de X et toute section $s \in H^0(U, L)$, la fonction

$$|s|_\varphi : (x \in U^{\mathrm{an}}) \longmapsto |s(x)|_\varphi(x)$$

est continue sur U^{an}. Dans le cas où le morphisme canonique de schémas $X \to \operatorname{Spec} k$ est propre, l'espace topologique X^{an} est compact. En particulier, pour toute section $s \in H^0(X, L)$, on a

$$\|s\|_{\varphi, \sup} := \sup_{x \in X^{\mathrm{an}}} |s|_\varphi(x) < +\infty,$$

et $\|\cdot\|_{\varphi, \sup}$ définit une norme sur $H^0(X, L)$, qui est ultramétrique lorsque $|\cdot|$ est non archimédienne. Cette norme est appelée *norme du supremum* associée à la métrique φ.

Si φ et φ' sont deux métriques sur le même $\mathscr{O}_X$-module inversible L, alors il existe une fonction strictement positive λ sur X^{an} telle que

$$\forall x \in X^{\mathrm{an}}, \quad |\cdot|_\varphi(x) = \lambda(x)|\cdot|_{\varphi'}(x).$$

On définit la *distance entre φ et φ'* comme

$$d(\varphi, \varphi') := \sup_{x \in X^{\mathrm{an}}} |\ln \lambda(x)|.$$

Si X est propre sur $\mathrm{Spec}\, k$ et si les métriques φ et φ' sont continues, la distance entre φ et φ' est finie.

Similairement, si V est un espace vectoriel de rang fini sur k et si $\|\cdot\|$ et $\|\cdot\|'$ sont deux normes sur V, on définit

$$d(\|\cdot\|, \|\cdot\|') := \sup_{0 \neq s \in V} \left| \ln \|s\| - \ln \|s\|' \right|.$$

Soit L un $\mathscr{O}_X$-module inversible, muni d'une métrique φ. Soit $i : Y \to X$ un morphisme de schémas. Si y est un point de Y^{an} et si x est l'image de y dans X^{an} par i^{an}, alors $(\widehat{\kappa}(y), |\cdot|_y)$ est une extension valuée de $(\widehat{\kappa}(x), |\cdot|_x)$. En particulier, la norme $|\cdot|_\varphi(x)$ sur $L(x) = L \otimes_{\mathscr{O}_X} \widehat{\kappa}(x)$ induit par extension de corps une norme sur

$$i^*(L)(y) = i^*(L) \otimes_{\mathscr{O}_Y} \widehat{\kappa}(y) \cong L(x) \otimes_{\widehat{\kappa}(x)} \widehat{\kappa}(y).$$

On obtient ainsi une métrique sur $i^*(L)$ que l'on note $i^*(\varphi)$, appelée *tirée en arrière* de φ par le morphisme i. Cette métrique est continue lorsque φ est continue.

Exemple 4.1 Soit $(V, \|\cdot\|)$ un espace vectoriel normé de rang fini sur k. Soit $\mathscr{O}_V(1)$ le faisceau inversible universel de l'espace projectif $\pi : \mathbb{P}(V) \to \mathrm{Spec}\, k$. Rappelons que l'on a un homomorphisme surjectif universel de $\mathscr{O}_{\mathbb{P}(V)}$-modules $f : \pi^*(V) \to \mathscr{O}_V(1)$, et tout k-point $\rho : \mathrm{Spec}\, A \to \mathbb{P}(V)$ de $\mathbb{P}(V)$ à valeurs dans une k-algèbre A correspond à un A-module quotient projectif de rang 1 de $V \otimes_k A$, donné par

$$\rho^*(f) : \rho^*(\pi^*(V)) \longrightarrow \rho^*(\mathscr{O}_V(1)).$$

En particulier, si x est un point de X^{an}, alors la norme $\|\cdot\|$ sur V induit par extension des scalaires une norme $\|\cdot\|_{\widehat{\kappa}(x)}$ sur $V \otimes_k \widehat{\kappa}(x)$ puis par quotient une norme sur $\mathscr{O}_V(1)(x)$. On obtient ainsi une métrique continue sur $\mathscr{O}_V(1)$, appelée *métrique de Fubini-Study* associée à $\|\cdot\|$, notée $|\cdot|_{\mathrm{FS}}$.

On peut montrer que, si $\|\cdot\|$ et $\|\cdot\|'$ sont deux normes sur le même espace vectoriel V qui est de rang fini sur k, alors on a

$$d(|\cdot|_{\mathrm{FS}}, |\cdot|'_{\mathrm{FS}}) \leqslant d(\|\cdot\|, \|\cdot\|'),$$

l'égalité est satisfaite si $|\cdot|$ est archimédienne (par le théorème de Hahn-Banach) ou si $\|\cdot\|$ et $\|\cdot\|'$ sont toutes deux ultramétriques.

Dans le cas où $|\cdot|$ est une valeur absolue non archimédienne et non triviale, une autre façon d'obtenir une métrique continue consiste à construire la métrique à partir d'un modèle entier. Soit $\mathfrak{o}_k$ l'anneau de valuation de $(k, |\cdot|)$. Si X est un schéma projectif sur $\operatorname{Spec} k$ et L est un $\mathscr{O}_X$-module inversible, on entend par *modèle* de (X, L) la donnée d'un $\mathfrak{o}_k$-schéma projectif et plat $\mathfrak{X}$ et d'un $\mathscr{O}_{\mathfrak{X}}$-module inversible $\mathfrak{L}$ tels que la fibre générique de $\mathfrak{X}$ soit isomorphe à X et que la restriction de $\mathfrak{L}$ à la fibre générique soit isomorphe à L. Si x est un point de X^{an}, on désigne par $\mathfrak{o}_x$ l'anneau de valuation de $\widehat{\kappa}(x)$. Par le critère valuatif de propreté, le morphisme canonique $p_x : \operatorname{Spec} \widehat{\kappa}(x) \to X$ s'étend de façon unique en un morphisme $\mathscr{P}_x : \operatorname{Spec} \mathfrak{o}_x \to \mathfrak{X}$. La tirée en arrière $\mathscr{P}_x^*(\mathfrak{L})$ induit une norme $|\cdot|_{\mathfrak{L}}(x)$ sur $L(x)$ telle que

$$\forall s \in L(x), \quad |s|_{\mathfrak{L}}(x) = \inf\{|a|_x \mid a \in \widehat{\kappa}(x)^\times, \ a^{-1}s \in \mathscr{P}_x^*(\mathfrak{L})\}.$$

On peut montrer que $|\cdot|_{\mathfrak{L}}$ définit une métrique continue sur L, appelée *métrique induite par le modèle* $(\mathfrak{X}, \mathfrak{L})$.

4.4 Version métrique du théorème de Hilbert-Samuel

Soient X un schéma projectif sur $\operatorname{Spec} k$ et L un $\mathscr{O}_X$-module inversible ample, et φ une métrique continue sur L. Pour tout $x \in X^{\mathrm{an}}$ et tout entier $n \in \mathbb{N}$, l'espace vectoriel $L^{\otimes n}(x)$ sur $\widehat{\kappa}(x)$ est la $n^{\mathrm{ième}}$ puissance tensorielle de $L(x)$. Ainsi la norme $|\cdot|_\varphi(x)$ induit par puissance tensorielle une norme $|\cdot|_{n\varphi}(x)$ sur $L^{\otimes n}(x)$ telle que, pour tout élément $\ell \in L(x)$, on ait

$$|\ell^{\otimes n}|_{n\varphi}(x) = |\ell|_\varphi(x)^n. \tag{4}$$

Ainsi on obtient une métrique $n\varphi$ sur $L^{\otimes n}$, qui est continue. On munit l'espace vectoriel $H^0(X, L^{\otimes n})$ de la norme $\|\cdot\|_{n\varphi,\sup}$.

Pour simplifier, on suppose que s est une section globale dans $H^0(X, L)$ qui induit une suite exacte

$$0 \longrightarrow H^0(X, L^{\otimes(n-1)}) \xrightarrow{\cdot s} H^0(X, L^{\otimes n}) \longrightarrow H^0(Y, i^*(L^{\otimes n})) \longrightarrow 0,$$

où n est un entier positif, Y est le sous-schéma fermé de X défini par l'annulation de s, et $i : Y \to X$ est le morphisme d'inclusion. On a un isomorphisme canonique entre des espaces vectoriels de rang un sur k :

$$\det(H^0(X, L^{\otimes(n-1)})) \otimes \det(H^0(Y, i^*(L^{\otimes n}))) \cong \det(H^0(X, L^{\otimes n})). \tag{5}$$

Si on munit ces espaces de déterminant des normes

$$\|\cdot\|_{(n-1)\varphi,\sup,\det}, \quad \|\cdot\|_{n\varphi,\sup,\det} \text{ et } \|\cdot\|_{i^*(n\varphi),\sup,\det},$$

respectivement, l'isomorphisme (5) n'est pas une isométrie en général. Cependant, si la métrique φ est semi-positive, le défaut d'isométrie peut être décrit par des invariants de la métrique.

Définition 4.2 Pour tout entier $n \geqslant 1$, soit φ_n la métrique continue sur L dont la $n^{\text{ième}}$ puissance symétrique s'identifie à la tirée en arrière de la métrique de Fubini-Study $|\cdot|_{n\varphi,\sup,\text{FS}}$ sur $\mathcal{O}_{\mathbb{P}(H^0(X,L^{\otimes n}))}(1)$ par l'immersion fermée $X \to \mathbb{P}(H^0(X,L^{\otimes n}))$. On dit que la métrique φ est *semi-positive* si

$$\lim_{n \to +\infty} d(\varphi_n, \varphi) = 0.$$

Dans le cas où la valeur absolue $|\cdot|$ est archimédienne, φ est semi-positive si et seulement si elle est pluri-sous-harmonique (voir [44]); dans le cas où la valeur absolue $|\cdot|$ est non archimédienne et non triviale, φ est semi-positive si et seulement si elle est limite uniforme d'une suite de métriques induites par des modèles amples sur l'anneau de valuation de $(k, |\cdot|)$ (voir [18, §3]). Cette condition est aussi équivalente à la pluri-sous-harmonicité non archimédienne (voir [30, §6] et [12, §6.8]).

Si la métrique φ est semi-positive, le défaut d'isométrie de l'isomorphisme (5) peut être estimé en fonction de la mesure de Monge-Ampère. On a

$$\frac{1}{\dim_k(H^0(X,L^{\otimes n}))} \ln \frac{\|\cdot\|_{n\varphi,\sup,\det}}{\|\cdot\|_{(n-1)\varphi,\sup,\det} \otimes \|\cdot\|_{i^*(n\varphi),\sup,\det}}$$

$$= -\int_{X^{\text{an}}} \ln|s|\, c_1(\varphi)^{\wedge d} + O(\ln(n)). \tag{6}$$

Dans le cas où $|\cdot|$ est archimédienne, la mesure $c_1(\varphi)^{\wedge d}$ est définie comme la puissance extérieure du courant. Dans le cas non archimédien, cette mesure est définie dans [11].

4.5 Méthode combinatoire

Similairement au cas géométrique, la méthode combinatoire peut aussi être appliquée aux systèmes linéaires gradués normés. Soient X un schéma projectif et intègre sur $\operatorname{Spec} k$ et $V_{\bullet}$ un système linéaire gradué sur X. Rappelons que $V_{\bullet} = \bigoplus_{n \in \mathbb{N}} V_n T^n$ est une sous-k-algèbre graduée de $K[T] = \bigoplus_{n \in \mathbb{N}} KT^n$, où K désigne le corps des fonctions rationnelles sur X. On suppose que chaque espace vectoriel V_n est muni d'une norme $\|\cdot\|_n$ de sorte que, pour tout $(m,n) \in \mathbb{N}^2$, et tout $(s_m, s_n) \in V_m \times V_n$, on ait

$$\|s_m \cdot s_n\|_{n+m} \leqslant \|s_m\|_m \cdot \|s_n\|_n. \tag{7}$$

Comme dans le cas géométrique, on suppose l'existence d'un point régulier $x \in X(k)$ et on fixe un système de paramètres $\{z_1, \ldots, z_d\}$ dans l'anneau local régulier $\mathcal{O}_{X,x}$. Ainsi l'homomorphisme de l'anneau des séries formelles $k[\![T_1, \ldots, T_d]\!]$ dans le complété $\widehat{\mathcal{O}}_{X,x}$, qui envoie T_i sur z_i, est un isomorphisme.

Soit $v : K \to \mathbb{Z}^d \cup \{+\infty\}$ la valuation comme dans le paragraphe précédent. Cette valuation définit une $\mathbb{Z}^d$-filtration sur K (comme espace vectoriel sur k) : pour tout $\alpha \in \mathbb{Z}^d$, soit

$$\mathscr{F}^\alpha(K) = \{s \in K \mid v(s) \geqslant \alpha\}.$$

Si V est un sous-espace vectoriel de rang fini de K, alors cette filtration induit par restriction une $\mathbb{Z}^d$-filtration sur V que l'on note encore $\mathscr{F}$ par abus de notation. On a vu que la somme directe des sous-quotients

$$\bigoplus_{n \in \mathbb{N}} \bigoplus_{\alpha \in \mathbb{Z}^d} \mathrm{gr}^\alpha(V_n)$$

forme une k-algèbre graduée qui est isomorphe à l'algèbre du semi-groupe $\Gamma(V_\bullet)$. Dans la suite, on identifie ces deux k-algèbres graduées. En outre, pour tout $n \in \mathbb{N}$ et tout $\alpha \in \mathbb{Z}^d$, la norme $\|\cdot\|_n$ sur V_n induit par passage au sous-quotient une norme $\|\cdot\|_{(n,\alpha),\mathrm{sq}}$ sur $\mathrm{gr}^\alpha(V_n)$. On déduit de l'hypothèse (7) que, si γ et γ' sont deux éléments de $\Gamma(V_\bullet)$, alors

$$\|e^{\gamma+\gamma'}\|_{\gamma+\gamma',\mathrm{sq}} \leqslant \|e^\gamma\|_{\gamma,\mathrm{sq}} \cdot \|e^{\gamma'}\|_{\gamma',\mathrm{sq}}.$$

Les normes déterminants $\|\cdot\|_{n,\mathrm{d\acute{e}t}}$ sont naturellement liées à ces normes sous-quotients. En effet, si on identifie V_n à

$$\bigoplus_{\alpha \in \Gamma_n} \mathrm{gr}^\alpha(V_n),$$

on a

$$\frac{1}{\mathrm{rg}_k(V_n)} \left| \ln \frac{\|\cdot\|_{n,\mathrm{d\acute{e}t}}}{\bigoplus_{\alpha \in \Gamma_n} \|\cdot\|_{(n,\alpha),\mathrm{sq}}} \right| \leqslant \ln(\mathrm{rg}_k(V_n)),$$

où $\Gamma_n = \{\alpha \in \mathbb{Z}^d \mid (n, \alpha) \in \Gamma(V_\bullet)\}$.

En utilisant la méthode dans [6] (voir aussi [45, 17]), on obtient le résultat suivant.

Théorème 4.3 *Soient X un schéma projectif et intègre de dimension $d \geqslant 1$ sur* Spec k *qui possède un point rationnel régulier, et $V_\bullet$ un système linéaire gradué sur*

V. On suppose que

$$\sup_{\substack{n \geqslant 1,\, \alpha \in \mathbb{Z}^d \\ (n,\alpha) \in \Gamma(V_\bullet)}} \left(-\frac{1}{n} \ln \|e^{(n,\alpha)}\|_{(n,\alpha),\mathrm{sq}} \right) < +\infty.$$

Alors la suite de mesures de probabilité

$$\frac{1}{\mathrm{rg}_k(V_n)} \sum_{\substack{\alpha \in \mathbb{Z}^d \\ (n,\alpha) \in \Gamma(V_\bullet)}} \delta_{-\ln \|e^{(n,\alpha)}\|_{(n,\alpha),\mathrm{sq}}}, \quad n \in \mathbb{N}(V_\bullet)$$

converge faiblement vers une mesure de probabilité borélienne sur $\mathbb{R}$, *où* δ_x *désigne la mesure de Dirac en* x. *En outre, si on désigne par* G *la fonction concave sur* $\Delta(V_\bullet)$ *dont le graphe est le bord supérieur de l'enveloppe convexe des points*

$$\frac{1}{n}((n,\alpha), -\ln \|e^{(n,\alpha)}\|_{(n,\alpha),\mathrm{sq}}), \quad (n,\alpha) \in \Gamma(V_\bullet), \ n \geqslant 1,$$

alors la mesure de probabilité limite s'identifie à l'image directe de la mesure uniforme sur $\Delta(V_\bullet)$ *par la fonction* G.

Remarque 4.4 Soient Σ un éventail tel que $|\Sigma| = \mathbb{R}^d$, $X = X_\Sigma$ la variété torique associée, et $\psi : \mathbb{R}^d \to \mathbb{R}$ une fonction de support virtuelle. On suppose que $V_\bullet = \bigoplus_{n \in \mathbb{N}} H^0(nD_\psi)$ est le système linéaire gradué total du $\mathbb{T}$-diviseur de Cartier D_ψ et que, pour tout $n \in \mathbb{N}$, la norme $\|\cdot\|_n$ est de la forme $\|\cdot\|_{n\varphi,\sup}$, où φ est une métrique continue semi-positive sur $\mathscr{O}(D_\psi)$ qui est invariante par l'action du tore. Il s'avère que la métrique φ correspond à une fonction concave $f_\varphi : \mathbb{R}^d \to \mathbb{R}$ telle que $|f_\varphi - \psi|$ soit une fonction bornée. Alors la fonction G décrite dans le théorème 4.3 s'identifie au dual au sens de Legendre-Fenchel de la fonction convexe f_φ. On renvoie les lecteurs à [8] pour plus de détails sur le point de vue de la géométrie convexe dans l'étude de l'arithmétique des variétés toriques, et à [35] pour la géométrie d'Arakelov (notamment la théorie de l'intersection arithmétique) des variétés toriques.

5 Cas arithmétique

Dans ce paragraphe, on suppose que k est un corps de nombres. On désigne par Σ_k l'ensemble des places de k. Pour toute place $\sigma \in \Sigma_k$, soit $|\cdot|_\sigma$ la valeur absolue sur k qui prolonge la valeur absolue usuelle sur $\mathbb{Q}$ (dans ce cas-là la place σ est dite *archimédienne*) ou la valeur absolue p-adique pour certain nombre premier p, où la valeur absolue de p est $1/p$ (dans ce cas-là la place σ est dite *non archimédienne*). On désigne par k_σ le complété de k par rapport à la valeur absolue $|\cdot|_\sigma$, sur lequel la valeur absolue $|\cdot|_\sigma$ s'étend de façon unique.

5.1   Fibré vectoriels adéliques sur un corps de nombres

On appelle *fibré vectoriel adélique* sur Spec k la donnée $\overline{V}$ d'un espace vectoriel de rang fini V sur k et une famille de normes $(\|\cdot\|_\sigma)_{\sigma\in\Sigma_k}$, où chaque $\|\cdot\|_\sigma$ est une norme sur $V_\sigma := V \otimes_k k_\sigma$, qui satisfait aux conditions suivantes :

(a) si la place σ est non archimédienne, alors la norme $\|\cdot\|_\sigma$ est ultramétrique ;
(b) il existe une base $(e_i)_{i=1}^r$ de V sur k telle que, pour toute place $\sigma \in \Sigma_k$ sauf un nombre fini, on a

$$\forall\,(\lambda_1,\ldots,\lambda_r) \in k_\sigma^r, \quad \|\lambda_1 e_1 + \cdots + \lambda_r e_r\|_\sigma = \max(|\lambda_1|_\sigma,\ldots,|\lambda_r|_\sigma).$$

Les fibrés vectoriels adéliques sont des cas particuliers des espaces adéliques rigides introduits dans le chapitre II. On renvoie les lecteurs à [25] pour une présentation détaillée de la théorie des fibrés vectoriels adéliques, voir aussi [4] pour le cadre classique de fibrés vectoriels hermitiens sur l'anneau des entiers d'un corps de nombres.

Soit $\overline{V}$ un fibré vectoriel adélique sur Spec k. Le rang de V sur k est appelé *rang* de $\overline{V}$. Si $\overline{V}$ est de rang 1, on définit son *degré d'Arakelov* comme

$$\widehat{\deg}(\overline{V}) = -\sum_{\sigma\in\Sigma_k} [k_\sigma : \mathbb{Q}_\sigma] \ln \|s\|_\sigma,$$

où s est un élément non nul de V. Rappelons que la formule du produit

$$\forall\, a \in k^\times, \quad \sum_{\sigma\in\Sigma_k} \ln |a|_\sigma = 0$$

montre que cette définition ne dépend pas du choix de s. Plus généralement, si $\overline{V} = (V, (\|\cdot\|_\sigma)_{\sigma\in\Sigma_k}))$ est un fibré vectoriel adélique de rang quelconque, on désigne par dét$(\overline{V})$ la donnée $(\det(V), (\|\cdot\|_{\sigma,\det})_{\sigma\in\Sigma_k})$ et on définit le degré d'Arakelov de $\overline{V}$ comme $\widehat{\deg}(\det(\overline{V}))$ (voir §2 du chapitre II *infra* pour la construction du degré d'Arakelov dans le cadre des espaces adéliques rigides).

5.2   Fibrés inversibles adéliques sur une variété arithmétique

Soit $\pi : X \to \operatorname{Spec} k$ un schéma projectif sur Spec k. On appelle *fibré inversible adélique* sur X la donnée $\overline{L}$ d'un $\mathcal{O}_X$-module inversible L et d'une famille $(\varphi_\sigma)_{\sigma\in\Sigma_k}$ de métriques, où chaque φ_σ est une métrique continue sur L_σ, la tirée en arrière de L par le changement de base $X_{k_\sigma} \to X$, qui satisfait à la condition suivante : il existe un modèle entier $(\mathscr{X}, \mathscr{L})$ de (X, L) tel que, pour toute place non archimédienne σ sauf un nombre fini, la métrique φ_σ soit induite par le modèle $(\mathscr{X}_{\mathfrak{o}_{k_\sigma}}, \mathscr{L}_{\mathfrak{o}_{k_\sigma}})$ de

(X_σ, L_σ), où $\mathfrak{o}_{k_\sigma}$ est l'anneau de valuation de k_σ. Rappelons qu'un modèle entier de (X, L) est par définition la donnée $(\mathscr{X}, \mathscr{L})$ d'un schéma projectif et plat $\mathscr{X}$ sur l'anneau des entiers algébriques dans k dont la fibre générique est isomorphe à X, ainsi qu'un $\mathscr{O}_{\mathscr{X}}$-module inversible $\mathscr{L}$ dont la restriction à la fibre générique est isomorphe à L. Il s'avère que

$$\pi_*(\overline{L}) := (H^0(X, L), (\|\cdot\|_{\varphi_\sigma, \sup})_{\sigma \in \Sigma_k})$$

est un fibré vectoriel adélique sur $\operatorname{Spec} k$.

5.3 Théorème de Hilbert-Samuel arithmétique

Si $\overline{L} = (L, (\varphi_\sigma)_{\sigma \in \Sigma_k})$ est un fibré inversible adélique sur X, pour tout entier $n \in \mathbb{N}$, on désigne par $\overline{L}^{\otimes n}$ le fibré inversible adélique $(L^{\otimes n}, (n\varphi_\sigma)_{\sigma \in \Sigma_k})$, où $n\varphi_\sigma$ désigne la $n^{\text{ième}}$ puissance tensorielle de φ_σ, définie dans la formule (4).

Comme le degré d'Arakelov peut être calculé par une somme par rapport aux places de k, on déduit de la version métrique du théorème de Hilbert-Samuel le résultat suivant.

Théorème 5.1 *Soient* $\pi : X \to \operatorname{Spec} k$ *un schéma projectif intègre de dimension d sur* $\operatorname{Spec} k$ *et* $\overline{L}$ *un fibré inversible adélique sur X. On suppose que L est ample et que les métriques φ_σ sont semi-positives, alors on a*

$$\lim_{n \to +\infty} \frac{\widehat{\deg}(\pi_*(\overline{L}^{\otimes n}))}{n^{d+1}/(d+1)!} = \widehat{\deg}(\widehat{c}_1(\overline{L})^{(d+1)}).$$

5.4 Cas sans hypothèse d'amplitude

Dans le cas où L n'est pas ample, la méthode de dévissage ne s'applique plus. Pour contourner cette difficulté, on applique le théorème 4.3 au cas où la valeur absolue est triviale. On désigne par $|\cdot|_0$ la valeur absolue triviale sur k. Si V est un fibré vectoriel adélique sur $\operatorname{Spec} k$, on définit une norme $\|\cdot\|_{\overline{V}}$ comme suit (où on considère la valeur absolue triviale sur k). Pour tout $s \in V \setminus \{0\}$, soit $\widehat{\deg}(s)$ le nombre

$$-\sum_{\sigma \in \Sigma_k} [k_\sigma : \mathbb{Q}_\sigma] \ln \|s\|_\sigma,$$

appelé *degré d'Arakelov* de s. Pour tout $s \in V$, soit $\|s\|_{\overline{V}}$ l'infimum de l'ensemble des $\lambda > 0$ tels que s puisse s'écrire comme une combinaison linéaire à coefficients dans k des vecteurs non nuls de V de degré d'Arakelov au moins $-\ln(\lambda)$. Comme

$\cdot|_0$ est la valeur absolue triviale, la restriction de $\|\cdot\|_{\overline{V}}$ à $V \backslash \{0\}$ ne prend qu'un nombre fini de valeurs, dont le cardinal ne dépasse pas le rang de V sur k. Ces valeurs sont précisément les minima successifs au sens de Thunder [43]. Ces minima sont naturellement liés aux invariants arithmétiques de $\overline{V}$ via le lemme de Siegel (voir [26]). Notamment on a

$$\frac{1}{\mathrm{rg}_k(V)} \left| \widehat{\deg}(\overline{V}) + \int_{\mathbb{R}} t \, \mathrm{d}\, \mathrm{rg}_k(B(V, e^{-t})) \right| = O(\ln(\mathrm{rg}_k(V))), \qquad (8)$$

où $B(V, e^{-t})$ désigne la boule fermée dans V de rayon e^{-t} et centrée à l'origine. En outre, si on désigne par $\widehat{h}^0(\overline{V})$ le logarithme du nombre des éléments $s \in V$ tels que $\sup_{\sigma \in \Sigma_k} \|s\|_\sigma \leqslant 1$, alors on a

$$\frac{1}{\mathrm{rg}_k(V)} \left| \widehat{h}^0(\overline{V}) + \int_0^{+\infty} t \, \mathrm{d}\, \mathrm{rg}_k(B(V, e^{-t})) \right| = O(\ln(\mathrm{rg}_k(V))). \qquad (9)$$

Soit $\overline{L}$ un fibré inversible adélique sur un k-schéma projectif et intègre X. On suppose que X possède un point rationnel régulier. Pour tout $n \in \mathbb{N}$, soit $\overline{V}_n := \pi_*(\overline{L}^{\otimes n})$. Il s'avère que la suite de normes $(\|\cdot\|_{\overline{V}_n})_{n \in \mathbb{N}}$ satisfait à la condition de sous-multiplicativité (7). Ainsi on déduit du théorème 4.3 le résultat suivant (voir [6, théorème 3.7]).

Théorème 5.2 *Soient* $\pi : X \to \mathrm{Spec}\, k$ *un schéma projectif et intègre, de dimension* d *sur* $\mathrm{Spec}\, k$ *et* $\overline{L}$ *un fibré inversible adélique sur* X *tel que* L *soit gros. Pour tout* $n \in \mathbb{N}$, *soit* $\overline{V}_n := \pi_*(\overline{L}^{\otimes n})$. *Il existe alors une fonction concave* G *sur le corps de Newton-Okounkov* $\Delta(L)$ *de l'algèbre graduée* $\bigoplus_{n \in \mathbb{N}} V_n$ *telle que*

$$\lim_{n \to +\infty} \frac{\widehat{\deg}(\overline{V}_n)}{n^{d+1}/(d+1)!} = \int_{\Delta(L)} G(x) \, \mathrm{d}x, \qquad (10)$$

$$\lim_{n \to +\infty} \frac{\widehat{h}^0(\overline{V}_n)}{n^{(d+1)}/(d+1)!} = \int_{\Delta(L)} \max(G(x), 0) \, \mathrm{d}x. \qquad (11)$$

Références

1. A. Abbes, T. Bouche, Théorème de Hilbert-Samuel "arithmétique". Ann. Inst. Fourier **45**, 375–401 (1995)
2. V.G. Berkovich, *Spectral Theory and Analytic Geometry over Non-Archimedean Fields*. Mathematical Surveys and Monographs, vol. 33 (American Mathematical Society, Providence, 1990)
3. R.J. Berman, G. Freixas i Montplet, An arithmetic Hilbert-Samuel theorem for singular hermitian line bundles and cusp forms. Compos. Math. **150**, 1703–1728 (2014)

4. J.-B. Bost, Périodes et isogénies des variétés abéliennes sur les corps de nombres (d'après D. Masser et G. Wüstholz), in *Séminaire Bourbaki, Vol. 1994/1995, Astérisque*, no. 237, p. Exp. No. 795, 4 (1996), pp. 115–161

5. S. Boucksom, Corps d'Okounkov (d'après Okounkov, Lazarsfeld-Mustaţă et Kaveh-Khovanskii), in *Astérisque*, no. 361, p. Exp. No. 1059, vii (2014), pp. 1–41

6. S. Boucksom, H. Chen, Okounkov bodies of filtered linear series. Compos. Math. **147**, 1205–1229 (2011)

7. N. Bourbaki, *Éléments de mathématique* (Masson, Paris, 1983). Algèbre commutative. Chapitre 8. Dimension. Chapitre 9. Anneaux locaux noethériens complets

8. J.I. Burgos Gil, P. Philippon, M. Sombra, Arithmetic geometry of toric varieties. Metrics, measures and heights, in *Astérisque*, no. 360 (2014), p. vi+222

9. J.I. Burgos Gil, P. Philippon, M. Sombra, Heights of toric varieties, entropy and integration over polytopes, in *Geometric Science of Information*. Lecture Notes in Comput. Sci., vol. 9389 (Springer, Cham, 2015), pp. 286–295

10. J.I. Burgos Gil, P. Philippon, M. Sombra, Successive minima of toric height functions. Ann. Inst. Fourier **65**, 2145–2197 (2015)

11. A. Chambert-Loir, Mesures et équidistribution sur les espaces de Berkovich. J. Reine Angew. Math. **595**, 215–235 (2006)

12. A. Chambert-Loir, A. Ducros, Formes différentielles réelles et courants sur les espaces de Berkovich. arXiv :1204.6277

13. H. Chen, Arithmetic Fujita approximation. Ann. Sci. École Norm. Sup. Quatrième Série **43**, 555–578 (2010)

14. H. Chen, Convergence des polygones de Harder-Narasimhan, in *Mémoires de la Société Mathématique de France. Nouvelle Série*, no. 120 (2010), p. 120

15. H. Chen, Okounkov bodies : an approach of function field arithmetic. J. Théor. Nombres Bordeaux **30**, 829–845 (2018)

16. H. Chen, H. Ikoma, On subfiniteness of graded linear series. Eur. J. Math. **6**, 367–399 (2020)

17. H. Chen, C. Maclean, Distribution of logarithmic spectra of the equilibrium energy. Manuscr. Math. **146**, 365–394 (2015)

18. H. Chen, A. Moriwaki, Extension property of semipositive invertible sheaves over a non-archimedean field. Ann. Scuola Norm. Sup. Pisa. Classe Sci. Serie V. **18**, 241–282 (2018)

19. H. Chen, A. Moriwaki, *Arakelov Geometry over Adelic Curves*. Lecture Notes in Mathematics, vol. 2258 (Springer, Singapore, 2020)

20. S.D. Cutkosky, Zariski decomposition of divisors on algebraic varieties. Duke Math. J. **53**, 149–156 (1986)

21. D. Eisenbud, *Commutative Algebra, with a View Toward Algebraic Geometry*. Graduate Texts in Mathematics, vol. 150 (Springer, New York, 1995)

22. G. Freixas i Montplet, An arithmetic Hilbert-Samuel theorem for pointed stable curves. J. Eur. Math. Soc. **14**, 321–351 (2012)

23. T. Fujita, Approximating Zariski decomposition of big line bundles. Kodai Math. J. **17**, 1–3 (1994)

24. W. Fulton, *Introduction to Toric Varieties*. Annals of Mathematics Studies, vol. 131 (Princeton University Press, Princeton, 1993). The William H. Roever Lectures in Geometry.

25. É. Gaudron, Pentes des fibrés vectoriels adéliques sur un corps global. Rend. Sem. Mat. Univ. Padova **119**, 21–95 (2008)

26. É. Gaudron, Géométrie des nombres adélique et lemmes de Siegel généralisés. Manuscr. Math. **130**, 159–182 (2009)

27. H. Gillet, C. Soulé, Arithmetic intersection theory. Inst. Hautes Études Sci. Publ. Math. (1990) **72**, 93–174 (1991)

28. H. Gillet, C. Soulé, An arithmetic Riemann-Roch theorem. Invent. Math. **110**, 473–543 (1992)

29. A. Grothendieck, Éléments de géométrie algébrique, rédigés avec la collaboration de J. Dieudonné. Inst. Hautes Études Sci. Publ. Math. **4**, **8**, **11**, **17**, **24**, **28**, **32**

30. W. Gubler, K. Künnemann, Positivity properties of metrics and delta-forms. J. Reine Angew. Math. **752**, 141–177 (2019)

31. K. Kaveh, A. G. Khovanskiĭ, Newton-Okounkov bodies, semigroups of integral points, graded algebras and intersection theory. Ann. Math. Second Series **176**, 925–978 (2012)
32. A.G. Khovanskiĭ, The Newton polytope, the Hilbert polynomial and sums of finite sets. Rossiĭskaya Akademiya Nauk. Funktsional'nyĭ Analiz i ego Prilozheniya **26**, 57–63, 96 (1992)
33. R. Lazarsfeld, Positivity in algebraic geometry. I, in *Ergebnisse der Mathematik und ihrer Grenzgebiete*. 3. Folge. A Series of Modern Surveys in Mathematics, vol. 48 (Springer, Berlin, 2004). Classical Setting : Line Bundles and Linear Series
34. R. Lazarsfeld, M. Mustaţă, Convex bodies associated to linear series. Ann. Sci. École Norm. Sup. Quatrième Série **42**, 783–835 (2009)
35. V. Maillot, Géométrie d'Arakelov des variétés toriques et fibrés en droites intégrables. Mém. Soc. Math. France. Nouv. Série **80**, p. vi+129 (2000)
36. A. Moriwaki, Continuity of volumes on arithmetic varieties. J. Algebraic Geom. **18**, 407–457 (2009)
37. A. Okounkov, Brunn-Minkowski inequality for multiplicities. Invent. Math. **125**, 405–411 (1996)
38. H. Randriambololona, Métriques de sous-quotient et théorème de Hilbert-Samuel arithmétique pour les faisceaux cohérents. J. Reine Angew. Math. **590**, 67–88 (2006)
39. R. Schneider, *Convex Bodies : The Brunn-Minkowski Theory*. Encyclopedia of Mathematics and Its Applications, vol. 151 (Cambridge University Press, Cambridge, 2014)
40. C. Soulé, *Lectures on Arakelov Geometry*. Cambridge Studies in Advanced Mathematics, vol. 33 (Cambridge University Press, Cambridge, 1992). With the collaboration of D. Abramovich, J.-F. Burnol and J. Kramer
41. S. Takagi, Fujita's approximation theorem in positive characteristics. J. Math. Kyoto Univ. **47**, 179–202 (2007)
42. B. Teissier, Du théorème de l'index de Hodge aux inégalités isopérimétriques. C. R. Hebd. Acad. Sci. Séries A B **288**, A287–A289 (1979)
43. J.L. Thunder, An adelic Minkowski-Hlawka theorem and an application to Siegel's lemma. J. Reine Angew. Math. **475**, 167–185 (1996)
44. G. Tian, On a set of polarized Kähler metrics on algebraic manifolds. J. Differ. Geom. **32**, 99–130 (1990)
45. D. Witt Nyström, Transforming metrics on a line bundle to the Okounkov body. Ann. Sci. École Norm. Sup. Quatrième Série **47**, 1111–1161 (2014)
46. X. Yuan, Big line bundles over arithmetic varieties. Invent. Math. **173**, 603–649 (2008)
47. X. Yuan, On volumes of arithmetic line bundles. Compos. Math. **145**, 1447–1464 (2009)
48. X. Yuan, Volumes of arithmetic Okounkov bodies. Math. Z. **280**, 1075–1084 (2015)

Chapter IV: Euclidean Lattices, Theta Invariants, and Thermodynamic Formalism

Jean-Benoît Bost

1 Introduction

1.1 My talks during the summer school *Arakelov Geometry and Diophantine Applications* were devoted to the formalism of infinite dimensional vector bundles over arithmetic curves and to the properties of their theta invariants studied in the monograph [9], and to some Diophantine applications of this formalism.

In this chapter, I will focus on the content of the first of these lectures, where I discussed various motivations for considering the theta invariants of (finite dimensional) Hermitian vector bundles over arithmetic curves, notably of Euclidean lattices.

Recall that a Euclidean lattice is defined as a pair

$$\overline{E} := (E, \|.\|),$$

where E is some free $\mathbb{Z}$-module of finite rank and $\|.\|$ is some Euclidean norm on the real vector space $E_{\mathbb{R}} := E \otimes \mathbb{R}$. The theta invariants of $\overline{E}$ are invariants defined by means of the theta series

$$\sum_{v \in E} e^{-\pi t \|x-v\|^2}, \tag{1.1}$$

J.-B. Bost (✉)
Département de Mathématique, Université Paris-Sud, Orsay Cedex, France
e-mail: jean-benoit.bost@math.u-psud.fr

E. Peyre, G. Rémond (eds.), *Arakelov Geometry and Diophantine Applications*,
Lecture Notes in Mathematics 2276, https://doi.org/10.1007/978-3-030-57559-5_5

where (t, x) belongs to $\mathbb{R}_+^* \times E_\mathbb{R}$, and of its special values. The most basic of these is the non-negative real number:

$$h_\theta^0(\overline{E}) := \log \sum_{v \in E} e^{-\pi \|v\|^2}.$$

My purpose in this chapter is to explain how they naturally arise when one investigates diverse basic questions concerning classical invariants of Euclidean lattices, such as their successive minima, their covering radius, or the number of lattice points in balls of a given radius.

1.2 The first part of this chapter consists in a self-contained introduction to the study of Euclidean lattices. (These are also discussed by E. Gaudron in Chapter II, from a more advanced perspective.)

In Sect. 2, we recall some basic definitions concerning Euclidean lattices and their basic invariants. We also introduce some less classical, although elementary, notions concerning Euclidean lattices, such as the admissible short exact sequences of Euclidean lattices. These notions naturally arise from the perspective of Arakelov geometry, but do not appear in classical introductions to Euclidean lattices. However this formalism should be appealing to geometrically minded readers, as it is specifically devised to emphasize the formal similarities between Euclidean lattices and vector bundles over varieties.

In Sect. 3, we discuss, in a simple guise, a central topic of the classical theory of Euclidean lattices, the so-called *reduction theory*. This will demonstrate the flexibility of the "geometric formalism" of Euclidean lattices previously introduced, and also exemplify one of the main features of the classical theory of Euclidean lattices: the occurrence, in diverse inequalities relating their classical invariants, of constants depending of the rank n of the Euclidean lattices under study.

1.3 The precise dependence on n of these constants is a formidable problem—already determining their asymptotic behavior when n grows to infinity is often delicate—and their occurence is a nuisance, both from a formal or aesthetic perspective and in applications, notably to Diophantine geometry. The use of more sophisticated invariants attached to Euclidean lattices, such as their slopes *à la* Stuhler-Grayson or their theta invariants, appears as a natural remedy to these difficulties.

In this chapter, we focus on the theta invariants, and the reader is refered to the survey article [10] for a discussion of these non-classical invariants with more emphasis on the role of slopes.[1] Our aim in the second part of this chapter will be

---

[1] Slopes are also studied in Chapter II, and some of their Diophantine applications are discussed by E. Peyre in Chapter V, notably in its Section V.6. There is some overlap between Sects. 1–3 and 5 of [10] and Sects. 1–4 of this chapter. The remaining sections of [10], devoted to some remarkable recent results of Regev, Dadush, and Stephens–Davidowitz ([20, 63]), provide some further illustrations of the relevance of theta invariants in the proofs of estimates relating invariants of Euclidean lattices.

to convince the reader of the significance of the theta invariants when investigating Euclidean lattices, by giving accessible presentations of diverse results involving their classical invariants, in the derivation or in the statement of which theta invariants play a key role.

In Sect. 4, after discussing some basic properties of the theta series (1.1), we give an introductory account of their use in the seminal article of Banaszczyk [5] for deriving *transference estimates*—namely, estimates comparing classical invariants of some Euclidean lattice $\overline{E}$ and of its dual $\overline{E}^{\vee}$—where the involved constants depending of $n := \mathrm{rk}\, E$ are basically optimal.

In Sect. 5, we discuss the occurrence of theta invariants of Euclidean lattices from a completely different perspective, namely when developing the classical analogy between number fields and function fields. In this analogy, the theta invariant $h_{\theta}^{0}(\overline{E})$ attached to a Euclidean lattice $\overline{E}$ appears as an arithmetic counterpart of the dimension

$$h^{0}(C, E) := \dim_{k} \Gamma(C, E)$$

of the k-vector space of sections $\Gamma(C, E)$ of some vector bundle E over a smooth projective geometrically irreducible curve C over some field k. The similarities between $h_{\theta}^{0}(\overline{E})$ and $h^{0}(C, E)$ may actually be pursued to a striking level of precision, and we survey several of them at the end of Sect. 5.

It turns out that, when dealing with the analogy between number fields and function fields, besides the invariant $h_{\theta}^{0}(\overline{E})$ attached to some Euclidean lattice $\overline{E} := (E, \|.\|)$, one also classically considers the non-negative real number

$$h_{\mathrm{Ar}}^{0}(\overline{E}) := \log |\{v \in E \mid \|v\| \leqslant 1\}|$$

—simply defined in terms of the number of lattice points in the unit ball of $(E_{\mathbb{R}}, \|.\|)$—as an arithmetic counterpart of $h^{0}(C, E)$.

The coexistence of two distinct invariants playing the role of an arithmetic counterpart of the basic geometric invariant $h^{0}(C, E)$ is intriguing. This puzzle has been solved in [9, Chapter 3], in two ways. Firstly, by establishing some comparison estimate, bounding the difference $h_{\theta}^{0}(\overline{E}) - h_{\mathrm{Ar}}^{0}(\overline{E})$ in terms of the rank of E, by means of Banaszczyk's methods discussed in Sect. 4. And secondly, by showing that the theta invariant $h_{\theta}^{0}(\overline{E})$ are related, by Fenchel-Legendre transform, to some "stable variant" $\tilde{h}_{\mathrm{Ar}}^{0}(\overline{E})$ of the invariant $h_{\mathrm{Ar}}^{0}(\overline{E})$ defined in terms of lattice points counting in the direct sums

$$\overline{E}^{\oplus n} := \overline{E} \oplus \ldots \oplus \overline{E} \quad (n\text{-times})$$

of copies of the Euclidean lattice $\overline{E}$, when the integer n goes to $+\infty$.

We present these relations between $h_\theta^0(\overline{E})$, $h_{\mathrm{Ar}}^0(\overline{E})$, and $\tilde{h}_{\mathrm{Ar}}^0(\overline{E})$ with some details in Sect. 5.4.

1.4 The "Legendre duality" between $\tilde{h}_{\mathrm{Ar}}^0(\overline{E})$ and $h_\theta^0(\overline{E})$ provides a striking motivation for considering the theta invariant $h_\theta^0(\overline{E})$. Somewhat surprisingly, this duality holds in a much more general context. It is indeed a special case of some general measure theoretic results, concerning a measure space $\mathscr{E}$ equipped with some measurable function H with values in $\mathbb{R}_+$, that describe the asymptotic behavior of the measure of the subsets

$$\{(e_1, \ldots, e_n) \in \mathscr{E}^n \mid H(e_1) + \ldots + H(e_n) \leqslant nx\} \tag{1.2}$$

of $\mathscr{E}^n$ when n goes to $+\infty$, for a given value of $x \in \mathbb{R}_+$. These results are actually closely related to the formalism of statistical thermodynamics.

The proof of these general measure theoretic results is arguably clearer than its specialization to the invariants $\tilde{h}_{\mathrm{Ar}}^0(\overline{E})$ and $h_\theta^0(\overline{E})$ associated to some Euclidean lattice $\overline{E} := (E, \|.\|)$.[2] In [9], these results were established by reduction to some classical theorems of the theory of large deviation, and their relations with the thermodynamic formalism was only alluded to. In the third part of this chapter, we provide a self-contained presentation of these results, requiring only some basic knowledge of measure theory and complex analysis (say, at the level of Rudin's classical textbook [65]). Our presentation also includes a discussion of the physical meaning of these results and of their relations with diverse classical techniques to derive estimates in probability and analytic number theory.

In Sect. 6, we state our general measure theoretic theorem (Theorem 6.2.1) and we discuss its interpretation in statistical physics and its application to the invariants $\tilde{h}_{\mathrm{Ar}}^0(\overline{E})$ and $h_\theta^0(\overline{E})$ of Euclidean lattices.

In Sect. 7, we give a proof of Theorem 6.2.1 that uses a few basic notions of measure theory only. The key point of this proof is a variation on a classical proof of Cramér's theorem, the starting point of the theory of large deviations.

Section 8 is devoted to some complements to Theorem 6.2.1 and its proof. Notably, we present Lanford's approach to the study of the asymptotic behavior of the measure of the sets (1.2) when n goes to infinity. We also discuss a mathematical interpretation of the second law of thermodynamics in our formalism and its application to Euclidean lattices.

Finally, in Sect. 9, we give an alternative derivation of the main assertion of Theorem 6.2.1, which originates in the works of Poincaré [59] and of Darwin and Fowler [21, 22, 23]. Instead of arguments from measure and probability theory, it relies on the theory of analytic functions and on the use of the saddle-point method.

We hope that this presentation will be suited to the arithmetically minded mathematicians for which the summer school was devised, and also to a wider circle

---

[2]This specialization arises from taking the measure space $\mathscr{E}$ to be the set E of lattice points of $\overline{E}$ equipped with the counting measure, and the function H to be a multiple of $\|.\|^2$.

of mathematicians and theoretical physicists with some interest in Euclidean lattices or in the mathematical foundations of statistical physics.

1.5 During the preparation of these notes, I benefited from the support of the ERC project AlgTateGro, supervised by François Charles (Horizon 2020 Research and Innovation Programme, grant agreement No 715747).

I warmly thank the referees for their careful reading of preliminary versions of these notes.

2 Euclidean Lattices

2.1 Un peu d'histoire

Let V be a finite dimensional vector space. A *lattice* Λ in V is a discrete subgroup of V such that the quotient topological group V/Λ is compact, or equivalently, such that there exists some $\mathbb{R}$-basis $(e_i)_{1 \leqslant i \leqslant n}$ of V such that $\Lambda = \bigoplus_{i=1}^{n} \mathbb{Z} e_i$. The $\mathbb{R}$ vector space V is then canonically isomorphic to $\Lambda_{\mathbb{R}} := \Lambda \otimes \mathbb{R}$.

A *Euclidean lattice* is the data $(V, \Lambda, \|.\|)$ of some finite dimensional $\mathbb{R}$-vector space V, equipped with some Euclidean norm $\|.\|$, and of some lattice Λ in V.

Equivalently, it is the data $\overline{E} := (E, \|.\|)$ of some free $\mathbb{Z}$-module of finite rank E, and of some Euclidean norm $\|.\|$ on the $\mathbb{R}$-vector space $E_{\mathbb{R}} := E \otimes \mathbb{R}$. (The $\mathbb{Z}$-module E will always be identified to its image by the injective morphism $(E \hookrightarrow E_{\mathbb{R}}, v \mapsto v \otimes 1)$. This image is a lattice in $E_{\mathbb{R}}$.)

Three-dimensional Euclidean lattices constitute a mathematical model for the spatial organization of atoms or molecules in a crystal and for this reason have been investigated since the seventeenth century (notably by Huyghens in his *Traité de la lumière*, published in 1690). At the end of the eighteenth century, the development of number theory led to the study of Euclidean lattices in a purely mathematical perspective: Lagrange, in his work on integral quadratic forms in two variables, considered two dimensional Euclidean lattices and their reduction properties; the investigation of integral quadratic forms in an arbitrary number of indeterminates led Gauss and then Hermite to study Euclidean lattices of rank three, and then of arbitrary rank.

At the beginning of the twentieth century, the study of Euclidean lattices had become a full fledged domain of pure mathematics, after major contributions of Korkin, Zolotarev, Minkowski (who introduced the terminology of *geometry of numbers* for the study of triples $(V, \Lambda, \|.\|)$ as above, with the norm $\|.\|$ non necessarily Euclidean), and Voronoi. We refer the reader to the books and surveys articles [13, 66, 45], and [49] for presentations of the classical results of this theory.

2.2   The Classical Invariants of Euclidean Lattices

There is an obvious notion of *isomorphism* between Euclidean lattices: an iso-morphism between $\overline{E}_1 := (E_1, \|.\|_1)$ and $\overline{E}_2 := (E_2, \|.\|_2)$ is an isomorphism $\varphi : E_1 \xrightarrow{\sim} E_2$ of $\mathbb{Z}$-modules such that the attached isomorphism of $\mathbb{R}$-vector spaces $\varphi_{\mathbb{R}} : E_{1,\mathbb{R}} \xrightarrow{\sim} E_{2,\mathbb{R}}$ is an isometry between $(E_{1,\mathbb{R}}, \|.\|_1)$ and $(E_{2,\mathbb{R}}, \|.\|_2)$.

To some Euclidean lattice $\overline{E} := (E, \|.\|)$ are classically attached the following invariants, which depend only of its isomorphism class:

– its *rank*:

$$\mathrm{rk}\, E = \dim_{\mathbb{R}} E_{\mathbb{R}} \in \mathbb{N};$$

– its *covolume*: if $m_{\overline{E}}$ denotes the Lebesgue measure[3] on the Euclidean vector space $(E_{\mathbb{R}}, \|.\|)$ and if Δ is a fundamental domain[4] for E acting by translation on $E_{\mathbb{R}}$, the covolume of $\overline{E}$ is defined as

$$\mathrm{covol}(\overline{E}) := m_{\overline{E}}(\Delta) \in \mathbb{R}_+^*.$$

Observe that $\mathrm{covol}(\overline{E}) = 1$ when $\mathrm{rk}\, E = 0$.
– its *first minimum*, when $\mathrm{rk}\, E > 0$:

$$\lambda_1(\overline{E}) := \min_{e \in E \setminus \{0\}} \|e\| \in \mathbb{R}_+^*.$$

More generally, one defines the *successive minima* $(\lambda_i(\overline{E}))_{1 \leqslant i \leqslant \mathrm{rk}\, E}$ of $\overline{E}$ as follows: $\lambda_i(\overline{E})$ is the minimum real number r such that $E \cap \overline{B}_{\|.\|}(0, r)$ contains i $\mathbb{R}$-linearly independent elements, where $\overline{B}_{\|.\|}(0, r)$ denotes the closed ball of center 0 and radius r in the Euclidean vector space $(E_{\mathbb{R}}, \|.\|)$.
– its *covering radius*, when $\mathrm{rk}\, E > 0$:

$$R_{\mathrm{cov}}(\overline{E}) := \max_{x \in E_{\mathbb{R}}} \min_{e \in E} \|x - e\| = \min\{r \in \mathbb{R}_+, E + \overline{B}_{\|.\|}(0, r) = E_{\mathbb{R}}\}.$$

Many results of the theory of Euclidean lattices may be stated as inequalities relating these divers invariants.

---

[3]It is defined as the unique translation invariant Borel measure on $E_{\mathbb{R}}$ such that $m_{\overline{E}}(\sum_{i=1}^{n}[0, 1[v_i) = 1$ for any orthonormal basis $(v_i)_{1 \leqslant i \leqslant n}$ of the Euclidean vector space $(E_{\mathbb{R}}, \|.\|)$. An equivalent normalization condition is the following one: $\int_{E_{\mathbb{R}}} e^{-\pi\|x\|^2} dm_{\overline{E}}(x) = 1$.
[4]Namely, a Borel subset of $E_{\mathbb{R}}$ such that $(\Delta+e)_{e \in E}$ is a partition of $E_{\mathbb{R}}$. One easily establishes that such a fundamental domain Δ exists and that the measure $m_{\overline{E}}(\Delta)$ does not depend of the choice of Δ.

For instance, a classical result, which goes back to Hermite and plays a central role in algebraic theory of numbers, is the following estimate for the first minimum of some Euclidean lattice in terms of its covolume:

Theorem 2.2.1 (Hermite, Minkowski) *For any integer $n > 0$, there exists $C(n)$ in $\mathbb{R}_+^*$ such that, for any Euclidean lattice $\overline{E}$ of rank n,*

$$\lambda_1(\overline{E}) \leqslant C(n)(\operatorname{covol}(\overline{E}))^{1/n}. \tag{2.1}$$

If we denote the Lebesgue measure of the unit ball in $\mathbb{R}^n$ by v_n, this holds with:

$$C(n) = 2v_n^{-1/n}. \tag{2.2}$$

Since $v_n = \pi^{n/2}/\Gamma(n/2 + 1)$, it follows from Stirling's formula that, when n goes to $+\infty$, this value of $C(n)$ admits the following asymptotics:

$$2v_n^{-1/n} \sim \sqrt{2n/e\pi}. \tag{2.3}$$

Hermite has proved this theorem by induction on the rank n, by developing what is known as *reduction theory* for Euclidean lattices of arbitrary rank. We present a modernized version of Hermite's arguments in Sect. 3 below. These arguments allowed him to establish the estimate (2.1) with

$$C(n) = (4/3)^{(n-1)/4}.$$

(see Sect. 3.1, *infra*).

In his *Geometrie der Zahlen* ([53], p. 73–76), Minkowski has given a new elegant proof of Hermite's estimate which leads to the value (2.2) for $C(n)$ and admits a simple physical interpretation. Let us think of the Euclidean lattice as a model for a crystal in the n-dimensional Euclidean space $(E_{\mathbb{R}}, \|.\|)$: the molecules in this crystal are represented by the points of the lattice E. As the open balls $\mathring{B}_{\|.\|}(v, \lambda_1(\overline{E})/2)$ of radius $\lambda_1(\overline{E})/2$ centered at these points are pairwise disjoint, the density of the crystal—defined as the number of its molecules per unit of volume—is at most the inverse of the volume of any of these balls, which is

$$v_n(\lambda_1(\overline{E})/2)^n.$$

This density is nothing but the inverse of the covolume of $\overline{E}$. Therefore:

$$\operatorname{covol}(\overline{E})^{-1} \leqslant [v_n(\lambda_1(\overline{E})/2)^n]^{-1}.$$

This estimate is precisely (2.1) with $C(n)$ given by (2.2).

Similarly, by observing that the ball $\overline{B}_{\|.\|}(0, R_{\mathrm{cov}}(\overline{E}))$ contains some fundamental domain for the action of E over $E_{\mathbb{R}}$, we obtain:

$$v_n R_{\mathrm{cov}}(\overline{E})^n \geqslant \mathrm{covol}(\overline{E}),$$

or equivalently:

$$R_{\mathrm{cov}}(\overline{E}) \geqslant v_n^{-1/n} \, \mathrm{covol}(\overline{E})^{1/n}. \tag{2.4}$$

The square $\gamma_n = C(n)^2$ of the best constant in Hermite's inequality (2.1) is classically known as the *Hermite constant*. Its exact value is known for small values of n only (see [17, 16]). However Minkowski has proved that the asymptotic estimate $\gamma_n = O(n)$, which follows (2.3), is essentially optimal—namely, when n goes to $+\infty$, we have:

$$\log \gamma_n = \log n + O(1).$$

By comparison, Hermite's arguments based on reduction theory lead to the weaker estimate:

$$\log \gamma_n \leqslant (n-1) \log \sqrt{4/3}.$$

The previous discussion exemplifies a major theme of the theory of Euclidean lattices, since the investigation by Hermite and his followers Korkin and Zolotarev of Euclidean lattices of arbitrary rank: the study of the "best constants" appearing in the estimates relating invariants of Euclidean lattices, and notably the determination of their asymptotic behavior when this rank goes to infinity.

2.3 Euclidean Lattices as Hermitian Vector Bundles Over Spec $\mathbb{Z}$

In this paragraph, we introduce a few additional definitions concerning Euclidean lattices, which are less classical than the ones discussed in Sect. 2.2 above, although they are still quite elementary. These definitions naturally arise from the perspective of Arakelov geometry, where Euclidean lattices occur as an instance of the so-called *Hermitian vector bundles* over some regular $\mathbb{Z}$-scheme of finite type $\mathscr{X}$, in the special case $\mathscr{X} = \mathrm{Spec}\,\mathbb{Z}$.

2.3.1 Short Exact Sequences and Duality
Let us consider a Euclidean lattice $\overline{E} := (E, \|.\|)$.

For any $\mathbb{Z}$-submodule F of E, the inclusion morphism $F \hookrightarrow E$ defines, by extension of scalars, a canonical injection $F_{\mathbb{R}} \hookrightarrow E_{\mathbb{R}}$. Equipped with the restriction to $F_{\mathbb{R}}$ of the norm $\|.\|$, the submodule F (which is also a free $\mathbb{Z}$-module of finite rank) defines some Euclidean lattice:

$$\overline{F} := (F, \|.\|_{|F_{\mathbb{R}}}).$$

If moreover F is *saturated* in E—namely, if the $\mathbb{Z}$-module E/F is torsion-free, or equivalently, if $F = F_{\mathbb{R}} \cap E$—then E/F is a free $\mathbb{Z}$-module of finite rank. Moreover the exact sequence

$$0 \longrightarrow F \xrightarrow{i} E \xrightarrow{p} E/F \longrightarrow 0$$

(where we denote by i and p the inclusion and quotient morphisms) becomes, by extension of scalars, a short exact sequence of $\mathbb{R}$-vector spaces:

$$0 \longrightarrow F_{\mathbb{R}} \xrightarrow{i_{\mathbb{R}}} E_{\mathbb{R}} \xrightarrow{p_{\mathbb{R}}} (E/F)_{\mathbb{R}} \longrightarrow 0.$$

Accordingly the $\mathbb{R}$-vector space $(E/F)_{\mathbb{R}}$ may be identified with the quotient of $E_{\mathbb{R}}$ by $F_{\mathbb{R}}$. In particular it may be equipped with the quotient Euclidean norm $\|.\|_{\text{quot}}$ induced by the Euclidean norm $\|.\|$ on $E_{\mathbb{R}}$. This defines the Euclidean lattice

$$\overline{E/F} := (E/F, \|.\|_{\text{quot}}).$$

With the previous notation, we shall say that the diagram

$$0 \longrightarrow \overline{F} \xrightarrow{i} \overline{E} \xrightarrow{p} \overline{E/F} \longrightarrow 0 \qquad (2.5)$$

is an *admissible short exact sequence* of Euclidean lattices.

Let us observe that any saturated $\mathbb{Z}$-submodule F in E is determined by the $\mathbb{R}$-vector subspace $F_{\mathbb{R}}$ in $E_{\mathbb{R}}$, and also by the $\mathbb{Q}$-vector subspace $F_{\mathbb{Q}} := F \otimes \mathbb{Q}$ of $E_{\mathbb{Q}} := E \otimes \mathbb{Q}$, since

$$F = F_{\mathbb{R}} \cap E = F_{\mathbb{Q}} \cap E.$$

The map $(F \mapsto F_{\mathbb{Q}})$ indeed establishes a bijection between the sets of saturated $\mathbb{Z}$-submodules of E and of $\mathbb{Q}$-vector subspaces of $E_{\mathbb{Q}}$.

Besides, to any Euclidean lattice $\overline{E} := (E, \|.\|)$ is attached its *dual Euclidean lattice*

$$\overline{E}^{\vee} := (E^{\vee}, \|.\|^{\vee})$$

defined as follows.

Its underlying $\mathbb{Z}$-module $E^{\vee}$ is the dual $\mathbb{Z}$-module dual of E,

$$E^{\vee} := \text{Hom}_{\mathbb{Z}}(E, \mathbb{Z}),$$

which a free $\mathbb{Z}$-module of the same rank as E. The $\mathbb{R}$-vector space $(E^\vee)_\mathbb{R} := E^\vee \otimes \mathbb{R}$ may be identified with $(E_\mathbb{R})^\vee := \operatorname{Hom}_\mathbb{R}(E_\mathbb{R}, \mathbb{R})$; we shall denote it by $E_\mathbb{R}^\vee$. The Euclidean norm $\|.\|^\vee$ is defined as the norm dual of the norm $\|.\|$ on $E_\mathbb{R}$. In other words, for any $\xi \in E_\mathbb{R}^\vee$,

$$\|\xi\|^\vee := \max\{|\xi(x)|; x \in \overline{B}_{\|.\|}(0, 1)\}.$$

There is a canonical biduality isomorphism:

$$\overline{E} \xrightarrow{\sim} \overline{E}^{\vee\vee}.$$

Moreover an admissible short exact sequence (2.5) of Euclidean lattices defines, by duality, a diagram

$$0 \longrightarrow \overline{E/F}^\vee \xrightarrow{{}^t i} \overline{E}^\vee \xrightarrow{{}^t p} \overline{F}^\vee \longrightarrow 0$$

which may be identified with the admissible short exact sequence

$$0 \longrightarrow \overline{F}^\perp \longrightarrow \overline{E}^\vee \longrightarrow \overline{E^\vee/F^\perp} \longrightarrow 0$$

attached to the saturated $\mathbb{Z}$-submodule

$$F^\perp := \{\xi \in E^\vee \mid \xi_{|F} = 0\}$$

in $E^\vee$. Actually the map $(F \mapsto F^\perp)$ establishes a bijection between the set of saturated submodules of E and of $E^\vee$.

2.3.2 Arakelov Degree and Slope Instead of its covolume, it is often more convenient to use the *Arakelov degree* of some Euclidean lattice $\overline{E}$, defined as the logarithm of its "density" $\operatorname{covol}(\overline{E})^{-1}$:

$$\widehat{\deg} \overline{E} := -\log \operatorname{covol}(\overline{E}), \tag{2.6}$$

and, when $\operatorname{rk} E > 0$, its *slope*

$$\widehat{\mu}(\overline{E}) := \frac{\widehat{\deg} \overline{E}}{\operatorname{rk} E} = \log(\operatorname{covol}(\overline{E})^{-1/\operatorname{rk} E}). \tag{2.7}$$

For instance, one easily sees that, for any admissible short exact sequence (2.5) of Euclidean lattices, the covolumes of $\overline{E}$, $\overline{F}$ and $\overline{E/F}$ satisfy :

$$\operatorname{covol}(\overline{E}) = \operatorname{covol}(\overline{F}) . \operatorname{covol}(\overline{E/F}). \tag{2.8}$$

Consequently their Arakelov degrees satisfy the additivity property:

$$\widehat{\deg}\,\overline{E} = \widehat{\deg}\,\overline{F} + \widehat{\deg}\,\overline{E/F}, \tag{2.9}$$

similar to the one satisfied by their rank:

$$\operatorname{rk} E = \operatorname{rk} F + \operatorname{rk} E/F.$$

In the same vein, the covolumes of some Euclidean lattice $\overline{E}$ and of its dual $\overline{E}^{\vee}$ satisfy the relation:

$$\operatorname{covol}(\overline{E}^{\vee}) = \operatorname{covol}(\overline{E})^{-1},$$

which may also be written as:

$$\widehat{\deg}\,\overline{E}^{\vee} = -\widehat{\deg}\,\overline{E}.$$

2.3.3 Operations on Euclidean Lattices

The operations of direct sum and of tensor product on $\mathbb{Z}$-modules on Euclidean $\mathbb{R}$-vector spaces allow one to define similar operations on Euclidean lattices.

For instance, if $\overline{E}_1 := (E_1, \|.\|_1)$ and $\overline{E}_2 := (E_2, \|.\|_2)$ are two Euclidean lattices, we let:

$$\overline{E}_1 \oplus \overline{E}_2 := (E_1 \oplus E_2, \|.\|_{\oplus}) \quad \text{et} \quad \overline{E}_1 \otimes \overline{E}_2 := (E_1 \otimes E_2, \|.\|_{\otimes}),$$

where the Euclidean norm $\|.\|_{\oplus}$ on $(E_1 \oplus E_2)_{\mathbb{R}} \simeq E_{1,\mathbb{R}} \oplus E_{2,\mathbb{R}}$ is defined by

$$\|x_1 \oplus x_2\|_{\oplus}^2 := \|x_1\|_1^2 + \|x_2\|_2^2,$$

and where the norm $\|.\|_{\otimes}$ sur $(E_1 \otimes E_2)_{\mathbb{R}} \simeq E_{1,\mathbb{R}} \otimes_{\mathbb{R}} E_{2,\mathbb{R}}$ is characterized by the following property: for any orthonormal basis $(e_{1\alpha})_{1 \leqslant \alpha \leqslant n_1}$ (resp. $(e_{2\beta})_{1 \leqslant \beta \leqslant n_2}$) of the Euclidean space $(E_{1,\mathbb{R}}, \|.\|_1)$ (resp. of $(E_{2,\mathbb{R}}, \|.\|_2)$), $(e_{1\alpha} \otimes e_{2\beta})_{1 \leqslant \alpha, \beta \leqslant n_1, n_2}$ is an orthonormal basis of $(E_{1,\mathbb{R}} \otimes_{\mathbb{R}} E_{2,\mathbb{R}}, \|.\|_{\otimes})$. (See Chapter II, Sect. 2, for a discussion of these constructions in the more general setting of "adelic spaces.")

The canonical inclusion $i : E_1 \longrightarrow E_1 \oplus E_2$ and projection $p : E_1 \oplus E_2 \longrightarrow E_2$ make the diagram

$$0 \longrightarrow \overline{E}_1 \overset{i}{\longrightarrow} \overline{E}_1 \oplus \overline{E}_2 \overset{p}{\longrightarrow} \overline{E}_2 \longrightarrow 0 \tag{2.10}$$

an admissible short exact sequence of Euclidean lattices.[5]

---

[5]One should beware that, in general, an admissible short exact sequence of Euclidean lattice *is not* isomorphic to an exact sequence of the form (2.10): the obstruction for the admissible short exact sequence (2.5) to be *split*, that is isomorphic to an admissible short exact sequence of the

In particular, as a special case of (2.9), we have:

$$\widehat{\deg}\,(\overline{E}_1 \oplus \overline{E}_2) = \widehat{\deg}\,\overline{E}_1 + \widehat{\deg}\,\overline{E}_2.$$

For any $t \in \mathbb{R}$, we define the Euclidean lattice of rank 1:

$$\overline{\mathscr{O}}(t) := (\mathbb{Z}, \|.\|_t),$$

where $\|.\|_t$ denotes the norm over $\mathbb{Z}_{\mathbb{R}} = \mathbb{R}$ defined by

$$\|x\|_t := e^{-t}|x|.$$

It is straightforward that

$$\widehat{\deg}\,\overline{\mathscr{O}}(t) = t$$

and that any Euclidean lattice $\overline{L}$ of rank 1 is isomorphic to $\overline{\mathscr{O}}(t)$ where $t := \widehat{\deg}\,\overline{L}$. Moreover, for any Euclidean lattice $\overline{E} := (E, \|.\|)$, the tensor product $\overline{E} \otimes \overline{\mathscr{O}}(t)$ may be identified to the Euclidean lattice $(E, e^{-t}\|.\|)$, deduced from $\overline{E}$ by scaling its norm by e^{-t}.

2.3.4 Example: Direct Sums of Euclidean Lattices of Rank 1 The invariants of Euclidean lattices direct sums of Euclidean lattices of rank 1 are easily computed. Let us indeed consider the Euclidean lattice

$$\overline{E} := \bigoplus_{i=1}^{n} \overline{\mathscr{O}}(t_i),$$

for some positive integer n and some non-increasing sequence $t_1 \geqslant \cdots \geqslant t_n$ of n real numbers. One easily computes:

$$\widehat{\deg}\,\overline{E} = t_1 + \ldots + t_n \quad \text{and} \quad \widehat{\mu}(\overline{E}) = \frac{t_1 + \ldots + t_n}{n},$$

$$\lambda_i(\overline{E}) = e^{-t_i} \quad \text{for any } i \in \{1, \ldots, n\}, \tag{2.11}$$

and

$$R_{\mathrm{cov}}(\overline{E}) = (1/2) \left(\sum_{i=1}^{n} e^{-2t_i} \right)^{1/2}.$$

---

form (2.10), is an element of some extension group attached to the Euclidean lattices $\overline{E}$ et $\overline{E/F}$, the properties of which are closely related to reduction theory; see [11].

This notably implies:

$$R_{cov}(\overline{E}) \in [(1/2)e^{-t_n}, (\sqrt{n}/2)e^{-t_n}]. \qquad (2.12)$$

Moreover, we have:

$$\overline{E}^{\vee} \simeq \bigoplus_{i=1}^{n} \overline{\mathscr{O}}(-t_i),$$

and this is easily seen to imply:

$$\lambda_i(\overline{E}^{\vee}) = e^{t_{n+1-i}} \quad \text{for every } i \in \{1, \ldots, n\}.$$

The relation (2.12) may therefore be written:

$$R_{cov}(\overline{E})\,\lambda_1(\overline{E}^{\vee}) \in [1/2, \sqrt{n}/2]. \qquad (2.13)$$

3 Reduction Theory for Euclidean Lattices

In this section, using the geometric language introduced in Sect. 2.3 above, we present a basic result of reduction theory. Namely we show that that any Euclidean lattice $\overline{E}$ of rank n may be "approximated" by some Euclidean lattice that is the direct sum $\overline{L}_1 \oplus \ldots \oplus \overline{L}_n$ of Euclidean lattices $\overline{L}_1, \cdots, \overline{L}_n$ of rank one, with an "error" controlled in terms of n, and accordingly is approximately determined by the n real numbers $\mu_i := \widehat{\deg}\,\overline{L}_i$, $1 \leqslant i \leqslant n$ (see Theorem 3.1.1 infra for a precise statement).

The derivation of this result in Sect. 3.1 below is nothing but a reformulation of some classical arguments that go back to Hermite, Korkin and Zolotarev. We believe that the geometric point of view used here—notably the notion of admissible short exact sequences of Euclidean lattices—makes the proof more transparent and demonstrates the conceptual interest of a more geometric approach.

As the successive minima or the covering radius of the direct sum $\overline{L}_1 \oplus \cdots \oplus \overline{L}_n$ and the dual Euclidean lattice $\overline{L}_1^{\vee} \oplus \cdots \oplus \overline{L}_n^{\vee}$ are simple functions of $(\mu_1, \cdots, \mu_n)$, our reduction theorem easily implies some "transference inequalities" that relates the above invariants of some Euclidean lattice $\overline{E}$ and of its dual $\overline{E}^{\vee}$.

The fact that the properties of some Euclidean lattice $\overline{E}$ of rank n are (approximately) controlled by the n real numbers $(\mu_1, \cdots, \mu_n)$, already demonstrated by this basic discussion of reduction theory, is a forerunner of the role of the so-called *slopes* $(\widehat{\mu}_1(\overline{E}), \ldots, \widehat{\mu}_n(\overline{E}))$, a non-increasing sequence of n real numbers associated by Stuhler to the Euclidean lattice $\overline{E}$ (see [72, 32], Chapter II and Chapter V). We refer the reader to [10] for a discussion and references concerning slopes of Euclidean lattices and recent advances on their properties.

3.1 A Theorem of Hermite, Korkin and Zolotarev

In substance, the following theorem appears in some letters of Hermite to Jacobi (see [37]). A streamlined version of Hermite's arguments appears in the work of Korkin and Zolotarev [43, pp. 370–373], and we gave below a geometric rendering of their proof, using the formalism introduced in the previous section.

Theorem 3.1.1 *For any positive integer n, there exists $D(n) \in \mathbb{R}_+^*$ such that, for any Euclidean lattice $\overline{E} := (E, \|.\|)$ of rank n, the $\mathbb{Z}$-module E admits some $\mathbb{Z}$-base $(v_1, \ldots, v_n)$ such that*

$$\prod_{i=1}^n \|v_i\| \leqslant D(n)\operatorname{covol}(\overline{E}). \tag{3.1}$$

This indeed holds with :

$$D(n) = (4/3)^{n(n-1)/4}. \tag{3.2}$$

Observe also that, with the notation of Theorem 3.1.1, we immediately obtain:

$$\lambda_1(\overline{E}) \leqslant (\prod_{i=1}^n \|v_i\|)^{1/n} \leqslant D(n)^{1/n}\operatorname{covol}(\overline{E})^{1/n}.$$

In this way, we recover Hermite's inequality (2.1), with

$$C(n) = D(n)^{1/n} = (4/3)^{(n-1)/4}.$$

(Compare with [37, pages 263–265 and 279–283]).

Proof The theorem is established by induction on the integer n.

Let $\overline{E}$ be a Euclidean lattice of rank $n > 0$. Let us choose some element $s \in E$ such that $\|s\| = \lambda_1(\overline{E})$. The submodule $\mathbb{Z}s$ is easily seen to be saturated in E.

If $n = 1$, then $E = \mathbb{Z}s$. In this case,

$$\operatorname{covol}(\overline{E}) = \lambda_1(\overline{E})$$

and the estimate (3.1) is satisfied by $v_1 := s$ and $D(1) = 1$.

When $n > 1$, we may consider the quotient Euclidean lattice

$$\overline{E/\mathbb{Z}s} := (E/\mathbb{Z}s, \|.\|_{\text{quot}}),$$

of rank $n - 1$. By induction, there exists some basis $(w_1, \ldots, w_{n-1})$ of $E/\mathbb{Z}s$ such that

$$\prod_{i=1}^{n-1} \|w_i\|_{\text{quot}} \leqslant D(n - 1) \operatorname{covol}(\overline{E/\mathbb{Z}s}).$$ (3.3)

If, for any $i \in \{0, \ldots, n - 1\}$, we choose some element v_i in the inverse image $p^{-1}(w_i)$ of w_i by the quotient map

$$p : E \longrightarrow E/\mathbb{Z}s$$

and if we let $v_n := s$, then $(v_1, \ldots, v_n)$ is a $\mathbb{Z}$-basis of E. Moreover, for any $i \in \{1, \ldots, n - 1\}$, we may choose for v_i an element of $p^{-1}(w_i)$ of minimal norm. Then we have:

$$\|v_i\| \leqslant \|v_i - s\| \quad \text{et} \quad \|v_i\| \leqslant \|v_i + s\|.$$ (3.4)

Besides, by the very definition of $\lambda_1(\overline{E})$, we also have:

$$\|v_i\| \geqslant \lambda_1(\overline{E}) = \|s\|.$$ (3.5)

Let us consider the element $v_i^{\perp}$ in $p_{\mathbb{R}}^{-1}(w_i)$ orthogonal to s. By definition of $\|.\|_{\text{quot}}$, we have:

$$\|v_i^{\perp}\| = \|w_i\|_{\text{quot}}.$$ (3.6)

Moreover we may write:

$$v_i = v_i^{\perp} + \eta_i s$$

for some $\eta_i \in \mathbb{R}$. Then we have:

$$\|v_i\|^2 = \|v_i^{\perp}\|^2 + \eta_i^2 \|s\|^2,$$

and similarly:

$$\|v_i - s\|^2 = \|v_i^{\perp}\|^2 + (\eta_i - 1)^2 \|s\|^2$$

and

$$\|v_i + s\|^2 = \|v_i^{\perp}\|^2 + (\eta_i + 1)^2 \|s\|^2$$

The conditions (3.4) may therefore be rephrased as

$$\eta_i^2 \leqslant \min((\eta_i - 1)^2, (\eta_i + 1)^2),$$

or equivalently as

$$|\eta_i| \leqslant 1/2.$$

This implies:

$$\|v_i\|^2 \leqslant \|v_i^\perp\|^2 + (1/4)\|s\|^2,$$

and finally, by taking (3.5) and (3.6) into account:

$$\|v_i\|^2 \leqslant (4/3)\|w_i\|_{\text{quot}}^2. \tag{3.7}$$

The estimates (3.7) and (3.3), together with the mutiplicativity (2.8) of the covolume, show that:

$$\prod_{i=1}^{n} \|v_i\| \leqslant (4/3)^{(n-1)/2} \prod_{i=1}^{n-1} \|w_i\|_{\text{quot}} \cdot \|s\|$$

$$\leqslant (4/3)^{(n-1)/2} D(n-1) \operatorname{covol}(\overline{E/\mathbb{Z}s}) \cdot \operatorname{covol}(\overline{\mathbb{Z}s})$$

$$= (4/3)^{(n-1)/2} D(n-1) \operatorname{covol}(\overline{E}).$$

This establishes the existence of some $\mathbb{Z}$-basis $(v_1, \ldots, v_n)$ of E that satisfies the inequality (3.1) with

$$D(n) = (4/3)^{(n-1)/2} D(n-1),$$

and finally with $D(n)$ given by (3.2).   □

The previous proof actually provides an algorithm[6] for constructing the basis $(v_1, \ldots, v_n)$. Bases obtained by this algorithm are called *Korkin-Zolotarev reduced* (see for instance [46]).

3.2 Complements

In applications, it is convenient to combine Theorem 3.1.1 with the following observations.

3.2.1 Non-isometric Isomorphisms and Invariant of Euclidean Lattices Let $\overline{E} := (E, \|.\|)$ and $\overline{E}' := (E', \|.\|')$ be two Euclidean lattices of the same rank

---

[6]Provided algorithms for finding some vector of shortest positive norm in a Euclidean lattice, etc., are known.

n and let

$$\varphi : E \xrightarrow{\sim} E'$$

be an isomorphism between the underlying $\mathbb{Z}$-modules.

The map

$$\varphi_{\mathbb{R}} := \varphi \otimes Id_{\mathbb{R}} : E_{\mathbb{R}} \longrightarrow E'_{\mathbb{R}}$$

is an isomorphism of $\mathbb{R}$-vector spaces, but is not necessary an isometry between the Euclidean vector spaces $(E_{\mathbb{R}}, \|.\|)$ and $(E'_{\mathbb{R}}, \|.\|')$. The "lack of isometry" of $\varphi_{\mathbb{R}}$ is measured by the operator norms $\|\varphi_{\mathbb{R}}\|$ and $\|\varphi_{\mathbb{R}}^{-1}\|$ defined by means of the norms $\|.\|$ and $\|.\|'$ on $E_{\mathbb{R}}$ and $E'_{\mathbb{R}}$. Indeed one easily sees, by unwinding the definitions, that the covolume, the successive minima, or the covering radius of $\overline{E}$ and $\overline{E}'$ may be compared, with some error terms controlled by these operator norms:

$$\|\varphi_{\mathbb{R}}^{-1}\|^{-n} \leqslant \frac{\mathrm{covol}(\overline{E}')}{\mathrm{covol}(\overline{E})} = \|\Lambda^n \varphi_{\mathbb{R}}\| \leqslant \|\varphi_{\mathbb{R}}\|^n,$$

$$\|\varphi_{\mathbb{R}}^{-1}\|^{-1} \leqslant \frac{\lambda_i(\overline{E}')}{\lambda_i(\overline{E})} \leqslant \|\varphi_{\mathbb{R}}\| \quad \text{for any } i \in \{1, \ldots, n\},$$

$$\|\varphi_{\mathbb{R}}^{-1}\|^{-1} \leqslant \frac{R_{\mathrm{cov}}(\overline{E}')}{R_{\mathrm{cov}}(\overline{E})} \leqslant \|\varphi_{\mathbb{R}}\|.$$

These estimates may be reformulated as follows:

Proposition 3.2.1 *If by ψ we denote any of the invariants $\widehat{\mu}$, $\log \lambda_i^{-1}$, or $\log R_{\mathrm{cov}}^{-1}$, we have:*

$$- \log \|\varphi_{\mathbb{R}}\| \leqslant \psi(\overline{E}') - \psi(\overline{E}) \leqslant \log \|\varphi_{\mathbb{R}}^{-1}\|. \tag{3.8}$$

Notably, for any $\lambda \in \mathbb{R}$, we have:

$$\psi(\overline{E} \otimes \overline{\mathcal{O}}(\lambda)) = \psi(\overline{E}) + \lambda. \tag{3.9}$$

3.2.2 Reduction Theory and Norms of Sum Maps Let us consider some Euclidean lattice $\overline{E}$ of rank $n > 0$, and let $L_1, \ldots, L_n$ be some $\mathbb{Z}$-submodules of rank 1 in E such that the $\mathbb{Z}$-module E is the direct sum of $L_1, \ldots, L_n$.

We may introduce the "sum map"

$$\Sigma : L_1 \oplus \ldots \oplus L_n \xrightarrow{\sim} E$$

and the Euclidean lattice $\overline{L}_1 \oplus \ldots \oplus \overline{L}_n$, and consider the operator norms $\|\Sigma_{\mathbb{R}}\|$, $\|\Lambda^n \Sigma_{\mathbb{R}}\|$ and $\|\Sigma_{\mathbb{R}}^{-1}\|$ defined by means of the Euclidean structures on $\overline{L}_1 \oplus \ldots \oplus \overline{L}_n$ and on $\overline{E}$.

Finally, we may define:

$$\delta(\overline{E}; L_1, \ldots, L_n) := \widehat{\mu}(\overline{E}) - \frac{1}{n} \sum_{i=1}^{n} \widehat{\deg}\, \overline{L}_i = \widehat{\mu}(\overline{E}) - \widehat{\mu}(\overline{L}_1 \oplus \ldots \oplus \overline{L}_n).$$

Proposition 3.2.2 *With the previous notation, we have:*

$$\delta(\overline{E}; L_1, \ldots, L_n) = -\frac{1}{n} \log \|\Lambda^n \Sigma_{\mathbb{R}}\| \geqslant 0, \tag{3.10}$$

$$\log \|\Sigma_{\mathbb{R}}\| \leqslant (1/2) \log n, \tag{3.11}$$

and

$$\log \|\Sigma_{\mathbb{R}}^{-1}\| \leqslant \frac{n-1}{2} \log n + n\delta(\overline{E}; L_1, \ldots, L_n). \tag{3.12}$$

Proof The estimates (3.10) and (3.11) easily follow from the definitions and Hadamard's inequality. They imply (3.12) thanks to "Cramer's formula" applied to Σ^{-1}. Indeed the latter identifies Σ^{-1} and $\Lambda^{n-1}\Sigma \otimes (\Lambda^n \Sigma)^{-1}$ and shows that:

$$\log \|\Sigma_{\mathbb{R}}^{-1}\| = \log \|\Lambda^{n-1}\Sigma_{\mathbb{R}}\| - \log \|\Lambda^n \Sigma_{\mathbb{R}}\|$$

$$\leqslant (n-1) \log \|\Sigma_{\mathbb{R}}\| + n\delta(\overline{E}; L_1, \ldots, L_n). \qquad \square$$

In the situation of Theorem 3.1.1, we may apply Proposition 3.2.2 with $L_i := \mathbb{Z}v_i$ for $1 \leqslant i \leqslant n$. Then we have:

$$n\delta(\overline{E}; L_1, \ldots, L_n) = -\log \operatorname{covol} \overline{E} + \sum_{i=1}^{n} \log \|v_i\| \leqslant \log D(n),$$

and therefore:

$$\log \|\Sigma_{\mathbb{R}}^{-1}\| \leqslant \frac{n-1}{2} \log n + \log D(n).$$

3.3 An Application to Transference Inequalities

Let us keep the previous notation. By applying Proposition 3.2.2 to $\varphi = \Sigma$, we obtain that, if ψ denotes any of the invariants $\log \lambda_i^{-1}$ or $\log R_{\text{cov}}^{-1}$, then the following estimate holds:

$$-\frac{n-1}{2}\log n - \log D(n) \leqslant \psi\left(\bigoplus_{i=1}^{n} \overline{\mathbb{Z}v_i}\right) - \psi(\overline{E}) \leqslant (1/2)\log n. \qquad (3.13)$$

We may also apply Proposition 3.2.2 to the isomorphism

$$^{t}\Sigma : E^{\vee} \xrightarrow{\sim} \bigoplus L_i^{\vee},$$

and thus we obtain:

$$-(1/2)\log n \leqslant \psi\left(\bigoplus_{i=1}^{n} \overline{\mathbb{Z}v_i}^{\vee}\right) - \psi(\overline{E}^{\vee}) \leqslant \frac{n-1}{2}\log n + \log D(n). \qquad (3.14)$$

The computations of paragraph 2.3.4 allow us to compute the invariants λ_i or R_{cov} of the Euclidean lattices $\bigoplus_{i=1}^{n} \overline{\mathbb{Z}v_i}$ and $\bigoplus_{i=1}^{n} \overline{\mathbb{Z}v_i}^{\vee}$ in terms of the sequence $(t_i)_{1\leqslant i\leqslant n} := (\log \|v_i\|^{-1})_{1\leqslant i\leqslant n}$, where the $\|v_i\|$ are ordered increasingly. Together with the estimates (3.13) and (3.14) above, these expressions allow one to relate invariants of the Euclidean lattice $\overline{E}$ and of its dual $\overline{E}^{\vee}$.

For instance, in this way, we may derive the following comparison estimate between the covering radius of $\overline{E}$ and the first minimum of $\overline{E}^{\vee}$:

Corollary 3.3.1 *For any Euclidean lattice $\overline{E}$ of positive rank n, we have:*

$$\left|\log R_{\mathrm{cov}}(\overline{E}) + \log \lambda_1(\overline{E}^{\vee})\right| \leqslant F(n), \qquad (3.15)$$

where

$$F(n) = \frac{n+1}{2}\log n + \log D(n). \qquad (3.16)$$

We leave the details of the proof as an exercise.

Statements like Corollary 3.3.1, which relates the invariants of geometry of numbers attached to some Euclidean lattice and to its dual, are classically known as *transference theorems*.[7] As demonstrated in the above derivation of Corollary 3.3.1, reduction theory allows one to give simple proofs of such estimates, by reducing to the easy case of Euclidean lattices direct sums of Euclidean lattices of rank 1.

However the constants depending on the rank n of the Euclidean lattices under study—such as the constant $E(n)$ in (3.15)—that occur in transference estimates derived in this way turn out to be "very large". For instance, as we shall see in the next section, the optimal constant $E(n)$ in (3.15) is actually $\log n + O(1)$, while its upper bound (3.16) derived from Theorem 3.1.1 is of the order of n^2.

---

[7]Originally, *Übertragungssätze*; see for instance [13, Chapter XI].

4 Theta Series and Banaszczyk's Transference Estimates

In this section, we discuss the basic properties of the theta series associated to Euclidean lattices and the remarkable applications of these theta series, due to Banaszczyk ([5]), to the study of their classical invariants.

4.1 Poisson Formula and Theta Series of Euclidean Lattices

The notion of dual lattice plays a central role in crystallography, since the development of the investigation of crystalline structures by X-ray diffraction: the diffraction pattern obtained from a crystal modeled by some three dimensional Euclidean lattice $\overline{E}$ produces a picture of the dual lattice $\overline{E}^\vee$ (Ewald, von Laue, Bragg, 1912; see for instance [3, Chapter 6]). This is a physical expression of the *Poisson formula* attached to the Euclidean lattice $\overline{E}$. Let us recall its formulation, for some Euclidean lattice $\overline{E} := (E, \|.\|)$ of arbitrary rank n.

Let us first recall a few basic facts concerning the Fourier transform and the Poisson formula. (The reader may refer to [38, Sections 7.1 and 7.2], for a concise and elegant presentation of the basic properties of the Fourier transform in the spaces $\mathscr{S}(\mathbb{R}^n)$ and $\mathscr{S}'(\mathbb{R}^n)$ that emphasizes the role of the Poisson formula.)

The Fourier transform provides an isomorphism of topological vector spaces

$$\mathscr{F} : \mathscr{S}(E_\mathbb{R}) \xrightarrow{\sim} \mathscr{S}(E_\mathbb{R}^\vee)$$

between the Schwartz spaces of $E_\mathbb{R}$ and of its dual $\mathbb{R}$-vector space $E_\mathbb{R}^\vee$, defined by the following formula, valid for any $f \in \mathscr{S}(E_\mathbb{R})$ and any $\xi \in E_\mathbb{R}^\vee$:

$$\mathscr{F}(f)(\xi) := \int_{E_\mathbb{R}} f(x) e^{-2\pi i \langle \xi, x \rangle} dm_{\overline{E}}(x).$$

It extends to an isomorphism of topological vector spaces between spaces of tempered distributions:

$$\mathscr{F} : \mathscr{S}'(E_\mathbb{R}) \xrightarrow{\sim} \mathscr{S}'(E_\mathbb{R}^\vee).$$

Poisson formula asserts that the counting measures $\sum_{v \in E} \delta_v$ and $\sum_{\xi \in E^\vee} \delta_\xi$— which are tempered distributions on $E_\mathbb{R}$ and $E_\mathbb{R}^\vee$—may be deduced from each other by Fourier transform:

$$\mathscr{F}(\sum_{v \in E} \delta_v) = (\mathrm{covol}(\overline{E}))^{-1} \sum_{\xi \in E^\vee} \delta_\xi. \tag{4.1}$$

Equivalently it asserts that, for any $f \in \mathscr{S}(E_\mathbb{R})$ and any $x \in E_\mathbb{R}$, the following equality holds:

$$\sum_{v \in E} f(x - v) = (\mathrm{covol}(\overline{E}))^{-1} \sum_{\xi \in E^\vee} \mathscr{F}(f)(\xi) e^{2\pi i \langle \xi, x \rangle}. \tag{4.2}$$

This equality is nothing but the Fourier series expansion of the function $\sum_{v \in E} f(. - v)$, which is E-periodic on $E_\mathbb{R}$.

For any $t \in \mathbb{R}_+^*$, we may apply (4.2) to the function $f_t \in \mathscr{S}(E_\mathbb{R})$ defined as

$$f_t(x) := e^{-\pi t \|x\|^2};$$

its Fourier transform is:

$$(\mathscr{F}f_t)(\xi) = t^{-n/2} e^{-\pi t^{-1} \|\xi\|^2}.$$

We thus obtain the following equality, for any $x \in E_\mathbb{R}$:

$$\sum_{v \in E} e^{-\pi t \|x - v\|^2} = (\mathrm{covol}(\overline{E}))^{-1} t^{-n/2} \sum_{\xi \in E^\vee} e^{-\pi t^{-1} \|\xi\|^2 + 2\pi i \langle \xi, x \rangle}. \tag{4.3}$$

In particular, when $x = 0$, the Poisson formula (4.3) becomes:

$$\theta_{\overline{E}}(t) = (\mathrm{covol}(\overline{E}))^{-1} t^{-n/2} \theta_{\overline{E}^\vee}(t^{-1}), \tag{4.4}$$

where the *theta function* $\theta_{\overline{E}}$ associated to the Euclidean lattice is defined, for any $t \in \mathbb{R}_+^*$, by the series:

$$\theta_{\overline{E}}(t) := \sum_{v \in E} e^{-\pi t \|v\|^2}. \tag{4.5}$$

4.2   Banaszczyk's Transference Estimates

In 1993, in his article [5], Banaszczyk has established some remarkable transference estimates, concerning the successive minima and the covering radius:

Theorem 4.2.1 (Banaszczyk) *For any Euclidean lattice $\overline{E}$ of positive rank n and for any integer i in $\{1, \ldots, n\}$, the following estimate holds :*

$$\lambda_i(\overline{E}).\lambda_{n+1-i}(\overline{E}^\vee) \leqslant n. \tag{4.6}$$

Moreover,

$$R_{\mathrm{cov}}(\overline{E}).\lambda_1(\overline{E}^{\vee}) \leqslant n/2. \tag{4.7}$$

As observed by Banaszczyk, these estimates are optimal, up to some multiplicative error term, uniformly bounded when n varies. This follows from the existence, establishes by Conway and Thompson, of a sequence of Euclidean lattices $\overline{\mathrm{CT}}_n$ such that

$$\mathrm{rk}\,\overline{\mathrm{CT}}_n = n,$$

$$\overline{\mathrm{CT}}_n{}^{\vee} \xrightarrow{\sim} \overline{\mathrm{CT}}_n \tag{4.8}$$

and:

$$\lambda_1(\overline{\mathrm{CT}}_n) \geqslant \sqrt{n/2\pi e}\,(1 + o(n)) \quad \text{when } n \longrightarrow +\infty.$$

(See [52, Chapter II, Theorem 9.5]. The lattices $\overline{\mathrm{CT}}_n$ are actually integral unimodular lattices, and their existence follows from the Smith–Minkowski–Siegel mass formula.) The lattices $\overline{\mathrm{CT}}_n$ satisfy:

$$\lambda_1(\overline{\mathrm{CT}}_n).\lambda_n(\overline{\mathrm{CT}}_n{}^{\vee}) \geqslant \lambda_1(\overline{\mathrm{CT}}_n)^2 \geqslant (n/2\pi e)(1 + o(n)) \quad \text{when } n \longrightarrow +\infty.$$

Moreover, according to (4.8), we have :

$$\mathrm{covol}(\overline{\mathrm{CT}}_n) = 1$$

and therefore, according to (2.4) :

$$R_{\mathrm{cov}}(\overline{\mathrm{CT}}_n) \geqslant v_n^{-1/n} = \sqrt{n/2\pi e}\,(1 + o(n)) \quad \text{when } n \longrightarrow +\infty.$$

Consequently,

$$\lambda_1(\overline{\mathrm{CT}}_n).R_{\mathrm{cov}}(\overline{\mathrm{CT}}_n{}^{\vee}) \geqslant (n/2\pi e)(1 + o(n)) \quad \text{when } n \longrightarrow +\infty.$$

To prove Theorem 4.2.1, Banaszczyk introduces an original method, which relies on the analytic properties of the theta series (4.5) associated to Euclidean lattices and on the Poisson formula (4.3). Previous approaches to transference inequalities such as (4.6) and (4.7) did rely on reduction theory and, in their best version, led to estimates with constants of the order of $n^{3/2}$ instead of n (see for instance [46]).

The role of the theta series $\theta_{\overline{E}}$ associated to *integral* Euclidean lattices—namely, the Euclidean lattices $\overline{E}$ defined by some Euclidean scalar product that is $\mathbb{Z}$-valued on $E \times E$—does not need to be emphasized: for such lattices, the functions $\theta_{\overline{E}}$

define modular forms and, through this construction, the theory of modular forms plays a key role in the study and in the classification of integral lattices (see for instance [25] for a modern presentation of this circle of ideas and for references).

Banaszczyk's method highlights the significance of the theta functions $\theta_{\overline{E}}$ when investigating the fine properties of *general* Euclidean lattices. We present it with some details in the next two subsections. For simplicity, we will focus on the second transference inequality (4.7) in Theorem 4.2.1; the proof of (4.6) relies on similar arguments, and we refer the reader to [5, pp. 631–632] for details.

Let us also emphasize that Banaszczyk's method has played a central role in the development of applications of Euclidean lattices to cryptography during the last decades. We refer the reader to [20] and [63] for some recent striking results inspired by these developments and to [10] for a survey intended to mathematicians and additional references.

4.3 The Key Inequalities

Let us consider a Euclidean lattice $\overline{E} := (E, \|.\|)$ of positive rank n.

Its theta function $\theta_{\overline{E}}$ clearly is a decreasing function. The same holds for $\theta_{\overline{E}^\vee}$ and the functional equation (4.4) relating $\theta_{\overline{E}}$ and $\theta_{\overline{E}^\vee}$ therefore shows that $t^{n/2}\theta_{\overline{E}}(t)$ is some increasing function of $t \in \mathbb{R}_+^*$.

Besides, Poisson formula (4.3) shows that, for any $x \in E_\mathbb{R}$ and any $t \in \mathbb{R}_+$, we have:

$$\sum_{v \in E} e^{-\pi t \|x-v\|^2} \leqslant \sum_{v \in E} e^{-\pi t \|v\|^2}, \tag{4.9}$$

and that the equality holds in (4.9) if and only if $x \in E$.

The starting point of Banaszczyk's technique is the following inequality, which easily follows from the previous observations:

Lemma 4.3.1 *For any $x \in E_\mathbb{R}$, any $r \in \mathbb{R}_+$ and any $t \in]0, 1]$, we have:*

$$\sum_{v \in E, \|v-x\| \geqslant r} e^{-\pi \|v-x\|^2} \leqslant t^{-n/2} e^{-\pi(1-t)r^2} \sum_{v \in E} e^{-\pi \|v\|^2}. \tag{4.10}$$

Proof This follows from the following chain of inequalities:

$$\sum_{v \in E, \|v-x\| \geqslant r} e^{-\pi \|v-x\|^2} = \sum_{v \in E, \|v-x\| \geqslant r} e^{-\pi(1-t)\|v-x\|^2} e^{-\pi t \|v-x\|^2}$$

$$\leqslant e^{-\pi(1-t)r^2} \sum_{v \in E, \|v-x\| \geqslant r} e^{-\pi t \|v-x\|^2}$$

$$\leqslant e^{-\pi(1-t)r^2} \sum_{v \in E} e^{-\pi t \|v\|^2} \tag{4.11}$$

$$\leqslant e^{-\pi(1-t)r^2} t^{-n/2} \sum_{v \in E} e^{-\pi \|v\|^2}. \tag{4.12}$$

Indeed, the estimate (4.11) is a consequence of (4.9), and (4.12) of the estimate $t^{n/2} \theta_{\overline{E}}(t) \leqslant \theta_{\overline{E}}(1)$.   □

The upper bound (4.9) shows that the estimate (4.10) is relevant only when r is such that

$$\inf_{t \in]0,1]} t^{-n/2} e^{-\pi(1-t)r^2} < 1.$$

An elementary computation, that we shall leave as an exercise, establishes that this inequality is satisfied precisely when $r > \sqrt{n/2\pi}$, and that, if this holds and if we define $\tilde{r} \in]1, +\infty[$ by the relation

$$r = \sqrt{n/2\pi}\, \tilde{r},$$

then the minimum of $t^{-n/2} e^{-\pi(1-t)r^2}$ on $]0, 1]$ is achieved at

$$t = t_{\min} := \tilde{r}^{-2}$$

and assumes the value:

$$t_{\min}^{-n/2} e^{-\pi(1-t_{\min})r^2} = \beta(\tilde{r})^n,$$

where

$$\beta(\tilde{r}) := \tilde{r} e^{-(1/2)(\tilde{r}^2 - 1)}. \tag{4.13}$$

These remarks show that Lemma 4.3.1 may be reformulated as the following proposition, better suited to applications:

Proposition 4.3.2 *Let* $\overline{E} := (E, \|.\|)$ *be a Euclidean lattice of positive rank n, and let x be some element of* $E_{\mathbb{R}}$. *For any* $\tilde{r} \in [1, +\infty[$, *if we let*

$$r := \sqrt{\frac{n}{2\pi}}\, \tilde{r},$$

then the following upper bound holds:

$$\sum_{v \in E, \|v-x\| \geqslant r} e^{-\pi \|v-x\|^2} \leqslant \beta(\tilde{r})^n \sum_{v \in E} e^{-\pi \|v\|^2}. \tag{4.14}$$

Observe that formula (4.13) defines a decreasing homeomorphism:

$$\beta : [1, +\infty) \xrightarrow{\sim} (0, 1].$$

Besides, Poisson formula (4.3) implies the following equalities:

$$\sum_{v \in E} e^{-\pi \|x-v\|^2} + \sum_{v \in E} e^{-\pi \|v\|^2} = (\operatorname{covol} \overline{E})^{-1} \sum_{\xi \in E^\vee} e^{-\pi \|\xi\|^2} [1 + \cos(2\pi \langle \xi, x \rangle)]$$

$$= 2(\operatorname{covol} \overline{E})^{-1} \sum_{\xi \in E^\vee} e^{-\pi \|\xi\|^2} \cos^2(\pi \langle \xi, x \rangle),$$

This implies:

Proposition 4.3.3 *For any Euclidean lattice $\overline{E}$ and for any x in $E_\mathbb{R}$, we have:*

$$\sum_{v \in E} e^{-\pi \|x-v\|^2} + \sum_{v \in E} e^{-\pi \|v\|^2} \geqslant 2(\operatorname{covol} \overline{E})^{-1}. \tag{4.15}$$

4.4 Proof of the Transference Inequality (4.7)

Let us first state two corollaries of Propositions 4.3.2 and 4.3.3.

By applying Proposition 4.3.2 to $x = 0$ et $r = \lambda_1(\overline{E})$, we get:

Corollary 4.4.1 *Let $\overline{E}$ be some Euclidean lattice of positive rank n and of first minimum $\lambda_1(\overline{E}) > \sqrt{n/2\pi}$, and let $\tilde{\lambda} \in]1, +\infty[$ be defined by*

$$\lambda_1(\overline{E}) = \sqrt{n/2\pi}\,\tilde{\lambda}.$$

Then the following upper-bound on $\theta_{\overline{E}}(1)$ holds:

$$\theta_{\overline{E}}(1) := \sum_{v \in E} e^{-\pi \|v\|^2} \leqslant (1 - \beta(\tilde{\lambda})^n)^{-1}. \tag{4.16}$$

Besides, by the very definition of the covering radius $R_{\operatorname{cov}}(\overline{E})$ of some Euclidean lattice $\overline{E}$, there exists $x \in E_\mathbb{R}$ such that $\|v - x\| \geqslant R_{\operatorname{cov}}(\overline{E})$ for any v in E. If we apply Proposition 4.3.2 to such a point x and to $r = R_{\operatorname{cov}}(\overline{E})$, we obtain the first assertion in the following corollary:

Corollary 4.4.2 *Let $\overline{E}$ be some Euclidean lattice of positive rank n and of covering radius*

$$R_{\operatorname{cov}}(\overline{E}) \geqslant \sqrt{n/2\pi},$$

and let $\tilde{R} \in [1, +\infty[$ be defined by $R_{\operatorname{cov}}(\overline{E}) = \sqrt{n/2\pi}\,\tilde{R}.$

Then there exists $x \in E_{\mathbb{R}}$ *such that*

$$\frac{\sum_{v \in E} e^{-\pi \|v - x\|^2}}{\sum_{v \in E} e^{-\pi \|v\|^2}} \leqslant \beta(\tilde{R})^n, \tag{4.17}$$

and consequently:

$$\beta(\tilde{R})^n \geqslant 2\theta_{\overline{E}^\vee}(1)^{-1} - 1. \tag{4.18}$$

Proof We are left to prove (4.18). To achieve this, observe that, according to Proposition 4.3.3,

$$\frac{\sum_{v \in E} e^{-\pi \|v - x\|^2}}{\sum_{v \in E} e^{-\pi \|v\|^2}} \geqslant 2 \operatorname{covol}(\overline{E})^{-1} \theta_{\overline{E}}(1)^{-1} - 1,$$

and use the functional equation (4.4) relating $\theta_{\overline{E}}$ et $\theta_{\overline{E}^\vee}$ for $t = 1$, which takes the form:

$$\theta_{\overline{E}}(1) = (\operatorname{covol}(\overline{E}))^{-1} \theta_{\overline{E}^\vee}(1). \qquad \square$$

We are now in position to establish the transference inequality (4.7), namely:

$$R_{\operatorname{cov}}(\overline{E}).\lambda_1(\overline{E}^\vee) \leqslant n/2.$$

Let us consider a Euclidean lattice $\overline{E}$ of positive rank n and let us define $\tilde{R}$ and $\tilde{\lambda}^\vee$ by the equalities

$$R_{\operatorname{cov}}(\overline{E}) = \sqrt{n/2\pi}\,\tilde{R} \quad \text{and} \quad \lambda_1(\overline{E}^\vee) = \sqrt{n/2\pi}\,\tilde{\lambda}^\vee.$$

Lemma 4.4.3 *If* $\min(\tilde{\lambda}^\vee, \tilde{R}) > 1$, *then :*

$$\beta(\tilde{R})^n + 2\beta(\tilde{\lambda}^\vee)^n \geqslant 1. \tag{4.19}$$

Proof Corollary 4.4.1, applied to $\overline{E}^\vee$, shows that:

$$1 - \beta(\tilde{\lambda}^\vee)^n \leqslant \theta_{\overline{E}^\vee}(1)^{-1}. \tag{4.20}$$

The inequality (4.19) follows from (4.18) and (4.20). $\qquad \square$

For any $n > 0$, we let:

$$t_n := \beta^{-1}(3^{-1/n}) \in\,]1, +\infty[.$$

Lemma 4.4.4 *When n goes to infinity,*

$$t_n = 1 + \sqrt{(\log 3)/n} + O(1/n). \tag{4.21}$$

Moreover,

$$t_n \leqslant \sqrt{\pi} \quad \text{for any } n \geqslant 3. \tag{4.22}$$

Proof An elementary computation shows that, when $x \in \mathbb{R}_+^*$ goes to 0,

$$\beta(1+x) = 1 - x^2 + O(x^3).$$

This implies that, when $y \in (0, 1)$ goes to zero,

$$\beta^{-1}(1-y) = 1 + \sqrt{y} + O(y).$$

Since

$$t_n = 1 - (\log 3)/n + O(1/n^2),$$

this proves (4.21).

Observe also that:

$$t_n \leqslant \sqrt{\pi} \iff \beta(t_n) \geqslant \beta(\sqrt{\pi})$$
$$\iff 3^{-1/n} \geqslant \sqrt{\pi} \exp(-(\pi - 1)/2)$$
$$\iff -(\log 3)/n \geqslant -(\pi - 1)/2 + (1/2)\log \pi.$$

As

$$\log 3 = 1.0986\ldots$$

and

$$(\pi - 1)/2 - (1/2)\log \pi = 0.4984\ldots,$$

the above inequalities hold for any integer $n \geqslant 3$. $\qquad\square$

Using Lemma 4.4.3, we may derive a slightly stronger version of Banaszczyk's inequality (4.7) for $n \geqslant 3$:

Proposition 4.4.5 *For any Euclidean lattice of positive rank n, the following inequality holds:*

$$R_{\text{cov}}(\overline{E}).\lambda_1(\overline{E}^\vee) \leqslant t_n^2 n/2\pi. \tag{4.23}$$

Actually (4.7) is trivial when $n = 1$ and, when $n = 2$, follows from elementary considerations, involving reduced bases of two dimensional Euclidean lattices, which we leave as an exercise for the reader.

Proof of Proposition 4.4.5 Let us first assume that

$$R_{\mathrm{cov}}(\overline{E}) = \lambda_1(\overline{E}^\vee) =: \sqrt{n/2\pi}\,\tilde{t}. \tag{4.24}$$

According to Lemma 4.4.3, if $\tilde{t} > 1$, then $\beta(\tilde{t}) \geqslant 3^{-1/n}$ and therefore $\tilde{t} \leqslant t_n$. Since $t_n > 1$, this inequality still holds when $\tilde{t} \leqslant 1$. The estimate (4.23) immediately follows.

The general validity of (4.23) follows from its validity under the additional assumption (4.24). Indeed, replacing the Euclidean lattice $\overline{E}$ by $\overline{E} \otimes \overline{\mathscr{O}}(\delta)$ for some $\delta \in \mathbb{R}$—that is, scaling the Euclidean norm of $\overline{E}$ by the positive $e^{-\delta}$—does not change the product $R_{\mathrm{cov}}(\overline{E}).\lambda_1(\overline{E}^\vee)$; moreover, by a suitable choice of δ, the condition

$$\rho(\overline{E} \otimes \overline{\mathscr{O}}(\delta)) = \lambda_1((\overline{E} \otimes \overline{\mathscr{O}}(\delta))^\vee)$$

may be achieved. Indeed, from the very definitions of the covering radius and of the first minimum, we obtain:

$$\rho(\overline{E} \otimes \overline{\mathscr{O}}(\delta)) = e^{-\delta}\rho(\overline{E}) \quad \text{and} \quad \lambda_1((\overline{E} \otimes \overline{\mathscr{O}}(\delta))^\vee) = e^{\delta}\lambda_1(\overline{E}^\vee). \qquad \square$$

5 Vector Bundles on Curves and the Analogy with Euclidean Lattices

5.1 Vector Bundles on Smooth Projective Curves and Their Invariants

In this subsection, we recall some basic facts concerning vector bundles on algebraic curves that play a key role in the analogy between vector bundles and Euclidean lattices.

Let C be a smooth, projective and geometrically connected curve over some field k. We shall denote the field of rational functions over C by

$$K := k(C).$$

5.1.1 A *vector bundle* E over C is a locally free coherent sheaf over C. Any coherent subsheaf F of E is again a vector bundle over C. We shall say that F is a *vector subbundle* of E when the coherent sheaf E/F is torsion-free, and therefore also defines a vector bundle over C.

The fiber E_K of E at the generic point of C—namely, the space of rational sections of E over C—is a finite dimensional K-vector space. When F is a coherent subsheaf of E, F_K is a K-vector subspace of E_K, and this construction establishes a bijection between vector subbundles of E and K-vector subspaces of E_K.

We may define tensor operations on vector bundles: to any vector bundle E over C, we may attach its dual vector bundle $E^\vee$ and, for any $n \in \mathbb{N}$, its tensor power $E^{\otimes n}$ and its exterior power $\Lambda^n E$; to any two vector bundles E and F over C, we may attach their tensor product $E \otimes F$ and the vector bundle

$$\mathrm{Hom}(E, F) \simeq E^\vee \otimes F.$$

To any vector bundle E over C are associated the following invariants:

– its rank

$$\mathrm{rk}\, E := \dim_K E_K \in \mathbb{N};$$

– its degree

$$\deg E \in \mathbb{Z}.$$

– when $\mathrm{rk}\, E > 0$, its slope:

$$\mu(E) := \frac{\deg E}{\mathrm{rk}\, E} \in \mathbb{Q}.$$

5.1.2 A reminder on the various definitions of $\deg E$ in the present setting may be in order.

When E has rank 1—that is when E is a *line bundle* or equivalently an *invertible sheaf*—hence isomorphic to the sheaf $\mathcal{O}_C(D)$ associated to the divisor

$$D = \sum_{i \in I} n_i P_i$$

of some non-zero rational section of E (defined by some family $(P_i)_{i \in I}$ of closed points of C and multiplicities $(n_i)_{i \in I} \in \mathbb{Z}^I$), it is defined as:

$$\deg E = \deg \mathcal{O}_C(D) = \deg D := \sum_{i \in I} n_i [\kappa(P_i) : k].$$

To define the degree of some vector bundle E of arbitrary rank, one reduces to the case of line bundles by considering its maximal exterior power:

$$\deg E := \deg \Lambda^{\mathrm{rk}\, E} E.$$

An alternative definition of the degree a vector bundle E involves its so-called Hilbert polynomial. Let us assume, for simplicity, that the curve C admits some divisor D of degree 1.[8] Then, when the integer n is large enough, we have:

$$\dim_k H^0(C, E \otimes \mathcal{O}_C(nD)) = n \operatorname{rk} E + (1 - g) \operatorname{rk} E + \deg E. \qquad (5.1)$$

where we denote by g the genus of C. This is a straightforward consequence of the Riemann-Roch formula for the vector bundle $E \otimes \mathcal{O}_C(nD)$, combined to the vanishing of $H^1(C, E \otimes \mathcal{O}_C(nD))$ when n is large enough, itself a consequence of the ampleness of $\mathcal{O}_C(D)$.

The right-hand side of (5.1), as a function of n, defines the Hilbert polynomial of E. In particular, when $g = 1$, its constant term is the degree of E.

5.1.3 The invariants defined above satisfy the following properties.

(i) For any vector bundle E over C and any vector subbundle F of E,

$$\deg E = \deg F + \deg E/F. \qquad (5.2)$$

(ii) For any vector bundle of positive rank E and any line bundle L over C,

$$\mu(E \otimes L) = \mu(E) + \deg L.$$

More generally, for any two vector bundles of positive rank E and F over C, we have:

$$\mu(E \otimes F) = \mu(E) + \mu(F).$$

(iii) If $\varphi : E \longrightarrow E'$ is a morphism of sheaves of $\mathcal{O}_C$-modules between two vector bundles which is an isomorphism at the generic point:

$$\varphi_K : E_K \xrightarrow{\sim} E'_K,$$

then

$$\deg E \leqslant \deg E',$$

and equality holds if and only if φ is an isomorphism.

(iv) For any vector bundle E over C, of dual $E^\vee := \operatorname{Hom}(E, \mathcal{O}_C)$, we have:

$$\deg E^\vee = -\deg E.$$

---

[8] Such a divisor exists when the base field k is algebraically closed (then the divisor D defined by any point in $C(k)$ will do), or when k is finite.

5.2   Euclidean Lattices as Analogues of Vector Bundles Over Projective Curves

The analogy between number fields and function fields has played a central role in the development of algebraic geometry and number theory since the end of the nineteenth century, starting with the works of Dedekind, Weber and Kronecker (see [24] and [44]).

Here we will be concerned with the version of this analogy that constitutes the framework of Arakelov geometry,[9] which originates in Hensel's idea that "all places of a number field K are on the same footing" and that, accordingly, besides the places of K defined by closed points of Spec $\mathscr{O}_K$, its archimedean places, associated to field extensions $\sigma : K \hookrightarrow \mathbb{C}$ (up to complex conjugation) play an equally important role.

An elementary but significant manifestation of the analogy between number fields and function fields is the analogy between Euclidean lattices and vector bundles over a smooth projective and geometrically irreducible curve C over some base field k.

In this analogy, the field $\mathbb{Q}$ takes the place of the field $K := k(C)$ of rational functions over k, and the set of places of $\mathbb{Q}$ (which may be identified with the disjoint union of the closed points of Spec $\mathbb{Z}$—in other words, the set of prime numbers—and of the archimedean place of $\mathbb{Q}$, defined by the usual absolute value) takes the place of the closed points of C.

Moreover the $\mathbb{Q}$-vector space $E_{\mathbb{Q}}$ associated to a Euclidean lattice $\overline{E}$ is the counterpart of the fibre E_K of a vector bundle E at the generic point of C. The Euclidean lattices $\overline{F}$ associated to some $\mathbb{Z}$-submodules F of E (resp. to saturated $\mathbb{Z}$-submodules) play the role of coherent subsheaves (resp. of sub-vector bundles) of E, and the admissible short exact sequences of euclidean lattices (2.5) the one of short exact sequences of vector bundles over C. Moreover, for any two Euclidean lattices $\overline{E}_1 := (E_1, \|.\|_1)$ and $\overline{E}_2 := (E_2, \|.\|_2)$, the set of morphisms of $\mathbb{Z}$–modules

$$\varphi : E_1 \longrightarrow E_2$$

such that, for any $v \in E_{1\mathbb{R}}$,

$$\|\varphi_{\mathbb{R}}(v)\|_2 \leqslant \|v\|_1$$

plays the role of the k-vector space

$$\mathrm{Hom}_{\mathscr{O}_C}(E_1, E_2) \simeq H^0(C, \mathrm{Hom}(E_1, E_2))$$

of the morphisms of $\mathscr{O}_C$-Modules between two vector bundles E_1 and E_2 over C.

---

[9]Rather different versions of this analogy have played a key role in some other areas of arithmetic geometry, for instance in Iwasawa theory.

These analogies, in their crudest form, have been pointed out for a long time (see notably [75] and [26, Chapter 1]). It turns out that the invariants $h^0_\theta(\overline{E})$ of Euclidean lattices allow one to pursue these classical analogies in diverse directions, with an unexpected level of precision. Notably diverse recent progresses in the study of Euclidean lattices that arose in the last decade in relation with their application to cryptography offer striking illustrations of this general philosophy. We refer the reader to [10] for a discussion of these developments, due notably to Micciancio, Regev, Dadush and Stephens-Davidowitz, which involve comparison estimates relating the slopes of Euclidean lattices and suitable invariants defined in terms of their theta series.

In this section, we discuss a few simple instances of this analogy.

5.3   The Invariants $h^0_{\mathrm{Ar}}(\overline{E})$, $h^0_\theta(\overline{E})$ and $h^1_\theta(\overline{E})$

In the literature devoted to the analogy between number fields and function fields and to Arakelov geometry are described *several* invariants of Euclidean lattices which play the role of the dimension

$$h^0(C, E) := \dim_k H^0(C, E)$$

of the space of sections of a vector bundle E over a curve C over some base field k, or of the dimension

$$h^1(C, E) := \dim_k H^1(C, E)$$

of its first cohomology group.

5.3.1 The Invariant $h^0_{\mathrm{Ar}}(\overline{E})$ With the notation of Sect. 5.1, the k-vector space $H^0(C, E)$ may be identified to the k-vector space $\mathrm{Hom}_{\mathcal{O}_C}(\mathcal{O}_C, E)$ of morphisms of sheaves of $\mathcal{O}_C$-modules from $\mathcal{O}_C$ to E. When the base field k is finite of cardinality q, it is a finite set and we have:

$$h^0(C, E) = \dim_k \mathrm{Hom}_{\mathcal{O}_C}(\mathcal{O}_C, E) = \frac{\log |\mathrm{Hom}_{\mathcal{O}_C}(\mathcal{O}_C, E)|}{\log q}.$$

This leads one to consider the set of morphisms from $\overline{\mathcal{O}}(0) = (\mathbb{Z}, |.|)$ to some Euclidean lattice $\overline{E} := (E, \|.\|)$—by mapping such a morphism φ to $\varphi(1)$, it may be identified with the finite set

$$E \cap \overline{B}_{\|.\|}(0, 1)$$

of the lattice points in the unit ball of $(E_\mathbb{R}, \|.\|)$—and then to consider the logarithm of its cardinality:

$$h^0_{Ar}(\overline{E}) := \log |E \cap \overline{B}_{\|.\|}(0, 1)|. \tag{5.3}$$

This definition appears implicitly in the works of Weil [75] and Arakelov [2], and more explicitly in the presentations of Arakelov geometry in [73] and [48]. See also [31] for some variation on this definition, and some definition in the same vein of an analogue of $h^1(C, E)$.

5.3.2 The Invariants $h^0_\theta(\overline{E})$ and $h^1_\theta(\overline{E})$ One may also introduce the theta series $\theta_{\overline{E}}$ associated to some Euclidean lattice $\overline{E} := (E, \|.\|)$, defined as:

$$\theta_{\overline{E}}(t) := \sum_{v \in E} e^{-\pi t \|v\|^2} \quad \text{for any } t \in \mathbb{R}^*_+$$

(see (4.5) *supra*), and then define :

$$h^0_\theta(\overline{E}) := \log \theta_{\overline{E}}(1) = \log \sum_{v \in E} e^{-\pi \|v\|^2} \in \mathbb{R}_+. \tag{5.4}$$

The fact that the so-defined invariant $h^0_\theta(\overline{E})$ of the Euclidean lattice $\overline{E}$ is an analogue of the invariant $h^0(C, E)$ attached to a vector bundle E over some projective curve C is a remarkable discovery of the German school of number theory, and goes back to F. K. Schmidt (at least). Indeed, if one compare the proofs, respectively by Hecke [36] and Schmidt [68] of the analytic continuation and of the functional equation of the zeta functions associated to a number field and to a function field $K := k(C)$ attached to some curve C (projective, smooth, and geometrically connected) over some finite field k of cardinality q, one sees that the sum

$$\sum_{v \in E} e^{-\pi \|v\|^2}$$

associated to some Euclidean lattice $\overline{E} := (E, \|.\|)$ plays the same role as the expression

$$q^{h^0(C,E)}.$$

A key feature of Schmidt's proof is actually that the Riemann-Roch formula for a (rank 1) vector bundle over a curve plays a role similar to the one of the Poisson formula (4.4) which relates $\theta_{\overline{E}}$ and $\theta_{\overline{E}^\vee}$. Indeed, at the point $t = 1$, this formula becomes:

$$\theta_{\overline{E}}(1) = (\text{covol}(\overline{E}))^{-1} \theta_{\overline{E}^\vee}(1)$$

and, by taking logarithms, may also be written:

$$h^0_\theta(\overline{E}) - h^0_\theta(\overline{E}^\vee) = \widehat{\deg}\,\overline{E}. \tag{5.5}$$

This equality is formally similar to the Riemann-Roch formula over a smooth projective curve C of genus 1 (hence of trivial canonical bundle), and leads one to define:

$$h^1_\theta(\overline{E}) := h^0_\theta(\overline{E}^\vee), \tag{5.6}$$

so that (5.5) becomes the "Poisson-Riemann-Roch" formula:

$$h^0_\theta(\overline{E}) - h^1_\theta(\overline{E}) = \widehat{\deg}\,\overline{E}. \tag{5.7}$$

During the last decades, the above definitions (5.4) and (5.6) have notably appeared in Quillen's mathematical diary [60] (see the entries on 12/24/1971, 04/26/1973 and 04/01/1983), in [64, 54], and more recently in the articles by van der Geer and Schoof [74] and Groenewegen [33].

5.4 How to Reconcile the Invariants $h^0_{\mathrm{Ar}}(\overline{E})$ and $h^0_\theta(\overline{E})$

It is comforting that, as shown in [10], the definitions (5.3) et (5.4) of the invariants $h^0_{\mathrm{Ar}}(\overline{E})$ and $h^0_\theta(\overline{E})$, both candidate for playing the role of $h^0(C, E)$ for Euclidean lattices, may be reconciled.

5.4.1 Comparing $h^0_{\mathrm{Ar}}(\overline{E})$ and $h^0_\theta(\overline{E})$ by Banaszczyk's Method

Proposition 5.4.1 ([9], Theorem 3.1.1) *For any Euclidean lattice of positive rank n, the following inequalities hold:*

$$-\pi \leqslant h^0_\theta(\overline{E}) - h^0_{\mathrm{Ar}}(\overline{E}) \leqslant (n/2)\log n + \log(1 - 1/2\pi)^{-1}. \tag{5.8}$$

Proof To prove the first inequality in (5.8), we simply observe that:

$$h^0_\theta(\overline{E}) = \log \sum_{v \in E} e^{-\pi\|v\|^2} \geqslant \log \sum_{\substack{v \in E \\ |v| \leqslant 1}} e^{-\pi\|v\|^2}$$

$$\geqslant \log(e^{-\pi}|\{v \in E \mid \|v\| \leqslant 1\}|) = -\pi + h^0_{\mathrm{Ar}}(\overline{E}).$$

The proof of second inequality in (5.8) will rely on the following assertions, which are variants of results in [5, Section 1]. □

Lemma 5.4.2

(1) The expression $\log \theta_{\overline{E}}(t)$ defines a decreasing function of t in $\mathbb{R}_+^*$, and the expression

$$\log \theta_{\overline{E}}(t) + \frac{1}{2}\mathrm{rk}\, E \cdot \log t \tag{5.9}$$

defines an increasing function of t in $\mathbb{R}_+^*$.

(2) We have:

$$\sum_{v \in E} \|v\|^2 e^{-\pi t \|v\|^2} \leqslant \frac{\mathrm{rk}\, E}{2\pi t} \sum_{v \in E} e^{-\pi t \|v\|^2}. \tag{5.10}$$

(3) For any t and r in $\mathbb{R}_+^*$, we have:

$$\sum_{v \in E, \|v\| < r} e^{-\pi t \|v\|^2} \geqslant \left(1 - \frac{\mathrm{rk}\, E}{2\pi t r^2}\right) \sum_{v \in E} e^{-\pi t \|v\|^2}. \tag{5.11}$$

Part (1) was already used in Sect. 4.3, as the starting point of the proof of Banaszczyk's transference estimates.

Proof The first assertion in (1) is clear. According to the functional equation (4.4) which relates $\theta_{\overline{E}}$ and $\theta_{\overline{E}^\vee}$, the expression (5.9) may also be written

$$\widehat{\deg}\, \overline{E} + \log \theta_{\overline{E}^\vee}(t^{-1}),$$

and consequently defines an increasing function of t.

The inequality (5.10) may also be written

$$-\frac{1}{\pi} \frac{d\theta_{\overline{E}}(t)}{dt} \leqslant \frac{\mathrm{rk}\, E}{2\pi t} \theta_{\overline{E}}(t),$$

and simply expresses that the derivative of (5.9) is non-negative.

To establish the inequality (5.11), we combine the straightforward estimate

$$\sum_{v \in E, \|v\| \geqslant r} e^{-\pi t \|v\|^2} \leqslant \frac{1}{r^2} \sum_{v \in E} \|v\|^2 e^{-\pi t \|v\|^2}$$

with (5.10). This yields:

$$\sum_{v \in E, \|v\| \geqslant r} e^{-\pi t \|v\|^2} \leqslant \frac{\mathrm{rk}\, E}{2\pi t r^2} \sum_{v \in E} e^{-\pi t \|v\|^2},$$

or equivalently,

$$\sum_{v \in E, \|v\| < r} e^{-\pi t \|v\|^2} \geq \left(1 - \frac{\mathrm{rk}\, E}{2\pi t r^2}\right) \sum_{v \in E} e^{-\pi t \|v\|^2}.$$

From (5.11) with $r = 1$, we obtain that, for any $t > \mathrm{rk}\, E/2\pi$, we have:

$$h^0_{\mathrm{Ar}}(\overline{E}) \geq \log(1 - \mathrm{rk}\, E/(2\pi t)) + \log \theta_{\overline{E}}(t).$$

Using also that, for any $t \geq 1$,

$$\log \theta_{\overline{E}}(t) \geq \log \theta_{\overline{E}}(1) - \frac{1}{2}\mathrm{rk}\, E . \log t,$$

we finally obtain that, for any $t \geq \min(1, \mathrm{rk}\, E/2\pi)$, the following inequality holds:

$$h^0_{\mathrm{Ar}-}(\overline{E}) \geq \log(1 - \mathrm{rk}\, E/(2\pi t)) - \frac{1}{2}\mathrm{rk}\, E . \log t + h^0_\theta(\overline{E}).$$

Notably we may choose $t = \mathrm{rk}\, E$, and then we obtain[10]:

$$h^0_{\mathrm{Ar}}(\overline{E}) \geq \log(1 - 1/2\pi) - \frac{1}{2}\mathrm{rk}\, E . \log \mathrm{rk}\, E + h^0_\theta(\overline{E}).$$

This completes the proof of the second inequality in (5.8).   □

5.4.2 The Stable Invariant $\tilde{h}^0_{\mathrm{Ar}}(\overline{E}, x)$ and the Legendre Transform of Theta Invariants

As before, we denote by $\overline{E}$ some Euclidean lattice of positive rank. As mentioned in the Introduction, it is also possible to relate $h^0_\theta(\overline{E})$ to some "stable version" of the invariant $h^0_{\mathrm{Ar}}(\overline{E})$.

For every $t \in \mathbb{R}^*_+$, we shall define:

$$h^0_{\mathrm{Ar}}(\overline{E}, x) := h^0_{\mathrm{Ar}}(\overline{E} \otimes \overline{\mathcal{O}}((\log x)/2) = \log |\{v \in E \mid \|v\|^2 \leq x\}|.$$

Theorem 5.4.3 ([9], Theorem 3.4.5) *For any $x \in \mathbb{R}^*_+$, the following limit exists in* $\mathbb{R}_+$:

$$\tilde{h}^0_{\mathrm{Ar}}(\overline{E}, x) := \lim_{k \to +\infty} \frac{1}{k} h^0_{\mathrm{Ar}}(\overline{E}^{\oplus k}, kx).$$

---

[10]The "optimal" choice of t in terms of $n := \mathrm{rk}\, E$ would be $t = (n+2)/2\pi$. This choice leads to the slightly stronger estimate: $h^0_{\mathrm{Ar}}(\overline{E}) \geq -\frac{n+2}{2}\log \frac{n+2}{2\pi} - \log \pi + h^0_\theta(\overline{E})$.

The functions $\log \theta_{\overline{E}}(\beta) = h_\theta^0(\overline{E} \otimes \overline{\mathcal{O}}((\log \beta^{-1})/2))$ *and* $\tilde{h}_{\mathrm{Ar}}^0(\overline{E}, x)$ *of* β *and* x *in* $\mathbb{R}_+^*$ *are real analytic, and respectively decreasing and strictly convex, and increasing and strictly concave.*

Moreover, they may be deduced from each other by Fenchel-Legendre duality. Namely, for any $x \in \mathbb{R}_+^*$, *we have:*

$$\tilde{h}_{\mathrm{Ar}}^0(\overline{E}, x) = \inf_{\beta > 0} (\log \theta_{\overline{E}}(\beta) + \pi \beta x),$$

and, for any $\beta \in \mathbb{R}_+^*$:

$$\log \theta_{\overline{E}}(\beta) = \sup_{x > 0} (\tilde{h}_{\mathrm{Ar}}^0(\overline{E}, x) - \pi \beta x).$$

When $\overline{E}$ is the "trivial" Euclidean lattice of rank one $\overline{\mathcal{O}}(0) := (\mathbb{Z}, |.|)$, Theorem 5.4.3 may be deduced from results of Mazo and Odlyzko [50, Theorem 1].

As mentioned in the introduction, the next sections of this chapter are devoted to the proof of Theorem 5.4.3. This proof will emphasize the relation between this theorem and the thermodynamic formalism. Notably, the function $\tilde{h}_{\mathrm{Ar}}^0(\overline{E}, x)$ will appear as some kind of "entropy function" associated to the Euclidean lattice $\overline{E}$. It satisfies the following additivity property, which constitutes an avatar concerning Euclidean lattices of the second principle of thermodynamics:

Corollary 5.4.4 *For any two Euclidean lattices* $\overline{E}_1$ *and* $\overline{E}_2$ *of positive rank, and any* $x \in \mathbb{R}_+^*$,

$$\tilde{h}_{\mathrm{Ar}}^0(\overline{E}_1 \oplus \overline{E}_2, x) = \max_{\substack{x_1, x_2 > 0 \\ x_1 + x_2 = x}} \left(\tilde{h}_{\mathrm{Ar}}^0(\overline{E}_1, x_1) + \tilde{h}_{\mathrm{Ar}}^0(\overline{E}_2, x_2) \right). \tag{5.12}$$

This will be proved in paragraph 8.4.3, in a more precise form.

5.4.3 Complements By elaborating on the proof of Proposition 5.4.1 above, it is possible to establish the following additional comparison estimates relating h_{Ar}^0, $\tilde{h}_{\mathrm{Ar}}^0$, and h_θ^0.

For any integer $n \geqslant 1$, we let:

$$C(n) := -\sup_{t > 1} [\log(1 - t^{-1}) - (n/2) \log t].$$

One easily shows that

$$C(n) = \log(n/2) + (1 + n/2) \log(1 + 2/n)$$

and that

$$1 \leqslant C(n) - \log(n/2) \leqslant (3/2) \log 3.$$

Proposition 5.4.5 *For every Euclidean lattice $\overline{E}$ of positive rank, we have:*

$$-C(\mathrm{rk}\,E) \leqslant h^0_{\mathrm{Ar}}(\overline{E}, \mathrm{rk}\,E/2\pi) - h^0_\theta(\overline{E}) \leqslant \mathrm{rk}\,E/2$$

and

$$0 \leqslant \tilde{h}^0_{\mathrm{Ar}}(\overline{E}, \mathrm{rk}\,E/2\pi) - h^0_\theta(\overline{E}) \leqslant \mathrm{rk}\,E/2.$$

We refer to [10, paragraph 3.4.4], for the details of the proof.

It may also be shown that, when rk E goes to infinity, the order of growth of the constants in the comparison estimates in Propositions 5.4.1 and 5.4.5 is basically optimal; see [10], Section 3.5.

5.5 Some Further Analogies Between $h^0_\theta(\overline{E})$ and $h^0(C, E)$

A major difference between the invariants $h^0_\theta(\overline{E})$ et $h^1_\theta(\overline{E})$ attached to Euclidean lattices and the dimensions $h^0(C, E)$ and $h^1(C, E)$ of the cohomology groups of vector bundles is that, while the latter are integers, the former are real numbers, and that, when $\overline{E}$ has positive rank, the former never vanish.

This being said, the analogies between the properties of the invariants $h^i_\theta(\overline{E})$ and $h^i(C, E)$ are especially striking. In the next paragraphs, we describe three of them, by order of increasing difficulty.

5.5.1 Asymptotic Behavior of $\log \theta_{\overline{E}}$ Starting from the equality

$$\int_{E_{\mathbb{R}}} e^{-\pi \|x\|^2}\, dm_{\overline{E}}(x) = 1,$$

by approximating this Gaussian integral by Riemann sums over the lattice $\sqrt{t}E$, where $t \in \mathbb{R}^*_+$ goes to zero, we obtain:

$$\lim_{t \longrightarrow 0_+} \sqrt{t}^{\mathrm{rk}\,E}\,\mathrm{covol}(\overline{E}) \sum_{v \in E} e^{-\pi t \|v\|^2} = 1,$$

or equivalently:

$$\log \theta_{\overline{E}}(t) = -(\mathrm{rk}\,E/2)\,\log t + \widehat{\deg}\,\overline{E} + o(1) \quad \text{when } t \longrightarrow 0_+. \tag{5.13}$$

If we let $\lambda = -(1/2)\log t$, we get:

$$h^0_\theta(\overline{E} \otimes \overline{\mathcal{O}}(\lambda)) = \mathrm{rk}\,E\,\lambda + \widehat{\deg}\,\overline{E} + \varepsilon(\lambda) \quad \text{where } \lim_{\lambda \to +\infty} \varepsilon(\lambda) = 0.$$

In this formulation, the asymptotic expression (5.13) for $\theta_{\overline{E}}(t)$ when t goes to 0_+ becomes the analogue of the expression (5.1) for the Hilbert polynomial of a vector bundle over some curve of genus $g = 1$.

The expression (5.13) is also a consequence of Poisson formula (4.4), which actually shows that the error term $\varepsilon(\lambda)$ decreases extremely fast at infinity. Namely, there exists $c \in \mathbb{R}_+^*$ such that:

$$\varepsilon(\lambda) = O(e^{-ce^{2\lambda}}) \quad \text{when } \lambda \longrightarrow +\infty.$$

5.5.2 Admissible Short Exact Sequences and Theta Invariants A further analogy between the properties of $h_\theta^0(\overline{E})$ and of $h^0(C, E)$ concerns their compatibility with direct sums and, more generally, with short exact sequences:

Proposition 5.5.1

(1) For any two Euclidean lattices $\overline{E}_1$ and $\overline{E}_2$, we have:

$$h_\theta^0(\overline{E}_1 \oplus \overline{E}_2) = h_\theta^0(\overline{E}_1) + h_\theta^0(\overline{E}_2). \tag{5.14}$$

(2) For any admissible short exact sequences of Euclidean lattices

$$0 \longrightarrow \overline{F} \xrightarrow{\ i\ } \overline{E} \xrightarrow{\ p\ } \overline{E/F} \longrightarrow 0,$$

we have :

$$h_\theta^0(\overline{E}) \leqslant h_\theta^0(\overline{F}) + h_\theta^0(\overline{E/F}). \tag{5.15}$$

The subadditivity inequality (5.15) has been observed by Quillen [60, entry of 04/26/1973] and Groenewegen [33, Lemma 5.3].

Proof

(1) The equality (5.14) follows from the relation:

$$\sum_{(v_1, v_2) \in E_1 \times E_2} e^{-\pi(\|v_1\|_{\overline{E}_1}^2 + \|v_2\|_{\overline{E}_2}^2)} = \sum_{v_1 \in E_1} e^{-\pi\|v_1\|_{\overline{E}_1}^2} \cdot \sum_{v_2 \in E_2} e^{-\pi\|v_2\|_{\overline{E}_2}^2}.$$

(2) Observe that the Poisson formula in the form (4.9) shows that, for any $\alpha \in E/F$,

$$\sum_{v \in p^{-1}(\alpha)} e^{-\pi\|v\|_{\overline{E}}^2} \leqslant e^{-\pi\|\alpha\|_{\overline{E/F}}^2} \sum_{f \in F} e^{-\pi\|f\|_{\overline{F}}^2}.$$

By summing over α and taking the logarithms, we obtain (5.15). $\qquad\square$

5.5.3 A Theorem of Regev and Stephens-Davidowitz Let C be a smooth projective curve over some base field k, as in Sect. 5.1. Let E be a vector bundle over

C, and F_1 and F_2 two coherent subsheaves of E. We may consider the following short exact sequence of vector bundles over C:

$$0 \longrightarrow F_1 \cap F_2 \longrightarrow F_1 \oplus F_2 \longrightarrow F_1 + F_2 \longrightarrow 0,$$

where the morphism from $F_1 \oplus F_2$ to $F_1 + F_2$ is the sum map, and the one from $F_1 \cap F_2$ to $F_1 \oplus F_2$ maps a section s to $(s, -s)$. It induces an exact sequence of finite dimensional k-vector spaces:

$$0 \longrightarrow H^0(C, F_1 \cap F_2) \longrightarrow H^0(C, F_1) \oplus H^0(C, F_2) \longrightarrow H^0(C, F_1 + F_2),$$

which yields the following inequality concerning their dimensions:

$$h^0(C, F_1) + h^0(C, F_2) \leqslant h^0(C, F_1 \cap F_2) + h^0(C, F_1 + F_2).$$

Answering a question by McMurray Price (see [51]), Regev and Stephens-Davidowitz have shown that this estimate holds without change for Euclidean lattices and their invariants h_θ^0:

Theorem 5.5.2 ([62]) *Let* $\overline{E} := (E, \|.\|)$ *be a Euclidean lattice and let* F_1 *and* F_2 *two* $\mathbb{Z}$-*submodules of* E. *Then the following estimate holds:*

$$h_\theta^0(\overline{F}_1) + h_\theta^0(\overline{F}_2) \leqslant h_\theta^0(\overline{F_1 \cap F_2}) + h_\theta^0(\overline{F_1 + F_2}).$$

Here we denote by $\overline{F}_1$, $\overline{F}_2$, $\overline{F_1 \cap F_2}$, and $\overline{F_1 + F_2}$ the Euclidean lattices defines by the free $\mathbb{Z}$-modules F_1, F_2, $F_1 \cap F_2$, and $F_1 + F_2$ equipped with the restrictions of the Euclidean norm $\|.\|$.

We refer the reader to [62] for the proof of this theorem, which is elementary but extremely clever.

5.6 Varia

As explained in [10], it is possible to understand the remarkable recent results of Dadush, Regev and Stephens-Davidowitz on the Kannan-Lovász conjecture [20, 63] as a further illustrations of the analogy between the invariants $h_\theta^i(\overline{E})$, $i \in \{0, 1\}$, associated to Euclidean lattices and the dimensions of cohomology groups $h^i(C, E)$ associated to vector bundles on curves (see notably *loc. cit.*, Section 5.4).

Let us also recall that Euclidean lattices are nothing but a special case, associated to the field $K = \mathbb{Q}$, of Hermitian vector bundles over the "arithmetic curve" Spec $\mathscr{O}_K$, attached to some number field K of ring of integers $\mathscr{O}_K$. The analogy between vector bundles over a curve and Euclidean lattices extends to an analogy between vector bundle over a curve and Hermitian vector bundles over Spec $\mathscr{O}_K$, where K is now an arbitrary number field.

This actually constitutes the natural framework for this analogy: considering arbitrary number fields is akin to considering curves C of arbitrary genus $g \geqslant 1$. The definitions of the invariants $h_\theta^i(\overline{E})$ extend to this setting, which already was, in substance, the one of [36] when $\mathrm{rk}_{\mathcal{O}_K} E = 1$. We refer to [9, Chapter 2], for their study in this more general framework.

Let us finally emphasize that the theta series (4.5) associated to Euclidean lattices appear in various areas of mathematics and mathematical physics and have led to multiple developments, from very diverse perspectives. We may notably mention the investigations of extremal values of theta functions, motivated by the classical theory of modular and automorphic forms (see for instance [67] and its references), the works on the "Gaussian core model", inspired by the study of sphere packing and statistical physics [15], and various developments in crystallography and solid state physics (see for instance [6]).

6 A Mathematical Model of the Thermodynamic Formalism

In this section, logically independent of the previous ones, we introduce a simple mathematical model of classical statistical physics and we establish some of its basic properties.

In this model, the central object of study is a pair $((\mathcal{E}, \mathcal{T}, \mu), H)$, consisting of a measure space $(\mathcal{E}, \mathcal{T}, \mu)$ equipped with some non-negative measurable function $H : \mathcal{E} \longrightarrow \mathbb{R}_+$ (see Sect. 6.1, *infra*). The measure space $(\mathcal{E}, \mathcal{T}, \mu)$ should be thought as the configuration space (or phase space) of some elementary physical system, and the function H as the energy function on this space.

For instance, when dealing with some physical system in the realm of classical mechanics, described by the Hamiltonian formalism, $(\mathcal{E}, \mathcal{T}, \mu)$ will be the measure space underlying a symplectic manifold (M, ω), of dimension $2n$, equipped with the Liouville measure defined by the top degree form

$$\mu := \frac{1}{n!} \omega^n,$$

and H will be the function in $C^\infty(M, \mathbb{R})$ such that the associated Hamiltonian vector field X_H on M, defined by

$$i_{X_H} \omega = dH,$$

describes the evolution of the system (see for instance [1] or [4]).

Besides such examples related to classical mechanics for the pair $((\mathcal{E}, \mathcal{T}, \mu), H)$, other examples, of physical and number theoretical origin, will turn out to be interesting (see Sects. 6.3.2 and 6.5, *infra*). However we shall still refer to the function H as the Hamiltonian or as the energy of the system under study.

6.1   Measure Spaces with a Hamiltonian: Basic Definitions

Let us consider a measure space $(\mathscr{E}, \mathscr{T}, \mu)$ defined by a set $\mathscr{E}$, a σ-algebra $\mathscr{T}$ of subsets of $\mathscr{E}$, and a non-zero σ-finite measure

$$\mu : \mathscr{T} \longrightarrow [0, +\infty].$$

Besides, let us consider some $\mathscr{T}$-measurable function

$$H : \mathscr{E} \longrightarrow \mathbb{R}_+.$$

We shall denote by $H_{\min}$ its essential infimum with respect to μ and introduce the $\mathscr{T}$-measurable subset of $\mathscr{E}$:

$$\mathscr{E}_{\min} := H^{-1}(H_{\min}).$$

We may introduce the following two conditions:

$\mathbf{T}_1$: $\mu(\mathscr{E}) = +\infty$,
   and:
$\mathbf{T}_2$: *For every $E \in \mathbb{R}_+$, the measure*

$$N(E) := \mu(H^{-1}([0, E]))$$

   is finite and, when E goes to $+\infty$,

$$\log N(E) = o(E).$$

The condition $\mathbf{T}_2$ on the finiteness and the subexponential growth of N is easily seen to be equivalent to following condition:
   $\mathbf{T}_2'$: *For every $\beta \in \mathbb{R}_+^*$, the function $e^{-\beta H}$ is μ-integrable.*
   When $\mathbf{T}_2$ and $\mathbf{T}_2'$ are satisfied, we may introduce the *partition function* (in German, *Zustandsumme*)

$$Z : \mathbb{R}_+^* \longrightarrow \mathbb{R}_+^*$$

and *Planck's characteristic function*

$$\Psi : \mathbb{R}_+^* \longrightarrow \mathbb{R}$$

defined by the relations, for any $\beta \in \mathbb{R}_+^*$:

$$Z(\beta) := \int_{\mathscr{E}} e^{-\beta H} \, d\mu \tag{6.1}$$

and

$$\Psi(\beta) := \log Z(\beta). \tag{6.2}$$

When the measure μ is a probability measure, the function $\Psi(-\beta)$ also appears in the literature as the *logarithmic moment generating function* (see for instance [71, Section 3.1.1]).

6.2 Main Theorem

For every positive integer n, we also consider the product $\mu^{\otimes n}$ of n copies of the measure μ on $\mathscr{E}^n$ equipped with the σ-algebra $\mathscr{F}^{\otimes n}$, and the $\mathscr{F}^{\otimes n}$-measurable function

$$H_n : \mathscr{E}^n \longrightarrow \mathbb{R}_+$$

defined by:

$$H_n(x_1, \ldots, x_n) := H(x_1) + \cdots + H(x_n)$$

for every $(x_1, \ldots, x_n) \in \mathscr{E}^n$.

For every $E \in \mathbb{R}$ and every integer $n \geqslant 1$, we may consider

$$A_n(E) := \mu^{\otimes n}(\{x \in \mathscr{E}^n \mid H_n(x) \leqslant nE\}). \tag{6.3}$$

Clearly, $A_n(E)$ is a non-decreasing function of E; moreover:

$$A_n(E) = 0 \quad \text{if} \quad E < H_{\min}$$

and

$$A_n(H_{\min}) = \mu^{\otimes n}(\mathscr{E}_{\min}{}^n) = \mu(\mathscr{E}_{\min})^n.$$

(Observe that the measure $\mu(\mathscr{E}_{\min})$ is finite when $\mathbf{T}_2$ holds.) Besides, when $E > H_{\min}$,

$$A_n(E) \geqslant \mu^{\otimes n}((H^{-1}([0, E]))^n) = N(E)^n > 0.$$

The following theorem, which constitutes the main result of the last part of this chapter, describes the asymptotic behavior of $A_n(E)$ when n goes to infinity. It shows that

$$A_n(E) = e^{(n+o(n))S(E)} \quad \text{when } n \to +\infty,$$

for some real valued function S on $(H_{\min}, +\infty)$ that is deduced from Ψ by Legendre-Fenchel transform.

Theorem 6.2.1 *Let us assume that Conditions* $\mathbf{T}_1$ *and* $\mathbf{T}_2$ *are satisfied.*

(1) For any $E \in (H_{\min}, +\infty)$ and any integer $n \geqslant 1$, $A_n(E)$ belongs to $\mathbb{R}_+^*$ and the limit

$$S(E) := \lim_{n \to +\infty} \frac{1}{n} \log A_n(E) \tag{6.4}$$

exists in $\mathbb{R}$, and actually coincides with $\sup_{n \geqslant 1}(1/n) \log A_n(E)$.
The function

$$S : (H_{\min}, +\infty) \longrightarrow \mathbb{R}$$

is real analytic, increasing and strictly concave[11], and satisfies:

$$\lim_{E \to (H_{\min})_+} S(E) = \log \mu(\mathcal{E}_{\min}) \tag{6.5}$$

and

$$\lim_{E \to +\infty} S(E) = +\infty. \tag{6.6}$$

Moreover its derivative establishes a real analytic decreasing diffeomorphism:

$$S' := (H_{\min}, +\infty) \xrightarrow{\sim} \mathbb{R}_+^*. \tag{6.7}$$

(2) The function $\Psi : \mathbb{R}_+^* \longrightarrow \mathbb{R}$ is real analytic, decreasing and strictly convex[12]. Its derivative up to a sign

$$U := -\Psi'$$

defines a real analytic decreasing diffeomorphism

$$U : \mathbb{R}_+^* \xrightarrow{\sim} (H_{\min}, +\infty) \tag{6.8}$$

and satisfies, for every $\beta \in \mathbb{R}_+^*$:

$$U(\beta) = \frac{\int_{\mathcal{E}} H e^{-\beta H} \, d\mu}{\int_{\mathcal{E}} e^{-\beta H} \, d\mu}. \tag{6.9}$$

---

[11] In other words, for any $E \in (H_{\min}, +\infty)$, $S'(E) > 0$ and $S''(E) < 0$.
[12] In other words, for any $\beta \in \mathbb{R}_+^*$, $\Psi'(\beta) < 0$ and $\Psi''(\beta) > 0$.

(3) The functions $-S(-.)$ *and* Ψ *are Legendre-Fenchel transforms of each other. Namely, for every* $E \in (H_{\min}, +\infty)$,

$$S(E) = \inf_{\beta \in \mathbb{R}_+^*} (\Psi(\beta) + \beta E), \tag{6.10}$$

and, for every $\beta \in \mathbb{R}_+^*$,

$$\Psi(\beta) = \sup_{E \in (H_{\min}, +\infty)} (S(E) - \beta E). \tag{6.11}$$

Moreover the diffeomorphisms S' *and* U *(see (6.7) and (6.8)) are inverse of each other. For any* $E \in (H_{\min}, +\infty)$ *and any* $\beta \in \mathbb{R}_+^*$, *the following inequality holds:*

$$S(E) \leqslant \Psi(\beta) + \beta E, \tag{6.12}$$

and (6.12) becomes an equality precisely when

$$\beta = S'(E), \quad \text{or equivalently} \quad E = U(\beta). \tag{6.13}$$

In [9, Appendix A], a more general form of this result is presented, with a strong emphasis on its relation with Cramér's fundamental theorem on large deviations (which deals with the situation where μ is a probability measure). In particular, Theorem 6.2.1 is established in [9] by some reduction to Cramér's theorem. (Theorem 6.2.1 appears as Theorem A.5.1 in *loc. cit.*; it is a special case of Theorem A.4.4, which extends Cramér's theorem by a means of a reduction trick discussed in Section A.4.1.)

The main purpose of the final part of this chapter is to present a self-contained derivation of Theorem 6.2.1, which hopefully will make clear the basic simplicity of the underlying arguments. These arguments will also show that there is a considerable flexibility in the definition (6.3) of $A_n(E)$ that ensures the validity of Theorem 6.2.1: diverse variants, where the conditions $H_n(x) \leqslant nE$ is replaced by stronger conditions—for instance $H_n(x) < nE$, or $(1 - \eta)nE \leqslant H_n(x) \leqslant nE$ for some fixed $\eta \in (0, 1)$—would still lead to to the convergence of $(1/n) \log A_n(E)$ towards the same limit $S(E)$.

For instance, in Sect. 7.2.4, we will derive the following result which notably covers the above variants of the condition $H_n(x) \leqslant nE$:

Proposition 6.2.2 *Let us keep the notation of Theorem 6.2.1.*

For any $E \in (H_{\min}, +\infty)$ *and any sequence* $(I_n)_{n \geqslant n_0}$ *of intervals in* $\mathbb{R}$ *such that*

$$\lim_{n \to +\infty} \sup I_n = E$$

and such that their lengths l_n (that is, their Lebesgue measure) satisfy

$$\lim_{n\to+\infty} (\sqrt{n}\,l_n) > 2\sqrt{\Psi''(\beta)} \quad \text{where } \beta := S'(E) = U^{-1}(E),$$

we have:

$$\lim_{n\to+\infty} \frac{1}{n} \log \mu^{\otimes n}(\{x \in \mathscr{E}^n \mid H_n(x) \in nI_n\}) = S(E).$$

6.3 Relation with Statistical Physics

In the next paragraphs, we want to explain briefly how the framework of measure spaces with Hamiltonian and the main theorem presented in this section are related to the formalism of thermodynamics.

These relations go back to the classical works of Boltzmann and Gibbs (see for instance [7, 8, 30]). However the present discussion is more specifically related to the approach to statistical thermodynamics presented in Schrödinger's seminar notes [69], and also, to a lesser extent, to Khinchin's exposition in [42]. We refer to [27] for another perspective on the relations between statistical mechanics and large deviations.

6.3.1 Let us consider a physical system consisting of a large number n of identical "elementary" systems, each of them described by a measure space equipped with some Hamiltonian function H. One assumes that these systems are "loosely" coupled: their coupling is assumed to allow the exchange of energy between these elementary systems; however, we suppose that these exchanges are negligible with respect to the internal dynamics of the elementary systems.

One is interested in the "average properties" of these elementary systems when the total energy of our composite system belongs to some "small interval" $[n(E - \delta E), nE]$, or in other words, when the average energy per elementary system is about E. Theorem 6.2.1 and its variant Proposition 6.2.2 provide the following answer to this type of question: they show that the measure

$$W_n(E) := \mu^{\otimes n}\left(\{x \in \mathscr{E}^n \mid E - \delta E \leqslant H_n(x)/n \leqslant E\}\right) \tag{6.14}$$

of the points in the phase space $\mathscr{E}^n$ describing states of our composite systems which satisfy the above energy condition grows like

$$\exp(n(S(E) + o(1))$$

when n goes to infinity.

In other words, when n is large,

$$n^{-1} \log W_n(E) = S(E) + o(1),$$

and $S(E)$ may be understood as the logarithmic volume of the phase space available "per elementary system" when their average energy is about E: it is the *Boltzmann entropy*, at the energy E, attached to the elementary system described by the measure space with Hamiltonian $((\mathscr{E}, \mathscr{T}, \mu), H)$.

Its expression (6.8) shows that $U(\beta)$ may be interpreted as the average value of the Hamiltonian of our elementary system computed by using Gibbs canonical distribution[13] at temperature β^{-1}—in brief, to its energy at temperature β^{-1}.

Part (3) of Theorem 6.2.1 relates the Boltzmann entropy of our system, defined as the limit (6.4), to its energy $U(\beta)$ as a function of β. As stated in (6.13), to every value of β in $\mathbb{R}_+^*$ is attached bijectively a value $E = U(\beta)$ in $(H_{\min}, +\infty)$ of this energy, and we then have:

$$\beta = S'(E).$$

If we let $\beta = 1/T$, this last relation takes the familiar form:

$$dS = \frac{dE}{T}.$$

The function $\Psi(\beta)$ satisfies:

$$\Psi(\beta) = S(E) - \beta E,$$

where, as before, $E = U(\beta)$. It coincides with the function initially introduced by Planck[14] as

$$\Psi := \frac{TS - U}{T} = -\frac{F}{T},$$

where $F := U - TS$ is the so-called Helmholtz free energy.

6.3.2 The general framework introduced in 6.1 also covers the thermodynamics of some quantum systems, namely of systems composed of some large number n of copies of some "elementary" quantum system described by a (non-negative selfadjoint) Hamiltonian operator $\mathscr{H}$ acting with a discrete spectrum on some Hilbert space, say the space of square integrable functions $L^2(X)$ on some measure space X.

---

[13]Namely, the probability measure $\nu_\beta := Z(\beta)^{-1} e^{-\beta H} \mu$; see paragraph 7.1.2, *infra*.

[14]Planck initially denoted this function by Φ. The notation Ψ seems to have been introduced in the English translation [56] of Planck's classical treatise on thermodynamics, and is also used by Schrödinger in [69].

To the data of such an elementary system are indeed associated the spectrum $\mathscr{E}$ of $\mathscr{H}$ (a discrete subset of $\mathbb{R}_+$), the spectral measure μ of $\mathscr{H}$—defined by the relation

$$\mathrm{Tr}_{L^2(X)}\, f(\mathscr{H}) = \int_{\mathscr{E}} f\, d\mu$$

for any finitely supported function f on $\mathscr{E}$—and the "tautological" function

$$H : \mathscr{E} \hookrightarrow \mathbb{R}_+,$$

defined by the inclusion of the spectrum $\mathscr{E}$ in $\mathbb{R}_+$. Then the partition function $Z(\beta)$ associated to the so-defined measure space with Hamiltonian is

$$Z(\beta) := \int_{\mathscr{E}} e^{-\beta H}\, d\mu = \mathrm{Tr}_{L^2(X)} e^{-\beta \mathscr{H}},$$

and the previous discussion still holds *mutatis mutandis*: $A_n(E)$ now represents the number of "quantum states" of our composite system of total energy at most nE, etc.

A remarkable instance of this situation is provided by the quantum harmonic oscillator, say of frequency v, described by the Hamiltonian operator

$$\mathscr{H} := -(h^2/2)\frac{d^2}{dx^2} + (v^2/2)x^2$$

acting on $L^2(\mathbb{R})$.

Then $\mathscr{E} = (\mathbb{N} + 1/2)hv$, and μ is the counting measure $\sum_{e \in \mathscr{E}} \delta_e$. The associated partition function is

$$Z(\beta) = \sum_{k \in \mathbb{N}} e^{-\beta(k+1/2)hv} = \frac{e^{-\beta hv/2}}{1 - e^{-\beta hv}}$$

and accordingly:

$$\Psi(\beta) := \log Z(\beta) = -(1/2)\,\beta hv - \log(1 - e^{-\beta hv}).$$

Consequently, we then have:

$$U(\beta) = \frac{hv}{2} + \frac{hv\, e^{-\beta hv}}{1 - e^{-\beta hv}}.$$

We recover Planck's formula for the energy of a quantum oscillator of frequency v at temperature β^{-1}, which constitutes the historical starting point of quantum physics.

6.4 Gaussian Integrals and Maxwell's Kinetic Gas Model

In this subsection, we discuss an instance of the formalism introduced in Sects. 6.1 and 6.2 which may be seen as a mathematical counterpart of Maxwell's statistical approach to the theory of ideal gases. It is included for comparison with the application in Sect. 6.5 of the above formalism to Euclidean lattices—the present example appears as a "classical limit" of the discussion of Sect. 6.5.

6.4.1 Euclidean Spaces and Gaussian Integrals We begin with a purely mathematical discussion.

Let V be a finite dimensional real vector space equipped with some Euclidean norm $\|.\|$.

We shall denote by λ the Lebesgue measure on V attached to this Euclidean norm. It may be defined as the unique translation invariant Radon measure on V which satisfies the following normalization condition: for any orthonormal base $(e_1, \cdots, e_N)$ of the Euclidean space $(V, \|.\|)$,

$$\lambda \left(\sum_{i=1}^{N} [0, 1)e_i \right) = 1.$$

This normalization condition may be equivalently expressed in terms of a Gaussian integral:

$$\int_V e^{-\pi \|x\|^2} \, d\lambda(x) = 1.$$

We may apply the formalism of Sects. 6.1 and 6.2 to the measure space $(V, \mathscr{B}, \lambda)$, defined by V equipped with the Borel σ-algebra $\mathscr{B}$ and with the Lebesgue measure λ, and to the function

$$H := (1/2m)\|.\|^2$$

where m denotes some positive real number.

Then, for every β in $\mathbb{R}_+^*$, we have:

$$\int_V e^{-\beta \|p\|^2 / 2m} \, d\lambda(p) = (2\pi m/\beta)^{\dim V/2}. \tag{6.15}$$

Therefore

$$\Psi(\beta) = (\dim V/2) \, \log(2\pi m/\beta) \tag{6.16}$$

and

$$U(\beta) = -\Psi'(\beta) = \dim V/(2\beta). \tag{6.17}$$

The relation (6.13) between the "energy" E and the "inverse temperature" β, takes the following form, for any $E \in (H_{\min}, +\infty) = \mathbb{R}^*_+$ and any $\beta \in \mathbb{R}^*_+$:

$$\beta E = (\dim V)/2. \tag{6.18}$$

The function $S(E)$ may be computed directly from its definition. Indeed, for any $E \in \mathbb{R}^*_+$ and any positive integer n, we have:

$$\lambda^{\otimes n}\left(\{(x_1, \ldots, x_n) \in V^n \mid (1/2m)(\|x_1\|^2 + \ldots + \|x_n\|^2) \leqslant nE\}\right)$$

$$= v_{n \dim V}(2mnE)^{n(\dim V)/2}. \tag{6.19}$$

Here $v_{n \dim V}$ denotes the volume of the unit ball in the Euclidean space of dimension $n \dim V$. It is given by:

$$v_{n \dim V} = \frac{\pi^{n(\dim V)/2}}{\Gamma(1 + n(\dim V)/2)}. \tag{6.20}$$

From (6.19) and (6.20), by a simple application of Stirling's formula, we get:

$$S(E) = \lim_{n \to +\infty} \frac{1}{n} \log\left[v_{n \dim V}(2mnE)^{n(\dim V)/2}\right] = \frac{\dim V}{2} \log \frac{4\pi em E}{\dim V}.$$

In particular

$$S'(E) = \frac{\dim V}{2E},$$

and we recover (6.18).

Conversely, combined with the expression (6.16) for the function Ψ, Part 3) of Theorem 6.2.1 allows one to recover the asymptotic behaviour of the volume v_n of the n-dimensional unit ball, in the form:

$$v_n^{1/n} \sim \sqrt{2e\pi/n} \quad \text{when } n \to +\infty.$$

Finally, observe that when $m = (2\pi)^{-1}$—the case relevant for the comparison with the application to Euclidean lattices in Sect. 6.5—the expressions for Ψ and S take the following simpler forms:

$$\Psi(\beta) = (\dim V/2) \log(1/\beta)$$

and

$$S(E) = \frac{\dim V}{2} \log \frac{2eE}{\dim V}.$$

6.4.2 Hamiltonian Dynamics on a Compact Riemannian Manifold Let X be a compact C^∞ manifold of (pure positive) dimension d, and let g be a C^∞ Riemannian metric on X.

We shall denote the tangent (resp. cotangent) bundle of X by T_X (resp. by $T_X^\vee$). The Riemannian metric g defines an isomorphism of C^∞ vector bundles

$$T_X \xrightarrow{\sim} T_X^\vee, \tag{6.21}$$

by means of which the Euclidean metric $\|.\|$ on the fibers of T_X defined by g may be transported into some Euclidean metric $\|.\|^\vee$ on the fibers of $T_X^\vee$.

The $2d$-dimensional C^∞ manifold $T_X^\vee$ is endowed with a canonical symplectic form ω, defined as the exterior differential $d\alpha$ of the tautological 1-form α on $T_X^\vee$, which is characterized by the following property (see for instance [1] or [4]): for any C^∞ function f on some open subset U of X, the differential of f defines a section Df over U of the structural morphism

$$\pi : T_X^\vee \longrightarrow X$$

of the cotangent bundle $T_X^\vee$, and the 1-form $Df^*\alpha$ over U, defined as the pull-back of α by this section, coincides with the differential df of f.

Let m be a positive real number. The Hamiltonian flow associated to the function

$$H := \frac{1}{2m}\|.\|^{\vee 2} : T_X^\vee \longrightarrow \mathbb{R} \tag{6.22}$$

on the symplectic manifold $(T_X^\vee, \omega)$ describes the dynamics of some particle of mass m moving freely on the Riemaniann manifold (X, g). When $m = 1$, this flow transported to T_X by means of (the inverse of) the diffeomorphism (6.21) is the geodesic flow of the Riemannian manifold (X, g) (see for instance [1, Section 3.7]).

The thermodynamics of a kinetic gas model composed of free particles of mass m on (X, g) is described by the formalism of Sects. 6.1 and 6.2 applied to $\mathscr{E} := T_X^\vee$ equipped with the Liouville measure $\mu := \omega^d/d!$ and with the Hamiltonian function H defined by (6.22).

More generally, one may consider a C^∞ function

$$V : X \longrightarrow \mathbb{R}$$

and introduce the Hamiltonian function

$$H_V : T_X^\vee \longrightarrow \mathbb{R}$$

defined by

$$H_V(p) := \frac{1}{2m}\|p\|^{\vee 2} + V(x)$$

for any point x in X and any p in the fiber $T^\vee_{X,x} := \pi^{-1}(x)$ of the cotangent bundle over x.

It describes a particle of mass m moving on the Riemannian manifold (X, g), submitted to the potential V, and our formalism applied to

$$(\mathscr{E}, \mu, H) := (T^\vee_X, \omega^d/d!, H_V)$$

describes the thermodynamics of a gas of such particles.

6.4.3 Euclidean Lattices and Flat Tori   Let $\overline{F}$ be a Euclidean lattice, and let (X, g) be the compact Riemannian manifold defined as the flat torus associated to the dual Euclidean lattice $\overline{F}^\vee := (F^\vee, \|.\|^\vee)$; namely,

$$X := F^\vee_{\mathbb{R}}/F^\vee$$

and g is the flat metric on

$$T_X \simeq (F^\vee_{\mathbb{R}}/F^\vee) \times F^\vee_{\mathbb{R}}$$

defined by the "constant" Euclidean norm $\|.\|^\vee$ on $F^\vee_{\mathbb{R}}$.

Then the function H on

$$T^\vee_X \simeq (F^\vee_{\mathbb{R}}/F^\vee) \times F_{\mathbb{R}}$$

is simply the composition

$$H : T^\vee_X \xrightarrow{\mathrm{pr}_2} F_{\mathbb{R}} \xrightarrow{\|.\|^2/2m} \mathbb{R},$$

where pr_2 denotes the projection of $(F^\vee_{\mathbb{R}}/F^\vee) \times F_{\mathbb{R}}$ onto its second factor. Moreover the Liouville measure μ on $T^\vee_X$ is nothing but the product of the translation invariant measure on $F^\vee_{\mathbb{R}}/F^\vee$ deduced from the Euclidean metric $\|.\|^\vee$—its total mass is

$$\mathrm{covol}\,\overline{F}^\vee = (\mathrm{covol}\,\overline{F})^{-1}$$

—and of the normalized Lebesgue measure on the Euclidean space $(E_{\mathbb{R}}, \|.\|)$.

Accordingly, in this situation, the triple $(T^\vee_X, \omega^d/d!, H)$ introduced in paragraph 6.4.2 coincides—up to some "trivial factor" $F^\vee_{\mathbb{R}}/F^\vee$—with the one associated in 6.4.1 to the Euclidean space $(V, \|.\|) = (F_{\mathbb{R}}, \|.\|)$.

In this way, we derive the following expressions for its partition and characteristic functions:

$$Z(\beta) = (\mathrm{covol}\,\overline{F})^{-1} (2\pi m/\beta)^{d/2}$$

and

$$\Psi(\beta) := (d/2)\log(2\pi m/\beta) + \widehat{\deg}\,\overline{F}, \qquad (6.23)$$

where

$$d := \dim F_{\mathbb{R}} = \mathrm{rk}\,F.$$

Consequently, the expression for the energy $U(\beta)$ is unchanged:

$$U(\beta) = d/(2\beta) \qquad (6.24)$$

and the entropy function is

$$S(E) = \frac{d}{2}\log(4\pi em E/d) + \widehat{\deg}\,\overline{F}.$$

We recover the classical formulae describing the kinetic theory of an ideal gas of particles of mass m in the "box with periodic boundary conditions" described by the flat torus associated to $\overline{F}^{\vee}$.

Observe that when $m = (2\pi)^{-1}$—the case relevant for the comparison with the application to Euclidean lattices discussed in Sect. 6.5—the above expressions for Ψ and S take the following simpler forms:

$$\Psi(\beta) = (d/2)\log\beta^{-1} + \widehat{\deg}\,\overline{F}$$

and

$$S(E) = (d/2)\log(2eE/d) + \widehat{\deg}\,\overline{F}.$$

6.4.4 Maxwell's Kinetic Gas Model on a Compact Riemannian Manifold Let us return to the situation of a compact Riemannian manifold (X, g), equipped with some potential function V, introduced in 6.4.2.

For simplicity, let us assume that X is oriented, and let us denote by λ_g the volume form on X associated to the Riemannian metric g (it is a C^{∞} form of degree d, everywhere positive). The expression (6.15) for the Gaussian integrals on some Euclidean vector space admits a straightforward "relative" version, concerning the projection map $\pi : T_X^{\vee} \longrightarrow X$, namely:

$$\pi_*(e^{-\beta\|.\|^{\vee 2}/2m}\,\omega^d/d!) = (2\pi m/\beta)^{d/2}\,\lambda_g.$$

(Here π_* denotes the operation of integration of differential forms along the fibers of π.)

This immediately implies that the partition function associated to $(\mathscr{E}, \mu, H) := (T_X^\vee, \omega^d/d!, H_V)$ is:

$$Z(\beta) := \int_{T_X^\vee} e^{-\beta H_V} \, \omega^d/d! = (2\pi m/\beta)^{d/2} \int_X e^{-\beta V} \lambda_g.$$

Therefore its characteristic function is

$$\Psi(\beta) = (d/2) \log(2\pi m/\beta) + \log \int_X e^{-\beta V} \lambda_g$$

and the energy function is given by:

$$U(\beta) = \frac{d}{2\beta} + \frac{\int_X e^{-\beta V} V \lambda_g}{\int_X e^{-\beta V} \lambda_g}.$$

When the potential V vanishes, we recover the same expression as the ones (6.23) and (6.24) previously derived for flat tori, with $\widehat{\deg} \, \overline{F}$ replaced by $\log \mathrm{vol}(X, g)$ in (6.23), where

$$\mathrm{vol}(X, g) := \int_X \lambda_g$$

denotes the volume of the Riemannian manifold (X, g). For a general potential V, we have:

$$\frac{d}{2\beta} + \min_X V \leqslant U(\beta) \leqslant \frac{d}{2\beta} + \max_X V.$$

6.5   Application to Euclidean Lattices: Proof of Theorem 5.4.3

We now explain how Theorem 5.4.3 follows from Theorem 6.2.1.

6.5.1 Let us consider a Euclidean lattice $\overline{E} := (E, \|.\|)$ of positive rank. To $\overline{E}$ is attached the measure space with Hamiltonian $((\mathscr{E}, \mathscr{T}, \mu), H)$ defined as follows:

$$\mathscr{E} := E, \tag{6.25}$$

$$\mathscr{T} := \mathscr{P}(E), \tag{6.26}$$

$$\mu := \sum_{v \in E} \delta_v \tag{6.27}$$

—in other words, $(\mathscr{E}, \mathscr{T}, \mu)$ is the set E underlying the Euclidean lattice $\overline{E}$ equipped with the counting measure—and:

$$H := \pi \|.\|^2. \tag{6.28}$$

The associated partition function is nothing but the theta function of $\overline{E}$:

$$Z(\beta) = \sum_{v \in E} e^{-\pi\beta\|v\|^2} = \theta_{\overline{E}}(\beta) \quad \text{for every } \beta \in \mathbb{R}_+^*. \tag{6.29}$$

Besides, for any $x \in \mathbb{R}_+^*$,

$$A_n(\pi x) = \mu^{\otimes n}(\{v \in \mathscr{E}^n \mid H_n(v) \leqslant n\pi x\})$$

$$= |\{(v_1, \ldots, v_n) \in E^{\oplus n} \mid \|v_1\|^2 + \ldots + \|v_n\|^2 \leqslant nx\}|.$$

In other words:

$$\log A_n(\pi x) = h_{\mathrm{Ar}}^0(\overline{E}^{\oplus n}, nx). \tag{6.30}$$

Using the relations (6.29) and (6.30), the content of Theorem 6.2.1 applied to the measure space with Hamiltonian defined by (6.25)–(6.28) translates into Theorem 5.4.3.

Indeed, we immediately obtain:

$$\Psi(\beta) = \log \theta_{\overline{E}}(\beta) \quad \text{for every } \beta \in \mathbb{R}_+^*,$$

and

$$S(\pi x) = \tilde{h}_{\mathrm{Ar}}^0(\overline{E}, x) \quad \text{for every } x \in \mathbb{R}_+^*. \tag{6.31}$$

6.5.2 From Theorem 6.2.1, we also derive some additional properties of the functions $\tilde{h}_{\mathrm{Ar}}^0(\overline{E}, .)$ and $\theta_{\overline{E}}$ that may have some interest.

Firstly:

$$\lim_{x \to 0_+} \tilde{h}_{\mathrm{Ar}}^0(\overline{E}, x) = 0.$$

Moreover, as functions of x and β,

$$\tilde{h}_{\mathrm{Ar}}^{0\prime}(\overline{E}, x) := \frac{d\tilde{h}_{\mathrm{Ar}}^0(\overline{E}, x)}{dx}$$

and $\theta'_{\overline{E}}(\beta)/\theta_{\overline{E}}(\beta)$ define real analytic decreasing diffeomorphisms of $\mathbb{R}_+^*$ to itself, and for any (x, β) in $\mathbb{R}_+^{*2}$,

$$\pi\beta = \tilde{h}^{0\prime}_{\mathrm{Ar}}(\overline{E}, x) \iff \pi x = -\theta'_{\overline{E}}(\beta)/\theta_{\overline{E}}(\beta). \tag{6.32}$$

Finally, for any $(x, \beta) \in \mathbb{R}_+^{*2}$, we have:

$$\tilde{h}^0_{\mathrm{Ar}}(\overline{E}, x) \leqslant \log \theta_{\overline{E}}(\beta) + \pi\beta x, \tag{6.33}$$

and equality holds in (6.33) if and only x and β are related by the equivalent conditions (6.32).

6.5.3 The measure space with Hamiltonian associated to $\overline{E}$ by the relations (6.25)–(6.28) may be seen as a quantum version of the one associated to a flat torus in paragraph 6.4.3.

Indeed, with the notation of this paragraph, the Hamiltonian operator that describes a non-relativistic particle of mass m that freely moves on the flat torus $X := F_{\mathbb{R}}^{\vee}/F^{\vee}$ associated to the Euclidean lattice $\overline{F}$ is

$$\mathscr{H} := -(h/2\pi)^2 \Delta/(2m)$$

acting on $L^2(X)$, where h denotes Planck's constant and Δ the usual Laplacian. This operator has a discrete spectrum, which may be parametrized by the lattice F: the eigenfunction associated to $v \in F$ is the function e_v on X defined by

$$e_v([x]) := e^{2\pi i \langle x, v \rangle} \quad \text{for every } x \in F_{\mathbb{R}}^{\vee};$$

it satisfies:

$$\mathscr{H}e_v := -\left(\frac{h}{2\pi}\right)^2 \frac{1}{2m} \Delta\, e_v = \frac{h^2}{2m} \|v\|_{\overline{F}}^2\, e_v.$$

The associated partition function is:

$$Z(\beta) = \sum_{v \in F} e^{-\pi\beta h^2 \|v\|_{\overline{F}}/2m} = \theta_{\overline{F}}(\beta h^2/(2\pi m)).$$

When $h = 1$ and $m = (2\pi)^{-1}$, this partition function Z coincides with θ_β (compare with paragraph 6.4.1 above).

7 Proof of the Main Theorem

In this section, we give a self-contained proof of Theorem 6.2.1, by "unfolding" the arguments in [9, Appendix A], and in the classical proofs of Cramér's theorem. This proof relies on some basic principles of measure and probability theory only.

In this proof, the assertions in Theorem 6.2.1 will *not* be established in the order in which they have been successively stated.

Actually, we shall first study the function Ψ and establish Part 2) of Theorem 6.2.1, and then define the function S as the Legendre-Fenchel transform of $-\Psi(-.)$ and establish its Part 3). This first part of the argument, in Sect. 7.1, will appear rather standard to any mathematician familiar with basic measure theory—except possibly for the introduction of the probability measures ν_β in paragraph 7.1.2, which however should be unsurprising to anybody familiar with the basic principles of statistical thermodynamics.

Then, in Sect. 7.2, we shall establish the expression (6.4) of $S(E)$ as the limit

$$\lim_{n \to +\infty} (1/n) \log A_n(E), \tag{7.1}$$

and thus establish the main assertions of Part 1) of Theorem 6.2.1. This step constitutes the key point of the proof Theorem 6.2.1, and will follow from applications of Markov and Chebyshev inequalities on the measure spaces defined as the product of n-copies of $(\mathscr{E}, \mathscr{T})$ equipped with the product measures $\mu^{\otimes n}$ and $\nu_\beta^{\otimes n}$. These arguments also lead to the strengthening of the limit formula (6.4) stated in Proposition 6.2.2.

At this stage, the proof of Theorem 6.2.1 will be completed, with the exception of the formula (6.5) for the entropy "at the zero temperature limit". This formula, of a more technical character, will be established in Sect. 7.3, which could be skipped at first reading.

Complements and variants to Theorem 6.2.1 and its derivation will be presented in the next sections.

Notably, in Sect. 8.3, we shall present a beautifully simple argument, due to Lanford [47], for the existence of the limit (7.1). Moreover, in Sect. 9, we shall discuss some alternative derivations of the key limit formula (6.4) which asserts the equality of this limit with $S(E)$ defined as the Legendre-Fenchel transform (6.10) of $-\Psi(-.)$. Instead of arguments from measure and probability theory, these derivations will rely on the theory of analytic functions and on the use of the saddle-point method. They originate in the work of Poincaré [59] and of Darwin and Fowler [21, 22, 23].

In this section, we consider a measure space $(\mathscr{E}, \mathscr{T}, \mu)$ and some $\mathscr{T}$-measurable function H from $\mathscr{E}$ to $\mathbb{R}_+$ as in Sect. 6, and we assume that Conditions $\mathbf{T}_1$ and $\mathbf{T}_2$ are satisfied.

7.1   The Functions Ψ, U and S

7.1.1 Analyticity Properties of Z and Ψ   For any a in $\mathbb{R}_+^*$ and any β in the half plane $[a, +\infty) + i\mathbb{R}$ in $\mathbb{C}$, we have:

$$|e^{-\beta H}| = e^{-\operatorname{Re}\beta . H} \leqslant e^{-aH}.$$

As the function e^{-aH} is μ-integrable on $\mathscr{E}$, we immediately derive from this estimate:

Proposition 7.1.1   *For any β in the open half plane*

$$\mathbb{C}_{>0} := \mathbb{R}_+^* + i\mathbb{R},$$

the integral (6.1) that defines $Z(\beta)$ is absolutely convergent. The so-defined function

$$Z : \mathbb{C}_{>0} \longrightarrow \mathbb{C}$$

is holomorphic and bounded on every half plane $[a, +\infty) + i\mathbb{R}$, where $a > 0$.

Moreover, for any $k \in \mathbb{N}$, the k-th derivative of Z is given by the absolutely convergent integral, for every $\beta \in \mathbb{C}_{>0}$:

$$Z^{(k)}(\beta) = \int_{\mathscr{E}} (-H)^k e^{-\beta H} \, d\mu. \tag{7.2}$$

This obviously yields the real analyticity of $Z : \mathbb{R}_+^* \longrightarrow \mathbb{R}$. Moreover, $Z(\beta)$ is clearly positive for any $\beta \in \mathbb{R}_+^*$ (since $\mu(\mathscr{E}) > 0$). Consequently $\Psi = \log Z$ is a well-defined real analytic function on $\mathbb{R}_+^*$.

Observe also that, for any $z \in \mathbb{C}_{>0}$,

$$|Z(z)| = \left| \int_{\mathscr{E}} e^{-zH} \, d\mu \right| \leqslant \int_{\mathscr{E}} |e^{-zH}| \, d\mu = Z(\operatorname{Re} z). \tag{7.3}$$

Consequently, for every $\beta \in \mathbb{R}_+^*$,

$$\Psi(\beta) = \max_{z \in \beta + i\mathbb{R}} \log |Z(z)|.$$

According to Hadamard's three-lines theorem (see for instance [70, Theorem 12.3]), this representation of Ψ implies its convexity on $\mathbb{R}_+^*$. It may also be derived from arguments of real analysis which we now present.

7.1.2 The Measures ν_β and the Convexity Properties of Ψ   For every $\beta \in \mathbb{R}_+^*$, we may introduce the probability measure

$$\nu_\beta := Z(\beta)^{-1} e^{-\beta H} \mu$$

on $(\mathscr{E}, \mathscr{D})$.

Clearly, for every $\varepsilon \in (0, \beta)$, the function $e^{\varepsilon H}$ is ν_β-integrable, and *a fortiori* H belongs to $L^p(\mathscr{E}, \nu_\beta)$ for every $p \in [1, +\infty)$. (However H is *not* essentially bounded with respect to μ—or equivalently to ν_β—as implied by $\mathbf{T}_1$ and $\mathbf{T}_2$.)

Actually, for any $k \in \mathbb{N}$ and every $\beta \in \mathbb{R}_+^*$, we have:

$$Z^{(k)}(\beta) = \int_{\mathscr{E}} (-H)^k e^{-\beta H} \, d\mu = (-1)^k Z(\beta) \int_{\mathscr{E}} H^k \, d\nu_\beta. \qquad (7.4)$$

Let us introduce the mean value m_β and the variance σ_β of H with respect to the probability ν_β, defined by the relations:

$$m_\beta := \int_{\mathscr{E}} H \, d\nu_\beta$$

and

$$\sigma_\beta^2 := \int_{\mathscr{E}} |H - m_\beta|^2 \, d\nu_\beta = \int_{\mathscr{E}} H^2 \, d\nu_\beta - m_\beta^2.$$

As a straightforward consequence of (7.4), we obtain the following formulae:

Proposition 7.1.2 *For every $\beta \in \mathbb{R}_+^*$, we have:*

$$m_\beta = Z(\beta)^{-1} \int_{\mathscr{E}} H e^{-\beta H} \, d\mu = -Z(\beta)^{-1} Z'(\beta) = -\Psi'(\beta), \qquad (7.5)$$

$$\int_{\mathscr{E}} H^2 \, d\nu_\beta = Z(\beta)^{-1} \int_{\mathscr{E}} H^2 e^{-\beta H} \, d\mu = Z(\beta)^{-1} Z''(\beta), \qquad (7.6)$$

and

$$\sigma_\beta^2 = Z(\beta)^{-1} Z''(\beta) - (Z(\beta)^{-1} Z'(\beta))^2 = \frac{d}{d\beta} \frac{Z'(\beta)}{Z(\beta)} = \Psi''(\beta). \qquad (7.7)$$

Corollary 7.1.3 *For every β in $\mathbb{R}_+^*$, $\Psi'(\beta) < 0$ and $\Psi''(\beta) > 0$.*

Proof The expression (7.5) (resp. (7.7)) of $-\Psi'(\beta)$ (resp., of $\Psi''(\beta)$) as m_β (resp., as σ_β^2) shows that it is positive, since H is non-negative and not (almost everywhere) constant.   □

7.1.3 The Function U Let us now consider the real analytic function

$$U := -\Psi' : \mathbb{R}_+^* \longrightarrow \mathbb{R}.$$

According to (7.5), for every $\beta \in \mathbb{R}_+^*$, we have:

$$U(\beta) = m_\beta.$$

As $H \geqslant H_{\min}$ ν_β-almost everywhere on $\mathscr{T}$, it satisfies:

$$U(\beta) \geqslant H_{\min}. \tag{7.8}$$

Besides, according to Corollary 7.1.3,

$$U'(\beta) = -\Psi''(\beta) < 0. \tag{7.9}$$

Proposition 7.1.4 *The limit behaviour of $Z(\beta)$, $\Psi(\beta)$ and $U(\beta)$ when $\beta \in \mathbb{R}_+^*$ goes to 0 and $+\infty$ is given by the following relations:*

$$\lim_{\beta \to 0_+} Z(\beta) = +\infty \quad and \quad \lim_{\beta \to +\infty} Z(\beta) = \mu(H^{-1}(0)) \, (\in \mathbb{R}_+), \tag{7.10}$$

$$\lim_{\beta \to 0_+} \Psi(\beta) = +\infty \quad and \quad \lim_{\beta \to +\infty} \Psi(\beta) = \log \mu(H^{-1}(0)) \, (\in [-\infty, +\infty)), \tag{7.11}$$

and

$$\lim_{\beta \to 0_+} U(\beta) = +\infty \quad and \quad \lim_{\beta \to +\infty} U(\beta) = H_{\min}. \tag{7.12}$$

Proof By monotone convergence (resp., by dominated convergence), as β goes to 0 (resp., to $+\infty$), $Z(\beta)$ goes to

$$\int_{\mathscr{E}} d\mu = \mu(\mathscr{E}) = +\infty \quad (\text{resp., to } \int_{\mathscr{E}} \mathbf{1}_{H^{-1}(0)} d\mu = \mu(H^{-1}(0))).$$

This establishes (7.10), and (7.11) immediately follows.

According to (7.8) and (7.9), the limit $l_0 := \lim_{\beta \to 0_+} U(\beta)$ and $l_\infty := \lim_{\beta \to +\infty} U(\beta)$ exist in $(H_{\min}, +\infty]$ and $[H_{\min}, +\infty)$ respectively.

If l_0 were not $+\infty$, then $U(\beta)$ would stay bounded when β goes to 0, and therefore its primitive $-\Psi(\beta)$ also. This would contradict the first part of (7.11).

Since U is decreasing, we have:

$$\Psi'(\beta) = -U(\beta) \leqslant -l_\infty$$

for every β in $\mathbb{R}_+^*$. Therefore, for every $\beta \geqslant 1$,

$$\Psi(\beta) \leqslant -l_\infty(\beta - 1) + \Psi(1).$$

In other words,

$$\int_{\mathscr{E}} e^{-\beta(H - l_\infty)} d\mu \leqslant e^{l_\infty + \Psi(1)} \quad \text{for } \beta \geqslant 1.$$

This immediately implies that

$$\mu(\{x \in \mathscr{E} \mid H(x) < l_\infty\}) = 0,$$

or equivalently:

$$H_{min} \geqslant l_\infty.$$

This shows that $l_\infty = H_{min}$.   □

From (7.9) and (7.12), we deduce:

Corollary 7.1.5 *The function U defines a decreasing real analytic diffeomorphism*

$$U := \mathbb{R}_+^* \xrightarrow{\sim} (H_{min}, +\infty).$$

7.1.4 The Entropy Function S For any $E \in (H_{min}, +\infty)$, the expression $\Psi(\beta) + \beta E$ defines a strictly convex function of $\beta \in \mathbb{R}_+^*$: its derivative $-U(\beta) + E$ is increasing on $\mathbb{R}_+^*$ and vanishes if (and only if) $\beta = U^{-1}(E)$. Therefore $\Psi(\beta) + \beta E$ attains its infimum over $\mathbb{R}_+^*$ precisely at $\beta = U^{-1}(E)$.

We shall define the entropy function

$$S : (H_{min}, +\infty) \longrightarrow \mathbb{R}$$

by

$$S(E) := \inf_{\beta \in \mathbb{R}_+^*} (\Psi(\beta) + \beta E) = \Psi(U^{-1}(E)) + U^{-1}(E)E.$$

In other words, the function $-S(-.)$ is the Legendre-Fenchel transform of the function Ψ, or equivalently the functions $-S(-.)$ and Ψ are dual in the sense of Young (see 7.1.5 *infra*).

The function S, like Ψ and U^{-1}, is clearly real analytic. Moreover the elementary properties of the Legendre-Fenchel-Young duality applied to Ψ and $-S(-.)$ show that $S'' > 0$ on $\mathbb{R}_+^*$, that S' defines a real analytic diffeomorphism

$$S' : (H_{min}, +\infty) \xrightarrow{\sim} \mathbb{R}_+^*$$

inverse of U, and that, for any $\beta \in \mathbb{R}_+^*$,

$$\Psi(\beta) = \sup_{E \in (H_{min}, +\infty)} (S(E) - \beta E); \tag{7.13}$$

moreover, the supremum in the right hand side of (7.13) is attained at a unique point E in (H_{min}, ∞), namely $E = U(\beta)$.

In other words, for any $E \in (H_{\min}, +\infty)$ and any $\beta \in \mathbb{R}_+^*$, the following inequality holds:

$$S(E) \leqslant \Psi(\beta) + \beta E, \qquad (7.14)$$

and it becomes an equality if and only if $E = U(\beta)$, or equivalently $\beta = S'(E)$.

Finally observe that, from the trivial lower bound

$$S(E) \geqslant \Psi(U^{-1}(E))$$

and the relations

$$\lim_{E \to +\infty} U^{-1}(E) = 0 \quad \text{and} \quad \lim_{\beta \to 0_+} \Psi(\beta) = +\infty$$

(see Proposition 7.1.4 and Corollary 7.1.5), immediately follows the relation (6.6):

$$\lim_{E \to +\infty} S(E) = +\infty.$$

7.1.5 A Reminder on Legendre Duality For the convenience of the reader, in this paragraph we establish the basic facts concerning the Legendre–Fenchel–Young duality of convex smooth functions of one variable used in 7.1.4. They are well known (see for instance [4, §14]), but usually not formulated in the precise form used here.[15]

Let I be a non empty interval in $\mathbb{R}$ and let $f : I \longrightarrow \mathbb{R}$ be a function of class C^2 that is strictly convex, namely that satisfies

$$f''(x) > 0 \quad \text{for every } x \in I.$$

The inverse function theorem applied to f' shows that $J := f(I)$ is a non empty open interval in $\mathbb{R}$ and that f' defines a C^1 diffeomorphism

$$f' : I \xrightarrow{\sim} J.$$

Moreover, for any $p \in J$, the function

$$F(x, p) := px - f(x)$$

---

[15]Legendre duality actually holds in a much more general setting, and we refer the reader to [39, Section 2.2] for a more general discussion of Legendre duality, concerning convex functions on finite dimensional vector spaces, with no smoothness assumptions, and to [70, Chapter 5], for its extension to convex functions on locally convex topological vector space.

of $x \in I$ is concave and attains its supremum at a unique point of I, namely $f'^{-1}(p)$.
(Indeed $\partial F(x, p)/\partial x = p - f'(x)$ and $\partial^2 F(x, p)/\partial x^2 = -f''(x) < 0$.)

The *Legendre-Fenchel transform* or *Young dual* of f is the function

$$g : J \longrightarrow \mathbb{R}$$

defined by

$$g(p) := \max_{x \in I} F(x, p) = F(f'^{-1}(p), p). \tag{7.15}$$

Proposition 7.1.6 *The function g is strictly convex of class C^2. Its derivative
defines a C^1 diffeomorphism, which coincides with the compositional inverse of
f':*

$$g' = f'^{-1} : J \xrightarrow{\sim} I.$$

Moreover, for any $x \in I$,

$$f(x) = \max_{p \in J} G(x, p) = G(x, g'^{-1}(x)) \tag{7.16}$$

where $G(x, p) := px - g(p)$.

The equality (7.16) precisely asserts that f coincides with the Legendre–Fenchel
transform of its Legendre–Fenchel transform g. In other words, the Legendre-
Fenchel transformation is involutive.

The symmetry between the two functions f and g may also be expressed by the
fact that, for any $(x, p) \in I \times J$,

$$px \leqslant f(x) + g(p) \tag{7.17}$$

and that the inequality (7.17) becomes an equality precisely when $p = f'(x)$,
or equivalently when $x = g'(p)$. (The inequality (7.17) is sometimes called the
inequality of Young; see for instance [35, § 4.8].)

Proof of Proposition 7.1.6 By definition of g, for any x in I,

$$g(f'(x)) = F(x, f'(x)) = f'(x)x - f(x).$$

As f' is a C^1 diffeomorphism from I onto J, this shows that g is of class C^1 on J
and that, for every $x \in I$,

$$g'(f'(x)) \, f''(x) = \frac{d}{dx} g(f'(x)) = \frac{d}{dx}[f'(x)x - f(x)] = f''(x) \, x.$$

As f'' does not vanish, this shows that

$$g' \circ f' = \mathrm{Id}_I.$$

Consequently $g'(J) = I$ and g' establishes a C^1 diffeomorphism from J to I, inverse to f'. In particular, like f', the function g' is C^1 with a positive derivative, and g is therefore of class C^2 and strictly convex.

For any $(x, y) \in I \times J$, we have:

$$g(f'(y)) = F(y, f'(y))) = f'(y)y - f(y)$$

and

$$G(x, f'(y)) = f'(y)x - g(f'(y)) = f(y) + f'(y)(x - y).$$

This is the ordinate at the point of abscissa x on the line tangent to the graph of f at the point $(y, f(y))$. As f is strictly convex, this tangent line lies below this graph, and we have

$$G(x, f'(y)) \leqslant f(x)$$

with equality if and only if $x = y$.

By letting $p := f'(y)$, this shows that, for any $(x, p) \in I \times J$,

$$G(x, p) \leqslant f(x),$$

with equality if and only if $p = f'(x)$.

This establishes (7.16) and completes the proof. $\square$

Observe finally that, if f and g are two strictly convex C^2 functions that are Young dual as in Proposition 7.1.6, then f is of class C^k for $k > 2$ (resp. of class C^∞, resp. real analytic) if and only g is. This directly follows from the expressions (7.15) and (7.16) for g and f in terms of each other.

7.2   The Convergence of $(1/n) \log A_n(E)$

7.2.1 The Markov and Chebyshev Inequalities and the Weak Law of Large Numbers
Let us start with a reminder of some basic results in probability theory, that we will formulate in a measure theoretic language adapted to the derivation of Theorem 6.2.1.

Let us consider a σ-finite measure on $\mathscr{E}$,

$$\nu : \mathscr{T} \longrightarrow [0, +\infty]$$

and some $\mathscr{F}$-measurable function

$$f : \mathscr{E} \longrightarrow \mathbb{R}.$$

Markov inequality is the observation that, when f is non-negative, then, for any $\varepsilon \in \mathbb{R}_+^*$,

$$\varepsilon\,\mu(f^{-1}([\varepsilon, +\infty))) = \int_{\mathscr{E}} \varepsilon\,\mathbf{1}_{f^{-1}([\varepsilon,+\infty))}\,dv \leqslant \int_{\mathscr{E}} f\,dv. \qquad (7.18)$$

Let us now assume that v is a probability measure and that f is square integrable, and therefore integrable, with respect to v, and let us introduce its "mean value"

$$m := \int_E f\,dv$$

and its "variance"

$$\sigma := \|f - m\|_{L^2(\mathscr{E},v)}^2.$$

In other words,

$$\sigma^2 = \int_{\mathscr{E}} |f - m|^2\,dv = \int_{\mathscr{E}} f^2\,dv - m^2. \qquad (7.19)$$

The Chebyshev inequality is obtained by applying Markov inequality (7.18) to the function $|f - m|^2$. It asserts that, for every $\varepsilon \in \mathbb{R}_+^*$,

$$\varepsilon^2\,v(\{x \in \mathscr{E} \mid |f(x) - m| \geqslant \varepsilon\}) \leqslant \int_{\mathscr{E}} |f - m|^2\,dv = \sigma^2.$$

For any integer $n \geqslant 1$, we may consider the n-functions

$$\varphi_i : \mathscr{E}^n \longrightarrow \mathbb{R}, \quad 1 \leqslant i \leqslant n,$$

defined by

$$\varphi_i(x_1, \ldots, x_n) := f(x_i) - m.$$

They are clearly square integrable on $(\mathscr{E}^n, \mathscr{F}^{\otimes n}, v^{\otimes n})$ and satisfy:

$$\|\varphi_i\|_{L^2}^2 := \int_{\mathscr{E}^n} |\varphi_i|^2\,dv^{\otimes n} = \int_{\mathscr{E}} |f - m|^2\,dv = \sigma^2.$$

Moreover, since the function $f - m$ satisfies

$$\int_{\mathscr{E}} (f - m)\, dv = 0,$$

or in other words, is orthogonal to the function $\mathbf{1}_{\mathscr{E}}$ in $L^2(\mathscr{E}, v)$, the functions $\varphi_1, \ldots, \varphi_n$ are pairwise orthogonal in $L^2(\mathscr{E}^n, v^{\otimes n})$. This implies that

$$\| (\varphi_1 + \ldots + \varphi_n)/n \|_{L^2}^2 = (\|\varphi_1\|_{L^2}^2 + \ldots + \|\varphi_n\|_{L^2}^2)/n^2 = \sigma^2/n.$$

This observation implies that, if we define the L^2-function $\tilde{f}_n$ on $(\mathscr{E}^n, \mathscr{T}^{\otimes n}, v^{\otimes n})$ by the equality

$$\tilde{f}_n(x_1, \ldots, x_n) := (f(x_1) + \ldots + f(x_n))/n,$$

the mean value of which is clearly

$$\tilde{m}_n := \int_{\mathscr{E}^n} \tilde{f}_n\, dv^{\otimes n} = m,$$

the variance of $\tilde{f}_n$ is given by:

$$\tilde{\sigma}_n^2 := \int_{\mathscr{E}^n} |\tilde{f}_n - \tilde{m}_n|^2\, dv^{\otimes n} = \| (\varphi_1 + \ldots + \varphi_n)/n \|_{L^2}^2 = \sigma^2/n.$$

Therefore Chebyshev inequality applied to the function $\tilde{f}_n$ establishes the following proposition, which constitutes a form of the weak law of large numbers:

Proposition 7.2.1 *With the above notation, for any integer $n \geqslant 1$ and any $\varepsilon \in \mathbb{R}_+^*$, we have:*

$$\varepsilon^2\, v^{\otimes n}(\{(x_1, \ldots, x_n) \in \mathscr{E}^n \mid |(f(x_1) + \ldots f(x_n))/n - m| \geqslant \varepsilon\}) \leqslant \sigma^2/n. \qquad (7.20)$$

7.2.2 Bounding $A_n(E)$ from Above The obvious relation

$$e^{-\beta H_n(x_1, \ldots, x_n)} = e^{-\beta H(x_1)} \ldots e^{-\beta H(x_n)}$$

and the very definition of the measure $\mu^{\otimes n}$ show that, for every integer $n \geqslant 1$,

$$Z(\beta)^n = \left(\int_{\mathscr{E}} e^{-\beta H}\, d\mu \right)^n = \int_{\mathscr{E}^n} e^{-\beta H_n}\, d\mu^{\otimes n}. \qquad (7.21)$$

Proposition 7.2.2 *For any integer $n \geqslant 1$, and any E and β in $\mathbb{R}_+^*$, we have:*

$$A_n(E) \leqslant e^{n\beta E}\, Z(\beta)^n. \qquad (7.22)$$

Proof This is the Markov inequality (7.18) applied to the function $f := e^{-\beta H_n}$ on the measure space $(\mathscr{E}^n, \mathscr{F}^{\otimes n}, \mu^{\otimes n})$ and to $\varepsilon := e^{-n\beta E}$. $\qquad\qquad\square$

By taking the logarithm of (7.22), we obtain:

$$\frac{1}{n} \log A_n(E) \leqslant \beta E + \Psi(\beta). \qquad (7.23)$$

(We define $\log 0$ to be $-\infty$.)

When $E > H_{\min}$, the infimum over $\beta \in \mathbb{R}_+^*$ of the right-and side of (7.23) is, by definition, $S(E)$, and the inequality (7.23) may be rephrased as follows:

Proposition 7.2.3 *For any E in $(H_{\min}, +\infty)$ and any integer $n \geqslant 1$,*

$$\frac{1}{n} \log A_n(E) \leqslant S(E). \qquad (7.24)$$

7.2.3 Bounding $A_n(E)$ from Below For every integer $n \geqslant 1$, and every (E, ε) in $\mathbb{R} \times \mathbb{R}_+^*$, we may consider the following $\mathscr{F}^{\otimes n}$-measurable subset of $\mathscr{E}^n$:

$$\mathscr{S}_n(E, \varepsilon) := \{x \in \mathscr{E}^n \mid |H_n(x) - nE| < n\varepsilon\}.$$

It describes the n-particle states of the system under study with average energy in the interval $(E - \varepsilon, E + \varepsilon)$.

We may also introduce its measure

$$\Sigma_n(E, \varepsilon) := \mu^{\otimes n}(\mathscr{S}_n(E, \varepsilon)).$$

The obvious inclusion

$$\mathscr{S}_n(E, \varepsilon) \subset H_n^{-1}((-\infty, E + \varepsilon])$$

yields the estimate:

$$\Sigma_n(E, \varepsilon) \leqslant A_n(E + \varepsilon). \qquad (7.25)$$

We shall derive a lower bound on $\Sigma_n(E, \varepsilon)$, which will immediately yield some lower bounds on $A_n(E + \varepsilon)$. It will be a consequence of the "weak law of large number" (7.20) applied to the function $f := H$ and to the probability measure ν_β.

Actually, using the expressions (7.5) and (7.7) for the mean value m and the variance σ in this special case, Proposition 7.2.1 then takes the form:

Lemma 7.2.4 *For every integer $n \geqslant 1$, and every β and ε in $\mathbb{R}_+^*$, we have:*

$$\nu_\beta^{\otimes n}(\mathscr{E}^n \backslash \mathscr{S}_n(U(\beta), \varepsilon)) \leqslant \varepsilon^{-2} \Psi''(\beta)/n. \qquad (7.26)$$

From the upper bound (7.26) on the measure of $\mathscr{E}^n \setminus \mathscr{S}_n(E, \varepsilon)$ with respect to $v_\beta^{\otimes n}$, we may derive some lower bound on the measure $\Sigma_n(E, \varepsilon)$ of $\mathscr{S}_n(E, \varepsilon)$ with respect to $\mu^{\otimes n}$:

Proposition 7.2.5 *For every integer $n \geqslant 1$, and every β and ε in $\mathbb{R}_+^*$, we have:*

$$\Sigma_n(U(\beta), \varepsilon) \geqslant e^{n(S(U(\beta)) - \varepsilon\beta)} (1 - \varepsilon^{-2} \Psi''(\beta)/n). \tag{7.27}$$

Proof From the very definition of v_β, we get:

$$\Sigma_n(U(\beta), \varepsilon) := \mu^{\otimes n}(\mathscr{S}_n(U(\beta), \varepsilon)) = Z(\beta)^n \int_{\mathscr{S}_n(U(\beta), \varepsilon)} e^{\beta H_n} \, dv_\beta^{\otimes n}. \tag{7.28}$$

Moreover the lower bound $H_n > n(U(\beta) - \varepsilon)$ holds over $\mathscr{S}_n(E, \varepsilon)$. Consequently:

$$\int_{\mathscr{S}_n(U(\beta), \varepsilon)} e^{\beta H_n} \, dv_\beta^{\otimes n} \geqslant e^{n\beta(U(\beta) - \varepsilon)} v_\beta^{\otimes n}(\mathscr{S}_n(U(\beta), \varepsilon)). \tag{7.29}$$

Besides, we have:

$$Z(\beta)^n = e^{n\Psi(\beta)}, \tag{7.30}$$

and, according to Lemma 7.2.4,

$$v_\beta^{\otimes n}(\mathscr{S}_n(U(\beta), \varepsilon)) \geqslant 1 - \varepsilon^{-2} \Psi''(\beta)/n. \tag{7.31}$$

The estimate (7.27) follows from (7.28)–(7.31) and from the relation $S(U(\beta)) = \beta U(\beta) + \Psi(\beta)$. $\square$

The estimate (7.27) is non-trivial only when $n > \varepsilon^{-2} \Psi''(\beta)$. When this holds, it may be written:

$$\frac{1}{n} \log \Sigma_n(U(\beta), \varepsilon) \geqslant S(U(\beta)) - \varepsilon\beta + (1/n) \log(1 - \varepsilon^{-2} \Psi''(\beta)/n). \tag{7.32}$$

This clearly implies:

Corollary 7.2.6 *For any β and ε in $\mathbb{R}_+^*$,*

$$\varliminf_{n \to +\infty} \frac{1}{n} \log \Sigma_n(U(\beta), \varepsilon) \geqslant S(\beta) - \varepsilon\beta.$$

We may now complete the proof of Theorem 6.2.1.

Together with the trivial estimate (7.25), Corollary 7.2.6 shows that, for any $E \in (H_{\min}, +\infty)$ and any $\varepsilon \in \mathbb{R}_+^*$,

$$\varliminf_{n \to +\infty} \frac{1}{n} \log A_n(E + \varepsilon) \geqslant S(E) - \varepsilon S'(E).$$

(We have performed the change of variable $E = U(\beta)$, or equivalently $\beta = S'(E)$.) Equivalently, for every $E \in (H_{\min}, +\infty)$ and any $\varepsilon \in (0, E - H_{\min})$,

$$\varliminf_{n \to +\infty} \frac{1}{n} \log A_n(E) \geqslant S(E - \varepsilon) - \varepsilon S'(E - \varepsilon).$$

By taking the limit when ε goes to 0_+, we get:

$$\varliminf_{n \to +\infty} \frac{1}{n} \log A_n(E) \geqslant S(E). \tag{7.33}$$

Together with Proposition 7.2.3, this proves that

$$\lim_{n \to +\infty} \frac{1}{n} \log A_n(E) = \sup_{n \geqslant 1} \frac{1}{n} \log A_n(E) = S(E).$$

7.2.4 Proof of Proposition 6.2.2 To establish the lower bound (7.33), we have used Proposition 7.2.5 through its Corollary 7.2.6. By using the full strength of the estimate (7.27) established in this Proposition, it is possible to derive stronger results. To illustrate this point, we now explain how to use it to establish Proposition 6.2.2.

With the notation of Proposition 6.2.2, we may clearly assume that, for every $n \geqslant n_0$, the interval I_n is bounded, contained in $(H_{\min}, +\infty)$, and has a non-empty interior (or equivalently, $l_n > 0$), and we may define

$$b_n := \sup I_n$$

and

$$a_n := b_n - \min(l_n, n^{-1/3}).$$

Then, for every $n \geqslant n_0$,

$$(a_n, b_n) \subset I_n.$$

Moreover,

$$\lim_{n \to +\infty} b_n = E,$$

and the positive real numbers

$$\varepsilon_n := (b_n - a_n)/2$$

satisfy

$$\lim_{n \to +\infty} \varepsilon_n = 0$$

and

$$\varliminf_{n \to +\infty} n\varepsilon_n^2 > \Psi''(\beta). \tag{7.34}$$

For every $n \geqslant n_0$, let us introduce:

$$\tilde{A}_n := \mu^{\otimes n}(\{x \in \mathscr{E}^n \mid H_n(x) \in nI_n\}).$$

To establish Proposition 6.2.2, we need to prove that

$$\lim_{n \to +\infty} \frac{1}{n} \log \tilde{A}_n = S(E). \tag{7.35}$$

The inclusion

$$\{x \in \mathscr{E}^n \mid H_n(x) \in nI_n\} \subseteq \{x \in \mathscr{E}^n \mid H_n(x) \leqslant nb_n\}$$

yields the upper bound:

$$\tilde{A}_n \leqslant A_n(b_n).$$

Together with the upper bound (7.24) on A_n, this implies:

$$\frac{1}{n} \log \tilde{A}_n \leqslant \frac{1}{n} \log A_n(b_n) \leqslant S(b_n),$$

and therefore:

$$\varlimsup_{n \to +\infty} \frac{1}{n} \log \tilde{A}_n \leqslant \lim_{n \to +\infty} S(b_n) = S(E). \tag{7.36}$$

For every $n \geqslant n_0$, we may also introduce

$$\beta_n := S'((a_n + b_n)/2) = U^{-1}((a_n + b_n)/2).$$

Clearly, when n goes to $+\infty$, $(a_n + b_n)/2$ converges to E, and β_n to $\beta = S'(E)$. Therefore the estimate (7.34) implies the existence of η in $(0, 1)$ such that, for any large enough integer n,

$$\varepsilon_n^{-2}\, \Psi''(\beta_n)/n \leqslant \eta.$$

The inclusion

$$\mathscr{S}((a_n + b_n)/2, \varepsilon_n) = \{x \in \mathscr{E}^n \mid H_n(x) \in (a_n, b_n)\}$$

$$\subseteq \{x \in \mathscr{E}^n \mid H_n(x) \in nI_n\}$$

yields the lower bound on $\tilde{A}_n$:

$$\Sigma_n(U(\beta_n), \varepsilon_n) := \mu^{\otimes n}(\mathscr{S}_n((a_n + b_n)/2, \varepsilon_n)) \leqslant \tilde{A}_n.$$

Besides, the lower bound on Σ_n established in Proposition 7.2.5, written in the form (7.32), shows that, when n is large enough:

$$\frac{1}{n} \log \Sigma_n(U(\beta_n), \varepsilon_n) \geqslant S(U(\beta_n)) - \varepsilon_n \beta_n + (1/n) \log(1 - \eta).$$

The last two estimates immediately imply that

$$\lim_{n \to +\infty} \frac{1}{n} \log \tilde{A}_n \geqslant \lim_{n \to +\infty} [S(b_n) - \varepsilon_n \beta_n + (1/n) \log(1 - \eta)] = S(E).$$

Together with (7.36), this establishes (7.35). $\qquad\qquad\qquad\qquad\qquad\qquad\square$

7.3 The Zero Temperature Limit

At this stage, all assertions in Theorem 6.2.1 have been established, but for the expression (6.5) for the limit of $S(E)$ when E decreases to $H_{\min}$.

We will establish it in this subsection, which turns out to be of a more technical character and logically independent of the proof of convergence of $(1/n) \log A_n(E)$ to $S(E)$ in the previous subsection, and could therefore be skipped at first reading.

In [9, Appendix A], the expression (6.5) is obtained as a consequence of the convexity and semi-continuity properties of $S(E)$ as a function of $E \in \mathbb{R}$ with values in $[-\infty, +\infty)$. Here we will derive it from a closer study of the asymptotic behavior of the function $Z(\beta)$—defined as the Laplace transform (6.1) of the measure μ— and of the associated functions $\Psi(\beta)$, and $U(\beta)$, when β goes to $+\infty$.

Physically, this corresponds to the limit where the temperature β^{-1} goes to zero, and the results of this paragraph may be seen as a mathematical interpretation of the third law of thermodynamics, which governs the behavior of the entropy and the heat capacity in this limit (see for instance [40, Sections 1.7 and 8.4]).

7.3.1 Asymptotics of Z and Its Derivatives at Zero Temperature Our study will
rely on the following asymptotic properties satisfied by the derivatives $Z^{(k)}(\beta)$ of
the partition function when β goes to $+\infty$.

Proposition 7.3.1 *When β goes to $+\infty$,*

$$(-1)^k Z^{(k)}(\beta) = H_{\min}^k e^{-\beta H_{\min}} \mu(\mathscr{E}_m) + o(e^{-\beta H_{\min}}) \qquad (7.37)$$

for every $k \in \mathbb{N}$, and

$$\beta^k \left(\frac{d}{d\beta} + H_{\min} \right)^k Z(\beta) := \beta^k \sum_{i=0}^{k} \binom{k}{i} H_{\min}^i Z^{(k-i)}(\beta) = o(e^{-\beta H_{\min}}) \qquad (7.38)$$

for every $k \in \mathbb{N}\backslash\{0\}$.

Proof To establish (7.37), observe that, for any $\beta \in \mathbb{R}_+^*$ and any $k \in \mathbb{N}$,

$$(-1)^k Z^{(k)}(\beta) \, e^{\beta H_{\min}} = \int_{\mathscr{E}} H^k e^{-\beta(H-H_{\min})} \, d\mu$$

and that, according to the dominated convergence theorem,

$$\lim_{\beta \to +\infty} \int_{\mathscr{E}} H^k e^{-\beta(H-H_{\min})} \, d\mu = \int_{\mathscr{E}} H^k \mathbf{1}_{\mathscr{E}_{\min}} \, d\mu = H_{\min}^k \, \mu(\mathscr{E}_{\min}).$$

To establish (7.38), we write, for any $\beta \in \mathbb{R}_+^*$ and any integer $k \geqslant 1$:

$$(-1)^k \beta^k \left(\frac{d}{d\beta} + H_{\min} \right)^k Z(\beta) = \int_{\mathscr{E}} \beta^k (H - H_{\min})^k e^{-\beta(H-H_{\min})} \, d\mu. \qquad (7.39)$$

Observe that, as a function of $t \in \mathbb{R}_+$, $t^k e^{-t}$ increases on $[0, k]$ and decreases
on $[k, +\infty)$; in particular, it is bounded from above by $k^k e^{-k}$. For any $E \in
(H_{\min}, +\infty)$, we let:

$$\mathscr{E}_{\leqslant E} := H^{-1}([H_{\min}, E]) \quad \text{and} \quad \mathscr{E}_{>E} := H^{-1}((E, +\infty)).$$

The non-negative function $\beta^k (H - H_{\min})^k \, e^{-\beta(H-H_{\min})}$ vanishes on $\mathscr{E}_{\min}$, is
bounded from above by $k^k e^{-k}$ and, over $\mathscr{E}_{>E}$, decreases as a function of β when
$\beta \geqslant k(E - H_{\min})^{-1}$. This shows that:

$$\int_{\mathscr{E}_{\leqslant E}} \beta^k (H - H_{\min})^k \, e^{-\beta(H-H_{\min})} \, d\mu \leqslant k^k e^{-k} \, \mu(\mathscr{E}_{\leqslant E}\backslash \mathscr{E}_{\min})$$

$$= k^k e^{-k} \, \mu(H^{-1}((H_{\min}, E]))$$

and, by dominated convergence again,

$$\lim_{\beta \to +\infty} \int_{\mathscr{E}_{>E}} \beta^k (H - H_{\min})^k \, e^{-\beta(H-H_{\min})} \, d\mu = 0.$$

This shows that

$$\overline{\lim_{\beta \to +\infty}} \int_{\mathscr{E}} \beta^k (H - H_{\min})^k \, e^{-\beta(H-H_{\min})} \, d\mu \leqslant k^k e^{-k} \, \mu(H^{-1}((H_{\min}, E])).$$

As E is arbitrary in $(H_{\min}, +\infty)$ and

$$\lim_{E \to H_{\min,+}} \mu(H^{-1}((H_{\min}, E])) = 0,$$

this shows that

$$\lim_{\beta \to +\infty} \int_{\mathscr{E}} \beta^k (H - H_{\min})^k \, e^{-\beta(H-H_{\min})} \, d\mu = 0.$$

Together with (7.39), this establishes (7.38). $\qquad\square$

7.3.2 The Asymptotics of S, U, U' at Zero Temperature and the Third Law of Thermodynamics When $k = 0$, the equality (7.37) shows that, if $\mu(\mathscr{E}_{\min}) = 0$, then

$$Z(\beta) = o(e^{-\beta H_{\min}}) \quad \text{when } \beta \to +\infty,$$

or equivalently, by taking logarithms:

$$\lim_{\beta \to +\infty} (\Psi(\beta) + \beta H_{\min}) = -\infty. \tag{7.40}$$

According to (7.14), for every $\beta \in \mathbb{R}_+^*$,

$$\overline{\lim_{E \to \Pi_{\min}}} S(E) \leqslant \Psi(\beta) + \beta H_{\min}'.$$

Together with (7.40), this shows that

$$\lim_{H \to H_{\min+}} S(E) = -\infty \quad \text{when} \quad \mu(\mathscr{E}_{\min}) = 0. \tag{7.41}$$

Let us now assume that $\mu(\mathscr{E}_{\min}) > 0$. Then (7.37) with $k = 0$ shows that, when β goes to $+\infty$,

$$Z(\beta) = \mu(\mathscr{E}_{\min})e^{-\beta H_{\min}} + o(e^{-\beta H_{\min}}) \sim \mu(\mathscr{E}_{\min})e^{-\beta H_{\min}},$$

or equivalently,

$$\Psi(\beta) := \log Z(\beta) = -H_{\min}\beta + \log \mu(\mathscr{E}_{\min}) + o(1).$$ (7.42)

Accordingly, the relations (7.37) and (7.38) may be written:

$$(-1)^k Z^{(k)}(\beta) = H_{\min}^k e^{-\beta H_{\min}} \mu(\mathscr{E}_{\min}) + o(Z(\beta))$$ (7.43)

and

$$\beta^k \sum_{i=0}^{k} \binom{k}{i} H_{\min}^i Z^{(k-i)}(\beta) = o(Z(\beta)).$$ (7.44)

The asymptotic relations (7.44) may be reformulated in a more convenient form, namely:

Corollary 7.3.2 *If* $\mu(\mathscr{E}_{\min}) > 0$, *then*

$$\lim_{\beta \to +\infty} \beta(U(\beta) - H_{\min}) = 0$$ (7.45)

and, for any integer $k \geqslant 1$,

$$\lim_{\beta \to +\infty} \beta^{k+1} U^{(k)}(\beta) = 0.$$ (7.46)

Proof *When* $k = 1$, *the relation (7.44) reads:*

$$\beta(Z'(\beta) + H_{\min} Z(\beta)) = o(Z(\beta)),$$

and may be written as (7.45), since $U(\beta) = -Z'(\beta)/Z(\beta)$.

From this last relation also follows, by a straightforward induction on the integer $n \geqslant 1$, *the existence of some polynomial* P_n *in* $\mathbb{Z}[X_0, \cdots, X_{n-2}]$ *such that*

$$\frac{Z^{(n)}(\beta)}{Z(\beta)} = -U^{(n-1)}(\beta) + P_n(U(\beta), \cdots, U^{(n-2)}(\beta)).$$

When $k = 1$, $P_1 = 0$. *Moreover* $P_n(X_0, \cdots, X_{n-2})$ *is homogeneous of weight* n *when each indeterminate* X_i *is given the weight* $i + 1$.

When $H_{\min} = 0$, *the relations (7.44) take the form:*

$$\beta^k Z^{(k)}(\beta) = o(Z(\beta)),$$

for every $k \geqslant 1$, and the relations (7.46) follow from (7.44) by induction on $k \geqslant 1$, by using the relation

$$\beta^n U^{(n-1)}(\beta) = -\beta^n \frac{Z^{(n)}(\beta)}{Z(\beta)} + P_n(\beta U(\beta), \beta^2 U'(\beta), \cdots, \beta^{n-1} U^{(n-2)}(\beta))$$

with $n = k + 1$.

One easily reduces to the situation where $H_{\min} = 0$, by considering the function $\tilde{H} := H - H_{\min}$ instead of H, which replaces $U(\beta)$ by $\tilde{U}(\beta) := U(\beta) - H_{\min}$. We leave the details to the reader.   □

Corollary 7.3.2 may be understood as a mathematical expression of the third law of thermodynamics, which notably asserts the existence of a finite limit of entropy at zero temperature.

Indeed, combined with the expression

$$S(U(\beta)) = \Psi(\beta) + \beta U(\beta)$$

for the entropy function S at the energy $U(\beta)$, the asymptotics (7.42) and (7.45) of $\Psi(\beta)$ and $U(\beta)$ at zero temperature show that, when $\mu(\mathscr{E}_{\min}) > 0$,

$$\lim_{E \to H_{\min,+}} S(E) = \lim_{\beta \to +\infty} S(U(\beta)) = \log \mu(\mathscr{E}_{\min}) \tag{7.47}$$

and is therefore finite.

As shown by (7.41), the relation (7.47) still holds when $\mu(\mathscr{E}_{\min}) = 0$ (with the convention $\log 0 = -\infty$). In our mathematical approach, the validity of the third law of thermodynamics—phrased as the existence of a finite limit of $S(U(\beta))$ when β goes to $+\infty$—is therefore equivalent to the positivity of $\mu(\mathscr{E}_{\min})$. It notably forbids "classical mechanical systems" for which $\mu(\mathscr{E}_{\min}) = 0$, like the ones discussed in Sect. 6.4 above. This gives a mathematical interpretation of the well-known fact that the third law of thermodynamics reflects the quantum nature of the physical world.

To interpret the relation (7.46), observe that the derivative of $U(\beta)$ with respect to the temperature β^{-1},

$$c(\beta^{-1}) := \frac{dU(\beta)}{d(\beta^{-1})} = -\beta^2 U'(\beta) = \beta^2 \Psi''(\beta) \tag{7.48}$$

represents the heat capacity of the system under study. Accordingly, (7.46) for $k = 1$ asserts that the heat capacity goes to zero with the temperature, a well known consequence of the third law of thermodynamics (see for instance [40, Section 1.7]).

More generally, the relation (7.46) for $k \geqslant 1$ arbitrary is easily seen to be equivalent to

$$c^{(k)}(T) = o(T^{-k}) \quad \text{when } T \longrightarrow 0_+.$$

8 Complements

In this section, we present some complements to Theorem 6.2.1 and its proof.

In Sects. 8.1 and 8.2, we begin with some remarks on its various possible formulations and on the relations between some of the estimates involved in its proof and various classical estimates in probability and analytic number theory.

Then, in Sect. 8.3, we present Lanford's direct approach to the asymptotic behavior of the measures $A_n(E)$ established in Theorem 6.2.1, approach based on elementary subadditivity estimates.

Finally, in Sect. 8.4, we discuss a mathematical interpretation of the second law of thermodynamics in our formalism and its application to Euclidean lattices.

8.1 The Main Theorem When $(\mathscr{T}, H) = (\mathbb{R}_+, \mathrm{Id}_{\mathbb{R}_+})$

8.1.1 In the special case where $(\mathscr{E}, \mathscr{T})$ is $(\mathbb{R}_+, \mathscr{B})$, the non-negative real numbers equipped with the σ-algebra of Borel subsets, and where H is the identity function:

$$H = \mathrm{Id}_{\mathbb{R}_+} : \mathbb{R}_+ \longrightarrow \mathbb{R}_+,$$

Theorem 6.2.1 boils down to a statement about positive Radon measures on $\mathbb{R}_+$ with finite Laplace transforms and their powers under convolution product.

Indeed, let us consider some non-negative Borel measure on $\mathbb{R}_+$. It is a Radon measure (that is, $\mu(K) < +\infty$ for every compact subset of $\mathbb{R}_+$) if and only if its distribution function, defined by:

$$N(E) := \mu([0, E]) < +\infty$$

is finite for every $E \in \mathbb{R}_+$. Then the function $N : \mathbb{R}_+ \longrightarrow \mathbb{R}_+$ is non-decreasing and right-continuous, and the measure μ is the Stieljes measure associated to the function N (extended by 0 on $\mathbb{R}_-^*$).[16]

When moreover $H = \mathrm{Id}_{\mathbb{R}_+}$, the partition function Z associated to

$$((\mathscr{E}, \mathscr{T}, \mu), H) = ((\mathbb{R}, \mathscr{B}, \mu), \mathrm{Id}_{\mathbb{R}_+})$$

---

[16]In other words, μ that is the distributional derivative of the distribution on $\mathbb{R}$ associated to the locally bounded function N. This construction establishes a bijection between non-negative Radon measures μ on $\mathbb{R}_+^*$ and non-decreasing right-continuous functions $N : \mathbb{R}_+^* \longrightarrow \mathbb{R}_+^*$, and one usually writes: $\int_{\mathbb{R}_+} f \, d\mu = \int_{\mathbb{R}_+} f(x) \, dN(x)$.

becomes the Laplace transform of μ, Ψ its logarithm, and U its logarithmic derivative. Namely, for every $\beta \in \mathbb{R}_+^*$, we have :

$$\Psi(\beta) = \log \int_{\mathbb{R}_+} e^{-\beta x} \, d\mu(x),$$

and

$$U(\beta) = \frac{\int_{\mathbb{R}_+} x e^{-\beta x} \, d\mu(x)}{\int_{\mathbb{R}_+} e^{-\beta x} \, d\mu(x)}.$$

Moreover, for any integer $n \geqslant 1$, we may consider the n-th power

$$\mu^{*n} := \mu * \cdots * \mu \quad (n \, \text{times})$$

of μ under the convolution product. It is the Borel measure on $\mathbb{R}$ defined by the equality:

$$\mu^{*n}(B) := \mu^{\otimes n}\{(x_1, \ldots, x_n) \in \mathbb{R}^n \mid x_1 + \ldots + x_n \in B\}$$

for any Borel subset B of $\mathbb{R}$. It is easily seen to be a Radon measure supported by $\mathbb{R}_+$. Moreover, for every $E \in \mathbb{R}_+$,

$$A_n(E) = \mu^{*n}([0, nE])$$

and therefore

$$S(E) = \lim_{n \to +\infty} \frac{1}{n} \log \mu^{*n}([0, nE]).$$

8.1.2 Let us return to Theorem 6.2.1, in its general formulation. We may introduce the Borel measure $\tilde{\mu}$ on $\mathbb{R}_+$ defined as the image of the measure μ by the measurable function H:

$$\tilde{\mu} := H_* \mu : B \longmapsto \mu(H^{-1}(B)).$$

It is then straightforward that the functions $Z(\beta)$, $\Psi(\beta)$ and $A_n(E)$ attached to the measure space $(\mathbb{R}_+, \mathcal{B}, \tilde{\mu})$ equipped with the function $\tilde{H} := \mathrm{Id}_{\mathbb{R}_+}$ coincides with the ones attached to $(\mathcal{E}, \mathcal{T}, \mu)$ equipped with H.

In particular, the validity of Theorem 6.2.1 in the special case discussed in 8.1.1 above implies its general validity. However this reduction does not lead to any actual simplification in the derivation of Theorem 6.2.1 presented in Sect. 7. One might even argue that the measure theoretic arguments in paragraphs 7.2.2 and 7.2.3 are clearer when presented in the general setting dealt with in Sect. 7.

8.2   Chernoff's Bounds and Rankin's Method

In paragraph 7.2.2, the first step in the proof of the convergence of $\log A_n(E)/n$ to

$$S(E) := \inf_{\beta \in \mathbb{R}_+^*} (\Psi(\beta) + \beta E) \tag{8.1}$$

has been to establish the upper bound

$$\frac{1}{n} \log A_n(E) \leqslant \inf_{\beta \in \mathbb{R}_+^*} (\Psi(\beta) + \beta E) \tag{8.2}$$

for any integer $n \geqslant 1$ and any $E > H_{\min}$. When $n = 1$, this upper bound reads:

$$\log \mu(H^{-1}([H_{\min}, E])) =: \log A_1(E) \leqslant \inf_{\beta \in \mathbb{R}_+^*} (\Psi(\beta) + \beta E). \tag{8.3}$$

This is the content of Proposition 7.2.2 when $n = 1$, itself a straightforward consequence of Markov inequality applied to the function $e^{-\beta H}$.

Inequalities of this type, which provides an upper bound for "tails probability" in terms of the "logarithmic moment generating function" Ψ, are classically known as *Chernoff's bounds*, by reference to Chernoff's seminal article [14], which constitutes, with the earlier article by Cramér [19], the starting point of the theory of large deviations. In [14], Chernoff establishes a basic theorem of large deviations, on which Theorem 6.2.1 is modeled, by considering in substance a framework similar to the one in Sect. 6.1, but were μ is a probability measure. Chernoff's theorem extends the earlier results in [19], established under more specific assumptions on the measure $H_*\mu$, and the arguments leading to (8.2) in paragraph 7.2.2 are direct adaptations of the ones in [14].

In spite of the simplicity of their derivation, Chernoff's bounds like (8.3) turn out to provide surprisingly sharp estimates[17] for tail probabilities, and have led to important inequalities, that play a key role in probability theory and its applications. We refer the reader to [12, Chapter 2], for a presentation of such inequalities, from the perspective of recent developments on concentration inequalities.

An avatar of Chernoff's bounds also appears in analytic number theory under the name of *Rankin's trick*. Let

$$f(s) := \sum_{n=1}^{+\infty} \frac{a_n}{n^s}$$

---

[17]The equality $\lim_{n \to +\infty} \frac{1}{n} \log A_n(E) = \inf_{\beta \in \mathbb{R}_+^*} (\Psi(\beta) + \beta E)$ somewhat explains this sharpness. See also [55] for related "Tauberian estimates".

be a Dirichlet series with non-negative coefficients, which admits 0 as abscissa of convergence. One is interested in bounding the partial sums $\sum_{1 \leqslant n \leqslant x} a_n$ from above, as a function of x in $[1, +\infty)$. To achieve this, one observes that, for any $\eta \in \mathbb{R}_+^*$,

$$\sum_{1 \leqslant n \leqslant x} a_n \leqslant x^\eta \sum_{n=1}^{+\infty} \frac{a_n}{n^\eta} = x^\eta f(\eta).$$

One often obtains a sharp upper bound on $\sum_{1 \leqslant n \leqslant x} a_n$ by choosing η in $\mathbb{R}_+^*$ that minimizes $x^\eta f(\eta)$.

Such estimates notably appear in Rankin's article [61] (see proof of Lemma II). Similar arguments had actually been used earlier by Hardy and Ramanujan (see [34, Section 4.1]). They constitute nothing but the special case of Chernoff's bound (8.3) when

$$\mathscr{E} = \mathbb{N}_{>0}, \quad \mu := \sum_{n=1}^{+\infty} a_n \delta_n, \quad \text{and} \quad H(n) = \log n.$$

8.3 Lanford's Estimates

It turns out that the existence of the limit (6.4):

$$\lim_{n \to +\infty} \frac{1}{n} \log A_n(E)$$

when E belongs to $(H_{\min}, +\infty)$, together with its concavity as a function of E, may be directly established, independently of the more sophisticated arguments[18] in paragraph 7.2.3.

In this paragraph, we briefly discuss this direct approach, which originates in Lanford's work [47] on the rigorous derivation of "thermodynamic limits" in statistical mechanics. We refer the reader to the original article [47] for developments of this approach, which emphasize the role of convexity in the thermodynamic formalism. One should also consult the long introduction[19] of [41] by Wightman for an enlightening discussion of this circle of ideas from a historical perspective.

---

[18]Which, of course, prove more, namely the equality of this limit with the function S defined by the Legendre transform of Ψ.

[19]Entitled *Convexity and the notion of equilibrium state in thermodynamics and statistical mechanics*.

We are going to present a simple proof of the following fragment of Theorem 6.2.1:

Proposition 8.3.1 *Let us consider a measure space with Hamiltonian $((\mathcal{E}, \mathcal{T}, \mu))$, $H)$, as in Sect. 6.1, and let us assume that the measure μ is non zero—or equivalently, that $H_{\min}$ is finite—and that Condition T_2 is satisfied.*

Then, for any $E \in (H_{\min}, +\infty)$, the limit $\lim_{n \to +\infty} \log A_n(E)/n$ exists in $\mathbb{R}$, and also equals $\sup_{n \geq 1} \log A_n(E)/n$. Moreover, it defines a continuous, non-decreasing, and concave function of $E \in (H_{\min}, +\infty)$.

Lanford's arguments to derive such a statement rely on the following subadditivity estimates:

Lemma 8.3.2 *For any (E_1, E_2) in $[H_{\min}, +\infty)^2$ and any two positive integers n_1 and n_2,*

$$A_{n_1}(E_1).A_{n_2}(E_2) \leqslant A_{n_1+n_2}\left(\frac{n_1 E_1 + n_2 E_2}{n_1 + n_2}\right). \tag{8.4}$$

Proof The following inclusion of subsets of $\mathcal{E}^{n_1+n_2}$ is a straightforward consequence of their definitions:

$$H_{n_1}^{-1}((-\infty, n_1 E_1]) \times H_{n_2}^{-1}((-\infty, n_2 E_2]) \subseteq H_{n_1+n_2}^{-1}((-\infty, n_1 E_1 + n_2 E_2]).$$

These subsets are $\mathcal{T}^{\otimes(n_1+n_2)}$-measurable, and by applying the measure $\mu^{\otimes(n_1+n_2)}$ to this inclusion, we get (8.4). $\qquad\square$

Proof of Proposition 8.3.1 Recall that, according to a well-known observation that goes back to Fekete [28], superadditive sequences of real numbers have a simple asymptotic behaviour: $\qquad\square$

Lemma 8.3.3 *Let $(a_n)_{n \geqslant 1}$ be a sequence in $(-\infty, +\infty]$ that is superadditive (namely, that satisfies $a_{n_1+n_2} \geqslant a_{n_1} + a_{n_2}$ for any two positive integers n_1 and n_2).*

Then the sequence $(a_n/n)_{n \geqslant 1}$ admits a limit in $(-\infty, +\infty]$. Moreover:

$$\lim_{n \to +\infty} a_n/n = \sup_{n \geqslant 1} a_n/n. \tag{8.5}$$

Proof For the sake of completeness, we recall how to establish (8.5).

The upper-bound

$$\overline{\lim_{n \to +\infty}} \, a_n/n \leqslant \sup_{n \geqslant 1} a_n/n$$

is clear. Moreover, for any $n \geqslant 1$ and any $i \in \{1, \ldots, n\}$, we have:

$$a_{kn+i} \geqslant k a_n + a_i \quad \text{for every } k \in \mathbb{N},$$

and therefore

$$\varliminf_{k \to +\infty} \frac{a_{kn+i}}{kn+i} \geqslant \lim_{k \to +\infty} \frac{k a_n + a_i}{kn+i} = \frac{a_n}{n}.$$

This shows that

$$\varliminf_{l \to +\infty} a_l / l \geqslant a_n / n,$$

and finally that

$$\varliminf_{l \to +\infty} a_l / l \geqslant \sup_{n \geqslant 1} a_n / n. \qquad \square$$

For any E in $(H_{\min}, +\infty)$, we define a sequence $(a_n)_{n \geqslant 1}$ in $(-\infty, +\infty]$ by letting:

$$a_n := \log A_n(E).$$

Indeed the estimates (8.4) with $E_1 = E_2$ yield the lower bound $A_n(E) \geqslant A_1(E)^n$, and this is positive by the very definition of $H_{\min}$.

These estimates also imply that the sequence $(a_n)_{n \geqslant 1}$ is superadditive. Moreover, the upper bound (7.23)—which, as explained in paragraph 7.2.2, easily follows from Condition $\mathbf{T}_2$, once it is expressed as the finiteness $\mathbf{T}_2'$ of the partition function $Z(\beta)$ for every $\beta \in \mathbb{R}_+^*$—shows that the sequence $(a_n/n)_{n \geqslant 1}$ is bounded from above.

According to Lemma 8.3.3, this already establishes the required convergence and finiteness:

$$\lim_{n \to +\infty} \log A_n(E)/n = \sup_{n \geqslant 1} \log A_n(E)/n \in \mathbb{R}.$$

As a function of $E \in (H_{\min}, +\infty)$, this limit

$$s(E) := \lim_{n \to +\infty} \log A_n(E)/n$$

is non-decreasing, like $\log A_n(E)$ for every $n \geqslant 1$. For any $E \in (H_{\min}, +\infty)$, we may define:

$$s(E)_- := \lim_{\tilde{E} \to E_-} s(\tilde{E}) \quad \text{and} \quad s(E)_+ := \lim_{\tilde{E} \to E_+} s(\tilde{E}).$$

Clearly, we have:

$$s(E)_- \leqslant s(E)_+, \qquad (8.6)$$

and the function s is continuous at the point E if and only if equality holds in (8.6).

Besides, for any E_1 and E_2 in $(H_{min}, +\infty)$, Lanford's estimates (8.4) may be written:

$$\frac{n_1}{n_1 + n_2} \frac{\log A_{n_1}(E_1)}{n_1} + \frac{n_2}{n_1 + n_2} \frac{\log A_{n_2}(E_2)}{n_2}$$

$$\leqslant \frac{\log A_{n_1+n_2}((n_1 E_1 + n_2 E_2)/(n_1 + n_2))}{n_1 + n_2}.$$

This implies that, for any E_1 and E_2 in $(H_{min}, +\infty)$ and any α_1 and α_2 in $\mathbb{Q} \cap [0, 1]$ such that $\alpha_1 + \alpha_2 = 1$, the following inequality holds:

$$\alpha_1 s(E_1) + \alpha_2 s(E_2) \leqslant s(\alpha_1 E_1 + \alpha_2 E_2). \tag{8.7}$$

For any E in $(H_{min}, +\infty)$, one may easily construct sequences $(E_{1,k})$ and $(E_{2,k})$ in $(H_{min}, +\infty)$ such that

$$\lim_{k \to +\infty} E_{1,k} = \lim_{k \to +\infty} E_{2,k} = E,$$

and such that $(E_{1,k})$ and $((E_{1,k} + E_{2,k})/2)$ are increasing and $(E_{2,k})$ is decreasing. Applying (8.7) to $E_1 = E_{1,k}$, $E_2 = E_{2,k}$, and $\alpha_1 = \alpha_2 = 1/2$, we obtain:

$$(1/2)s(E)_- + (1/2)s(E)_+ \leqslant s(E)_-,$$

and therefore:

$$s(E)_+ \leqslant s(E)_-.$$

This establishes the continuity of s.

Using this continuity, we immediately derive that the estimates (8.7) still holds for any α_1 and α_2 in $[0, 1]$ such that $\alpha_1 + \alpha_2 = 1$. This establishes the concavity of s.

8.4　Products and Thermal Equilibrium

The formalism developed in Sects. 6 and 7—that attaches functions Ψ and S to a measure space $(\mathscr{E}, \mathscr{T}, \mu)$ and to a non-negative function H on $\mathscr{E}$ satisfying $\mathbf{T_2}$—satisfies a simple but remarkable compatibility with finite products, that we briefly discuss in this subsection.

8.4.1 Products of Measures Spaces with a Hamiltonian Assume that, for any element i in some non-empty finite set I, we are given a measure space $(\mathscr{E}_i, \mathscr{T}_i, \mu_i)$ and a measurable function $H_i : \mathscr{E}_i \longrightarrow \mathbb{R}_+$ as in Sect. 6.1 above.

Then we may form the product measure space $(\mathcal{E}, \mathcal{T}, \mu)$ defined by the set $\mathcal{E} :=$ $\prod_{i \in I} \mathcal{E}_i$ equipped with the σ-algebra $\mathcal{T} := \bigotimes_{i \in I} \mathcal{T}_i$ and the product measure $\mu :=$ $\bigotimes_{i \in I} \mu_i$.

We may also define a measurable function

$$H : \mathcal{E} \longrightarrow \mathbb{R}_+$$

by the formula

$$H := \sum_{i \in I} \mathrm{pr}_i^* H_i,$$

where $\mathrm{pr}_i : \mathcal{E} \longrightarrow \mathcal{E}_i$ denotes the projection on the i-th factor.

Let us assume that, for every $i \in I$, the measure μ_i is not zero and that $(\mathcal{E}_i, \mathcal{T}_i, \mu_i)$ and H_i satisfy the condition $\mathbf{T_2}$, or equivalently that the functions $e^{-\beta H_i}$ is μ_i-integrable for every $\beta \in \mathbb{R}_+^*$.

Then $(\mathcal{E}, \mathcal{T}, \mu)$ and H are easily seen to satisfy $\mathbf{T_2}$ also, as a consequence of Fubini's Theorem. Actually Fubini's Theorem shows that the function $Z : \mathbb{R}_+^* \longrightarrow$ $\mathbb{R}_+^*$ and $\Psi : \mathbb{R}_+^* \longrightarrow \mathbb{R}$ attached to the above data, defined as in Sect. 6.1 by the formulae

$$Z(\beta) = \int_{\mathcal{E}} e^{-\beta H} d\mu \quad \text{and} \quad \Psi(\beta) := \log Z(\beta)$$

and the "partial functions" Z_i and Ψ_i, $i \in I$, attached to the measured space $(\mathcal{E}_i, \mathcal{T}_i, \mu_i)$ equipped with the function H_i by the similar formulae

$$Z_i(\beta) := \int_{\mathcal{E}_i} e^{-\beta H_i} d\mu_i \quad \text{and} \quad \Psi_i(\beta) := \log Z_i(\beta),$$

satisfy the relations:

$$Z(\beta) = \prod_{i \in I} Z_i(\beta) \quad \text{and} \quad \Psi(\beta) = \sum_{i \in I} \Psi_i(\beta). \tag{8.8}$$

In particular, the functions U and U_i, $i \in I$, defined by (6.9), satisfy the additivity relation:

$$U(\beta) = \sum_{i \in I} U_i(\beta). \tag{8.9}$$

8.4.2 The Entropy Function Associated to a Product and the Second Law of Thermodynamics From now on, let us also assume that, for every $i \in I$, Condition $\mathbf{T_1}$ holds, namely that $\mu_i(\mathcal{E}_i) = +\infty$. Then $\mu(\mathcal{E}) = +\infty$—in other words, $(\mathcal{E}, \mathcal{T}, \mu)$ also satisfies $\mathbf{T_1}$—and we may apply Theorem 6.2.1 to the data $(\mathcal{E}_i, \mathcal{T}_i, \mu_i, H_i)$, $i \in I$, and $(\mathcal{E}, \mathcal{T}, \mu, H)$.

Notably, if $H_{i,\min}$ (resp. $H_{\min}$) denotes the essential infimum of the function H_i on the measure space $(\mathscr{E}_i, \mathscr{T}_i, \mu_i)$ (resp., of H on $\mathscr{E}, \mathscr{T}, \mu$)), we may define some concave functions:

$$S_i : (H_{i,\min}, +\infty) \longrightarrow \mathbb{R}, \quad \text{for } i \in I,$$

and

$$S : (H_{\min}, +\infty) \longrightarrow \mathbb{R}.$$

Observe also that, as a straightforward consequence of the definitions, we have:

$$H_{\min} = \sum_{i \in I} H_{i,\min}.$$

The expression (8.8) of Ψ as sum of the Ψ_i's translates into the following description of the entropy function S in terms of the S_i's:

Proposition 8.4.1

(1) For each $i \in I$, let E_i be a real number in $(H_{i,\min}, +\infty)$. Then the following inequality is satisfied:

$$\sum_{i \in I} S_i(E_i) \leqslant S(\sum_{i \in I} E_i). \tag{8.10}$$

Moreover equality holds in (8.10) if and only if the positive real numbers $S'(E_i)$, $i \in I$, are all equal. When this holds, if β denotes their common value, we also have:

$$\beta = S'(\sum_{i \in I} E_i).$$

(2) Conversely, for any $E \in (H_{\min}, +\infty)$, there exists a unique family $(E_i)_{i \in I} \in \prod_{i \in I}(H_{i,\min}, +\infty)$ such that

$$E = \sum_{i \in I} E_i \quad \text{and} \quad S(E) = \sum_{i \in I} S_i(E_i).$$

Indeed, if $\beta = S'(E)$, it is given by

$$(E_i)_{i \in I} = (U_i(\beta))_{i \in I},$$

where $U_i = -\Psi_i'$.

Proof Let $(E_i)_{i\in I}$ be an element of $\prod_{i\in I}(H_{i,\min}, +\infty)$. According to Theorem 6.2.1, (3), we have, for every $i \in I$:

$$S(E_i) = \inf_{\beta>0}(\beta E_i + \Psi_i(\beta)). \tag{8.11}$$

Moreover, the infimum is attained for a unique β in $\mathbb{R}_+^*$, namely $S'(E_i)$.
Similarly, for $E := \sum_{i\in I} E_i$,

$$S(E) = \inf_{\beta>0}(\beta E + \Psi(\beta)), \tag{8.12}$$

and the infimum is attained for a unique positive β, namely $S'(E)$.

Besides, the additivity relation (8.8) shows that, for every β in $\mathbb{R}_+^*$,

$$\beta E + \Psi(\beta) = \sum_{i\in I}(\beta E_i + \Psi_i(\beta)).$$

Part (1) of the proposition directly follows from these observations. Part (2) follows from Part (1) and from the relation $\Psi' = \sum_{i\in I} \Psi_i'$. $\qquad\square$

Proposition 8.4.1 notably asserts that, for any E in $(H_{\min}, +\infty)$,

$$S(E) = \max\left\{\sum_{i\in I} S(E_i); (E_i)_{i\in I} \in \prod_{i\in I}(H_{i,\min}, +\infty), \sum_{i\in I} E_i = E\right\}.$$

In other words, the function S is the "tropical convolution" of the functions $(S_i)_{i\in I}$. (Recall that, in tropical mathematics, products are replaced by sums, and sums and integrals by maxima and suprema.)

The above results admit the following physical interpretation, in line with the discussion in Sect. 6.3.

The product $((\mathscr{E}, \mathscr{T}, \mu), H)$ of the measure spaces with Hamiltonian $((\mathscr{E}_i, \mathscr{T}_i, \mu_i), H_i)$ represents an elementary system composed of subsystems (indiced by I). The relation (8.9) expresses the fact that the energy is an extensive quantity. Proposition 8.4.1 shows that the entropy $S(E)$ of the composite system may be computed as the sum of the entropies $S_i(E_i)$ of its subsystems for the (unique) values of the energies $(E_i)_{i\in I}$ of its subsystems which add up to E, and maximize the sum of these partial entropies, or equivalently that gives each of the subsystems the same temperature as the total system.

In this way, Proposition 8.4.1 appears as a mathematical interpretation of the second law of thermodynamics.

8.4.3 Application to Euclidean Lattices In Sect. 6.5, in order to derive Theorem 5.4.3 from Theorem 6.2.1, we have associated a measure space with Hamiltonian to any Euclidean lattice, defined by the relations (6.25)–(6.28).

It directly follow from these relations that this construction is compatible with direct sums of Euclidean lattices: for any two Euclidean lattices $\overline{E}_1$ and $\overline{E}_2$, the measure space with Hamiltonian associated to $\overline{E} := \overline{E}_1 \oplus \overline{E}_2$ may be identified with the product of the measure spaces with Hamiltonian associated to $\overline{E}_1$ and $\overline{E}_2$.

Taking into account the relation (6.31) between the invariant $\tilde{h}^0_{Ar}$ attached to Euclidean lattices and the entropy function of the associated measure spaces with Hamiltonian, Proposition 8.4.1 (with $I = \{1, 2\}$) applied to this product decomposition immediately establishes Corollary 5.4.4. Using (6.32), it actually shows that the maximum in the right-hand side of (5.12) is achieved at unique pair (x_1, x_2), namely when

$$x_1 = -\pi^{-1} \theta'_{\overline{E}_1}(\beta)/\theta_{\overline{E}_1}(\beta) \quad \text{and} \quad x_2 = -\pi^{-1} \theta'_{\overline{E}_2}(\beta)/\theta_{\overline{E}_2}(\beta),$$

where $\beta \in \mathbb{R}^*_+$ is defined by the equality:

$$x = -\pi^{-1} \theta'_{\overline{E}}(\beta)/\theta_{\overline{E}}(\beta).$$

9   The Approaches of Poincaré and of Darwin-Fowler

In this section, we consider a measure space equipped with some Hamiltonian $((\mathscr{E}, \mathscr{T}, \mu), H)$ as in Sect. 6, and we use the notation introduced in Sect. 6.1 and 6.2. Our aim will be to give, under suitable assumptions on the measure $H_* \mu$, some asymptotic expression for

$$A_n(E) := \mu^{\otimes n}(\{x \in \mathscr{E}^n \mid H_n(x) \leqslant nE\})$$

when n goes to infinity. These expressions will be refined versions of the limit formula

$$\lim_{n \to +\infty} \frac{1}{n} \log A_n(E) = S(E), \tag{9.1}$$

valid for every $E \in (H_{\min}, +\infty)$, established by probabilistic arguments in Sect. 6.2.

Our derivation of these asymptotic expressions will rely on arguments involving Fourier and Laplace transforms in the complex domain. The measure $A_n(E)$ will be expressed as a weighted integral along a suitable complex path of the function

$$\left[Z(s)e^{Es} \right]^n$$

of the complex variable s in the right half-plane defined by $\operatorname{Re} s > 0$. The asymptotic expression of $A_n(E)$ when n goes to infinity will be obtained as an application of Laplace's method to this integral.

This derivation may be seen as an application of the saddle-point method (see for instance [18, Chapters 7 and 8]) and is a modern rendering of arguments in the articles [59] by Poincaré and [21, 22], and [23] by Darwin and Fowler, devoted to the statistical mechanics of classical and quantum systems.

We will rely on the analyticity and convexity properties of the functions Z and Ψ and on the construction of the entropy function S presented in Sect. 7.1, but not on the results in Sect. 7.2. Actually, from the asymptotic expressions for $A_n(E)$ established in Sect. 9.2 *infra* under some additional assumptions on the measure $H_*\mu$, one may recover the validity of the limit formula (9.1) under the general assumptions of Theorem 6.2.1—which constituted the main result of Sect. 7.2—by some simple approximation arguments that we present in Sect. 9.3.

Needless to say, to comply with the change of standards of rigor during the last century, we have been led to formulate the asymptotic results in this section with more precision than in the original articles by Poincaré and Darwin and Fowler.[20] The informal character of Poincaré's arguments[21] appears to have led to divergent appreciations of its significance (compare for instance the discussions by Planck in [57, Appendix II], and [58], and the less laudatory comments by Fowler in [29, Section 6.7]). It is however remarkable that now routine analytic techniques are enough to transform the arguments in [59] and [21], which are either informal or of limited scope, into rigorous derivations of the general limit formula (9.1).

9.1   Preliminaries

9.1.1 Laplace Transforms of Measures on $\mathbb{R}_+^*$ Let μ be a complex valued Radon measure on $\mathbb{R}$, supported by $\mathbb{R}_+$. For any $\gamma \in \mathbb{R}$, we shall say that *the measure μ satisfies the condition Σ_γ* when

$$\int_{\mathbb{R}} e^{-\gamma E} d|\mu|(E) < +\infty. \tag{9.2}$$

---

[20]Notably by the introduction of Condition $\mathbf{L}_\varepsilon^2$ in 9.2.1 and 9.2.2 *infra*. Observe that Theorem 9.2.2 requires such an additional assumption on the measure μ to be valid, as shown by a comparison with Theorem 9.2.6.

[21]Poincaré sketches an argument, based on the use of Laplace transforms, to derive an asymptotic formula of the kind of the one established in Theorem 9.2.2, but is concerned with applications of his results in situations where the measure $H_*\mu$ may have a discrete support, or even satisfy Condition **DF** of 9.2.1, in which case this derivation actually fails, as demonstrated by the difference between the asymptotic expressions (9.15) and (9.26) for $A_n(E)$.

When this holds, for any s in the right half-plane

$$\mathbb{C}_{\geqslant \gamma} := \{s \in \mathbb{C} \mid \operatorname{Re} s \geqslant \beta\},$$

we may consider the integral

$$\mathscr{L}\mu(s) := \int_{\mathbb{R}} e^{-sE} d\mu(E).$$

The function

$$\mathscr{L}\mu : \mathbb{C}_{\geqslant \gamma} \longrightarrow \mathbb{C}$$

so-defined—the *Laplace transform* of μ—is continuous on $\mathbb{C}_{\geqslant \gamma}$ and holomorphic on its interior $\mathbb{C}_{>\gamma}$. Its restriction to any vertical line $\beta + i\mathbb{R}$ is, up to a normalization, the Fourier transform of the measure of finite mass $e^{-\beta E} d\mu(E)$, and consequently uniquely determines μ. Moreover, for any $s \in \mathbb{C}_{\geqslant \gamma}$,

$$|\mathscr{L}\mu(s)| \leqslant \mathscr{L}|\mu|(\gamma) := \int_{\mathbb{R}} e^{-\gamma E} d|\mu|(E).$$

For any two Radon measures μ_1 and μ_2 on $\mathbb{R}_+$, we may consider their convolution product $\mu_1 * \mu_2$, namely the Radon measure on $\mathbb{R}_+$ defined by

$$\mu_1 * \mu_2(E) := (\mu_1 \otimes \mu_2)\left(\Sigma^{-1}(E)\right)$$

for any bounded Borel subset of $\mathbb{R}_+$, where

$$\Sigma : \mathbb{R}_+ \times \mathbb{R}_+ \longrightarrow \mathbb{R}_+$$

denotes the sum map. From basic measure theory, it follows that, if μ_1 and μ_2 both satisfy the integrability condition Σ_γ, then $\mu_1 * \mu_2$ satisfies it also, and that, for any $s \in \mathbb{C}_{\geqslant \gamma}$,

$$\mathscr{L}(\mu_1 * \mu_2)(s) = \mathscr{L}\mu_1(s)\,\mathscr{L}\mu_2(s). \tag{9.3}$$

Proposition 9.1.1 *Let μ be a Radon measure on $\mathbb{R}$, supported by $\mathbb{R}_+$, that satisfies Σ_γ for some $\gamma \in \mathbb{R}$.*

For any $\beta \in [\gamma, +\infty)$, the following two conditions are equivalent:

(i) the measure $e^{-\beta E}\, d\mu(E)$ on $\mathbb{R}_+$ is defined by some L^2-function on $\mathbb{R}_+$;
(ii) $\int_{-\infty}^{+\infty} |\mathscr{L}(\beta + i\xi)|^2\, d\xi < +\infty$.
When they are satisfied and when $\beta > 0$, then, for every $E \in \mathbb{R}$,

$$\mu((-\infty, E]) = \frac{1}{2\pi} \int_{\mathbb{R}} \mathscr{L}\mu(\beta + i\xi)\, e^{E(\beta + i\xi)}\, \frac{d\xi}{\beta + i\xi}. \tag{9.4}$$

Condition (i) precisely means that the measure μ is absolutely continuous with respect to the Lebesgue measure λ on $\mathbb{R}$ and that, if f denotes the Radon-Nikodym derivative[22] $d\mu/d\lambda$—the function $(E \longmapsto e^{-\beta E} f(E))$ is square integrable on $\mathbb{R}_+$.

Observe also that, when (ii) holds, the integral in the right-hand side of (9.4) is absolutely convergent, since $e^{E(\beta+i\xi)}/(\beta + i\xi)$ is, like $\mathscr{L}\mu(\beta + i\xi)$, a function of ξ in $L^2(\mathbb{R})$. This integral may be seen as an integral along the "infinite vertical path" in the complex plane defined by the map $(\mathbb{R} \longrightarrow \mathbb{C}, t \longmapsto \beta + it)$, and the equality (9.4) may be written more suggestively as:

$$\mu([0, E]) = \frac{1}{2\pi i} \int_{\beta+i\mathbb{R}} \mathscr{L}\mu(s)\, e^{Es}\, s^{-1} ds. \tag{9.5}$$

Proof of Proposition 9.1.1 For any $\beta \in [\gamma, +\infty)$, the Radon measure μ_β defined as

$$d\mu_\beta(x) = e^{-\beta x}\, d\mu(x)$$

has a finite mass, and its Fourier transform $\mathscr{F}\mu_\beta$ is a continuous function, defined by:

$$\mathscr{F}\mu_\beta(\xi) := \int_{\mathbb{R}} e^{-ix\xi}\, d\mu_\gamma(x) = \int_{\mathbb{R}_+} e^{-(\beta+i\xi)x}\, d\mu(x) = \mathscr{L}\mu(\beta + i\xi), \tag{9.6}$$

for every $\xi \in \mathbb{R}$.

By Parseval's Theorem, the Radon measure μ_β belongs to $L^2(\mathbb{R})$ if and only if $\mathscr{F}\mu_\beta$ belongs to L^2. This proves the equivalence of (i) and (ii).

Moreover, when $\beta > 0$, for every $E \in \mathbb{R}$, we may consider the function $\varphi_{E,\beta}$ in $L^1(\mathbb{R}) \cap L^\infty(\mathbb{R})$ defined by

$$\varphi_{E,\beta}(x) := e^{\beta x} \mathbf{1}_{(-\infty, E]}(x).$$

Its Fourier transform $\mathscr{F}\varphi_{E,\beta}$ is easily computed; namely, for every $\xi \in \mathbb{R}$, we have:

$$\mathscr{F}\varphi_{E,\beta}(\xi) := \int_{\mathbb{R}} e^{-ix\xi} \varphi_{E,\beta}(x)\, dx = \int_{-\infty}^{E} e^{-i(\xi+i\beta)x}\, dx$$

$$= (\beta - i\xi)^{-1} e^{E(\beta - i\xi)}. \tag{9.7}$$

---

[22]Namely, a non-negative Borel function supported by $\mathbb{R}_+$ such that $\mu = f\lambda$, or more exactly, its class modulo equality λ-almost everywhere.

The functions $\varphi_{E,\beta}$ and $\mathscr{F}\varphi_{E,\beta}$ belong to $L^2(\mathbb{R})$, and Parseval's formula applied to $\varphi_{E,\beta}$ and μ_β shows that

$$\int_\mathbb{R} \overline{\varphi_{E,\beta}}(x)\,d\mu_\beta(x) = \frac{1}{2\pi}\int_\mathbb{R} \overline{\mathscr{F}\varphi_{E,\beta}}(\xi)\,\mathscr{F}\mu_\beta(\xi)\,d\xi.$$

This establishes (9.4). Indeed, according to the very definitions of $\varphi_{E,\beta}$ and μ_β, we have:

$$\int_\mathbb{R} \overline{\varphi_{E,\beta}}(x)\,d\mu_\beta(x) = \mu((-\infty, E]),$$

and (9.6) and (9.7) imply:

$$\frac{1}{2\pi}\int_\mathbb{R} \overline{\mathscr{F}\varphi_{E,\beta}}(\xi)\,\mathscr{F}\mu_\beta(\xi)\,d\xi = \frac{1}{2\pi}\int_\mathbb{R} (\beta + i\xi)^{-1} e^{E(\beta+i\xi)}\,\mathscr{L}(\beta + i\xi)\,d\xi. \quad \square$$

9.1.2 Asymptotics of Complex Integrals by Laplace's Method Let I be an interval in $\mathbb{R}$ that contains 0 in its interior, and let g and F be two complex valued Borel functions on I.

Let assume that they satisfy the following condition:

L$_1$: *The function F is bounded on I and there exists $N_0 \in \mathbb{N}$ such that $g.F^{N_0}$ is integrable on I.*

Then, for any integer $N \geqslant N_0$, the function $g.F^N$ is integrable on I and we may consider its integral:

$$I_N := \int_I g(t)\,F(t)^N\,dt.$$

Let us introduce some further conditions on g and F, that will allow us to use Laplace's method to obtain the asymptotic behavior of these integrals when N goes to infinity.

L$_2$: *The functions g and F do not vanish and are continuous at 0. Moreover, there exists $\alpha \in \mathbb{R}_+^*$ such that*

$$F(t) = F(0)(1 - \alpha t^2) + o(t^2) \quad \text{when } t \longrightarrow 0.$$

L$_3$: *For any $\eta \in \mathbb{R}_+^*$,*

$$M_\eta := \sup_{t \in I \setminus (-\eta, \eta)} |F(t)| < |F(0)|.$$

For instance, when $I = \mathbb{R}$, the function F is bounded on $\mathbb{R}$ and satisfies **L$_3$** when it is continuous and satisfies

$$\lim_{|t| \longrightarrow +\infty} F(t) = 0 \tag{9.8}$$

and

$$|F(t)| < |F(0)| \quad \text{for any } t \in \mathbb{R}_+. \tag{9.9}$$

Proposition 9.1.2 *With the above notation, when Conditions* L_1, L_2 *and* L_3 *are satisfied, we have:*

$$I_N \sim g(0) F(0)^N \sqrt{\frac{2\pi}{\alpha N}} \quad \text{when } N \longrightarrow +\infty. \tag{9.10}$$

Proof For any $\eta \in \mathbb{R}_+^*$ and any integer $N \geqslant N_0$, we may define:

$$I_N(\eta) := \int_{I \cap [-\eta, \eta]} g(t) \, F(t)^N \, dt.$$

We have:

$$|I_N(\eta) - I_N| \leqslant \int_{I \setminus [-\eta, \eta]} |g(t)| \, |F(t)|^N \, dt \leqslant M_\eta^{N-N_0} \int_I |g(t)| \, |F(t)|^{N_0} \, dt,$$

and therefore:

$$|I_N(\eta) - I_N| = O(M_\eta^N) \quad \text{when } N \longrightarrow +\infty. \tag{9.11}$$

Besides, for $t \in I$ close enough to 0, we may write:

$$F(0)^{-1} F(t) = e^{-\alpha t^2 + \varepsilon(t) t^2},$$

with

$$\lim_{t \to 0} \varepsilon(t) = 0.$$

Let us choose $\eta \in \mathbb{R}_+^*$ small enough, so that I contains the interval $[-\eta, \eta]$ and

$$\tilde{\alpha} := \sup_{|t| \leqslant \eta} \operatorname{Re} \varepsilon(t) < \alpha \quad \text{and} \quad M := \sup_{|t| \leqslant \eta} |g(t)| < +\infty.$$

Then, for any integer $N \geqslant N_0$, we have:

$$g(0)^{-1} F(0)^{-N} I_N(\eta) = \int_{-\eta}^{\eta} g(0)^{-1} g(t) \, [F(0)^{-1} F(t)]^N \, dt$$

$$= \int_{-\eta}^{\eta} g(0)^{-1} g(t) \, e^{-(\alpha - \varepsilon(t)) N t^2} \, dt$$

$$= \sqrt{N}^{-1} \int_{-\sqrt{N}\eta}^{\sqrt{N}\eta} g(0)^{-1} g(u/\sqrt{N}) \, e^{-(\alpha - \varepsilon(u/\sqrt{N})) u^2} \, du.$$

(We have performed the change of variables $u = \sqrt{N}t$.) When N goes to infinity, the last integral converges to

$$\int_{-\infty}^{+\infty} e^{-\alpha u^2}\, du = \sqrt{2\pi/\alpha}.$$

Indeed, for any fixed $u \in \mathbb{R}$, its integrand converges to $e^{-\alpha u^2}$ when N goes to $+\infty$, and its absolute value is bounded from above by $|g(0)|^{-1} M e^{-(\alpha - \tilde{\alpha})u^2}$, which is integrable over $\mathbb{R}$. This proves that

$$I_N(\eta) \sim g(0) F(0)^N \sqrt{\frac{2\pi}{\alpha N}} \qquad \text{when } N \longrightarrow +\infty. \tag{9.12}$$

Since $M_\eta < |F(0)|$, the asymptotic equivalent (9.10) for I_N follows from (9.11) and (9.12). $\square$

9.2 Asymptotics of $A_n(E)$ by the Saddle-Point Method

We return to the notation introduced at the beginning of this section. Namely, we consider a measure space equipped with some Hamiltonian $((\mathscr{E}, \mathscr{T}, \mu), H)$ as in Sect. 6, and we freely use the notation introduced in Sect. 6.1 and 6.2.

9.2.1 The Conditions $\mathbf{L}_\varepsilon^2$ and DF Let us introduce the following additional conditions on the measure μ and on the function H:

$\mathbf{L}_\varepsilon^2$: *For any $\beta \in \mathbb{R}_+^*$, the measure*

$$H_*(e^{-\beta H}\, \mu) = e^{-\beta \mathrm{Id}_{\mathbb{R}_+}}\, H_* \mu$$

on $\mathbb{R}_+$ is absolutely continuous with square integrable density[23];
and:

DF : *There exists η in $\mathbb{R}_+^*$ such that, μ-almost everywhere on $\mathscr{E}$, the function H takes its values in $H_{\min} + \mathbb{N}\eta$, or equivalently such that the measure $H_*\mu$ is supported by $H_{\min} + \mathbb{N}\eta$.*

We are going to derive some asymptotic representation of $A_n(E)$ when n goes to infinity when, besides Conditions $\mathbf{T}_1$ and $\mathbf{T}_2$, one of these conditions holds.

Clearly the conditions $\mathbf{L}_\varepsilon^2$ and DF are never simultaneously satisfied, unless $\mu = 0$.

Let us also indicate that the pairs $((\mathscr{E}, \mathscr{T}, \mu), H)$ that arise from classical mechanics, as discussed in the introduction of Sect. 6 and in paragraph 6.4.2, often

---

[23]In other words, this measure may be written $f\, d\lambda$, where $d\lambda$ denotes the Lebesgue measure and f some non-negative L^2 function on $\mathbb{R}_+$.

satisfy Condition $\mathbf{L}_\varepsilon^2$. This is related to the following observation, that we leave as an exercise for the reader:

If $\mathscr{E}$ is a C^∞ manifold of pure dimension $n \geqslant 2$, if μ is defined by some C^∞ density on this manifold, and if the function $H : M \longrightarrow \mathbb{R}_+$ is C^∞, proper, and a Morse function, then the measure $H_*\mu$ is locally L^2.

Clearly such pairs never satisfy Condition **DF**, except in trivial cases where H is locally constant.

9.2.2 The Approach of Poincaré

Proposition 9.2.1 *Let us assume that Condition $\mathbf{L}_\varepsilon^2$ is satisfied.*

Then Condition $\mathbf{T_2}$ holds, and for any $\beta \in \mathbb{R}_+^*$, the function $(\xi \mapsto Z(\beta + i\xi))$ belongs to $\mathscr{C}_0(\mathbb{R}) \cap L^2(\mathbb{R})$. Moreover, for any $E \in \mathbb{R}$ and any integer $n \geqslant 1$,

$$A_n(E) = \frac{1}{2\pi} \int_{\mathbb{R}} \left[Z(\beta + i\xi)\, e^{E(\beta + i\xi)} \right]^n \frac{d\xi}{\beta + i\xi}$$

$$=: \frac{1}{2\pi i} \int_{\beta + i\mathbb{R}} \left[Z(s)\, e^{Es} \right]^n s^{-1} ds. \qquad (9.13)$$

Observe that, since the function $(\xi \mapsto Z(\beta + i\xi))$ is both L^∞ and L^2, the function

$$\left(\xi \longmapsto \left[Z(\beta + i\xi)\, e^{E(\beta + i\xi)} \right]^n \right)$$

is L^2 for any positive integer n. The integrals in (9.13) are therefore absolutely convergent.

Proof If Condition $\mathbf{L}_\varepsilon^2$ is satisfied, then, for any $\beta \in \mathbb{R}_+^*$, the measure $e^{-\beta \mathrm{Id}_{\mathbb{R}_+}}\, H_*\mu$ on $\mathbb{R}_+$ is defined by some L^1 function on $\mathbb{R}_+^*$. Indeed, the equality

$$e^{-\beta \mathrm{Id}_{\mathbb{R}_+}}\, H_*\mu = e^{-(\beta/2)\mathrm{Id}_{\mathbb{R}_+}} . e^{-(\beta/2)\mathrm{Id}_{\mathbb{R}_+}}\, H_*\mu$$

shows that it is the product of two L^2-functions on $\mathbb{R}_+^*$. This implies that $e^{-\beta H}\mu$ has a finite mass for every $\beta \in \mathbb{R}_+^*$, that is, that Condition $\mathbf{T_2}$ holds.

By the very definitions of the image measure $H_*\mu$ and of the partition function Z (see (6.1) and Proposition 7.1.1), we have:

$$A_1(E) := \mu(H^{-1}((-\infty, E])) = H_*\mu((-\infty, E]) \quad \text{for every } E \in \mathbb{R},$$

and:

$$Z(s) = \mathscr{L}(H_*\mu)(s) \quad \text{for every } s \in \mathbb{C}_{>0}.$$

Using (9.6) with $H_*\mu$ instead of μ, this shows that the function $(\xi \mapsto Z(\beta + i\xi))$ is the Fourier transform of the measure $e^{-\beta \mathrm{Id}_{\mathbb{R}_+}}\, H_*\mu$ on $\mathbb{R}_+$, which is defined by some function both in L^2 and L^1, and therefore belongs to $\mathscr{C}_0(\mathbb{R}) \cap L^2(\mathbb{R})$. Moreover, when $n = 1$, the equality (9.13) follows from Proposition 9.1.1 applied to the measure $H_*\mu$ on $\mathbb{R}_+$.

Observe that Proposition 9.1.1 also shows that, conversely, when $\mathbf{T}_2$ holds and Z is L^2 on the vertical line $\beta + i\mathbb{R}$, then the measure $e^{-\beta \mathrm{Id}_{\mathbb{R}_+}} H_*\mu$ on $\mathbb{R}_+$ is defined by some L^2-function.

Let us now consider an arbitrary positive integer n, and let us introduce the pair

$$((\mathscr{E}^n, \mathscr{T}^{\otimes n}, \mu^{\otimes n}), H_n). \tag{9.14}$$

It is nothing but the product, in the sense of Sect. 8.4, of n-copies of the given measure space with Hamiltonian $((\mathscr{E}, \mathscr{T}, \mu), H)$.

As observed in *loc. cit.*, it follows from Fubini theorem that it still satisfies Condition $\mathbf{T}_2$ and its partition function is Z^n. This function is L^2 on the vertical line $\beta + i\mathbb{R}$ for any $\beta \in \mathbb{R}_+^*$, and the above observation shows that the pair (9.14) also satisfies Condition $\mathbf{L}_\varepsilon^2$ and that we may apply to it the equality (9.13) with $n = 1$.

This shows that, for any $E \in \mathbb{R}$,

$$\mu^{\otimes n}(H_n^{-1}((-\infty, nE])) = \frac{1}{2\pi i} \int_{\beta+i\mathbb{R}} Z(s)^n \, e^{nEs} \, s^{-1} ds$$

and establishes (9.13) in general.   □

For any given $E \in (H_{\min}, +\infty)$, we may derive an asymptotic expression for $A_n(E)$ when n goes to $+\infty$ from the integral formulae (9.13), by choosing

$$\beta := S'(E)$$

and then applying Laplace's method. In this way, we shall establish:

Theorem 9.2.2 *Let us assume that conditions* $\mathbf{T}_1$ *and* $\mathbf{L}_\varepsilon^2$ *are satisfied. Then, for any* $E \in (H_{\min}, +\infty)$, *we have:*

$$A_n(E) \sim \left[\pi\beta^2\Psi''(\beta)n\right]^{-1/2} e^{nS(E)}, \tag{9.15}$$

with $\beta := S'(E)$, *when the integer* n *goes to infinity.*

Proof For any $E \in \mathbb{R}$ and $\beta \in \mathbb{R}_+^*$, the right-hand side of the integral expression (9.13) for $A_n(E)$ may be written as

$$\frac{1}{2\pi} \int_{\mathbb{R}} \left[Z(\beta + i\xi) \, e^{E(\beta+i\xi)}\right]^n \frac{d\xi}{\beta + i\xi} = \int_I g_\beta(t) \, F_\beta(t)^n \, dt, \tag{9.16}$$

where

$$I := \mathbb{R}, \quad g_\beta(t) := [2\pi(\beta + it)]^{-1}, \quad \text{and} \quad F_\beta(t) := Z(\beta + it)e^{E(\beta + it)}.$$

The functions g_β and F_β satisfy the conditions $\mathbf{L_1}$ and $\mathbf{L_3}$ on the functions g and F introduced in our discussion of Laplace's method in paragraph 9.1.2.

Indeed, F_β is continuous on $I := \mathbb{R}$ and satisfies (9.8), as shown in Proposition 9.2.1; it also satisfies (9.9), as a consequence of the estimates (7.3) on $|Z|$ and of the fact that e^{-itH} is not μ-almost everywhere constant for any $t \in \mathbb{R}_+^*$, as a straightforward consequence of $\mathbf{L}_\varepsilon^2$. This implies that F_β is bounded and satisfies $\mathbf{L_3}$. Moreover, as observed after Proposition 9.2.1, the function $g_\beta F_\beta$ is integrable, and this establishes $\mathbf{L_1}$.

Let us now assume that $E > H_{\min}$, and let us choose

$$\beta := S'(E),$$

where the function S has been introduced in paragraph 7.1.4. By the very definition of S, the function S' is the compositional inverse of the function $U := -\Psi' = -Z'/Z$, and β is the unique zero in $\mathbb{R}_+^*$ of the derivative

$$\frac{d}{ds}(\Psi(s) + Es) = \Psi'(s) + E.$$

Moreover,

$$S(\beta) = \Psi(\beta) + E\beta.$$

Accordingly, when $s \in \mathbb{C}$ goes to β, we may write:

$$Z(s)\, e^{Es} = e^{\Psi(s)+Es} = e^{\Psi(\beta)+E\beta+\Psi''(\beta)(s-\beta)^2/2+o((s-\beta)^2)}$$

$$= e^{S(\beta)+\Psi''(\beta)(s-\beta)^2/2+o((s-\beta)^2)}. \qquad (9.17)$$

(The analytic function $\Psi := \log Z$ is well defined on some open neighbourhood of $\mathbb{R}_+^*$ in $\mathbb{C}_{>0}$.)

This immediately shows that F_β satisfies Condition $\mathbf{L_2}$ with

$$F_\beta(0) = e^{S(\beta)} \quad \text{and} \quad \alpha = (1/2)\Psi''(\beta).$$

Moreover,

$$g_\beta(0) = (2\pi\beta)^{-1}.$$

We may therefore apply Proposition 9.1.2 to the integrals (9.16). The asymptotic expression (9.10) for these integrals given by Laplace's method takes the form (9.15). □

9.2.3 The Approach of Darwin-Fowler Let us assume in this paragraph that Condition T_2 is satisfied.

The following proposition follows from the fact that a Radon measure on $\mathbb{R}_+$ which satisfies Condition Σ_0 (see paragraph 9.1.1) is uniquely determined by its Laplace transform on the half-plane $\mathbb{C}_{>0}$.

Proposition 9.2.3 *For any $\eta \in \mathbb{R}_+^*$, the following three conditions are equivalent:*

$\mathbf{DF}_\eta^1$: *For μ-almost every $x \in \mathscr{E}$, $H(x)$ belongs to $H_{\min} + \mathbb{N}\eta$;*

$\mathbf{DF}_\eta^2$: *the measure $H_*\mu$ is supported by $H_{\min} + \mathbb{N}\eta$;*

$\mathbf{DF}_\eta^3$: *the function $e^{H_{\min}s} Z(s)$ of $s \in \mathbb{C}_{>0}$ is $2\pi i /\eta$-periodic.*

Condition $\mathbf{DF}$ is equivalent to the existence of η in $\mathbb{R}_+^*$ such that these conditions $\mathbf{DF}_\eta^1$ - $\mathbf{DF}_\eta^3$ are satisfied.

Actually, the (obviously equivalent) conditions $\mathbf{DF}_\eta^1$ and $\mathbf{DF}_\eta^2$ are satisfied if and only if we may write the measure $H_*\mu$ as:

$$H_*\mu = \sum_{k\in\mathbb{N}} h_k \, \delta_{H_{\min}+k\eta} \tag{9.18}$$

for some sequence $(h_k)_{k\in\mathbb{N}}$ in $\mathbb{R}_+^*$. When this holds, we have:

$$Z(s) = e^{-H_{\min}s} \sum_{k\in\mathbb{N}} h_k \, e^{-k\eta s} \quad \text{for any } s \in \mathbb{C}_{>0}.$$

Then the series with non-negative coefficients

$$f(X) := \sum_{k\in\mathbb{N}} h_k X^k \tag{9.19}$$

has radius of convergence at least 1 (this is a reformulation of Condition T_2), and we have:

$$Z(s) = e^{-H_{\min}s} f(e^{-\eta s}) \quad \text{for any } s \in \mathbb{C}_{>0}. \tag{9.20}$$

This makes clear the validity of Condition $\mathbf{DF}_\eta^3$.

Observe also that

$$\mu(\mathscr{E}) = \sum_{k\in\mathbb{N}} h_k = \lim_{q\to 1_-} f(q).$$

Condition T_1 holds if and only if this limit is $+\infty$.

These observations show that, when conditions $\mathbf{T_1}$, $\mathbf{T_2}$, and $\mathbf{DF}$ are satisfied, there exists a largest $\eta \in \mathbb{R}_+^*$ such that the conditions $\mathbf{DF}_\eta^{1-3}$ are satisfied. We shall denote it by η_H. Then the analytic function f on $D(0, 1)$ such that

$$Z(s) = e^{-H_{\min} s} f(e^{-\eta_H s}) \quad \text{for any } s \in \mathbb{C}_{>0}. \tag{9.21}$$

is defined by the series (9.19) where the $(h_k)_{k\in\mathbb{N}}$ is defined by the relation (9.18) with $\eta = \eta_H$. Moreover, by the very definition of η_H, for every integer $m > 1$, we have:

$$\Sigma := \{k \in \mathbb{N} \mid h_k \neq 0\} \not\subseteq m\mathbb{N}.$$

(Otherwise, $e^{H_{\min} s} Z(s)$ would be $2\pi i / m\eta_H$-periodic.)

This implies:

Lemma 9.2.4 *For every $s \in \mathbb{C}_{>0}$, the inequality (7.3)*

$$|Z(s)| \leqslant Z(\operatorname{Re} s)$$

is an equality if and only if s belongs to $\operatorname{Re} s + (2\pi i / \eta_H)\, \mathbb{Z}$, or equivalently if and only if the element

$$q := e^{-\eta_H s}$$

of the pointed unit disc $D(0, 1)\backslash\{0\}$ belongs to the interval $(0, 1)$.

Proof We have to show that the inequality

$$\left| \sum_{k\in\mathbb{N}} h_k q^k \right| \leqslant \sum_{k\in\mathbb{N}} h_k |q|^k$$

is an equality (if and) only if $q = |q|$. As

$$h_0 = \mu(\mathscr{E}_{\min})$$

does not vanish and the h_k are non-negative, this holds if and only if $q^k / |q|^k = 1$ for every k in Σ. As Σ is infinite, this implies that $q / |q|$ is a root of unity. Let m be its order. The set Σ is contained in $m\mathbb{N}$; therefore $m = 1$ and $q / |q| = 1$. $\qquad\square$

For any $r \in (0, 1)$, we shall denote by $C(r)$ the closed path in the complex plane

$$([0, 1] \longrightarrow \mathbb{C}, t \longmapsto re^{2\pi i t}).$$

Proposition 9.2.5 *Let us assume that Conditions* T_1*,* T_2*, and* **DF** *hold. Let* $r \in (0, 1)$ *and let*

$$\beta := \eta_H^{-1} \log r^{-1}.$$

Then, for every $E \in \mathbb{R}$ *and any integer* $n \geqslant 1$ *such that* $n(E - H_{\min}) \in \mathbb{Z}\eta_H$*, we have:*

$$A_n(E) = \frac{1}{2\pi i} \int_{C(r)} (1 - q)^{-1} q^{-nE/\eta_H} f(q)^n \, q^{-1} \, dq \tag{9.22}$$

$$= \frac{1}{2\pi i} \int_{\beta - \pi i/\eta_H}^{\beta + \pi i/\eta_H} \frac{\eta_H}{1 - e^{-\eta_H s}} \left[Z(s) e^{Es} \right]^n \, ds \tag{9.23}$$

$$= \frac{1}{2\pi} \int_{-\pi/\eta_H}^{\pi/\eta_H} \frac{\eta_H}{1 - e^{-\eta_H(\beta + it)}} \left[Z(\beta + it) e^{E(\beta + it)} \right]^n \, dt. \tag{9.24}$$

Proof Let us write, as above:

$$H_* \mu = \sum_{k \in \mathbb{N}} h_k \delta_{H_{\min} + k\eta_H}.$$

Then the measure

$$H_{n*}(\mu^{\otimes n}) = (H_* \mu)^{*n} := (H_* \mu) * \ldots * (H_* \mu) \quad (n\text{-times})$$

may be written

$$H_{n*}(\mu^{\otimes n}) = \sum_{k \in \mathbb{N}} h_k^{[n]} \delta_{nH_{\min} + k\eta_H},$$

where the sequence $(h_k^{[n]})_{k \in \mathbb{N}}$ satisfies:

$$\sum_{k \in \mathbb{N}} h_k^{[n]} X^k = \left(\sum_{k \in \mathbb{N}} h_k X^k \right)^n = f(X)^n.$$

When $\tilde{n} := n(E - H_{\min})/\eta_H$ is an integer, we have:

$$A_n(E) = H_{n*}(\mu^{\otimes n})([0, nE]) = \sum_{0 \leqslant k \leqslant \tilde{n}} h_k^{[n]}$$

$$= \mathrm{Res}_0[(1 - X)^{-1} X^{-(\tilde{n}+1)} f(X)^n], \tag{9.25}$$

where we denote by Res_0 the residue at 0. By the residue formula, this coincides with the right-hand side of (9.22).

We deduce (9.23) and (9.24) from (9.22) by the changes of variables

$$q = e^{-\eta_H s} \quad \text{and} \quad s = \beta + it.$$ $\square$

By applying Laplace's method to the integral formulae established in Proposition 9.2.5, it is possible to derive an asymptotic expression for $A_n(E)$ similar to the one in Theorem 9.2.2:

Theorem 9.2.6 *Let us assume that Conditions* $\mathbf{T_1}$, $\mathbf{T_2}$, *and* $\mathbf{DF}$ *hold. Let us consider*

$$E \in H_{\min} + \mathbb{Q}_+^* \eta_H$$

and let us define:

$$\mathcal{M}(E) := \{n \in \mathbb{Z}_{>0} \mid n(E - H_{\min}) \in \mathbb{Z}_{>0}\,\eta_H\}.$$

Then, when the integer $n \in \mathcal{M}(E)$ *goes to infinity, we have:*

$$A_n(E) \sim \frac{\eta_H \beta}{1 - e^{-\eta_H \beta}} \left[\pi \beta^2 \Psi''(\beta) n\right]^{-1/2} e^{nS(E)},   \tag{9.26}$$

with $\beta := S'(E)$.

If $E = H_{\min} + \eta_H a/b$ *for coprime integers* a *and* b, *then:*

$$\mathcal{M}(E) = |b|\mathbb{Z}_{>0}.$$

Proof For any n in $\mathcal{M}(E)$ and any β in $\mathbb{R}_+^*$, according to (9.24), $A_n(E)$ admits the expression (9.24):

$$A_n(E) = \frac{1}{2\pi} \int_{-\pi/\eta_H}^{\pi/\eta_H} \frac{\eta_H}{1 - e^{-\eta_H(\beta+it)}} \left[Z(\beta + it)e^{E(\beta+it)}\right]^n dt$$

$$= \int_I g_\beta(t) F_\beta(t)^n \, dt,   \tag{9.27}$$

where:

$$I := [-\pi/\eta_h, \pi/\eta_H], \quad g_\beta(t) := \frac{\eta_H}{1 - e^{-\eta_H(\beta+it)}},$$

and:

$$F_\beta(t) := Z(\beta + it)e^{E(\beta+it)}.$$

The functions g_β and F_β clearly satisfy the condition $\mathbf{L}_1$ on the functions g and F in our discussion of Laplace's method in paragraph 9.1.2. According to Lemma 9.2.4, the function F_β also satisfies Condition $\mathbf{L}_3$. Moreover, if we choose $\beta := S'(E)$, F_β also satisfies Condition $\mathbf{L}_2$, as already shown in the proof of Theorem 9.2.2.

The asymptotic expression (9.15) therefore follows from Proposition 9.1.2 applied to the integrals (9.27).   □

Besides the original articles ([21, 22, 23]) the results of Darwin-Fowler are presented in the reference text by Fowler [29, Chapter 2], in the beautiful seminar notes by Schrödinger [69, Chapter 4], and in the textbook of Huang [40, Section 9.1].

9.3   Approximation Arguments

The asymptotic equivalents (9.15) and (9.26) for $A_n(E)$, established under the additional assumptions $\mathbf{L}_\varepsilon^2$ and $\mathbf{DF}$ on the pair $((\mathscr{E}, \mathscr{T}, \mu), H)$ in Theorems 9.2.2 and 9.2.6, both imply the limit formula

$$\lim_{n \to +\infty} \frac{1}{n} \log A_n(E) = S(E), \qquad (9.28)$$

which was the key point in our derivation of Theorem 6.2.1.

Indeed, when $\mathbf{L}_\varepsilon^2$ holds, (9.15) clearly implies this limit formula. When $\mathbf{DF}$ holds, it follows from the asymptotic expression (9.26) combined with the existence in $(-\infty, +\infty]$ of the limit $\lim_{n \to +\infty}(1/n)\log A_n(E)$, established in Sect. 8.3 by Lanford's method, and the fact that this limit is a non-decreasing function of E.

In this subsection, we present some simple approximation arguments that will allow us to derive (9.28) in the general context of Theorem 6.2.1 from its special cases implied by Theorem 9.2.2 or Theorem 9.2.6.

As already observed in 8.1.2, to prove the validity of Theorem 6.2.1, one immediately reduces to the case where $(\mathscr{E}, \mathscr{T})$ is $(\mathbb{R}_+, \mathscr{B})$—where $\mathscr{B}$ denotes the σ-algebra of Borel subsets of $\mathbb{R}_+$—and where the Hamiltonian function H is $\mathrm{Id}_{\mathbb{R}_+}$.

From now one, we place ourselves in this framework; namely, we consider a positive Radon measure μ on $\mathbb{R}_+$ which satisfies the conditions

$$\mathbf{T}_1 : \quad \mu(\mathbb{R}_+) = +\infty,$$

and

$$\mathbf{T}_2 : \quad Z(\beta) := \mathscr{L}\mu(\beta) := \int_{\mathbb{R}_+} e^{-\beta E} \, d\mu(E) < +\infty \quad \text{for any } \beta \in \mathbb{R}_+^*.$$

9.3.1 Approximation by Convolution Let us choose χ in $C_c^\infty(\mathbb{R})$ such that

$$\chi(\mathbb{R}) \subset \mathbb{R}_+, \quad \text{supp} \, \chi \subset [0, 1], \quad \text{and} \quad \int_{\mathbb{R}} \chi(x) \, dx = 1.$$

For any $\eta \in \mathbb{R}_+^*$, we define the function in $C_c^\infty(\mathbb{R})$:

$$\chi_\eta := \eta^{-1} \chi(\eta^{-1}.)$$

and the Radon measure on $\mathbb{R}_+$:

$$\mu_\eta := \mu * \chi_\eta.$$

The validity of $\mathbf{T}_1$ and $\mathbf{T}_2$ for μ immediately implies their validity for μ_η.

To the pair $((\mathbb{R}_+, \mathscr{B}, \mu), \text{Id}_{\mathbb{R}_+})$ are associated the non-negative real number $H_{\min}$, the functions Z, Ψ and S, and the sequence of functions $(A_n)_{n \geq 1}$.

Similarly, for any $\eta \in \mathbb{R}_+^*$, to $((\mathbb{R}_+, \mathscr{B}, \mu), \text{Id}_{\mathbb{R}_+})$ are associated the essential minimum $H_{\min,\eta}$ of $\text{Id}_{\mathbb{R}_+}$ with respect to μ_η (or equivalently, the minimum of $\text{supp} \, \mu_\eta$), and the functions Z_η, Ψ_η and S_η and the sequence of functions $(A_{n,\eta})_{n \geq 1}$ defined as follows: for any $s \in \mathbb{C}_{>0}$,

$$Z_\eta(s) := \int_{\mathbb{R}_+} e^{-sE} \, d\mu_\eta(E);$$

for any $\beta \in \mathbb{R}_+^*$,

$$\Psi_\eta(\beta) := \log Z_\eta(\beta);$$

for any $E \in (H_{\min,\eta}, +\infty)$,

$$S_\eta(E) := \sup_{\beta \in \mathbb{R}_+^*} (\Psi_\eta(\beta) + \beta E);$$

and for any positive integer n and any $E \in \mathbb{R}_+$,

$$A_{n,\eta}(E) := \mu_\eta^{\otimes n}(\{(x_1, \ldots, x_n) \in \mathbb{R}_+^n \mid x_1 + \ldots + x_n \leq nE\})$$

$$= \mu_\eta^{*n}([0, nE]).$$

Lemma 9.3.1 *For any $\eta \in \mathbb{R}_+^*$ and any $E \in \mathbb{R}_+$, we have:*

$$H_{\min} \leq H_{\min,\eta} \leq H_{\min} + \eta \tag{9.29}$$

and

$$A_{n,\eta}(E) \leqslant A_n(E) \leqslant A_{n,\eta}(E+\eta). \tag{9.30}$$

Proof The estimates (9.29) are straightforward. The estimates (9.30) follow from the identities:

$$A_{n,\eta}(E) = \mu_\eta^{*n}([0, nE]) = \int_{x,y \geqslant 0, x+y \leqslant nE} d\mu^{*n}(x) . \chi_\eta^{*n}(y) \, dy,$$

$$A_n(E) = \mu^{*n}([0, nE]) = \int_{0 \leqslant x \leqslant nE} d\mu^{*n}(x),$$

and

$$A_{n,\eta}(E+\eta) = \mu_\eta^{*n}([0, nE+n\eta]) = \int_{x,y \geqslant 0, x+y \leqslant nE+n\eta} d\mu^{*n}(x) . \chi_\eta^{*n}(y) \, dy,$$

and from the fact that χ_η^{*n} is non-negative, supported by $[0, n\delta]$, and of integral 1.

$\square$

Lemma 9.3.2 *For any* $(\eta, \beta) \in \mathbb{R}_+^{*2}$,

$$\Psi(\beta) - \beta\eta \leqslant \Psi_\eta(\beta) \leqslant \Psi(\beta). \tag{9.31}$$

Proof From the multiplicativity property (9.3) of the Laplace transform, we obtain that, for any $s \in \mathbb{C}_{>0}$:

$$Z_\eta(s) = \mathscr{L}\mu_\eta(s) = \mathscr{L}\mu(s) \mathscr{L}\chi(s) = Z(s) \int_0^{+\infty} e^{-sE} \chi_\eta(E) \, dE.$$

Therefore, for any $\beta \in \mathbb{R}_+^*$:

$$\Psi_\eta(\beta) = \Psi(\beta) + \log \int_0^{+\infty} e^{-\beta E} \chi_\eta(E) \, dE. \tag{9.32}$$

Besides, as χ_η is non-negative, supported by $[0, \eta]$ and of integral 1, we have:

$$e^{-\beta\eta} \leqslant \int_0^{+\infty} e^{-\beta E} \chi_\eta(E) \, dE \leqslant 1. \tag{9.33}$$

The estimates (9.31) follow from (9.32) and (9.33). $\square$

From Lemma 9.3.1 and 9.3.2, one easily derives:

Lemma 9.3.3 *If, for any $\eta \in \mathbb{R}_+^*$ and any $E \in (H_{\min,\eta}, +\infty)$,*

$$\lim_{n \to +\infty} \frac{1}{n} \log A_{n,\eta}(E) = S_\eta(E), \tag{9.34}$$

then, for any $E \in (H_{\min}, +\infty)$,

$$\lim_{n \to +\infty} \frac{1}{n} \log A_n(E) = S(E). \tag{9.35}$$

Proof From Lemma 9.3.2, we immediately obtain:

$$S(E - \eta) \leqslant S_\eta(E) \quad \text{for any } \eta > 0 \text{ and any } E > H_{\min} + \eta, \tag{9.36}$$

and

$$S_\eta(E) \leqslant S(E) \quad \text{for any } \eta > 0 \text{ and any } E > H_{\min,\eta}. \tag{9.37}$$

Let us now consider $E \in (H_{\min}, +\infty)$. For any $\eta \in (E - H_{\min})$, from (9.30) and (9.34), we get:

$$S_\eta(E) = \lim_{n \to +\infty} \frac{1}{n} \log A_{n,\eta}(E) \leqslant \varliminf_{n \to +\infty} \frac{1}{n} \log A_n(E)$$

and

$$\varlimsup_{n \to +\infty} \frac{1}{n} \log A_n(E) \leqslant \lim_{n \to +\infty} \frac{1}{n} \log A_{n,\eta}(E + \eta) = S_\eta(E + \eta).$$

Together with (9.36) and (9.37), this shows:

$$S(E - \eta) \leqslant \varliminf_{n \to +\infty} \frac{1}{n} \log A_n(E) \leqslant \varlimsup_{n \to +\infty} \frac{1}{n} \log A_n(E) \leqslant S(E + \eta).$$

Since the function S is continuous on $(H_{\min}, +\infty)$, this establishes (9.35) by taking the limit when η goes to zero. $\qquad\square$

Lemma 9.3.3 allows one to derive the validity of the limit formula (9.35) from its validity when furthermore Condition $\mathbf{L}_\varepsilon^2$ holds. Indeed, for any $\eta \in \mathbb{R}_+^*$, the pair $((\mathbb{R}_+, \mathscr{B}, \mu_\eta), \mathrm{Id}_{\mathbb{R}_+})$ satisfies this condition, which ensures the validity of (9.34); in other words, the measure μ_η is absolutely continuous with respect to the Lebesgue measure λ and may be written

$$\mu_\eta = f_\eta \, \lambda$$

where, for every $\varepsilon \in \mathbb{R}_+^*$, the function $(x \longmapsto e^{-\varepsilon x} f_\eta(x))$ is in $L^2(\mathbb{R}_+)$.

Actually, the density f_η is a C^∞ function on $\mathbb{R}_+$, and we have, for any $x \in \mathbb{R}_+^*$:

$$f_\eta(x) = \int_\mathbb{R} \chi_\eta(x - t)\, d\mu(t) = \eta^{-1} \int_{x-\eta}^x \chi(\eta^{-1}(x - t))\, d\mu(t)$$

$$\leqslant \eta \|\chi\|_{L^\infty}\, \mu([0, x]).$$

This shows that, for any $\delta \in \mathbb{R}_+^*$,

$$f_\eta(x) = O(e^{\delta x}) \quad \text{when } x \longrightarrow +\infty,$$

and therefore

$$e^{-\varepsilon x} f_\eta(x) = O(e^{-\varepsilon' x}) \quad \text{when } x \longrightarrow +\infty$$

for any $\varepsilon' \in (0, \varepsilon)$.

9.3.2 Approximation by Discretization For any $\eta \in \mathbb{R}_+^*$, we may also introduce the Radon measure

$$\mu_\eta := \sum_{k \in \mathbb{Z}_{>0}} \mu([(k - 1)\eta, k\eta))\, \delta_{k\eta}.$$

As in the previous paragraph, to the measure μ_η are associated its essential minimum $H_{\min,\eta}$, and the functions Z_η, Ψ_η, S_η, and $A_{n,\eta}$. Then Lemma 9.3.1 and 9.3.2 remain valid (we leave the details as an exercise for the interested reader), and consequently Lemma 9.3.3 also.

By construction, μ_η is supported by $\mathbb{N}\eta$, and therefore the pair $((\mathbb{R}_+, \mathscr{B}, \mu_\eta), \mathrm{Id}_{\mathbb{R}_+})$ satisfies Condition **DF**. Thus the fact that Lemma 9.3.3 holds in the present context shows that the validity of the limit formula (9.35) follows from its validity when furthermore Condition **DF** holds.

Let us finally remark that, when we consider a Euclidean lattice $\overline{E}$ of positive rank, the associated measure space with Hamiltonian $((\mathscr{E}, \mathscr{T}, \mu), H)$ associated to $\overline{E}$ by the construction in Sect. 6.5 never satisfies Condition $\mathbf{L}_\varepsilon^2$. It satisfies Condition **DF** if and only if some positive real multiple of $\|.\|^2$ is an integral quadratic form on the free $\mathbb{Z}$-module E. Accordingly, in this case, the validity of Theorem 5.4.3 directly follows from the asymptotics à la Darwin-Fowler established in paragraph 9.2.3, combined with Lanford's arguments presented in Sect. 8.2 (see [50, Section 3], for a related discussion).

References

1. R. Abraham, J.E. Marsden, *Foundations of Mechanics*, 2nd edn. (Benjamin/Cummings Publishing Co., Inc., Advanced Book Program, Reading, MA, 1978)

2. S.J. Arakelov, Theory of intersections on the arithmetic surface, in *Proceedings of the International Congress of Mathematicians (Vancouver, BC, 1974)*, Montreal, 1975, vol. 1, pp 405–408
3. N.W. Ashcroft, N.D. Mermin, *Solid State Physics* (Brooks/Cole, Cengage Learning, Boston, 1976)
4. V.I. Arnol'd, *Mathematical Methods of Classical Mechanics*. Graduate Texts in Mathematics, vol. 60, 2nd edn. (Springer, New York, 1989)
5. W. Banaszczyk, New bounds in some transference theorems in the geometry of numbers. Math. Ann. **296**, 625–635 (1993)
6. L. Bétermin, M. Petrache, Dimension reduction techniques for the minimization of theta functions on lattices. J. Math. Phys. **58**, 071902 (2017)
7. L. Boltzmann, Weitere Studien über Wärmegleichgewicht unter Gasmolekülen. Wien. Ber. **66**, 275–370 (1872)
8. L. Boltzmann, Über die Beziehungen zwischen dem zweiten Hauptsatz der Wärmetheorie und der Warscheinlichkeitsrechnung respektive den Sätzen über das Wärmegleichgewicht. Wien. Ber. **76**, 373–435 (1877)
9. J.-B. Bost, *Theta Invariants of Euclidean Lattices and Infinite-Dimensional Hermitian Vector Bundles over Arithmetic Curves*. Progress in Mathematics, vol. 334 (Birkhaüser, Basel, 2020)
10. J.-B. Bost, Réseaux euclidiens, séries thêta et pentes (d'après W. Banasczyk, O. Regev, D. Dadush, S. Stephens–Davidowitz, ...). Séminaire N. Bourbaki, Exposé 1152, Octobre 2018, http://www.bourbaki.ens.fr/TEXTES/Exp1152-Bost.pdf; to appear in Astérisque **422** (2020)
11. J.-B. Bost, K. Künnemann, Hermitian vector bundles and extension groups on arithmetic schemes. I. Geometry of numbers. Adv. Math. **223**, 987–1106 (2010)
12. S. Boucheron, G. Lugosi, P. Massart, *Concentration Inequalities – A Nonasymptotic Theory of Independence* (Oxford University Press, Oxford, 2013)
13. J.W.S. Cassels, *An Introduction to the Geometry of Numbers*. Grundlehren der Mathematischen Wissenschaften, vol. 99, second corrected printing (Springer, Berlin/New York, 1971)
14. H. Chernoff, A measure of asymptotic efficiency for tests of a hypothesis based on the sum of observations. Ann. Math. Stat. **23**, 493–507 (1952)
15. H. Cohn, M. de Courcy-Ireland, The Gaussian core model in high dimensions (2016). arXiv:1603.09684
16. H. Cohn, A. Kumar, Optimality and uniqueness of the Leech lattice among lattices. Ann. Math. (2) **170**, 1003–1050 (2009)
17. J.H. Conway, N.J.A. Sloane, *Sphere Packings, Lattices and Groups*. Grundlehren der Mathematischen Wissenschaften, vol. 290, 3rd edn. (Springer, New York, 1999)
18. E.T. Copson, *Asymptotic Expansions*. Cambridge Tracts in Mathematics and Mathematical Physics, vol. 55 (Cambridge University Press, Cambridge, 1965)
19. H. Cramér, Sur un nouveau théorème-limite de la théorie des probabilités, in *Conférences internationales de sciences mathématiques (Université de Genève 1937). Théorie des probabilités. III: Les sommes et les fonctions de variables aléatoires*. Actualités scientifiques et industrielles, vol. 736 (Hermann, Paris, 1938), pp. 5–23
20. D. Dadush, O. Regev, Towards strong reverse Minkowski-type inequalities for lattices, in *57th Annual IEEE Symposium on Foundations of Computer Science—FOCS 2016* (IEEE Computer Society, Los Alamitos, CA, 2016), pp. 447–456
21. C.G. Darwin, R.H. Fowler, On the partition of energy. Philos. Mag. **44**, 450–479 (1922)
22. C.G. Darwin, R.H. Fowler, On the partition of energy. – Part II. Statistical principles and thermodynamics. Philos. Mag. **44**, 823–842 (1922)
23. C.G. Darwin, R.H. Fowler, Fluctuations in an assembly in statistical equilibrium. Proc. Camb. Philos. Soc. **21**, 391–404 (1923)
24. R. Dedekind, H. Weber, Theorie der algebraischen Functionen einer Veränderlichen. J. Reine Angew. Math. **92**, 181–291 (1882)

25. W. Ebeling, *Lattices and Codes – A Course Partially Based on Lectures by Friedrich Hirzebruch*. Advanced Lectures in Mathematics, 3rd edn. (Springer Spektrum, Wiesbaden, 2013)
26. M. Eichler, *Introduction to the Theory of Algebraic Numbers and Functions*. Pure and Applied Mathematics, vol. 23 (Academic, New York/London, 1966)
27. R.S. Ellis, *Entropy, Large Deviations, and Statistical Mechanics*. Grundlehren der Mathematischen Wissenschaften, vol. 271 (Springer, New York, 1985)
28. M. Fekete, Über die Verteilung der Wurzeln bei gewissen algebraischen Gleichungen mit ganzzahligen Koeffizienten. Math. Z. **17**, 228–249 (1923)
29. R.H. Fowler, *Statistical Mechanics: The Theory of the Properties of Matter in Equilibrium*, 2nd edn. (Cambridge University Press, Cambridge, 1936)
30. J.W. Gibbs, *Elementary Principles in Statistical Mechanics: Developed with Especial Reference to the Rational Foundation of Thermodynamics*. Yale Bicentennial Publications (Scribner and Sons, New York, 1902)
31. H. Gillet, B. Mazur, C. Soulé, A note on a classical theorem of Blichfeldt. Bull. Lond. Math. Soc. **23**, 131–132 (1991)
32. D.R. Grayson, Reduction theory using semistability. Comment. Math. Helv. **59**, 600–634 (1984)
33. R.P. Groenewegen, An arithmetic analogue of Clifford's theorem. J. Théor. Nombres Bordeaux **13**, 143–156 (2001). 21st Journées Arithmétiques (Rome, 2001)
34. G.H. Hardy, S. Ramanujan, Asymptotic formulae for the distribution of integers of various types. Proc. Lond. Math. Soc. (2) **16**, 112–132 (1917)
35. G.H. Hardy, J.E. Littlewood, G. Pólya, *Inequalities*, 2nd edn. (Cambridge University Press, Cambridge, 1952)
36. E. Hecke, Über die Zetafunktion beliebiger algebraischer Zahlkörper. Nachr. Ges. Wiss. Göttingen, Math.-Phys. Kl. **1917**, 77–89 (1917)
37. C. Hermite, Extraits de lettres de M. Ch. Hermite à M. Jacobi sur différents objets de la théorie des nombres. J. Reine Angew. Math. **14**, 261–315 (1850)
38. L. Hörmander, *The Analysis of Linear Partial Differential Operators. I*. Grundlehren der Mathematischen Wissenschaften, vol. 256, 2nd edn. (Springer, Berlin, 1990)
39. L. Hörmander, *Notions of Convexity* (Birkhäuser Boston Inc., Boston, MA, 1994)
40. K. Huang, *Statistical Mechanics*, 2nd edn. (Wiley, New York, 1987)
41. R.B. Israel, *Convexity in the Theory of Lattice Gases*, with an introduction by Arthur S. Wightman. Princeton Series in Physics (Princeton University Press, Princeton, NJ, 1979)
42. A.I. Khinchin, *Mathematical Foundations of Statistical Mechanics*. Translated by G. Gamow (Dover Publications, Inc., New York, NY, 1949)
43. A. Korkine, G. Zolotareff, Sur les formes quadratiques. Math. Ann. **6**, 366–389 (1873)
44. L. Kronecker, Grundzüge einer arithmetischen Theorie der algebraischen Grössen. (Festschrift zu Herrn Ernst Eduard Kummers fünfzigjährigem Doctor-Jubiläum, 10 September 1881). J. Reine Angew. Math. **92**, 1–122 (1882)
45. J.C. Lagarias, Point lattices, in *Handbook of combinatorics*, vols. 1, 2 (Elsevier Sci. B. V., Amsterdam, 1995), pp. 919–966
46. J.C. Lagarias, H.W. Lenstra Jr., C.-P. Schnorr, Korkin-Zolotarev bases and successive minima of a lattice and its reciprocal lattice. Combinatorica **10**, 333–348 (1990)
47. O.E. Lanford, Entropy and equilibrium states in classical statistical mechanics, in *Statistical mechanics and mathematical problems (Battelle Seattle 1971 Rencontres)*, ed. by A. Lenard. Lecture Notes in Physics, vol. 20 (Springer, Berlin, Heidelberg, 1973), pp. 1–113
48. Yu.I. Manin, New dimensions in geometry, in *Workshop Bonn 1984 (Bonn, 1984)*. Lecture Notes in Mathematics, vol. 1111 (Springer, Berlin, 1985), pp. 59–101
49. J. Martinet, *Perfect Lattices in Euclidean Spaces*. Grundlehren der Mathematischen Wissenschaften, vol. 327 (Springer, Berlin, 2003)
50. J.E. Mazo, A.M. Odlyzko, Lattice points in high-dimensional spheres. Monatsh. Math. **110**, 47–61 (1990)
51. T. McMurray Price, Numerical cohomology. Algebr. Geom. **4**, 136–159 (2017)

52. J. Milnor, D. Husemoller, *Symmetric Bilinear Forms* (Springer, New York/Heidelberg, 1973). Ergebnisse der Mathematik und ihrer Grenzgebiete, Band 73
53. H. Minkowski, *Geometrie der Zahlen* (Teubner-Verlag, Leipzig, Berlin, 1896)
54. M. Morishita, Integral representations of unramified Galois groups and matrix divisors over number fields. Osaka J. Math. **32**, 565–576 (1995)
55. A.M. Odlyzko, Explicit Tauberian estimates for functions with positive coefficients. J. Comput. Appl. Math. **41**, 187–197 (1992)
56. M. Planck, *Treatise on Thermodynamics* (Longman, Green, and Co., London, New York, Bombay, 1903)
57. M. Planck, The theory of heat radiation. Authorised translation by M. Masius (1914)
58. M. Planck, Henri Poincaré und die Quantentheorie. Acta Math. **38**, 387–397 (1921)
59. H. Poincaré, Sur la théorie des quanta. Journal de Physique théorique et appliquée **2**, 5–34 (1912) (= Oeuvres, Tome IX, pp. 606–659)
60. D. Quillen, *Quillen Notebooks 1968–2003*, ed. by G. Luke, G. Segal. Published online by the Clay Mathematics Institute. http://www.claymath.org/publications/quillen-notebooks
61. R.A. Rankin, The difference between consecutive prime numbers. J. Lond. Math. Soc. **11**, 242–245 (1936)
62. O. Regev, N. Stephens-Davidowitz, An inequality for Gaussians on lattices. SIAM J. Discrete Math. **31**, 749–757 (2017)
63. O. Regev, N. Stephens-Davidowitz, A reverse Minkowski theorem, in *STOC'17—Proceedings of the 49th Annual ACM SIGACT Symposium on Theory of Computing* (ACM, New York, 2017), pp. 941–953
64. D. Roessler, The Riemann-Roch theorem for arithmetic curves. Diplomarbeit, ETH Zürich, 1993
65. W. Rudin, *Real and Complex Analysis*, 3rd edn. (McGraw-Hill Book Co., New York, 1987)
66. S.S. Ryškov, E.P. Baranovskiĭ, Classical methods of the theory of lattice packings. Uspekhi Mat. Nauk **34**, 3–63, 256 (1979)
67. P. Sarnak, A. Strömbergsson, Minima of Epstein's zeta function and heights of flat tori. Invent. Math. **165**, 115–151 (2006)
68. F.K. Schmidt, Analytische Zahlentheorie in Körpern der Charakteristik *p*. Math. Z. **33**, 1–32 (1931)
69. E. Schrödinger, *Statistical Thermodynamics*, 2nd edn. A course of seminar lectures delivered in January-March 1944, at the School of Theoretical Physics, Dublin Institute for Advanced Studies (Cambridge University Press, Cambridge, 1952)
70. B. Simon, *Convexity*. Cambridge Tracts in Mathematics, vol. 187 (Cambridge University Press, Cambridge, 2011)
71. D.W. Stroock, *Probability Theory. An Analytic View*, 2nd edn. (Cambridge University Press, Cambridge, 2011)
72. U. Stuhler, Eine Bemerkung zur Reduktionstheorie quadratischer Formen. Arch. Math. (Basel) **27**, 604–610 (1976)
73. L. Szpiro, Degrés, intersections, hauteurs. Astérisque **127**, 11–28 (1985)
74. G. van der Geer, R. Schoof, Effectivity of Arakelov divisors and the theta divisor of a number field. Selecta Math. (N.S.) **6**, 377–398 (2000)
75. A. Weil, Sur l'analogie entre les corps de nombres algébriques et les corps de fonctions algébriques. Rev. Sci. **77**, 104–106 (1939)

Part B
Distribution of Rational Points and Dynamics

Chapter V: Beyond Heights: Slopes and Distribution of Rational Points

Emmanuel Peyre

1 Introduction

For varieties with infinitely many rational points, one may equip the variety with a height and study asymptotically the finite set of rational points with a bounded height. The study of many examples shows that the distribution of rational points of bounded height on algebraic varieties is far from uniform. Indeed the points tend to accumulate on thin subsets which are images of non-trivial finite morphisms. It is natural to look for new invariants to characterise the points in these thin subsets. First of all, it is natural to consider all possible heights, instead of one relative to a fixed line bundle. But the geometric analogue described in Sect. 5 suggests to go beyond heights to find a property similar to being very free for rational curves. The slopes introduced by Jean-Benoît Bost give the tool for such a construction. In Sect. 6, we describe the notion of freeness which measures how free a rational point is. This section will present several cases in which this approach is fruitful. In Sect. 7, we also describe its use in connection with the notion of locally accumulating subvarieties which arises when one considers rational points of bounded height near a fixed rational point.

The author thanks D. Loughran for a discussion which led to a crucial improvement of this paper.

E. Peyre (✉)
Institut Fourier, Université Grenoble Alpes, Grenoble, France
e-mail: Emmanuel.Peyre@univ-grenoble-alpes.fr

E. Peyre, G. Rémond (eds.), *Arakelov Geometry and Diophantine Applications*, Lecture Notes in Mathematics 2276, https://doi.org/10.1007/978-3-030-57559-5_6

2 Norms and Heights

2.1 Adelic Metric

In this chapter, I am going to use heights defined by an adelic metric, which I use in a more restrictive sense than in the rest of the volume. In fact, an adelic metric will be an analogue of the notion of Riemannian metric in the adelic setting. Let me fix some notation for the remaining of this chapter.

Notation 2.1 The letter **K** denotes a number field. The set of places of **K** is denoted by Val(**K**), and its set of finite places by Val(**K**)$_f$. Let w be a place of **K**. We denote by $\mathbf{K}_w$ the completion of **K** at w. For an ultrametric place, $\mathscr{O}_w$ is the ring of integers of $\mathbf{K}_w$ and $\mathfrak{m}_w$ its maximal ideal. Let $v \in$ Val(**Q**) denote the restriction of w to **Q**. We consider the map $|\cdot|_w : \mathbf{K}_w \to \mathbf{R}_{\geqslant 0}$ defined by

$$|x|_w = |N_{\mathbf{K}_w/\mathbf{Q}_v}(x)|_v$$

for $x \in \mathbf{K}_w$, where $N_{\mathbf{K}_w/\mathbf{Q}_v}$ denotes the norm map. The Haar measure on the locally compact field $\mathbf{K}_w$ is normalized as follows:

(a) $\int_{\mathscr{O}_w} \mathrm{d}x_w = 1$ for a non-archimedean place w;
(b) $\mathrm{d}x_w$ is the usual Lebesgue measure if w is real;
(c) $\mathrm{d}x_w$ is twice the usual euclidean measure for a complex place.

Remark 2.2 The map $|\cdot|_w$ is an absolute value if w is ultrametric or real, it is the square of the modulus for a complex place. This choice of notation is motivated by the fact that $|\lambda|_w$ is the multiplier of the Haar measure for the change of variables $y = \lambda x$:

$$\mathrm{d}y_w = |\lambda|_w \mathrm{d}x_w$$

and we have the product formula:

$$\prod_{w \in \mathrm{Val}(\mathbf{K})} |x|_w = 1$$

for any $x \in \mathbf{K}^*$.

Terminology 2.3 The varieties we consider are integral separated schemes of finite type over a field. We shall say that a variety V is *nice* if it is smooth, projective, and geometrically integral.

Notation 2.4 Let X be a variety over **K**. For any commutative **K**-algebra A, we denote by X_A the product $X \times_{\mathrm{Spec}(\mathbf{K})} \mathrm{Spec}(A)$ and by $X(A)$ the set of A-points which is defined as $\mathrm{Mor}_{\mathrm{Spec}(\mathbf{K})}(\mathrm{Spec}(A), X)$. For any place w of **K**, we equip $X(\mathbf{K}_w)$ with the w-adic topology.

For the rest of this chapter, we denote by V a nice variety on the number field $\mathbf{K}$. The Picard group of V, denoted by $\mathrm{Pic}(V)$, is thought as the set of isomorphism classes of line bundles on V. A line bundle L is said to be *big* if a multiple of its class may be written as a sum of an ample class and an effective one.

Definition 2.5 Let $\pi : E \to V$ be a vector bundle on V. For any extension $\mathbf{L}$ of $\mathbf{K}$ and any $\mathbf{L}$-point P of V, we denote by $E_P \subset E(\mathbf{L})$ the $\mathbf{L}$-vector space corresponding to the fiber $\pi^{-1}(P)$ of π at P. In this text, a *classical adelic norm on E* is a family $(\|\cdot\|_w)_{w \in \mathrm{Val}(\mathbf{K})}$ of continuous maps

$$\|\cdot\|_w : E(\mathbf{K}_w) \to \mathbf{R}_{\geqslant 0}$$

such that:

  (i) If w is non-archimedean, for any $P \in V(\mathbf{K}_w)$, the restriction $\|\cdot\|_{w|E_P}$ is an ultrametric norm with values in $\mathrm{Im}(|\cdot|_w)$;
 (ii) If $\mathbf{K}_w$ is isomorphic to $\mathbf{R}$, then, for any P in $V(\mathbf{K}_w)$, the restriction $\|\cdot\|_{w|E_P}$ is a euclidean norm;
(iii) If $\mathbf{K}_w$ is isomorphic to $\mathbf{C}$, then, for any P in $V(\mathbf{K}_w)$, there exists a positive definite hermitian form ϕ_P on E_P such that

$$\forall y \in E_P, \quad \|y\|_w = \phi_P(y, y);$$

 (iv) There exists a finite set of places $S \subset \mathrm{Val}(\mathbf{K})$ containing the set of archimedean places and a model $\mathscr{E} \to \mathscr{V}$ of $E \to V$ over $\mathscr{O}_S$ such that for any place w in $\mathrm{Val}(\mathbf{K}) \backslash S$ and any $P \in \mathscr{V}(\mathscr{O}_w)$

$$\mathscr{E}_P = \{\, y \in E_P \mid \|y\|_w \leqslant 1 \,\},$$

where $\mathscr{E}_P$ denotes the $\mathscr{O}_w$-submodule of E_P defined by $\mathscr{E}$.

In the rest of this chapter, we shall say adelic norm for classical adelic norm. An *adelically normed vector bundle* is a vector bundle equipped with an adelic norm. We call *adelic metric* an adelic norm on the tangent bundle TV.

The point of using this type of norms is that you can do all the usual constructions:

Examples 2.6

  (i) If E and F are vector bundles equipped with classical adelic norms, then we can define adelic norms on the dual $E^\vee$, the direct sum $E \oplus F$ and the tensor product $E \otimes F$.
 (ii) If E is a vector bundle equipped with a classical norm, then we define a classical norm on the exterior product $\Lambda^m E$ in the following manner (see also Chapter II, Sect. 2.2). Let $P \in V(\mathbf{K}_w)$. If w is an ultrametric space, then let

$$\mathscr{E}_P = \{\, y \in E_P \mid \|y\|_w \leqslant 1 \,\}.$$

The set $\mathscr{E}_P$ is a $\mathscr{O}_w$-submodule of E_P of maximal rank. Then we take on $\Lambda^m E_P$ the norm defined by the module $\Lambda^m \mathscr{E}_P$. In the archimedean case, we choose the norm on $\Lambda^m E_P$ so that if $(e_1, \ldots, e_r)$ is an orthonormal basis of E_P then the family $(e_{k_1} \wedge e_{k_2} \wedge \ldots \wedge e_{k_m})_{1 \leqslant k_1 < k_2 < \cdots < k_m \leqslant r}$ is an orthonormal basis of $\Lambda^m E_P$.

(iii) It is possible to define pull-backs for morphisms of nice varieties over $\mathbf{K}$.

(iv) If $V = \mathrm{Spec}(\mathbf{K})$, then we may consider a vector bundle on V as a $\mathbf{K}$-vector space. Let E be a $\mathbf{K}$ vector space of dimension r equipped with an adelic norm $(\| \cdot \|)_{w \in \mathrm{Val}(\mathbf{K})}$. Then

$$\mathscr{E} = \{ y \in E \mid \forall w \in \mathrm{Val}(\mathbf{K})_f, \|y\|_w \leqslant 1 \}$$

is a projective $\mathscr{O}_\mathbf{K}$ module of constant rank r.

If $r = 1$, by the product formula, the product

$$\prod_{w \in \mathrm{Val}(\mathbf{K})} \|y\|_w$$

is constant for $y \in E \backslash \{0\}$. So we can define

$$\widehat{\deg}(E) = - \sum_{w \in \mathrm{Val}(\mathbf{K})} \log(\|y\|_w).$$

Let $\widehat{\mathrm{Pic}}(\mathrm{Spec}(\mathbf{K}))$ be the set of isomorphism classes of line bundles with an adelic norm on $\mathrm{Spec}(\mathbf{K})$. Let r_1 be the number of real places and r_2 the number of complex places. Let $\mathrm{Val}(\mathbf{K})_\infty \subset \mathrm{Val}(\mathbf{K})$ be the set of archimedean places. Let $H \subset \mathbf{R}^{\mathrm{Val}(\mathbf{K})_\infty}$ be the hyperplane given by the equation $\sum_{w \in \mathrm{Val}(\mathbf{K})_\infty} X_w = 0$. Then the map

$$x \mapsto (\log(|x|_w))_{w \in \mathrm{Val}(\mathbf{K})_\infty}$$

induces a map from $\mathscr{O}_\mathbf{K}^*$ to H. Let T be the quotient of H by the image of this map. The group T is a compact torus of dimension $r_1 + r_2 - 1$ and we get an exact sequence

$$0 \longrightarrow T \longrightarrow \widehat{\mathrm{Pic}}(\mathrm{Spec}(\mathbf{K})) \longrightarrow \mathrm{Pic}(\mathrm{Spec}(\mathscr{O}_\mathbf{K})) \times \mathbf{R} \longrightarrow 0$$

where a line bundle E equipped with an adelic norm $(\| \cdot \|_w)_{w \in \mathrm{Val}(\mathbf{K})}$ is sent on the pair $([\mathscr{E}], \widehat{\deg}(E))$ where $[\mathscr{E}]$ is the class of $\mathscr{E}$ in the ideal class group of $\mathscr{O}_\mathbf{K}$.

For arbitrary rank r, we may define:

$$\widehat{\deg}(E) = \widehat{\deg}(\Lambda^r(E)).$$

2.2 Arakelov Heights

Definition 2.7 For any vector bundle E over V equipped with an adelic norm, the corresponding *logarithmic height* is defined as the map $h_E : V(\mathbf{K}) \to \mathbf{R}$ given by $P \mapsto \widehat{\deg}(E_P)$, where E_P is the pull-back of E by the map $P : \mathrm{Spec}(\mathbf{K}) \to V$. The corresponding *exponential height* is defined by $H_E = \exp \circ h_E$.

Remark 2.8 If $r = \mathrm{rk}(E)$, we have that $h_E = h_{\wedge^r E} = h_{\det(E)}$. Therefore we do not get more than the heights defined by line bundles.

Example 2.9 For any $w \in \mathrm{Val}(\mathbf{K})$, we may consider the map $\| \cdot \|_w : \mathbf{K}_w^{N+1} \to \mathbf{R}$ defined by

$$\|(y_0, \ldots, y_N)\|_w = \max_{0 \leqslant i \leqslant N} |y_i|_w.$$

This does not define a classical norm on $\mathbf{K}_w^{N+1}$ in the sense above, however it defines a norm on the tautological line bundle as follows. Let $w \in \mathrm{Val}(\mathbf{K})$. The fibre of the tautological $\mathscr{O}_{\mathbf{P}_\mathbf{K}^N}(-1)$ over a point $P \in \mathbf{P}^N(\mathbf{K}_w)$ may be identified with the line corresponding to the point. By restricting $\| \cdot \|_w$ to these lines, we obtain an adelic norm $(\| \cdot \|_w)_{w \in \mathrm{Val}(\mathbf{K})}$ on $\mathscr{O}_{\mathbf{P}_\mathbf{K}^N}(-1)$ and by duality on $\mathscr{O}_{\mathbf{P}_\mathbf{K}^N}(1)$. If $(y_0, \ldots, y_N) \in \mathbf{K}^{N+1} \backslash \{0\}$, let P, also denoted by $[y_0 : \ldots : y_N]$, be the corresponding point in $\mathbf{P}^N(\mathbf{K})$. Then $y = (y_0, \ldots, y_n) \in \mathscr{O}(-1)_P$ and we get the formula

$$H_{\mathscr{O}(-1)}(P) = \prod_{w \in \mathrm{Val}(K)} \|y\|_w^{-1}.$$

Thus $H_{\mathscr{O}(1)}(P) = \prod_{w \in \mathrm{Val}(\mathbf{K})} \|y\|_w$. In the case where $\mathbf{K} = \mathbf{Q}$ and $y_0, \ldots, y_N$ are coprime integers, we have $\|(y_0, \ldots, y_N)\|_v = 1$ for any finite place v and the height may be written as

$$H_{\mathscr{O}(1)}(P) = \max_{0 \leqslant i \leqslant N} |y_i|$$

which is one of the naïve heights for the projective space.

Notation 2.10 For any function $H : V(\mathbf{K}) \to \mathbf{R}$, any subset $W \subset V(\mathbf{K})$ and any positive real number B, we consider the set

$$W_{H \leqslant B} = \{ P \in V(\mathbf{K}) \mid H(P) \leqslant B \}.$$

Our aim is to study such sets for heights H as B goes to infinity. Let us motivate this study with a few pictures of such sets.

Examples 2.11 Figure 1 represents rational points of bounded height in the projective plane. More precisely this drawing represents

$$\{\, (x, y) \in \mathbf{Q}^2 \mid H_{\mathcal{O}(1)}(x : y : 1) < 40,\ |x| \leqslant 1 \text{ and } |y| \leqslant 1 \,\}.$$

Figure 2 represents rational points of bounded height in the one-sheeted hyperboloid defined by the equation $xy = zt$ in $\mathbf{P}_{\mathbf{Q}}^3$:

$$\{\, P = (x, y) \in \mathbf{Q}^2 \mid H_{\mathcal{O}(1)}(xy : 1 : x : y) \leqslant 50,\ |x| \leqslant 1 \text{ and } |y| \leqslant 1 \,\}.$$

This quadric is the image of the Segre embedding

$$([u_1 : v_1], [u_2 : v_2]) \longmapsto [u_1 u_2 : v_1 v_2 : u_1 v_2 : v_1 u_2]$$

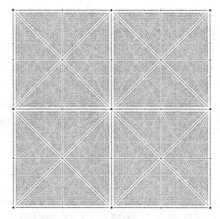

Fig. 1 Projective plane

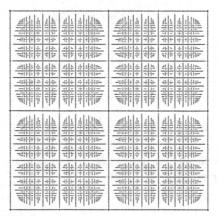

Fig. 2 Hyperboloid

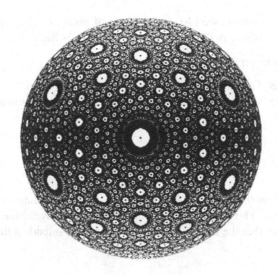

Fig. 3 The sphere

and therefore isomorphic to the product $\mathbf{P}_{\mathbf{Q}}^1 \times \mathbf{P}_{\mathbf{Q}}^1$. The last picture represents rational points of bounded height on the sphere (Fig. 3):

$$\{\, P = [x : y : z : t] \in \mathbf{P}^3(\mathbf{Q}) \mid H_{\mathscr{O}(1)}(P) \leqslant B \text{ and } x^2 + y^2 + z^2 = t^2 \,\}.$$

Proposition 2.12 *If L is a big line bundle and H a height relative to L, then there exists a dense open subset $U \subset V$ for Zariski topology such that for any $B \in \mathbf{R}_{\geqslant 0}$, the set $U(\mathbf{K})_{H \leqslant B}$ is finite.*

Proof It is enough to prove the result for a multiple of L. Thus we may assume that we can write L as $E + A$ where E is effective and A very ample. Taking U as the complement of the base locus of E, and choosing a basis $(s_0, \ldots, s_N)$ of $\Gamma(V, L)$, we get an embedding

$$\varphi : U \longrightarrow \mathbf{P}_{\mathbf{K}}^N.$$

Using the height of Example 2.9 on $\mathbf{P}_{\mathbf{K}}^N$, we get that

$$\frac{H(\varphi(x))}{H(x)} = \prod_{w \in \mathrm{Val}(\mathbf{K})} \max_{0 \leqslant i \leqslant N} \|s_i(x)\|_w.$$

Thus there exists a constant $C \in \mathbf{R}_{>0}$ such that

$$\forall x \in V(\mathbf{K}), \quad H(\varphi(x)) \leqslant C H(x).$$

Using Northcott theorem (see [40, 41]), the set of points of bounded height in the projective space is finite. A fortiori, the set $U(\mathbf{K})_{H \leqslant B}$ is finite.   □

The height depends on the metric, but in a bounded way:

Proposition 2.13 *Let H and H' be heights defined by adelic norms on a line bundle L then the quotient H/H' is bounded: there exist real constants $0 < C < C'$ such that*

$$\forall P \in V(\mathbf{K}), \quad C \leqslant \frac{H'(P)}{H(P)} < C'.$$

Proof The quotient of the norms $\frac{\|\cdot\|'_w}{\|\cdot\|_w}$ induces a continuous map from the compact set $V(\mathbf{K}_w)$ to $\mathbf{R}_{>0}$. Thus it is bounded from below and above. Moreover the adelic condition imposes that the norms coincide for all places outside a finite set.   □

3 Accumulation and Equidistribution

In this chapter, I shall first consider the distribution of rational points of bounded height on the variety.

3.1 Sandbox Example: The Projective Space

First, I have to explain what I mean by distribution. Let us for example consider the picture in Fig. 4. We have selected a "simple" open subset W in $\mathbf{P}^n(\mathbf{R})$, which is drawn in grey. We may then study asymptotically the proportion of rational points of bounded height in this open set. More precisely, one may formulate the following question:

Question 3.1 *Does the quotient*

$$\frac{\sharp(W \cap \mathbf{P}^n(\mathbf{Q}))_{H \leqslant B}}{\sharp \mathbf{P}^n(\mathbf{Q})_{H \leqslant B}}.$$

have a limit as B goes to $+\infty$ and how can we interpret its value?

Similarly, let us fix some integer $M > 0$ and consider the reduction modulo M of the points. More precisely, let A be a commutative ring. The set of A-points of the projective space, denoted by $\mathbf{P}^n(A)$, is the set of morphisms from $\mathrm{Spec}(A)$ to $\mathbf{P}^n_{\mathbf{Z}}$. This defines a covariant functor from the category of rings to the category of sets. A $(n+1)$-tuple $(a_0, \ldots, a_n)$ in A^{n+1} is said to be primitive if the generated ideal $(a_0, \ldots, a_n)$ is A itself; this is equivalent to the existence of $(u_0, \ldots, u_n) \in A^{n+1}$ such that $\sum_{i=0}^{n} u_i a_i = 1$. The group of invertible elements acts by multiplication

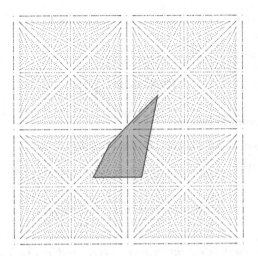

Fig. 4 Open subset

on the set of primitive elements in A^{n+1}. Then the $\mathbf{Z}/M\mathbf{Z}$ points of the projective space $\mathbf{P}^n_{\mathbf{Z}}$ may be described as the orbits for the action of $(\mathbf{Z}/M\mathbf{Z})^*$ on the set of primitive elements in $(\mathbf{Z}/M\mathbf{Z})^{n+1}$. For any point P in $\mathbf{P}^n(\mathbf{Q})$, we may choose homogeneous coordinates $[y_0 : \ldots : y_n]$ so that $y_0, \ldots, y_n$ are coprime integers. The reduction modulo M of P, is the point of $\mathbf{P}^n(\mathbf{Z}/M\mathbf{Z})$ defined by the primitive element $(\overline{y_0}, \ldots, \overline{y_n})$, where $\overline{y}$ denotes the reduction modulo M of the integer y. This define a map

$$r_M : \mathbf{P}^n(\mathbf{Q}) \longrightarrow \mathbf{P}^n(\mathbf{Z}/M\mathbf{Z}).$$

This description of the reduction map generalises easily to any quotient of a principal ring. Then for any subset W of $\mathbf{P}^n(\mathbf{Z}/M\mathbf{Z})$, we may consider the question

Question 3.2 *Does the quotient*

$$\frac{\sharp(r_M^{-1}(W))_{H \leqslant B}}{\sharp \mathbf{P}^N(\mathbf{Q})_{H \leqslant B}}$$

converges as B goes to infinity?

With the adelic point of view, we can see Questions 3.1 and 3.2 as particular cases of the following more general question:

Question 3.3 *Let $\mathbf{K}$ be a number field. Let*

$$\mathbf{P}^N(A_{\mathbf{K}}) = \prod_{w \in \mathrm{Val}(\mathbf{K})} \mathbf{P}^n(\mathbf{K}_w)$$

be the adelic projective space and let $f : \mathbf{P}^N(A_{\mathbf{K}}) \to \mathbf{R}$ be a continuous function. Does the quotient

$$S_B(f) = \frac{1}{\sharp \mathbf{P}^n(\mathbf{K})_{H \leqslant B}} \sum_{P \in \mathbf{P}^n(\mathbf{K})_{H \leqslant B}} f(P)$$

have a limit as B goes to infinity?

The answer is positive and we shall state it as a proposition:

Proposition 3.4 *With the notations introduced in Question 3.3,*

$$S_B(f) \xrightarrow[B \to +\infty]{} \int_{\mathbf{P}^n_{\mathbf{K}}(A_{\mathbf{K}})} f \mu_{\mathbf{P}^n}$$

where $\mu_{\mathbf{P}^n}$ is the probability measure given as the product $\prod_{w \in \mathrm{Val}(\mathbf{K})} \mu_w$ where μ_w is the borelian probability measure on $\mathbf{P}^n(\mathbf{K}_w)$ defined by:

- *If w is a non-archimedean place, let $\pi_k : \mathbf{P}^n(\mathbf{K}_w) \to \mathbf{P}^n(\mathscr{O}_w/\mathfrak{m}_w^k)$ be the reduction modulo $\mathfrak{m}_w^k$ then we equip $\mathbf{P}^n(\mathbf{K}_w)$ with the natural probability measure:*

$$\mu_w(\pi_k^{-1}(W)) = \frac{\sharp W}{\sharp \mathbf{P}^n(\mathscr{O}_w/\mathfrak{m}_w^k)}$$

for any subset W of $\mathbf{P}^n(\mathscr{O}_w/\mathfrak{m}_w^k)$;
- *If w is archimedean, let $\pi : \mathbf{K}_w^{n+1} \setminus \{0\} \to \mathbf{P}^n(\mathbf{K}_w)$ be the natural projection. Than μ_w is defined by*

$$\mu_w(U) = \frac{\mathrm{Vol}(\pi^{-1}(U) \cap B_{\|\cdot\|_w}(1))}{\mathrm{Vol}(B_{\|\cdot\|_w}(1))},$$

for any borelian subset U in $\mathbf{P}^n(\mathbf{K}_w)$, where $B_{\|\cdot\|_w}(1)$ denotes the ball of radius 1 for $\|\cdot\|_w$.

As a consequence, we may give a precise answer to Questions 3.1 and 3.2:

Corollary 3.5 *If W is an open subset of $\mathbf{P}^n(A_{\mathbf{K}})$ such that $\mu_{\mathbf{P}^n}(\partial W) = 0$ then*

$$\frac{\sharp(W \cap \mathbf{P}^n(\mathbf{K}))_{H \leqslant B}}{\sharp \mathbf{P}^n(\mathbf{K})_{H \leqslant B}} \xrightarrow[B \to +\infty]{} \mu_{\mathbf{P}^n}(W).$$

Sketch of the Proof of Proposition 3.4 for $\mathbf{K} = \mathbf{Q}$ Take an open cube $\mathscr{C} = \prod_{i=0}^n]a_i, b_i[$ where a_i and b_i are real numbers with $a_i < b_i$ for $i \in \{0, \ldots, n\}$, an integer $M \geqslant 1$ and an element $P_0 \in \mathbf{P}^n(\mathbf{Z}/M\mathbf{Z})$. We imbed $\mathbf{R}^n$ in $\mathbf{P}^n(\mathbf{R})$ and consider $\mathscr{C}$ as an open subset of the projective space. We choose a primitive element y_0 in $(\mathbf{Z}/M\mathbf{Z})^{n+1}$ representing P_0. We then want to estimate

$$\sharp\{\, P \in \mathbf{P}^n(\mathbf{Q}) \mid H(P) \leqslant B,\ P \in \mathscr{C} \text{ and } \pi_M(P) = P_0 \,\}$$

$$= \frac{1}{2} \sum_{\lambda \in (\mathbf{Z}/M\mathbf{Z})^*} \sharp\{\, y \in \mathbf{Z}^{n+1} \mid y \text{ primitive}, \|y\|_\infty \leqslant B, y \in \pi^{-1}(\mathscr{C})$$

$$\text{and } y \equiv \lambda y_0\, [M] \,\}$$

$$= \frac{1}{2} \sum_{\substack{d>0 \\ \lambda \in \mathbf{Z}/M\mathbf{Z}^*}} \mu(d) \sharp\{\, y \in (d\mathbf{Z})^{n+1} \backslash \{0\} \mid \|y\|_\infty \leqslant B, y \in \pi^{-1}(\mathscr{C})$$

$$\text{and } y \equiv \lambda y_0\, [M] \,\}$$

where $\mu : \mathbf{N}\backslash\{0\} \to \{-1, 0, 1\}$ denotes the Möbius function. As y_0 is primitive, the set we obtained in the sum is empty if M and d are not coprime. Otherwise it is the intersection of the translation of a lattice of covolume $(dM)^{n+1}$, the cone $\pi^{-1}(\mathscr{C})$ and the ball $B_{\|\cdot\|_\infty}(B)$. Thus its cardinal may be approximated by

$$\frac{\operatorname{Vol}(\pi^{-1}(\mathscr{C}) \cap B_{\|\cdot\|_\infty}(1)) B^{n+1}}{(dM)^{n+1}}$$

with an error term which is bounded up to a constant by $\left(\frac{B}{d} + 1\right)^n$. Up to an error term left to the reader, we get that the sum is equivalent to

$$\frac{1}{2} \operatorname{Vol}(\pi^{-1}(\mathscr{C}) \cap B_{\|\cdot\|_\infty}(1)) \times \frac{\varphi(M)}{M^{n+1} \prod_{p|M} \left(1 - \frac{1}{p^{n+1}}\right)} \times \frac{1}{\zeta_{\mathbf{Q}}(n+1)} B^{n+1}.$$

In this product, the term

$$\frac{\varphi(M)}{M^{n+1} \prod_{p|M} \left(1 - \frac{1}{p^{n+1}}\right)}$$

is $\sharp(\mathbf{P}^n(\mathbf{Z}/M\mathbf{Z}))^{-1}$. In particular, this implies that

$$\frac{1}{2} \operatorname{Vol}(B_{\|\cdot\|_\infty}(1))/\zeta_{\mathbf{Q}}(n+1)$$

is the limit of $\sharp \mathbf{P}^n(\mathbf{Q})_{H \leqslant B}/B^{n+1}$ as B goes to infinity. $\qquad \square$

3.2 Adelic Measure

By choosing different norms on the anticanonical line bundle, and thus different heights on a variety, one realizes that the measure which gives the asymptotic distribution as B goes to infinity may be directly defined from the adelic norm on ω_V^{-1}, exactly as a Riemannian metric defines a volume form. This construction in fact applies to any nice variety equipped with an adelic metric.

Construction 3.6 Let V be a nice variety with a rational point. We fix an adelic norm $(\| \cdot \|_w)_{w \in \text{Val}(\mathbf{K})}$ on $\omega_V^{-1} = \det(TV)$. The formula for the change of variables (see [56, §2.2.1]) proves that the local measures

$$\left\| \frac{\partial}{\partial x_1} \wedge \frac{\partial}{\partial x_2} \wedge \ldots \wedge \frac{\partial}{\partial x_n} \right\|_w dx_{1,w} \, dx_{2,w} \ldots dx_{n,w}, \tag{1}$$

where $(x_1, \ldots, x_n) : \Omega \to \mathbf{K}_w^n$ is a local system of coordinates defined on an open subset Ω of $V(\mathbf{K}_w)$, does not depend on the choice of coordinates; therefore by patching together these measures, we get a measure $\omega_{V,w}$ on $V(\mathbf{K}_w)$, which induces a probability measure

$$\mu_{V,w} = \frac{1}{\omega_{V,w}(V(\mathbf{K}_w))} \omega_{V,w}.$$

Then the product

$$\mu_V = \prod_{w \in \text{Val}(\mathbf{K})} \mu_{V,w}$$

is a probability measure on the adelic space $V(A_\mathbf{K})$.

Remark 3.7 For the projective space, this construction gives the right asymptotic distribution for the points of bounded height. So it is natural to try to generalise to other varieties. To state precisely our question, we introduce the counting measure defined by the set of points of bounded height.

Definition 3.8 For any non-empty subset $W \subset V(\mathbf{K})$ we define, for B a real number bigger than the smallest height of a point of W,

$$\delta_{W_{H \leqslant B}} = \frac{1}{\sharp W_{H \leqslant B}} \sum_{P \in W_{H \leqslant B}} \delta_P,$$

where δ_P denotes the Dirac measure at P on the adelic space.

Naïve Equidistribution 3.9 We shall say that the naïve equidistribution (NE) holds if the measure $\delta_{V(\mathbf{K})_{H \leqslant B}}$ converges to μ_V as B goes to infinity for the weak topology.

Remark 3.10 In other words, the naïve equidistribution holds if for any continuous function $f : V(A_K) \to \mathbf{R}$, one has the convergence

$$\int_{V(A_K)} f \delta_{V(K)_{H \leqslant B}} \xrightarrow[B \to +\infty]{} \int_{V(A_K)} f \mu_V.$$

This equidistribution may seem to be overoptimistic and one may wonder whether there exists any case besides the projective space for which it is valid.

Theorem 3.11 *If V is a generalized flag variety, that is a quotient G/P where G is a linear algebraic group over $\mathbf{K}$ and P a parabolic subgroup of G, then* (NE) *is true.*

Example 3.12 Grassmannian are examples of such flag varieties. Any smooth quadric with a rational point is a generalized flag variety for the orthogonal group. Therefore any smooth quadric with a rational point satisfies the naïve equidistribution.

Tools of the Proof of Theorem 3.11 To prove this result one may use harmonic analysis on the adelic space $G/P(A_K)$ and apply Langland's work on Eisenstein series (see [45, corollaire 6.2.17], [33]). □

So we have solved the case of hypersurfaces of degree 2. In higher degrees, the equidistribution, when the number of variables is large enough, is an easy consequence of the very general result of Birch [7] based on the circle method. His result implies the following theorem:

Theorem 3.13 *Let $V \subset \mathbf{P}_{\mathbf{Q}}^n$ be a smooth hypersurface of degree d such that $V(A_Q) \neq \emptyset$ with $n > (d-1)2^d$, then V satisfies* (NE).

Remark 3.14 In fact, it applies to all the cases considered by Birch, that is for smooth complete intersection of m hypersurfaces of the same degree d if $n > m(m+1)(d-1)2^{d-1}$.

3.3 Weak Approximation

The first indications of the naïveté of (NE) appear when one considers obvious consequences of it. Let us recall the definition of weak approximation:

Definition 3.15 A nice variety V satisfies *weak approximation* if the rational points of V are dense in the adelic space $V(A_K)$.

Remarks 3.16

(i) Let V be a nice variety with a rational point. If it satisfies the naïve equidistribution, then it satisfies weak approximation and therefore $V(\mathbf{K})$ is dense for Zariski topology.

This follows from the fact that for any real number B, the support of the measure $\delta_{V(\mathbf{K})_{H \leqslant B}}$ in $V(A_{\mathbf{K}})$ is contained in the closure $\overline{V(\mathbf{K})}$ of the set of rational points. But the support of the measure μ_V is the whole adelic space. Thus (NE) implies that $\overline{V(\mathbf{K})} = V(A_{\mathbf{K}})$. Let then U be a non-empty open subset for Zariski topology. If V has an adelic point, the implicit function theorem ensures that for any place w, the set $U(\mathbf{K}_w)$ is a non-empty open subset of $V(\mathbf{K}_w)$. If $V(\mathbf{K})$ is dense in $V(\mathbf{K}_w)$, it follows that U contains a rational point and the rational points are Zariski dense.

(ii) So (NE) has to fail for any variety in which the rational points are not Zariski dense. In that case, one may consider the desingularisation of the closure of the rational points for Zariski topology and ask wether the principle holds for that variety. But even such a modified question fails because examples are known where rational points are dense for Zariski topology but the variety does not satisfy weak approximation.

Convention 3.17 From now on, we assume that V is a nice variety in which the set of rational points $V(\mathbf{K})$ is Zariski dense.

About weak approximation, we are going to give a quick overview of the Brauer–Manin obstruction, which was introduced by Y. Manin in [37] to explain the previously known counterexamples to weak approximation (see also [47] for a survey).

Construction 3.18 For a nice variety V, we define its *Brauer group* as the cohomology group

$$\mathrm{Br}(V) = H^2_{\text{ét}}(V, \mathbf{G}_m)$$

which defines a contravariant functor from nice varieties to the category of abelian groups. In the case of the spectrum of a field of characteristic 0, we get the Brauer group of $\mathbf{L}$, which is defined in terms of Galois cohomology by

$$\mathrm{Br}(\mathbf{L}) = H^2(\mathrm{Gal}(\overline{\mathbf{L}}/\mathbf{L}), \mathbf{G}_m),$$

where $\overline{\mathbf{L}}$ is an algebraic closure of $\mathbf{L}$. Class field theory gives for any place w an injective morphism

$$\mathrm{inv}_w : \mathrm{Br}(\mathbf{K}_w) \longrightarrow \mathbf{Q}/\mathbf{Z}$$

which is an isomorphism if w is not archimedean, so that the sequence

$$0 \to \mathrm{Br}(\mathbf{K}) \to \bigoplus_{w \in \mathrm{Val}(\mathbf{K})} \mathrm{Br}(\mathbf{K}_w) \xrightarrow{\sum_w \mathrm{inv}_w} \mathbf{Q}/\mathbf{Z} \to 0 \qquad (2)$$

is an exact sequence. Therefore we may define a pairing

$$\mathrm{Br}(V) \times V(A_{\mathbf{K}}) \longrightarrow \mathbf{Q}/\mathbf{Z}$$

$$(\alpha, (P_w)_{w \in \mathrm{Val}(\mathbf{K})}) \longmapsto \sum_{w \in \mathrm{Val}(\mathbf{K})} \mathrm{inv}_w(\alpha(P_w))$$

where $\alpha(P_w)$ denotes the pull-back of α by the morphism $\mathrm{Spec}(\mathbf{K}_w) \to V$ defined by P_w. Let us denote by $\mathrm{Br}(V)^\vee$ the group $\mathrm{Hom}(\mathrm{Br}(V), \mathbf{Q}/\mathbf{Z})$ then the above pairing may be seen as a map

$$\eta : V(A_{\mathbf{K}}) \longrightarrow \mathrm{Br}(V)^\vee.$$

If $P \in V(\mathbf{K})$ then the fact that (2) is a complex implies that

$$\sum_{w \in \mathrm{Val}(\mathbf{K})} \mathrm{inv}_w(\alpha(P)) = 0;$$

in other words, $\eta(P) = 0$. By arguments of continuity, one gets that

$$\overline{V(\mathbf{K})} \subset V(A_{\mathbf{K}})^{\mathrm{Br}} = \{ P \in V(A_{\mathbf{K}}) \mid \eta(P) = 0 \}.$$

The element $\eta(P)$ is called the *Brauer–Manin obstruction* to weak approximation at P.

Remark 3.19 Let $\overline{\mathbf{K}}$ be an algebraic closure of $\mathbf{K}$ and $\overline{V} = V_{\overline{\mathbf{K}}}$. Since we assume V to have a rational point, there is an exact sequence

$$0 \to \mathrm{Br}(\mathbf{K}) \to \ker(\mathrm{Br}(V) \to \mathrm{Br}(\overline{V})) \to H^1(\mathrm{Gal}(\overline{\mathbf{K}}/\mathbf{K}), \mathrm{Pic}(\overline{V})) \to 0.$$

Also the exponential map gives an exact sequence

$$H^1(V_{\mathbf{C}}, \mathscr{O}_V) \to \mathrm{Pic}(\overline{V}) \to H^2(V(\mathbf{C}), \mathbf{Z}) \to$$

$$H^2(V_{\mathbf{C}}, \mathscr{O}_V) \to \mathrm{Br}(\overline{V}) \to H^3(V(\mathbf{C}), \mathbf{Z})_{\mathrm{tors}}$$

Thus assuming that $H^i(V, \mathscr{O}_V) = \{0\}$ for $i = 1$ and $i = 2$, which is automatic for Fano varieties by Kodaira's vanishing theorem, we get first that the geometric Picard of the variety is finitely generated. Thus the action of the Galois group on the Picard group is trivial over a finite extension of the ground field. Therefore, in this case, the groups $H^1(\mathrm{Gal}(\overline{\mathbf{K}}/\mathbf{K}), \mathrm{Pic}(\overline{V}))$ and $\mathrm{Br}(\overline{V})$ are finite. Hence the cokernel of the morphism $\mathrm{Br}(\mathbf{K}) \to \mathrm{Br}(V)$ is finite, which implies that $V(A_{\mathbf{K}})^{\mathrm{Br}}$ is open and closed in the adelic space.

If one hopes that the Brauer–Manin obstruction to the weak approximation is the only one, then it is natural to define the measure induced by the probability measure

μ_V on the space on which the obstruction is 0. Since we assume that the variety V has a rational point, the space $V(A_{\mathbf{K}})^{\mathrm{Br}}$ is not empty. In that setting, we may give the following definition:

Definition 3.20 The measure μ_V^{Br} is defined as follows: for any Borelian subset W of $V(A_{\mathbf{K}})$

$$\mu_V^{\mathrm{Br}}(W) = \frac{\mu_V(W \cap V(A_{\mathbf{K}})^{\mathrm{Br}})}{\mu_V(V(A_{\mathbf{K}})^{\mathrm{Br}})}.$$

The following question then takes into account the Brauer–Manin obstruction to weak approximation:

Global Equidistribution 3.21 *We shall say that global equidistribution holds if the measure* $\delta_{V(\mathbf{K})_{H \leqslant B}}$ *converges weakly to* μ_V^{Br} *as B goes to infinity.*

Potential counterexamples to global equidistribution have been known for quite a long time (see for example [54]), but Y. Manin was the first to consider accumulating subsets, which we will study at length in the next section.

3.4 Accumulating Subsets

In fact, the support of the limit of the measure $\delta_{V(\mathbf{K})_{H \leqslant B}}$ is, in general, much smaller than the closure $\overline{V}(\mathbf{K})$ of the set of rational points. Let me give a few examples.

The Plane Blown Up in One Point

The blowing up of the projective plane at the point $P_0 = [0 : 0 : 1]$ may be described as the hypersurface V in the product $\mathbf{P}_{\mathbf{Q}}^2 \times \mathbf{P}_{\mathbf{Q}}^1$ defined by the equation $XV = YU$, where X, Y, Z denote the coordinates on the first factor and U, V the coordinates on the second one. Let π be the projection on the first factor. Then $E = \pi^{-1}(P_0)$ is an exceptional divisor on V and the second projection pr_2 defines an isomorphism from E to $\mathbf{P}_{\mathbf{Q}}^1$. Let U be the complement of E in V. The projection π induces an isomorphism from U to $\mathbf{P}_{\mathbf{Q}}^2 \backslash \{P_0\}$. As an exponential height, we may use the map

$$H : V(\mathbf{Q}) \longrightarrow \mathbf{R}_{>0}$$

$$(P, Q) \longmapsto H_{\mathcal{O}_{\mathbf{P}_{\mathbf{Q}}^2}(1)}(P)^2 H_{\mathcal{O}_{\mathbf{P}_{\mathbf{Q}}^1}(1)}(Q).$$

This example as been used as a sandbox case for the study of rational points of bounded height by many people, including J.-P. Serre [54, §2.12], as well as V. V. Batyrev and Y. I. Manin [3, proposition 1.6] and the results may be summarized as follows:

Proposition 3.22 *On the exceptional line, the number of points of bounded height is given by*

$$\sharp E(\mathbf{Q})_{H \leqslant B} \sim \frac{2}{\zeta_{\mathbf{Q}}(2)} B^2$$

as B goes to infinity, whereas on its complement it is given by

$$\sharp U(\mathbf{Q})_{H \leqslant B} \sim \frac{8}{3\zeta_{\mathbf{Q}}(2)^2} B \log(B)$$

as B goes to infinity.

Remark 3.23 Thus there are much more rational points on the exceptional line E than on the dense open subset U. In fact, since the points on the exceptional line are distributed as on $\mathbf{P}_{\mathbf{Q}}^1$, we get that the measure $\delta_{V(\mathbf{Q})_{H \leqslant B}}$ converges to μ_E for the weak topology.

On the other hand, if we only consider the rational points on the open set U, we get the right limit:

Proposition 3.24 *The measure $\delta_{U(\mathbf{Q})_{H \leqslant B}}$ converges to μ_V for the weak topology as B goes to infinity.*

Remarks 3.25

(i) Let W be an infinite subset of $V(K)$. If the measure $\delta_{W_{H \leqslant B}}$ converges to μ_V for the weak topology, then, for any strict closed subvariety F in V, we have that

$$\sharp(W \cap F(\mathbf{Q}))_{H \leqslant B} = o(\sharp W_{H \leqslant B})$$

since we have $\mu_V(F(A_{\mathbf{K}})) = 0$. Thus any strict closed subset with a strictly positive contribution to the number of points has to be removed to get equidistribution.

(ii) It may seem counterintuitive that by removing points, we get a measure with a larger support. But this comes from the fact that we divide the counting measure on U by a smaller term. From this example, it follows that it is natural to consider only the points outside a set of "bad" points. The problem is that this set of bad points might be quite big.

The Principle of Manin

The principle suggested by Manin and his collaborators in the founding papers [3, 24] is that, on Fano varieties, there should be an open subset on which the points of bounded height behave as expected. Let us give a precise expression for this principle, in a slightly more general setting. Since this principle deals with the number of points of bounded height rather than their distribution, we have to introduce another normalisation of the measures to get a conjectural value for the constant, which is defined as a volume.

Notation 3.26 Let $\mathrm{NS}(V)$ be the Néron-Severi group of V, that is the quotient of the Picard group by the connected component of the neutral element. We put $\mathrm{NS}(V)_{\mathbf{R}} = \mathrm{NS}(V) \otimes_{\mathbf{Z}} \mathbf{R}$ and denote by $C_{\mathrm{eff}}(V)$ the closed cone in $\mathrm{NS}(V)_{\mathbf{R}}$ generated by the classes of effective divisors. We write $C_{\mathrm{eff}}(V)^{\vee}$ for the dual of the effective cone in the dual space $\mathrm{NS}(V)_{\mathbf{R}}^{\vee}$:

$$C_{\mathrm{eff}}(V)^{\vee} = \{\, y \in \mathrm{NS}(V)_{\mathbf{R}}^{\vee} \mid \forall x \in C_{\mathrm{eff}}(V), \langle y, x \rangle \geqslant 0 \,\}.$$

To construct the constant, we shall restrict ourselves to a setting in which the local measures can be normalized using the action of the Galois group of $\mathbf{K}$ on the Picard group. Therefore, we make the following hypothesis:

Hypotheses 3.27 From now on, V is a nice variety, which satisfies the following conditions:

   (i) A multiple of the class of ω_V^{-1} is the sum of an ample divisor and a divisor with normal crossings;
  (ii) The set $V(\mathbf{Q})$ is Zariski dense;
 (iii) The groups $H^i(V, \mathcal{O}_V)$ are $\{0\}$ if $i \in \{1, 2\}$;
 (iv) The geometric Brauer group $\mathrm{Br}(\overline{V})$ is trivial and the geometric Picard group $\mathrm{Pic}(\overline{V})$ has no torsion;
  (v) The closed cone $C_{\mathrm{eff}}(\overline{V})$ is generated by the classes of a finite set of effective divisors.

Remark 3.28 The first four conditions are satisfied by Fano varieties, that is varieties for which ω_V^{-1} is ample. The fifth has been conjectured by V. V. Batyrev for these varieties [2].

Construction 3.29 We choose a finite set S of places containing the archimedean places and the places of bad reduction for V. Let $\mathbf{L}$ be a finite extension of $\mathbf{K}$ such that the Picard group $\mathrm{Pic}(V_{\mathbf{L}})$ is isomorphic to the geometric Picard group $\mathrm{Pic}(\overline{V})$. We assume that S contains all the places which ramify in the extension $\mathbf{L}/\mathbf{K}$. With this assumption, for any place $w \in \mathrm{Val}(\mathbf{K}) \backslash S$, let $\mathbf{F}_w$ be the residual field at w. The Frobenius lifts to an element $(w, \mathbf{L}/\mathbf{K})$ in $\mathrm{Gal}(\mathbf{L}/\mathbf{K})$ which is well defined up to conjugation (see [53, §1.8]). Then we can consider the local factors of the L function defined by the Picard group:

$$L_w(s, \mathrm{Pic}(\overline{V})) = \frac{1}{\det(1 - \sharp \mathbf{F}_w^{-s}(w, \mathbf{L}/\mathbf{K}) | \mathrm{Pic}(\overline{V}))},$$

where s is a complex number with $\Re(s) > 0$. If the real part of s satisfies $\Re(s) > 1$, then a theorem of Artin [1, Satz 3] implies that the eulerian product

$$L_S(s, \mathrm{Pic}(\overline{V})) = \prod_{w \in \mathrm{Val}(\mathbf{K}) \setminus S} L_w(s, \mathrm{Pic}(\overline{V}))$$

converges. For $w \in \mathrm{Val}(\mathbf{K})$, we define $\lambda_w = L_w(1, \mathrm{Pic}(\overline{V}))^{-1}$ if w does not belong to S and $\lambda_w = 1$ otherwise. We put $t = \mathrm{rk}(\mathrm{Pic}(V))$. It follows from the Weil's conjecture proven by P. Deligne [20] that the product of measures

$$\omega_V = \frac{\lim_{s \to 1} (s-1)^t L_S(s, \mathrm{Pic}(\overline{V}))}{\sqrt{d_{\mathbf{K}}}^{\dim(V)}} \prod_{w \in \mathrm{Val}(\mathbf{K})} \lambda_w \omega_{V,w} \qquad (3)$$

converges (see [45, §2.1]), where $d_{\mathbf{K}}$ denotes the absolute value of the discriminant of $\mathbf{K}$. We may then define the *Tamagawa–Brauer–Manin volume of V* as

$$\tau^{\mathrm{Br}}(V) = \omega_V(V(A_{\mathbf{K}})^{\mathrm{Br}}).$$

We also introduce the constant

$$\alpha(V) = \frac{1}{(t-1)!} \int_{C^1_{\mathrm{eff}}(V)^\vee} e^{-\langle \omega_V^{-1}, y \rangle} \mathrm{d}y$$

which is a rational number under the Hypothesis 3.27(v), and the integer

$$\beta(V) = \sharp(\mathrm{Br}(V)).$$

Then the *empirical constant* associated to the chosen metric on V is the constant

$$C(V) = \alpha(V)\beta(V)\tau^{\mathrm{Br}}(V).$$

Batyrev–Manin Principle 3.30 *Let V be a variety which satisfies the conditions 3.27. We say that V satisfies the refined Batyrev–Manin principle if there exists a dense open subset U of V such that*

$$\sharp U(\mathbf{K})_{H \leqslant B} \sim C(V) B \log(B)^{t-1} \qquad (4)$$

as B goes to infinity.

For equidistribution, we may introduce the following notion

Relative Equidistribution 3.31 *Let W be an infinite subset of $V(\mathbf{K})$, we say that the points of W are equidistributed in V if the counting measure $\delta_{W_{H \leqslant B}}$ converges to μ_V.*

Remark 3.32 The relation between the Batyrev-Manin principle as stated here and the equidistribution may be described as follows: if the principle holds for a given open subset U for any metric on V, then the points of $U(\mathbf{K})$ are equidistributed on V. Conversely if the principle holds for a particular choice of the metric and an open subset U and if the points of $U(\mathbf{K})$ are equidistributed, then the principle holds for any choice of the metric (see [45, §3]).

The Counterexample of V. V. Batyrev and Y. Tschinkel

This example was described in [4]. We consider the hypersurface V in $\mathbf{P}_{\mathbf{Q}}^3 \times \mathbf{P}_{\mathbf{Q}}^3$ defined by the equation

$$\sum_{i=0}^{3} X_i Y_i^3 = 0.$$

We denote by $\mathcal{O}_V(a, b)$ the restriction to V of the line bundle $\mathrm{pr}_1^*(\mathcal{O}_{\mathbf{P}_{\mathbf{Q}}^3}(a)) \otimes \mathrm{pr}_2^*(\mathcal{O}_{\mathbf{P}_{\mathbf{Q}}^3}(b))$ Then the anticanonical line bundle on V is given by $\mathcal{O}_V(3, 1)$ and therefore the function $H : V(\mathbf{Q}) \to \mathbf{R}$ defined by

$$H(P, Q) = H_{\mathcal{O}_{\mathbf{P}_{\mathbf{Q}}^3}(1)}(P)^3 H_{\mathcal{O}_{\mathbf{P}_{\mathbf{Q}}^3}(1)}(Q)$$

defines a height relative to the anticanonical line bundle on V. Let π be the projection on the first factor and for any $P \in \mathbf{P}^3(\mathbf{Q})$, let $V_P = \pi^{-1}(P)$ the fibre over P. If $P = [x_0 : x_1 : x_2 : x_2]$ with $\prod_{i=0}^3 x_i \neq 0$, then the fibre V_P is a smooth cubic surface which contains 27 projective lines. The complement U_P of these 27 lines is defined over $\mathbf{Q}$. For cubic surfaces, it is expected that the Batyrev-Manin principle holds for any dense open subset contained in U_P. For any P as above, let $t_P = \mathrm{rk}(\mathrm{Pic}(V_P))$ be the rank of the Picard group of the cubic surface corresponding to P. Thus, according to (4), one expects that for any $U \subset U_P$, one has

$$\sharp U(\mathbf{Q})_{H \leqslant B} \sim C(V_P) B \log(B)^{t_P - 1}$$

as B goes to infinity. One can show that $t_P \in \{1, 2, 3, 4\}$ and that $t_P = 4$ if all the quotients x_i/x_j are cubes, that is if P is in the image of the morphism c from $\mathbf{P}_{\mathbf{Q}}^3$ to $\mathbf{P}_{\mathbf{Q}}^3$ defined by $[x_0 : x_1 : x_2 : x_3] \mapsto [x_0^3 : x_1^3 : x_2^3 : x_3^3]$. But, on the other hand, by Lefschetz theorem, the application $(a, b) \mapsto \mathcal{O}_V(a, b)$ induces an isomorphims of groups from $\mathbf{Z}^2$ to $\mathrm{Pic}(V)$. Therefore, the principle of Batyrev and Manin would be satisfied for V if and only if there existed an open subset U of V such that

$$\sharp U(\mathbf{Q})_{H \leqslant B} \sim C(V) B \log(B)$$

as B goes to infinity. Since the rational points in the image of c are dense for Zariski topology, the open set U has to intersect an open set U_P for some P. Thus the principle can not hold for both the cubic surfaces and V itself.

Remarks 3.33

(i) In fact, V. V. Batyrev and Y. Tschinkel proved in [4] that any dense open set of V contains too many rational points over $\mathbf{Q}(j)$, where j is a primitive third root of unity. More recently, C. Frei, D. Loughran, and E. Sofos proved in [25] that it is in fact the case over any number field.

(ii) One may look at the set

$$T = \{\, P \in \mathbf{P}^3(\mathbf{Q}) \mid \mathrm{rk}(\mathrm{Pic}(V_P)) > 1 \,\}$$

that is the set of points for which the rank of the Picard group is bigger than the generic one. As we are about to explain,

$$\sharp T_{H \leqslant B} = o(\sharp \mathbf{P}^3(\mathbf{Q})_{H \leqslant B})$$

which means that most of the fibers have a Picard group of rank one.

This example led to the introduction of a new kind of accumulating subsets, namely thin subsets (see J.-P. Serre [55, §3.1]).

Definition 3.34 Let V be a nice variety over the number field $\mathbf{K}$. A subset $T \subset V(\mathbf{K})$ is said to be *thin*, if there exists a morphism of varieties $\varphi : X \to V$ which satisfies the following conditions:

(i) The morphism φ is generically finite;
(ii) The morphism φ has no rational section;
(iii) The set T is contained in the image of φ.

Remarks 3.35

(i) If E is an elliptic curve, the group $E(\mathbf{K})/2E(\mathbf{K})$ is a finite group. Let $(P_i)_{i \in I}$ be a finite family of points of $E(\mathbf{K})$ containing a representative for each element of $E(\mathbf{K})/2E(\mathbf{K})$. Then the morphism $\varphi : \coprod_{i \in I} E \to E$ which maps a point P in the i-th component to $P_i + 2P$ gives a surjective map onto the sets of rational points. This shows that $E(\mathbf{K})$ itself is thin.

(ii) In the example of Batyrev and Tschinkel, as T is a thin subset in $\mathbf{P}^3(\mathbf{Q})$, it follows from [54, §13, theorem 3] that

$$\sharp T_{H \leqslant B} = o(\sharp \mathbf{P}^3(\mathbf{Q})_{H \leqslant B}).$$

The set

$$V_T = \bigcup_{P \in T} V_P(\mathbf{Q})$$

is itself a thin subset of $V(\mathbf{Q})$. Conjecturally we may hope that

$$\sharp(V(\mathbf{Q})\backslash V_T)_{H\leqslant B} \sim C_H(V)B\log(B)$$

as B goes to infinity. In other words, the points on the complement of the accumulating subset should behave as expected. We shall explain below how a result of this kind was proven by C. Le Rudulier for a Hilbert scheme of the projective plane [36]. More recently, T. Browning and D.R. Heath-Brown [11] proved that for the hypersurface of $\mathbf{P}_{\mathbf{Q}}^3 \times \mathbf{P}_{\mathbf{Q}}^3$ defined by the equation

$$\sum_{i=0}^{3} X_i Y_i^2$$

the number of points on the complement of an accumulating thin subset behaves as expected.

(iii) The work of B. Lehmann, S. Tanimoto and Y. Tschinkel [34] shows how common varieties with accumulating thin subsets probably are.

(iv) We may assume that φ is a proper morphism. Then $\varphi(X(A_{\mathbf{K}})) \subset V(A_{\mathbf{K}})$ is a closed subset. Under mild hypotheses, T. Browning and D. Loughran proved in [12] that

$$\mu_V(\varphi(X(A_{\mathbf{K}}))) = 0.$$

Thus the existence of such a thin subset with a positive contribution to the asymptotic number of points is an obstruction to the global equidistribution of points.

The Example of C. Le Rudulier

C. Le Rudulier considers the Hilbert scheme V which parametrizes the points of degree 2 in $\mathbf{P}_{\mathbf{Q}}^2$ [36]. To describe this scheme, let us consider the scheme Y defined as the second symmetric product of $\mathbf{P}_{\mathbf{Q}}^2$:

$$Y = \mathrm{Sym}^2(\mathbf{P}_{\mathbf{Q}}^2) = (\mathbf{P}_{\mathbf{Q}}^2)^2/\mathfrak{S}_2.$$

More precisely, we may define it as the projective scheme associated to the ring of invariant polynomials $\mathbf{Q}[X_1, Y_1, Z_1, X_2, Y_2, Z_2]^{\mathfrak{S}_2}$. Let us denote by Δ_Y the image of the diagonal Δ in Y. The scheme Y is singular along this diagonal and V may be seen as the blowing up of Y along the diagonal Δ_Y. From this point of view, the variety V is a desingularization of Y. Let us define P as the blowing up of $(\mathbf{P}_\mathbf{Q}^2)^2$ along the diagonal. We get a cartesian square

$$
\begin{array}{ccc}
P & \longrightarrow & (\mathbf{P}_\mathbf{Q}^2)^2 \\
\tilde{\pi} \downarrow & \square & \downarrow \pi \\
V & \xrightarrow{\ b\ } & Y
\end{array}
$$

We put $\Delta_V = b^{-1}(\Delta_Y)$ and $U_0 = V \backslash \Delta_V$. Then the set

$$
T = \tilde{\pi}(P(\mathbf{Q})) \cap U_0(\mathbf{Q})
$$

is a Zariski dense thin accumulating subset. More precisely, C. Le Rudulier proves the following theorem:

Theorem 3.36 (C. Le Rudulier)

(a) *Asymptotically the points of T give a positive contribution to the total number of points:*

$$
\frac{\sharp T_{H \leqslant B}}{\sharp U_0(\mathbf{Q})_{H \leqslant B}} \xrightarrow[B \to +\infty]{} c
$$

for a real number $c > 0$. But for any strictly closed subset $F \subset V$, one has

$$
\sharp(F(\mathbf{Q}) \cap T)_{H \leqslant B} = o(U_0(\mathbf{Q})_{H \leqslant B}).
$$

(b) *On the complement of T, one has*

$$
\sharp(U_0(\mathbf{Q}) \backslash T)_{H \leqslant B} \sim C(V) B \log(B)
$$

as $B \to +\infty$.

Remarks 3.37

(i) It follows from this theorem that the set T is a thin subset which is not the union of accumulating subvarieties but which gives a positive contribution to the total number of points of bounded height on the variety. In the adelic space the closure of the points of T are contained in a closed subset F with a volume $\mu_V(F)$ equal to 0. Therefore this thin accumulating subset is an obstruction to the equidistribution of the points on V.

(ii) Hopefully, in general, if ω_V^{-1} is "big enough", there should be a natural "small" subset T such that the points of bounded height on $W = V(\mathbf{K}) \backslash T$ should behave as expected. The problem is to describe this subset T.

(iii) W. Sawin recently proved that, in the example described in Theorem 3.36, the empiric formula [48, formule empirique 6.13] is *false* [51]. However the approach described below involving several heights might still be correct.

(iv) In these notes, so far, we did not go into the distribution of the rational points of bounded height for a height associated to an ample line with a class which is not a multiple of ω_V^{-1}. The description in that case requires to introduce more complicated measures and we refer the interested reader to the work of V. V. Batyrev and Y. Tschinkel (see [5]).

4 All the Heights

4.1 Heights Systems

A natural approach to select the points we wish to keep is to introduce more invariants. The rest of this chapter is devoted to such invariants. Let us start by considering other heights. Traditionally, most authors in arithmetic geometry consider only one height given by a given ample line bundle. However there are no reason to do so, and we may consider the whole information given by heights. In order to do this, let us introduce the notion of family of heights.

Definition 4.1 Let L and L' be adelically normed line bundles on a nice variety V. Let $(\| \cdot \|_w)_{w \in \mathrm{Val}(\mathbf{K})}$ be the adelic norm on L. We say that L and L' are *equivalent* if there is an integer $M > 0$, a family $(\lambda_w)_{w \in \mathrm{Val}(\mathbf{K})}$ in $\mathbf{R}_{>0}^{(\mathrm{Val}(\mathbf{K}))}$, such that its support $\{ w \in \mathrm{Val}(\mathbf{K}) \mid \lambda_w \neq 1 \}$ is finite and $\prod_{w \in \mathrm{Val}(\mathbf{K})} \lambda_w = 1$, and an isomorphism of adelically normed line bundles from the line bundle $L^{\otimes M}$ equipped with the adelic norm $(\lambda_w \| \cdot \|_w^{\otimes M})_{w \in \mathrm{Val}(\mathbf{K})}$ to the adelically normed line bundle $L'^{\otimes M}$. We denote by $\mathscr{H}(V)$ the set of equivalence classes of adelically normed line bundles. It has a structure of group induced by the tensor product of line bundles, we call this group the group of *Arakelov heights on V*.

Remark 4.2 The height introduced in Definition 2.7 depends only on the equivalence class of the adelically normed line bundle $\det(E)$. From that point of view, the group $\mathscr{H}(V)$ does parametrize the heights on V. If V satisfies weak approximation and has an adelic point, then two distinct elements of $\mathscr{H}(V)$ define heights which differ at least at one rational point.

Example 4.3 If V is a point, that is the spectrum of a field, then the height defines an isomorphism from $\mathscr{H}(V)$ to $\mathbf{R}_{>0}$. Indeed, it is surjective and if we take a representative L of an element of $\mathscr{H}(\mathrm{Spec}(\mathbf{K}))$ of height 1, then let y be an nonzero element of L. The unique morphism of vector spaces from $\mathbf{K}$ to L which maps 1 to y then induces an isomorphism from $\mathbf{K}$ equipped with the adelic norm $(\|y\|_w \mid \cdot \mid_w)_{w \in \mathrm{Val}(\mathbf{K})}$ to L.

Definition 4.4 A *system of Arakelov heights* on our nice variety V is a section s of the forgetful morphism of groups

$$o : \mathcal{H}(V) \longrightarrow \text{Pic}(V).$$

Such a system defines a map

$$h : V(\mathbf{K}) \to \text{Pic}(V)_{\mathbf{R}}^{\vee}$$

constructed as follows: for any $P \in V(\mathbf{K})$ and any $L \in \text{Pic}(V)$, the real number $h(P)(L)$ is the logarithmic height of the point P relative to the Arakelov height $s(L)$ (see Definition 2.7). We shall call $h(P)$ the *multiheight* of the point P. By abuse of language, a function of the form $P \mapsto \exp(\langle u, h(P)\rangle)$ for some $u \in \text{Pic}(V)_{\mathbf{R}}$ will also be called an exponential height on V.

Since $\text{Pic}(V)$ is finitely generated, we may fix a system of Arakelov heights on our nice variety V. We still assume that V satisfies the Hypotheses 3.27. Then one can study the multiheights of rational points.

Lemma 4.5 *Under the Hypotheses 3.27, there is a dense open subset U of V and an element $c \in \text{Pic}(V)_{\mathbf{R}}^{\vee}$ such that*

$$\forall P \in U(\mathbf{K}), \quad h(P) \in c + C_{\text{eff}}(V)^{\vee}.$$

Proof Let $L_1, \ldots, L_m$ be line bundles the classes of which generate the effective cone in $\text{Pic}(V)_{\mathbf{R}}$. We may assume that they have nonzero sections. Let U be the complement of the base loci of these line bundles. Let $i \subset \{1, \ldots, m\}$. Then choosing a basis $(s_0, \ldots, s_{N_i})$ of the space of sections of the line bundle L_i, we get a morphism from U to a projective space $\mathbf{P}_{\mathbf{K}}^{N_i}$. For any place w, there exist a constant c_w such that $\|s_j(x)\|_w \leqslant c_w$ for any $x \in V(\mathbf{K}_w)$ and any $j \in \{0, \ldots, N_i\}$. Moreover we may take $c_w = 1$ outside a finite set of places. Therefore there exists a constant C such that for any $x \in U(\mathbf{K})$ there is an $j \in \{0, \ldots, N_i\}$ with

$$0 < \prod_{w \in \text{Val}(\mathbf{K})} \|s_j(x)\| \leqslant C.$$

It follows that there exists a constant $c_i \in \mathbf{R}$ such that $h_i(P) \geqslant c_i$ for any $P \in U(\mathbf{K})$. The statement of the lemma follows. $\square$

Remark 4.6 Let $C_{\text{eff}}^{\circ}(V)^{\vee}$ be the interior of the dual cone $C_{\text{eff}}(V)^{\vee}$. This lemma shows that it is quite natural to count the number of rational points in $V(K)$ such that $h(P) \in \mathscr{D}_B$ for some compact domain $\mathscr{D}_B \subset C_{\text{eff}}^{\circ}(V)^{\vee}$ depending on a parameter $B \in \mathbf{R}_{>0}$. In the following, we shall consider domains of the form

$$\mathscr{D}_B = \mathscr{D}_1 + \log(B)u$$

where $u \in C^{\circ}_{\mathrm{eff}}(V)^{\vee}$ and $\mathscr{D}_1$ is a compact polyhedron in $\mathrm{Pic}(V)^{\vee}_{\mathbf{R}}$. In other words, we get a finite number of conditions of the form

$$aB \leqslant H(P) \leqslant bB$$

where H is an exponential height on V, in the sense of Definition 4.4, and $a, b \in \mathbf{R}_{>0}$.

Notation 4.7 We define the measure ν on $\mathrm{Pic}(V)^{\vee}_{\mathbf{R}}$ as follows: for a compact subset $\mathscr{D}$ of $\mathrm{Pic}(V)^{\vee}_{\mathbf{R}}$,

$$\nu(\mathscr{D}) = \int_{\mathscr{D}} e^{\langle \omega_V^{-1}, y \rangle} \mathrm{d}y \,,$$

where the Haar measure $\mathrm{d}y$ on $\mathrm{Pic}(V)^{\vee}_{\mathbf{R}}$ is normalised so that the covolume of the dual of the Picard group is one.

For any domain $\mathscr{D} \subset \mathrm{Pic}(V)^{\vee}_{\mathbf{R}}$, we define

$$V(\mathbf{K})_{h \in \mathscr{D}} = \{ P \in V(\mathbf{K}) \mid h(P) \in \mathscr{D} \}$$

With these notations, we may ask the following question:

Question 4.8 We assume that our nice variety V satisfies the conditions of the Hypothesis 3.27. Let $\mathscr{D}_1$ be a compact polyhedron of $\mathrm{Pic}(V)^{\vee}_{\mathbf{R}}$ and u be an element of the open cone $C^{\circ}_{\mathrm{eff}}(V)^{\vee}$. For a real number $B > 1$, let $\mathscr{D}_B = \mathscr{D}_1 + \log(B)u$. Can we find a "small" subset T so that we have an equivalence of the form

$$\sharp(V(\mathbf{K}) - T)_{h \in \mathscr{D}_B} \sim \beta(V)\nu(\mathscr{D}_1)\omega_V(V(A_{\mathbf{K}})^{\mathrm{Br}})B^{\langle \omega_V^{-1}, u \rangle} \tag{5}$$

as B goes to infinity?

Remarks 4.9

(i) One may note that in the right hand side of (5), one may use $\nu(\mathscr{D}_B) = \nu(\mathscr{D}_1)B^{\langle \omega_V^{-1}, u \rangle}$.

(ii) One can easily imagine variants of this question. For example, some methods from analytic number theory give much better error terms if ones use smooth functions instead of characteristic functions of sets. So it would be natural to consider a smooth function $\varphi : \mathrm{Pic}(V)^{\vee}_{\mathbf{R}} \to \mathbf{R}$ with compact support and ask whether we have

$$\sum_{P \in V(\mathbf{K})} \varphi(h(P) - Bu) \sim \beta(V) \int_{\mathrm{Pic}(V)^{\vee}_{\mathbf{R}}} \varphi \mathrm{d}\nu \, \omega_V(V(A_{\mathbf{K}})^{\mathrm{Br}})B^{\langle \omega_V^{-1}, u \rangle}$$

as B goes to infinity.

(iii) Let us compare formula (5) with formula (4). First we may note that

$$\nu(\{y \in C_{\mathrm{eff}}(V)^\vee \mid \langle y, \omega_V^{-1} \rangle \leqslant \log(B)\}) \sim \alpha(V) B \log(B)^{t-1}$$

Thus using Remark 4.9, formula (4) may be seen as integrating formula (5) over

$$\mathscr{D}_B = \{y \in C_{\mathrm{eff}}(V)^\vee \mid \langle y, \omega_V^{-1} \rangle \leqslant \log(B)\}.$$

In this context in which we consider all the possible heights, we may consider again the question of the global equidistribution.

Global Equidistribution 4.10 *We shall say that the global equidistribution holds for **h** if, for any compact polyhedron $\mathscr{D}_1$ in $\mathrm{Pic}(V)_{\mathbf{R}}^\vee$ and any u in the open cone $C_{\mathrm{eff}}^\circ(V)^\vee$, the measure $\delta_{V(K)_{h \in \mathscr{D}_B}}$ converges weakly to μ_V^{Br} as B goes to infinity.*

Note that the expected limit probability measure is the same as before and does not depend on u.

4.2 Compatibility with the Product

A positive answer to Question 4.8 is compatible with the product of varieties in the following sense:

Proposition 4.11 *Let V_1 and V_2 be nice varieties equipped with system of heights which satisfy the conditions 3.27. If the sets $V_1(\mathbf{K}) - T_1$ and $V_2(\mathbf{K}) - T_2$ satisfy the equivalences (5) for any compact polyhedra, then this is also true for the product*

$$(V_1(\mathbf{K}) - T_1) \times (V_2(\mathbf{K}) - T_2),$$

equipped with the induced system of heights.

If these varieties satisfy the global equidistribution 4.10, then so does their product.

Proof We put $W_i = V_i(\mathbf{K}) - T_i$ for $i \in \{1, 2\}$. Let W be the product $W_1 \times W_2$. For $i \in \{1, 2\}$, we denote by h_i the multiheight on V_i, and fix a compact polyhedron $\mathscr{D}_{i,1}$ in $\mathrm{Pic}(V_i)_{\mathbf{R}}^\vee$, as well as an element $u_i \in C_{\mathrm{eff}}^\circ(V_i)^\vee$. Let us first note that by Hartshorne [28, exercise III.12.6], the natural morphism induced by pull-backs $\mathrm{Pic}(V_1) \times \mathrm{Pic}(V_2) \to \mathrm{Pic}(V)$ is an isomorphism which maps the product $C_{\mathrm{eff}}(V_1) \times C_{\mathrm{eff}}(V_2)$ onto $C_{\mathrm{eff}}(V)$ and $(\omega_{V_1}^{-1}, \omega_{V_2}^{-1})$ on ω_V^{-1} (see [28, exercise II.8.3]). Therefore we identify these groups and consider $\mathscr{D}_1 = \mathscr{D}_{1,1} \times \mathscr{D}_{2,1}$ as a subset of

$\mathrm{Pic}(V)^{\vee}_{\mathbf{R}}$ and $u = (u_1, u_2)$ as an element of $C^{\circ}_{\mathrm{eff}}(V)^{\vee}$. If we put $\mathscr{D}_B = \log(B)u + \mathscr{D}_1$, we have

$$\sharp W_{h \in \mathscr{D}_B} = \sharp (W_1)_{h_1 \in \mathscr{D}_{1,B}} \times \sharp (W_2)_{h_2 \in \mathscr{D}_{2,B}}$$

and the result follows from the compatibility of equivalence with products. To extend the result to an arbitrary polyhedra $\mathscr{D}$, we find domains $\mathscr{D}'$ and $\mathscr{D}''$ which are finite unions of products of polyhedra with disjoint interiors such that $\mathscr{D}' \subset \mathscr{D} \subset \mathscr{D}''$ and use the fact that the equivalence is valid for such a finite union.

Similarly for the equidistribution, it is enough to count the points in open subsets U of $V(A_{\mathbf{K}})^{\mathrm{Br}}$ which are of the form $U = U_1 \times U_2$ for open subsets U_1 and U_2 such that $\omega_{V_1}(\partial U_1) = 0$ and $\omega_{V_2}(\partial U_2) = 0$. But in that case,

$$\sharp (W \cap U)_{h \in \mathscr{D}_B} = \sharp (W_1 \cap U_1)_{h_1 \in \mathscr{D}_{1,B}} \times \sharp (W_2 \cap U_2)_{h_2 \in \mathscr{D}_{2,B}}$$

and we may conclude in the same way. $\square$

It is worthwhile to note that this proof is much simpler than the proof of the compatibility of the principle of Batyrev and Manin for products (see [24, §1.1]). It illustrates the fact that in Question 4.8 we cut out the "spikes" where the heights of the components of the points are very different.

4.3   Lifting to Versal Torsors

Following Salberger [50], we shall now explain how the question lifts naturally to versal torsors (see also [46]). Let us start by a quick reminder on versal torsors. In our setting, the geometric Picard is supposed to be without torsion, thus we shall restrict ourselves to torsors under algebraic tori.

Definition 4.12 Let $\mathbf{L}$ be a field and $\mathbf{L}^s$ be a separable closure of $\mathbf{L}$. For any scheme X over $\mathbf{L}$, we write X^s for the product $X \times_{\mathrm{Spec}(\mathbf{L})} \mathrm{Spec}(\mathbf{L}^s)$.

An algebraic group G over a field $\mathbf{L}$ is said to be *of multiplicative type* if there exists an integer n such that G^s is isomorphic to a closed subgroup of $\mathbf{G}^n_{m,\mathbf{L}^s}$. A *torus* $\mathbf{T}$ over $\mathbf{L}$ is an algebraic group $\mathbf{T}$ over $\mathbf{L}$ such that $\mathbf{T}^s$ is isomorphic to a power $\mathbf{G}^n_{m,\mathbf{L}^s}$ of the multiplicative group.

The *group of characters* of an algebraic group G, denoted by $X^*(G^s)$ is the group of group homomorphisms from G^s to $\mathbf{G}_{m,\mathbf{L}^s}$. If G is of multiplicative type, it is a finitely generated $\mathbf{Z}$-module. If G is a torus, it is a free $\mathbf{Z}$-module of rank n. In both cases, it is equipped with an action of the absolute Galois group of $\mathbf{L}$, that is $\mathscr{G}_{\mathbf{L}} = \mathrm{Gal}(\mathbf{L}^s/\mathbf{L})$, which splits over a finite separable extension of $\mathbf{L}$.

Conversely, let us define a *Galois module L over* $\mathbf{L}$ (resp. a *Galois lattice L over* $\mathbf{L}$) as a finitely generated $\mathbf{Z}$-module (resp. a free $\mathbf{Z}$-module of finite rank) equipped with an action of the Galois group $\mathscr{G}_{\mathbf{L}}$ which splits over a finite extension. To a Galois module L, we may associate the monoid algebra $\mathbf{L}^s[L]$ and thus the algebraic variety

$$\mathbf{T} = \mathrm{Spec}(\mathbf{L}^s[L]^{\mathscr{G}_\mathbf{L}})$$

equipped with the algebraic group structure induced by the coproduct ∇ on $\mathbf{L}^s[L]$ defined by $\nabla(\lambda) = \lambda \otimes \lambda$ for any $\lambda \in L$. This algebraic group is an algebraic group of multiplicative type, which we shall say to be *associated to L*.

Example 4.13 As a basic example, the group of characters of $\mathbf{G}_{m,\mathbf{L}}^n$ is $\mathbf{Z}^n$ with a trivial action of the Galois group and the torus associated with $\mathbf{Z}^n$ is isomorphic to $\mathbf{G}_{m,\mathbf{L}}^n$.

Remark 4.14 These constructions are functorial and we get a contravariant equivalence of categories between the category of tori (resp. groups of multiplicative type) over $\mathbf{L}$ and the category of Galois lattices (resp. Galois modules) over $\mathbf{L}$.

Notation 4.15 We shall denote by $\mathbf{T}_{\mathrm{NS}}$ the torus associated to the Galois lattice $\mathrm{Pic}(\overline{V})$.

We are going to use pointed torsors, that is torsors in the category of pointed schemes.

Definition 4.16 Let G be an algebraic group over a field $\mathbf{L}$ and let X be an algebraic variety over $\mathbf{L}$. A *G-torsor* T over X is an algebraic variety T over $\mathbf{L}$ equipped with a faithfully flat morphism $\pi : T \to X$ and an action $\mu : G \times T \to T$ of G such that $\pi \circ \mu = \pi \circ \mathrm{pr}_2$ and the morphism given by $(g, y) \mapsto (gy, y)$ is an isomorphism from $G \times T$ to $T \times_X T$.

A *pointed variety* over $\mathbf{L}$ is a variety X over $\mathbf{L}$ equipped with a chosen rational point $x \in X(\mathbf{L})$. A *pointed torsor* over the pointed variety X is a torsor T over X equipped with a rational point $t \in T(\mathbf{L})$ such that $\pi(t) = x$.

Example 4.17 For any line bundle L over X, we can define a $\mathbf{G}_{m,\mathbf{L}}$ torsor by considering $L^\times$ which is the complement of the zero section in L. Conversely for a nice variety X, given a $\mathbf{G}_m$ torsor T, we get a line bundle by considering the contracted product $T \times^{\mathbf{G}_{m,\mathbf{L}}} \mathbf{\Lambda}_\mathbf{L}^1$ which is the quotient $(T \times \mathbf{\Lambda}_\mathbf{L}^1)/\mathbf{G}_{m,\mathbf{L}}$ where $\mathbf{G}_{m,\mathbf{L}}$ acts by $t.(y, a) = (t.y, t^{-1}.a)$. We get in that way the equivalence of category between the line bundles and the $\mathbf{G}_{m,\mathbf{L}}$-torsors over X.

Versal and Universal Torsors

The versal torsors were introduced by J.-L. Colliot-Thélène and J.-J. Sansuc in the study of the Brauer-Manin obstruction for Hasse principle and weak approximation (see [14, 15, 16]). For a survey on versal torsors, the reader may also look at [47].

In topology, universal coverings for an unlaceable pointed space X answers a universal problem for coverings: it is a pointed covering E over X such that for any pointed covering $C \to X$ there exists a unique morphism $E \to C$ of pointed spaces over X (see [9, TA IV, §1, n°3]). We could in fact restrict ourselves to

Galois coverings, that is connected coverings with an automorphism group which acts transitively on the fibre over the marked point of X. Fixing a point in the space X is necessary to guarantee the unicity, up to a unique isomorphism, of the universal covering. The universal torsor is the answer to a similar problem for torsor under groups of multiplicative type.

Definition 4.18 Let $\mathbf{L}$ be a field and $\overline{\mathbf{L}}$ be an algebraic closure of $\mathbf{L}$. Let X be a smooth and geometrically integral variety over $\mathbf{L}$ with a rational point such that all invertible functions on $\overline{X}$ are constant: $\Gamma(\overline{X}, \mathbf{G}_m) = \overline{\mathbf{L}}^*$. We see X as a pointed space by fixing a rational point $x \in X(\mathbf{L})$. Then a *universal torsor* is a pointed torsor T_u over the pointed space X under a group of multiplicative type $\mathbf{T}_u$ such that for any pointed torsor T over X under a group of multiplicative type $\mathbf{T}$, there is a unique morphism of group $\varphi : \mathbf{T}_u \to \mathbf{T}$ and a unique morphism $\psi : T_u \to T$ over X, compatible with the actions of $\mathbf{T}_u$ and $\mathbf{T}$ and the marked points.

Remarks 4.19

(i) If such a torsor exists it is by definition unique up to a unique isomorphism.
(ii) Using the cohomological characterisation of universal torsors [16, §2], one may show that the extension of scalars of a universal torsor is also a universal torsor.
(iii) Let us assume that there exists a universal torsor T_u. Let x be the chosen point of X. For any line bundle L over X, we can consider the $\mathbf{G}_m$-torsor $L^\times$ and fix a point in its fibre over x. Thus there exists a unique morphism of pointed torsors from T_u to $L^\times$ compatible with a morphism $\mathbf{T}_u \to \mathbf{G}_m$. By duality, it corresponds to a homomorphism of groups from $\mathbf{Z}$ to the group of characters of $\mathbf{T}_u$. Moreover if $L^{\otimes n}$ is isomorphic to the trivial line bundle, the image of $n \in \mathbf{Z}$ in $X^*(\mathbf{T}_u)$ is trivial. Therefore, over $\mathbf{L}^s$, we get a homomorphism of groups from $\mathrm{Pic}(X^s)$ to $X^*(\mathbf{T}_u^s)$, which is compatible with the Galois actions.

Conversely, for any torsor T under a multiplicative group $\mathbf{T}$ and any group character $\chi : \mathbf{T} \to \mathbf{G}_m$, the contracted product $T \times^{\mathbf{T}} \mathbf{G}_{m,\mathbf{L}}$ is a $\mathbf{G}_m$ torsor over $\mathbf{L}$. We get a homomorphism of groups from $X^*(\mathbf{T}^s)$ to $\mathrm{Pic}(X^s)$. It is possible to deduce from such arguments that the character group of $\mathbf{T}_u$ over $\mathbf{L}^s$ has to be isomorphic to $\mathrm{Pic}(X^s)$.

Construction 4.20 Let us now explain how it is possible to construct such universal torsors. We shall assume again Hypothesis 3.27, and fix a rational point $x \in V(\mathbf{K})$. In that case the group $\mathbf{T}_u$ is canonically isomorphic to the Néron-Severi torus $\mathbf{T}_{\mathrm{NS}}$. Over $\overline{\mathbf{K}}$, the construction of Remark 4.19 gives an isomorphism of $\mathbf{T}_{\mathrm{NS}}$-torsor from a universal torsor $\overline{T}_u$ to the product $L_1^\times \times_V \cdots \times_V L_t^\times$ where $([L_1], \ldots, [L_t])$ is a basis of $\mathrm{Pic}(\overline{V})$. But the unicity of the universal torsor shows that, by marking $\overline{T}_u$ with a point in the fibre of x, there exists no non-trivial automorphism of $\overline{T}_u$ as a pointed torsor over X. By descent theory, $\overline{T}_u$ comes from a unique pointed $\mathbf{T}_{\mathrm{NS}}$-torsor T_u over X.

Remark 4.21 In particular, as a non-pointed $\overline{\mathbf{T}}_{NS}$-torsor over $\overline{V}$, the torsor $\overline{T}_u$ does not depend on the choice of the point x in $V(\mathbf{K})$. This is not true over $\mathbf{K}$.

Definition 4.22 A *versal torsor* over V is a $\mathbf{K}$-form of the $\overline{\mathbf{T}}_{NS}$-torsor $\overline{T}_u$.

Remark 4.23 The automorphisms of $\overline{T}_u$ as a $\overline{\mathbf{T}}_{NS}$-torsor over $\overline{V}$ are given by the action of $\mathbf{T}_{NS}(\overline{\mathbf{K}})$. It follows that if we fix a rational point, and therefore a universal torsor T_u, the versal torsors are classified by the group of Galois cohomology $H^1(\mathbf{K}, \mathbf{T}_{NS})$ and we get a map from $V(\mathbf{K})$ to $H^1(\mathbf{K}, \mathbf{T}_{NS})$ which maps a point to the class of the corresponding universal torsor. In general this cohomology group is infinite. But Colliot-Thélène and Sansuc proved in [15, proposition 2] that the image of the map is finite. In other words, there exists a finite family $(T_i)_{i \in I}$ of non-isomorphic versal torsors over V with a rational point such that

$$V(\mathbf{K}) = \coprod_{i \in I} \pi_i(T_i(\mathbf{K})),$$

where $\pi_i : T_i \to V$ is the structural morphism.

Structures on Versal Torsors

Let T_u be a universal torsor over V. By definition of the torsors, there is a natural isomorphism

$$T_u \times_V T_u \xrightarrow{\sim} \mathbf{T}_{NS} \times T_u$$

which shows that the pull-back of T_u to T_u is trivial. But from the universality of T_u it is possible to show that the pull-back of any pointed torsor under a group of multiplicative type is trivial [16, proposition 2.1.1]. By this proposition, we also have that invertible functions on T_u are constant: $\Gamma(T_u, \mathbf{G}_m) = \mathbf{K}^*$. Moreover, by Peyre [46, lemme 2.1.10], ω_{T_u} is isomorphic to the pull-back of ω_V. We get the following assertion concerning *volume forms*, that is non-vanishing sections of ω_{T_u}.

Proposition 4.24 *Let T be a versal torsor over V. Then up to multiplication by a constant there exists a unique volume form on T.*

Construction 4.25 Let T be a versal torsor on V with a rational point. By the proposition, we may take a non-vanishing section ω of ω_T. For any place w of $\mathbf{K}$, the expression

$$\left| \left\langle \omega, \frac{\partial}{\partial x_1} \wedge \frac{\partial}{\partial x_2} \wedge \ldots \wedge \frac{\partial}{\partial x_n} \right\rangle \right|_w \mathrm{d}x_{1,w}\, \mathrm{d}x_{2,w} \ldots \mathrm{d}x_{n,w},$$

defines a local measure, which, like in construction 3.6, we may patch together to get a measure $\omega_{T_u,w}$ on $T_u(\mathbf{K}_w)$.

We then choose a finite set S of places containing all the places of bad reduction for V, the archimedean places, as well as the ramified places in a extension splitting the action of the Galois group on the Picard group of V. Moreover, we may assume that any isomorphism class of versal torsors with a rational point has a model over the ring of S-integers $\mathcal{O}_S$ and that the projection maps $T(\mathbf{K}_w) \to V(\mathbf{K}_w)$ are surjective for $w \notin S$ [16, lemme 3.2.3]. Let us fix such a model $\mathcal{T}$ of our versal torsor T. Then for any place w outside a finite set of places, one can prove (see the proof of Theorem 4.33 below) that

$$\omega_{T,w}(\mathcal{T}(\mathcal{O}_w)) = L_w(1, \mathrm{Pic}(\overline{V}))^{-1}\omega_{V,w}(V(\mathbf{K}_w)).$$

Using the arguments of Construction 3.29, it follows that we can define the product of the measures

$$\omega_T = \frac{1}{\sqrt{d_{\mathbf{K}}}^{\dim T}} \prod_{w \in \mathrm{Val}(\mathbf{K})} \omega_{T,w}.$$

on the adelic space $T(A_{\mathbf{K}})$. By the product formula, this measure does not change if we multiply ω by a nonzero constant. Thus we may call ω_T the *canonical measure* on the adelic space of the versal torsor T.

Example 4.26 For a smooth hypersurface V of degree d in $\mathbf{P}_{\mathbf{K}}^N$, with $N \geqslant 4$, any versal torsor is isomorphic to the cone over the hypersurface in $\mathbf{A}_{\mathbf{K}}^{N+1}\backslash\{0\}$, and the canonical measure is given by the Leray form [35, chapter IV, §1]. If F is a homogeneous equation for V, then locally the measure may be defined as

$$\omega_{T,w} = \frac{1}{\left|\frac{\partial F}{\partial x_0}(1, x_1, \ldots, x_N)\right|_w}\mathrm{d}x_{1,w} \ldots \mathrm{d}x_{N,w}.$$

Let us now turn to the lifting of heights to versal torsors. We have to take into account that the rank of the Picard group at a place w depends on w.

Construction 4.27 We choose a system of representatives $(T_i)_{i \in I}$ of the isomorphism classes of versal torsors over V which have a rational points over $\mathbf{K}$. For each $i \in I$, we also fix a point $y_i \in T_i(\mathbf{K})$. Let $\mathbf{L}$ be a Galois extension of $\mathbf{K}$ which splits the Picard group of V. Let $s_{\mathbf{L}} : \mathrm{Pic}(V_{\mathbf{L}}) \to \mathcal{H}(V_{\mathbf{L}})$ be a system of heights over $\mathbf{L}$. We also fix a place w_0 of $\mathbf{K}$. Let $i \in I$. For any line bundle L over $\mathrm{Pic}(V_{\mathbf{L}})$ there exists a morphism $\phi_L : T_i \to L^\times$ over V, which is compatible with the character $\chi_L : (\mathbf{T}_{\mathrm{NS}})_{\mathbf{L}} \to \mathbf{G}_{m,\mathbf{L}}$ defined by L. This morphism is unique up to multiplication by a constant. Let us choose a representant $(\|\cdot\|_v)_{v \in \mathrm{Val}(\mathbf{L})}$ of $s_{\mathbf{L}}([L])$ defining the exponential height H_L on $V(\mathbf{L})$. For any $v \in \mathrm{Val}(\mathbf{L})$, we may then consider the map from $T_i(\mathbf{L}_v)$ to $\mathbf{R}$ given by

$$y \longmapsto \|y\|_v^L = \begin{cases} \frac{\|\phi_L(y)\|_v}{\|\phi_L(y_i)\|_v} & \text{if } v \nmid w_0 \\ \frac{\|\phi_L(y)\|_v}{\|\phi_L(y_i)\|_v} H_L(\pi_i(y_i))^{-\frac{[\mathbf{L}_v:\mathbf{K}_{w_0}]}{[\mathbf{L}:\mathbf{K}]}} & \text{otherwise.} \end{cases}$$

This map does not depend on the choice of ϕ_L nor on the choice of the representant of $s_L([L])$ and satisfies

$$\forall y \in T_i(\mathbf{L}), \quad H_L(\pi_i(y)) = \prod_{v \in \mathrm{Val}(\mathbf{L})} (\|y\|_v^L)^{-1}.$$

Moreover it satisfies the formula $\|t.y\|_v^L = |\chi_L(t)|_v \|y\|_v^L$, for $t \in \mathbf{T}_{\mathrm{NS}}(\mathbf{L}_v)$ and y in $T_i(\mathbf{L}_v)$. We get a map

$$\tilde{h}_v : T_i(\mathbf{L}_v) \longrightarrow (\mathrm{Pic}(V_{\mathbf{L}_v}))_{\mathbf{R}}^{\vee}$$

defined by the relations

$$\|y\|_v^L = q_v^{-\langle \tilde{h}_v(y), [L] \rangle}$$

for $y \in T_i(\mathbf{L}_v)$ and $[L] \in \mathrm{Pic}(V_{\mathbf{L}_v})$, with q_v the cardinal of the residue field $\mathbf{F}_v$ if v is ultrametric, $q_v = e$ for a real place and $q_v = e^2$ for a complex one. Let us now write V_w for $V_{\mathbf{K}_w}$. Using the inclusion $T_i(\mathbf{K}_w) \to \prod_{v|w} T_i(\mathbf{L}_v)$ and the projection $\mathrm{pr} : \prod_{v|w} \mathrm{Pic}(V_{L_v})_{\mathbf{R}}^{\vee} \to \mathrm{Pic}(V_w)_{\mathbf{R}}^{\vee}$, we define a map

$$\tilde{h}_w : T_i(\mathbf{K}_w) \longrightarrow \mathrm{Pic}(V_w)_{\mathbf{R}}^{\vee}.$$

so that the diagram

$$
\begin{array}{ccc}
\prod_{v|w} T_i(\mathbf{L}_v) & \longrightarrow & \prod_{v|w} \mathrm{Pic}(V_{L_v})_{\mathbf{R}}^{\vee} \\
\uparrow & & \downarrow {\scriptstyle \frac{1}{[\mathbf{L}:\mathbf{K}]} \mathrm{pr}} \\
T_i(\mathbf{K}_w) & \xrightarrow{\tilde{h}_w} & \mathrm{Pic}(V_{\mathbf{K}_w})_{\mathbf{R}}^{\vee}
\end{array}
\tag{6}
$$

commutes.

If L is line bundle over X and if $(\|\cdot\|_v)_{v \in \mathrm{Val}(\mathbf{L})}$ is an adelic norm for the extension of scalars $L_{\mathbf{L}}$, then it induces an adelic norm on L defined by

$$\forall w \in \mathrm{Val}(\mathbf{K}), \forall y \in L(\mathbf{K}_w), \|y\|_w = \left(\prod_{v|w} \|y\|_v \right)^{\frac{1}{[\mathbf{L}:\mathbf{K}]}}.$$

Therefore the system of heights $s_{\mathbf{L}}$ induces a system of heights $s : \mathrm{Pic}(V) \to \mathcal{H}(V)$. For any point $y \in T_i(\mathbf{K})$ we have the formula

$$h(\pi_i(y)) = \sum_{w \in \mathrm{Val}(\mathbf{K})} \log(q_w) \tilde{h}_w(y).$$

These construction enables us to lift a system of heights to versal torsors with a rational point.

Lifting of the Asymptotic Formula

We now wish to express the asymptotic formula (5) at the torsor level. The fibre of the projection map $\pi_i : T_i(\mathbf{K}) \to V(\mathbf{K})$ is either empty or a principal homogeneous space under $\mathbf{T}_{NS}(\mathbf{K})$. Therefore we now need to use the description of the rational points of the torus $\mathbf{T}_{NS}$, as described in the work of Ono [42, 43].

Definition 4.28 Let $\mathbf{T}$ be an algebraic torus over $\mathbf{K}$. We denote by $W(\mathbf{T})$ the torsion subgroup of $\mathbf{T}(\mathbf{K})$. By an abuse of notation, for any place w of $\mathbf{K}$, we denote by $\mathbf{T}(\mathscr{O}_w)$ the maximal compact subgroup of $\mathbf{T}(\mathbf{K}_w)$. Let us put $K_{\mathbf{T}} = \prod_{w \in \mathrm{Val}(\mathbf{K})} \mathbf{T}(\mathscr{O}_w)$ which is a compact subgroup of $\mathbf{T}(A_{\mathbf{K}})$. We also have that $W(\mathbf{T}) = K_{\mathbf{T}} \cap \mathbf{T}(\mathbf{K})$. For any place w, there is an injective morphism of groups

$$\log_w : \mathbf{T}(\mathbf{K}_w)/\mathbf{T}(\mathscr{O}_w) \longrightarrow X^*(\mathbf{T}_w)_{\mathbf{R}}^{\vee}$$

so that for any $t \in \mathbf{T}(\mathbf{K}_w)$ and any $\chi \in X^*(\mathbf{T}_w)$, we have $q_w^{\langle \log_w(t), \chi \rangle} = |\chi(t)|_w$. For almost all places w the image of $\log_w$ coincide with $X^*(\mathbf{T}_w)^{\vee}$. In fact, by Ono [42, theorem 4] and [43, §3] there exists a finite set of places S_T such that the induced map gives an exact sequence

$$1 \longrightarrow \mathbf{T}(\mathscr{O}_{S_T}) \longrightarrow \mathbf{T}(\mathbf{K}) \longrightarrow \bigoplus_{w \in \mathrm{Val}(\mathbf{K}) - S_T} X^*(\mathbf{T})_w^{\vee} \longrightarrow 0 \tag{7}$$

and there is an exact sequence

$$1 \longrightarrow W(\mathbf{T}) \longrightarrow \mathbf{T}(\mathscr{O}_{S_T}) \xrightarrow{\log_{S_T}} \bigoplus_{w \in S_T} X^*(\mathbf{T}_w)_{\mathbf{R}}^{\vee}, \tag{8}$$

where $\log_{S_T}$ is the map defined by taking $\log_w$ for $w \in S_T$. For any $w \in S_T$, the extension of scalars defines a linear map $\pi_w : X^*(T_w)_{\mathbf{R}}^{\vee} \to X^*(T)_{\mathbf{R}}^{\vee}$. We then consider the linear map $\pi = \sum_{w \in S_T} \log(q_w)\pi_w$:

$$\bigoplus_{w \in S_{\mathbf{T}}} X^*(\mathbf{T}_w)_{\mathbf{R}}^{\vee} \longrightarrow X^*(\mathbf{T})_{\mathbf{R}}^{\vee}.$$

By the product formula, the image of $\mathbf{T}(\mathscr{O}_{S_T})$ is contained in $\ker(\pi)$. The image $M = \pi(\mathbf{T}(\mathscr{O}_{S_T}))$ is a lattice in the $\mathbf{R}$-vector space $\ker(\pi)$. Let $(e_1, \ldots, e_m)$ be a basis for this lattice and let

$$\Delta = \left\{ \sum_{i=1}^m t_i e_i, \ (t_i)_{1 \leqslant i \leqslant m} \in [0, 1[^m \right\}.$$

By construction, Δ is a fundamental domain for the action of $\mathbf{T}(\mathscr{O}_{S_T})$ on $\ker(\pi)$.

Construction 4.29 By increasing the finite set of places S introduced in Construction 3.6, we assume that we may take $S_{T_{NS}} = S$ for the finite set of places considered in the last definition. In particular, we get that outside of S, the map

$$\mathbf{T}_{NS}(\mathbf{K}_w) \longrightarrow \mathrm{Pic}(V_w)^\vee$$

is surjective. For each of the chosen torsors we may also fix models $\mathscr{T}_i$ over $\mathscr{O}_S$. We may further assume that, for a family of line bundles which generates $\mathrm{Pic}(V_L)$ and is invariant under the action of the Galois group $\mathrm{Gal}(\mathbf{L}/\mathbf{K})$, the heights are given by models of the corresponding line bundles and that the maps ϕ_L from the chosen versal torsors to a line bundle L of the family are defined over $\mathscr{O}_S$. We may also assume that the adelic metrics outside S are compatible with the action of the Galois group. For any i in I, we define the set $\Delta(T_i)$ as

$$\{ y \in T_i(\mathbf{A_K}) \mid \mathrm{pr}((\widetilde{\pmb{h}}_w(y_w))_{w \in S}) \in \Delta \text{ and } \forall w \notin S, \ y_w \in \mathscr{T}_i(\mathscr{O}_w) \},$$

where pr is a linear projection on $\ker(\pi)$.

Lemma 4.30 *For any place $w \notin S$, the projection map $\mathscr{T}_i(\mathscr{O}_w) \to V(\mathbf{K}_w)$ is surjective and the map $\widetilde{\pmb{h}}_w$ is characterized by the following two conditions:*

(i) *We have the relation $\widetilde{\pmb{h}}_w(t.y) = -\log_w(t) + \widetilde{\pmb{h}}_w(y)$ for any $t \in \mathbf{T}_{NS}(\mathbf{K}_w)$ and any $y \in T_i(\mathbf{K}_w)$;*

(ii) *The integral points of $\mathscr{T}_i$ are given by*

$$\mathscr{T}_i(\mathscr{O}_w) = \{ y \in T_i(\mathbf{K}_w) \mid \widetilde{\pmb{h}}_w(y) = 0 \}.$$

Proof Relation (i) follows from the formula for $\|t.y\|_w^L$ and the description in (ii) from the fact that all maps are compatible with the models. By the choice of S, for any place $w \notin S$, the projection $\pi_i : T_i(\mathbf{K}_w) \to V(\mathbf{K}_w)$ is surjective. Moreover the functions $\| \cdot \|_w^L$ are compatible with the action of the Galois group. By the diagram (6), it follows that $\widetilde{\pmb{h}}_w(y)$ belongs to $\mathrm{Pic}(V_w)^\vee$. Since the map $\log_w$ is surjective, we may find in any fibre an element y such that $\widetilde{\pmb{h}}_w(y) = 0$. By (ii), this element is an integral point. Since the map $\pi_i : \mathscr{T}_i(\mathscr{O}_w) \to V(\mathbf{K}_w)$ is surjective, conditions (i) and (ii) characterize $\widetilde{\pmb{h}}_w$. $\qquad\square$

Theorem 4.31 *The set* $\Delta(T_i) \cap T_i(\mathbf{K})$ *is a fundamental domain for the action of* $\mathbf{T}_{NS}(\mathbf{K})$ *modulo* $W(\mathbf{T}_{NS})$. *In other words, it satisfies the following conditions:*

(i) *We have* $T_i(\mathbf{K}) = \cup_{t \in \mathbf{T}_{NS}(\mathbf{K})} t.\big(\Delta(T_i) \cap T_i(\mathbf{K})\big);$
(ii) *For any* $t \in \mathbf{T}_{NS}(\mathbf{K})$, *we have*

$$\big(\Delta(T_i) \cap T_i(\mathbf{K})\big) \cap t.\big(\Delta(T_i) \cap T_i(\mathbf{K})\big) \neq \emptyset$$

if and only if $t \in W(\mathbf{T}_{NS})$.
(iii) *For* $t \in W(\mathbf{T}_{NS})$, *we have*

$$t.\big(\Delta(T_i) \cap T_i(\mathbf{K})\big) = \Delta(T_i) \cap T_i(\mathbf{K}).$$

Proof Let $y \in T_i(\mathbf{K})$. By the lemma, for any $w \notin S$, $\tilde{h}_w(y) \in \mathrm{Pic}(V_w)^{\vee}$. Thus, using the exact sequence (7), we get an element $t \in \mathbf{T}_{NS}(\mathbf{K})$ such that $t.y \in \mathscr{T}_i(\mathscr{O}_w)$ for $w \notin S$. Using the exact sequence (8) and the definition of Δ, there is an element t' in $\mathbf{T}_{NS}(\mathscr{O}_S)$ such that $(t't).y \in \Delta(T_i)$. Assertions (ii) and (iii) follow from the definition of Δ.   □

Notation 4.32 For any $i \in I$, we define the map

$$\tilde{h} : \Delta(T_i) \longrightarrow \mathrm{Pic}(V)_{\mathbf{R}}^{\vee}$$

by the relation $\tilde{h}(y) = \pi\big((\tilde{h}_w(y_w))_{w \in S}\big)$.

Theorem 4.33 *We assume conditions 3.27. Let W be a borelian subset of $V(\mathbf{A_K})$. Let $\mathscr{D}$ be a borelian subset of* $\mathrm{Pic}(V)_{\mathbf{R}}^{\vee}$. *Then*

$$\beta(V)\nu(\mathscr{D})\boldsymbol{\omega}_V(W \cap V(\mathbf{A_K})^{\mathrm{Br}})$$

$$= \frac{1}{W(\mathbf{T}_{NS})} \sum_{i \in I} \boldsymbol{\omega}_{T_i}(\{\, y \in \Delta(T_i) \cap \pi_i^{-1}(W) \mid \tilde{h}(y) \in \mathscr{D}\}).$$

Proof This proof follows the ideas of Salberger [50] (see also [46, §3.5] for more details). If $(\xi_1, \ldots, \xi_r)$ is a basis of $X^*(\mathbf{T}_{NS}) = \mathrm{Pic}(V_{\mathbf{L}})$, then $\bigwedge_{i=1}^{r} \xi_i^{-1} \mathrm{d}\xi_i$ is a section of $\omega_{\mathbf{T}_{NS}}$, which, up to sign, does not depend on the choice of the basis. This defines a canonical Haar measure $\omega_{\mathbf{T}_{NS}, w}$ on $\mathbf{T}_{NS}(\mathbf{K}_w)$ for any place w of $\mathbf{K}$. Let $w \in \mathrm{Val}(\mathbf{K}) \backslash S$. Locally for w-adic topology, we may choose a section of $\pi_i : \mathscr{T}_i(\mathscr{O}_w) \to V(\mathbf{K}_w)$ and the measure $\omega_{T_i, w}$ on $\mathscr{T}_i(\mathbf{K}_w)$ is locally isomorphic to the measure

$$L_w(1, X^*(\mathbf{T}_{NS}))|\omega_V(t)|_w \omega_{\mathbf{T}_{NS}, w} \times \lambda_w \omega_{V, w}.$$

where ω_V is seen as a character of $\mathbf{T}_{NS}$. Let us also consider the groups $\mathbf{T}_{NS}(A_{\mathbf{K}})^1$, defined as

$$\left\{ (t_w)_{w \in \mathrm{Val}(\mathbf{K})} \in \mathbf{T}_{NS}(A_{\mathbf{K}}) \,\middle|\, \forall \xi \in X^*(\mathbf{T}_{NS}), \prod_{w \in \mathrm{Val}(\mathbf{K})} |\xi(t_w)|_w = 1 \right\},$$

and $\mathbf{T}_{NS}(\mathbf{K}_S)^1$, defined as

$$\left\{ (t_w)_{w \in S} \in \prod_{w \in S} \mathbf{T}_{NS}(\mathbf{K}_w) \,\middle|\, \forall \xi \in X^*(\mathbf{T}_{NS}), \prod_{w \in S} |\xi(t_w)|_w = 1 \right\}.$$

The lattice $X^*(\mathbf{T}_{NS})^\vee$ normalises the Haar measure on $X^*(\mathbf{T}_{NS})_{\mathbf{R}}^\vee$ and therefore on the quotient $\prod_{w \in S} \mathbf{T}_{NS}(\mathbf{K}_w)/\mathbf{T}_{NS}(\mathbf{K}_S)^1$. Using the measure $\prod_{w \in S} \omega_{\mathbf{T}_{NS},w}$ on the product, we get a normalised Haar measure ω_{T^1} on $\mathbf{T}_{NS}(\mathbf{K}_S)^1$. We consider the fibration

$$\widetilde{h} \times \pi_i : \prod_{w \in S} T_i(\mathbf{K}_w) \longrightarrow \mathrm{Pic}(V)_{\mathbf{R}}^\vee \times \prod_{w \in S} V(\mathbf{K}_w),$$

which, over its image, is a principal homogeneous space under $\mathbf{T}_{NS}(\mathbf{K}_S)^1$. By choosing a local adequate section of this fibration, we get that the measure $\prod_{w \in S} \omega_{T_i,w}$ on $\prod_{w \in S} T_i(\mathbf{K}_w)$ is the measure induced by the product measure $\nu \times \prod_{w \in S} \omega_{V,w}$ on the image and the measure ω_{T^1} on $\mathbf{T}_{NS}(\mathbf{K}_S)^1$. Taking the product over all places, and multiplying by the normalisation terms, we get that

$$\frac{1}{\sharp W(\mathbf{T}_{NS})} \omega_{T_i}(\{ y \in \Delta(T_i) \cap \pi_i^{-1}(W) \mid \widetilde{h}(y) \in \mathscr{D}\})$$

$$= \tau(\mathbf{T}_{NS})\nu(\mathscr{D})\omega_V(\pi_i(T_i(A_{\mathbf{K}})) \cap W),$$

where $\tau(\mathbf{T}_{NS})$ is the Tamagawa number of $\mathbf{T}_{NS}$, that is the normalized volume of the compact quotient $\mathbf{T}_{NS}(A_{\mathbf{K}})^1/\mathbf{T}_{NS}(\mathbf{K})$ which is isomorphic to the product

$$\mathbf{T}_{NS}(\mathbf{K}_S)^1/\mathbf{T}_{NS}(\mathscr{O}_S) \times \prod_{w \notin S} \mathbf{T}_{NS}(\mathscr{O}_w).$$

By Ono's theorem [44, §3], the Tamagawa number of $\mathbf{T}_{NS}$ is given by

$$\tau(\mathbf{T}_{NS}) = \frac{\sharp H^1(\mathbf{K}, X^*(\mathbf{T}_{NS}))}{\sharp \mathrm{III}^1(\mathbf{K}, \mathbf{T}_{NS})}$$

where $\mathrm{III}^1(\mathbf{K}, \mathbf{T}_{NS}) = \ker(H^1(\mathbf{K}, \mathbf{T}_{NS}) \to \prod_{w \in \mathrm{Val}(\mathbf{K})} H^1(\mathbf{K}_w, \mathbf{T}_{NS}))$. By definition, $\beta(V) = \sharp H^1(\mathbf{K}, X^*(\mathbf{T}_{NS}))$. To conclude the proof, we use the crucial fact, first proven by Salberger [50], that for any $x \in V(A_{\mathbf{K}})^{\mathrm{Br}}$, the number of $i \in I$ such that $x \in \pi_i(T_i(A_{\mathbf{K}}))$ is precisely equal to $\sharp \mathrm{III}^1(\mathbf{K}, \mathbf{T}_{\mathrm{NS}})$. □

Remarks 4.34

(i) Using Theorems 4.31 and 4.33, we see that the equivalence formula (5) of Question 4.8, reduces to an equivalence of the form

$$\sharp\{ y \in T_i(\mathbf{K}) \cap \Delta(T_i) \mid \widetilde{h}(y) \in \mathscr{D}_B \} \sim \omega_{T_i}(\{ y \in \Delta(T_i) \mid \widetilde{h}(y) \in \mathscr{D}_B \})$$

as $B \to +\infty$.

(ii) The conditions $y \in T_i(\mathscr{O}_w)$ for $w \in \mathrm{Val}(\mathbf{K}) \backslash S$ correspond to an integrality condition combined with a gcd condition. For example, if V is a smooth complete intersection of dimension $\geqslant 3$ in the projective space $\mathbf{P}_{\mathbf{Q}}^N$, then the unique versal torsor T is the corresponding cone in $\mathbf{A}_{\mathbf{Q}}^{N+1} \backslash \{0\}$ and the condition $(y_0, \ldots, y_N) \in T(\mathbf{Z}_p)$ corresponds to $(y_0, \ldots, y_N) \in \mathbf{Z}_p^{N+1}$ and $\gcd(y_0, \ldots, y_N) = 1$. Therefore to reduce to counting integral points in a bounded domain, the next step is to use a Moebius inversion formula to remove the gcd condition. Such an inversion formula is described in [46, §2.3].

(iii) In the preceding description, we were not very careful about the choice of the finite set S of bad places. For practical reasons, to use this method, it is in fact more efficient to use a small set of bad primes.

(iv) The lifting to the versal torsors has been used in many cases, see for example [18] or [19]. For practical reasons, it is often simpler to consider an intermediate torsor corresponding to the Picard group Pic(V) (see for example the work of K. Destagnol [22]). The main difference in the new approach described in this section is that the domain obtained after lifting does not have "spikes". In other words, the area of the boundary has a smaller rate of growth, which should remove some of the problems encountered when using a single height relative to the anticanonical line bundle.

4.4 Varieties of Picard Rank One

If the rank of Pic(V) is one, then without loss of generality formula (5) is reduced to estimating a difference of the form

$$\sharp(V(\mathbf{K}) - T)_{H \leqslant bB} - \sharp(V(\mathbf{K}) - T)_{H \leqslant aB} \tag{9}$$

as B goes to infinity, where H is a height relative to the anticanonical line bundle and a, b are real numbers with $0 < a < b$. Therefore, in that case, a positive answer

to Question 4.8 is true if the principle of Batyrev and Manin is valid for $V(\mathbf{K}) -$
T. Similarly the global equidistribution in the sense of 4.10, follows from global
equidistribution 3.21. However the knowledge of estimates for the difference (9)
does not gives an estimate for $(V(\mathbf{K}) - T)_{H \leqslant B}$, unless we have a uniform upper
bound for the error term.

But several examples of Fano varieties of Picard rank one with acccumulating
subvarieties are known in dimension $\geqslant 3$ (see the list given in [12]). For example,
if we consider a cubic volume, the projective lines it contains are parametrized by
the Fano surface, which is of general type. Each of these rational lines has degree
2 and as we shall explain in section "Rational Curves of Low Degree", these lines
give a non negligible contribution to the total number of points thus contradicting
the global equidistribution. In the case of a smooth complete intersection of two
quadrics in $\mathbf{P}^5$, the situation is even worse since the projective lines it contains may
be Zariski dense.

This shows that in higher dimension, even in the case of varieties with a Picard
group of rank one, there might be accumulating subvarieties of codimension $\geqslant 2$
which are not detected by heights on line bundles. Thus one needs to go beyond
heights. To help us in that direction we shall first consider the geometric analogue
of this problem.

5   Geometric Analogue

The geometric analogue of the study of rational points of bounded height is the
study of rational curves of bounded degree. This is a very active subject in algebraic
geometry, and we are going to give a very superficial survey of some particular
aspects of this subject in this section. In fact, there is a very classical dictionary
between number fields, global fields of positive characteristic and function fields of
curves. To simplify the description, we shall mostly restrict ourselves to morphisms
from $\mathbf{P}_k^1$ to a variety V defined over k.

Notation 5.1 Let k be a field and let $\mathscr{C}$ be a smooth geometrically integral projective
curve over k. In this section, we denote by $\mathbf{K} = k(\mathscr{C})$ the function field of $\mathscr{C}$. Let V
be a nice variety over k. The image of the generic point gives a bijection between
the set of rational point $V(\mathbf{K})$ and the set of morphisms $f : \mathscr{C} \to V$. From now
on, we shall identify these sets. Let $f : \mathscr{C} \to V$ be a point of this space. Then the
pull-back map is a morphism of groups $f^* : \operatorname{Pic}(V) \to \operatorname{Pic}(\mathscr{C})$. The composition
$\deg \circ f^*$ is an element of $\operatorname{Pic}(V)^{\vee}$, which we call the *multidegree of* f and denote
by $\mathbf{deg}(f)$.

The constructions of Grothendieck [27, §4.c] prove that for any $d \in \operatorname{Pic}(V)^{\vee}$,
there exists a variety $\mathscr{H}\!om^d(\mathscr{C}, V)$ defined over k, which parametrizes the morphisms from $\mathscr{C}$ to V of multidegree d.

In that geometric setting, we want to describe *asymptotically* the geometric properties of the variety $\mathcal{H}om^d(\mathscr{C}, V)$ as the distance from d to the boundary of the dual of the effective cone goes to infinity. The problem is to give a framework for the asymptotic study of a variety. We shall use the framework given by the ring of integration which was introduced by Kontsevich (see also [21]).

5.1 The Ring of Motivic Integration

Of course, the dimension of the variety $\mathcal{H}om^d(\mathscr{C}, V)$ goes to infinity as the multidegree d grows. But, as suggested by the work of J. Ellenberg [23], we could consider the stabilisation of cohomology groups. The ring of motivic integration enables us to consider the limit of a class associated to the variety.

Construction 5.2 We denote by $\mathcal{M}_k$ the Grothendieck ring of varieties over k: as a group it is generated by the isomorphism classes of varieties over k, where the class of a variety V is denoted by $[V]$, with the relations

$$[V] = [F] + [U]$$

for any closed subvariety F of V, with $U = V \backslash F$. We can then extend the definition of a class to non reduced schemes. Then $\mathcal{M}_k$ is equipped with the unique ring structure such that

$$[V_1] \times [V_2] = [V_1 \times_k V_2],$$

for any varieties V_1 and V_2 over k. We define the Tate symbol as $L = [\mathbf{A}_k^1]$ and consider the localized ring $\mathcal{M}_{k,\mathrm{loc}} = \mathcal{M}_k[L^{-1}]$. We then introduce a decreasing filtration on this ring where, for $i \in \mathbf{Z}$,

$$F^i \mathcal{M}_{k,\mathrm{loc}}$$

is the subgroup of $\mathcal{M}_{k,\mathrm{loc}} = \mathcal{M}_k[L^{-1}]$ generated by symbols of the form $[V]L^{-n}$ if $\dim(V) - n \leqslant -i$. We have the inclusion

$$F^i \mathcal{M}_{k,\mathrm{loc}}.F^j \mathcal{M}_{k,\mathrm{loc}} \subset F^{i+j} \mathcal{M}_{k,\mathrm{loc}},$$

for $i, j \in \mathbf{Z}$. Thus the inverse limit $\widehat{\mathcal{M}_k} = \varprojlim_i \mathcal{M}_{k,\mathrm{loc}}/F^i \mathcal{M}_{k,\mathrm{loc}}$ comes equipped with a structure of topological ring so that the natural map $\mathcal{M}_{k,\mathrm{loc}} \to \widehat{\mathcal{M}_k}$ is a morphism of rings.

Remark 5.3 The morphism $\mathcal{M}_k \to \mathcal{M}_{k,\mathrm{loc}}$ is not injective (see [8]), so we loose information by looking at classes in $\widehat{\mathcal{M}_k}$.

With this ring we may formulate the analogue of Question 4.8:

Question 5.4 *We assume that the nice variety V over k is rationally connected, satisfies conditions (i) and (iii) to (v) of Hypotheses 3.27 and that the rational points over k(T) are Zariski dense. Does the symbol*

$$\left[\mathcal{H}\!om^d(\mathbf{P}_k^1, V)\right] L^{-\langle \omega_V^{-1}, d\rangle}$$

converges in $\widehat{\mathcal{M}_k}$ *for* $d \in \mathrm{Pic}(V)^\vee \cap C_{\mathrm{eff}}^\circ(V)^\vee$ *as* $\mathrm{dist}(d, \partial C_{\mathrm{eff}}(V)^\vee)$ *goes to infinity and can we interpret the limit as some adelic volume?*

5.2   A Sandbox Example: The Projective Space

In the case of the projective space, it turns out that the symbol in fact stabilizes, and thus converges:

Proposition 5.5 *If* $d \geqslant 1$, *then*

$$\left[\mathcal{H}\!om^d(\mathbf{P}_k^1, \mathbf{P}_k^n)\right] L^{-(n+1)d} = \frac{L^{n+1} - 1}{L - 1}(1 - L^{-n}).$$

Proof In this proof, we shall describe the sets of k-points of our varieties and gloss over the description of the varieties themselves. So if we consider the set $W_d(k)$ of $(P_0, \ldots, P_n) \in k[T]^{n+1}$ such that $\gcd_{0 \leqslant i \leqslant n}(P_i) = 1$ and $\max_{0 \leqslant i \leqslant n}(\deg(P_i)) = d$ then W_d is a $\mathbf{G}_m$ torsor over the space $\mathcal{H}\!om^d(\mathbf{P}_k^1, \mathbf{P}_k^n)$ which is locally trivial for Zariski topology. Hence

$$(L - 1)\left[\mathcal{H}\!om^d(\mathbf{P}_k^1, \mathbf{P}_k^n)\right] = [W_d]. \tag{10}$$

But if we consider the space of $(n + 1)$-tuples of polynomials $(P_0, \ldots, P_n)$ such that $\max_{0 \leqslant i \leqslant n}(\deg(P_i)) = d$, then it is naturally isomorphic to $\mathbf{A}^{(n+1)(d+1)} - \mathbf{A}^{(n+1)d}$ and we may decompose it as a disjoint union according to the degree of the gcd of the polynomials. The piece corresponding to the families with $\deg(\gcd_{0 \leqslant i \leqslant n}(P_i)) = k$ is isomorphic to $[W_{d-k}] \times \mathbf{A}^k$ where $\mathbf{A}^k$ parametrizes the gcd which is a unitary polynomial of degree k. We get the formula

$$L^{(n+1)(d+1)} - L^{(n+1)d} = \sum_{k=0}^{d} L^k [W_{d-k}].$$

We may introduce formal series in $\widehat{\mathcal{M}_k}[[T]]$ to get the formula

$$\sum_{d \geqslant 0}(L^{n+1} - 1)L^{(n+1)d}T^d = \left(\sum_{k \geqslant 0} L^k T^k\right)\left(\sum_{d \geqslant 0}[W_d]T^d\right).$$

From which we deduce

$$\sum_{d \geqslant 0}[W_d]T^d = (1 - LT)(L^{n+1} - 1)\sum_{d \geqslant 0}L^{(n+1)d}T^d.$$

Therefore, if $d \geqslant 1$, we get

$$[W_d] = (L^{n+1} - 1)(L^{(n+1)d} - LL^{(n+1)(d-1)})$$
$$= (L^{n+1} - 1)L^{(n+1)d}(1 - L^{-n}).$$

Combining with formula (10) gives the formula of the proposition.   □

Remarks 5.6

(i) Let us quickly explain how the constant obtained might be interpreted as an adelic volume. First, for the projective space the L function associated to the Picard group coincide with the usual zeta function. This has a motivic analogue decribed by M. Kapranov in [32]:

$$Z_{\mathbf{C}(T)}(U) = \sum_{d \geqslant 0}[(\mathbf{P}_k^1)^{(d)}]U^d$$

where $(\mathbf{P}_k^1)^{(d)}$ is the symmetric product $(\mathbf{P}_k^1)^d/\mathfrak{S}_d$ and is isomorphic to $\mathbf{P}_k^d$. The parameter U should be understood as L^{-s}. The residue of the zeta function at $s = 1$ corresponds to

$$((1 - LU)Z_{\mathbf{C}(T)}(U))(L^{-1})$$

$$= \left((1 - LU)\sum_{d \geqslant 0}\frac{L^{d+1} - 1}{L - 1}U^d\right)(L^{-1})$$

$$= \frac{1}{L - 1}\left((1 - LU)\left(\frac{L}{1 - LU} - \frac{1}{1 - U}\right)\right)(L^{-1})$$

$$= \frac{1}{L - 1}\left(\frac{L - 1}{1 - U}\right)(L^{-1})$$

$$= \frac{1}{1 - L^{-1}}.$$

By translating the formula (3), the expected constant should formally have the form

$$C = \frac{L^n}{1 - L^{-1}} \prod_{P \in \mathbf{P}_k^1} (1 - L^{-\deg(P)})[\mathbf{P}_{\kappa(P)}^n]L^{-n\deg(P)},$$

where L^{-1} plays the rôle of the square root of the discriminant. The term appearing in the product may be simplified as $1 - L^{-(n+1)\deg(P)}$. However this formal constant involves a product over a possibly uncountable set $\mathbf{P}_k^1$. Nevertheless, in this very particular case, we may consider the *inverse* of this product. Then, we get

$$\prod_{P \in \mathbf{P}_k^1} \sum_{m \geq 0} L^{-(n+1)m\deg(P)},$$

where the product is taken over closed points of $\mathbf{P}_k^1$. If we admit that it makes sense to develop this product, we get, noting that we get a sum over all divisors of $\mathbf{P}_k^1$,

$$\sum_{m \geq 0} \sum_{P \in (\mathbf{P}_k^1)^{(m)}(k)} L^{-(n+1)m}.$$

But we may now interpret each interior sum as a motivic integral and get, using the fact that $(\mathbf{P}_k^1)^{(m)}$ is isomorphic to the projective space $\mathbf{P}_k^m$,

$$\sum_{m \geq 0} [\mathbf{P}_k^m]L^{-(n+1)m} = \sum_{m \geq 0} \frac{1 - L^{m+1}}{1 - L}L^{-(n+1)m}$$

$$= \frac{1}{1 - L}\left(\frac{1}{1 - L^{-n-1}} - \frac{L}{1 - L^{-n}}\right)$$

$$= \frac{1}{1 - L} \times \frac{1 - L}{(1 - L^{-n})(1 - L^{-n-1})}$$

Finally we get

$$C = \frac{L^{n+1} - 1}{L - 1}(1 - L^{-n})$$

as wanted.

(ii) This type of result is compatible with products and we get a result for products of projective spaces for free. D. Bourqui has more general results for toric varieties [10].

(iii) M. Bilu in [6] has defined an Euler product giving a precise meaning for the expected constant in this setting.[1]

5.3 Equidistribution in the Geometric Setting

In the geometric setting equidistribution may be described as follows.

Construction 5.7 Let $\mathscr{S}$ be a subscheme of dimension 0 of $\mathscr{C}$, then we may consider the moduli space $\mathcal{H}\!om(\mathscr{S}, V)$ which parametrizes the morphisms from $\mathscr{S}$ to V. For any subvariety W of $\mathcal{H}\!om(\mathscr{S}, V)$, we may then consider the set of morphisms $f : \mathbf{P}_k^1 \to V$ of multidegree d such that the restriction $f_{|\mathscr{S}}$ belongs to W. This is parametrized by a variety $\mathcal{H}\!om_W^d(\mathscr{C}, V)$ contained in $\mathcal{H}\!om^d(\mathscr{C}, V)$.

Naïve Geometric Equidistribution 5.8 *We shall say that naïve equidistribution holds for V if for any subscheme $\mathscr{S}$ of dimension 0 in $\mathscr{C}$ and any subvariety W of $\mathcal{H}\!om(\mathscr{S}, V)$, the symbol*

$$\left(\left[\mathcal{H}\!om_W^d(\mathscr{C}, V) \right]\left[\mathcal{H}\!om(\mathscr{S}, V) \right] - \left[\mathcal{H}\!om^d(\mathscr{C}, V) \right][W] \right) \mathbf{L}^{-\langle \omega_V^{-1}, d \rangle}$$

converges to 0 in $\widehat{\mathcal{M}_k}$ for $d \in \mathrm{Pic}(V)^\vee \cap C_{\mathrm{eff}}^\circ(V)^\vee$ as the distance from d to $\partial C_{\mathrm{eff}}(V)^\vee$ goes to infinity.

Remark 5.9 This statement gives a precise meaning to the idea of a convergence

$$\frac{\left[\mathcal{H}\!om_W^d(\mathscr{C}, V) \right]}{\left[\mathcal{H}\!om^d(\mathscr{C}, V) \right]} \longrightarrow \frac{[W]}{\left[\mathcal{H}\!om(\mathscr{S}, V) \right]}.$$

5.4 Crash Course about Obstruction Theory

Obstruction theory gives a sufficient condition for the moduli spaces to have the expected dimension. Let us give a very short introduction to these tools, the interested reader may turn to the book of O. Debarre [17] for a more serious introduction to this subject.

---

[1]The construction of M. Bilu and the work of D. Bourqui suggest that the filtration described here is not the correct one to get the expected limit. In fact, one may need a filtration such that if X and Y are geometrically irreducible varieties then $[\mathbf{L}^{-\dim(X)} X] - [\mathbf{L}^{-\dim(Y)} Y]$ belongs to $F^1 \mathcal{M}_{k,\mathrm{loc}}$.

Let $f : \mathbf{P}_k^1 \to V$ be a morphism of multidegree d then we may consider the tangent space at f and the dimension at f. There is a natural isomorphism

$$T_f \mathcal{H}\mathrm{om}^d(\mathbf{P}_k^1, V) \xrightarrow{\sim} H^0(\mathbf{P}_k^1, f^*(T_V))$$

and

$$\dim_f \left(\mathcal{H}\mathrm{om}^d(\mathbf{P}_k^1, V) \right) \geqslant h^0(\mathbf{P}_k^1, f^*(T_V)) - h^1(\mathbf{P}_k^1, f^*(T_V)).$$

On the other hand, on $\mathbf{P}_k^1$, any vector bundle splits into a direct sum of line bundles. In other words, there exists an isomorphism

$$f^*(TV) \xrightarrow{\sim} \bigoplus_{i=1}^n \mathcal{O}_{\mathbf{P}_k^1}(a_i)$$

with $a_1 \geqslant a_2 \geqslant \cdots \geqslant a_n$ and $(a_1, \ldots, a_n)$ is uniquely determined. If $a_n \geqslant 0$, then we get that $h^1(\mathbf{P}_k^1, f^*(TV)) = 0$ and

$$\dim_f \left(\mathcal{H}\mathrm{om}(\mathbf{P}_k^1, V) \right) = h^0(\mathbf{P}_k^1, f^*(T_V)) = \sum_{i=1}^n h^0(\mathcal{O}_{\mathbf{P}_k^1}(a_i))$$

$$= \sum_{i=1}^n a_i + 1 = n + \langle d, \omega_V^{-1} \rangle,$$

which is the expected dimension. Thus a sufficient condition to get the expected dimension is $a_n \geqslant 0$.

But let us now add some conditions related to equidistribution. Let $\mathscr{S}$ be a subscheme of $\mathbf{P}_k^1$ of dimension 0. Then $\mathscr{S}$ corresponds to a divisor $D = \sum_{P \in I} n_P P$ on $\mathbf{P}_k^1$ and may described as $\mathrm{Spec}(\times_{P \in I} \mathcal{O}_{\mathbf{P}_k^1, P} / \mathfrak{m}_P^{n_P})$, where $\mathfrak{m}_P$ is the maximal ideal of the local ring $\mathcal{O}_{\mathbf{P}_k^1, P}$. Let s be the degree of D, that is $\sum_{P \in I} n_P [\kappa(P) : k]$. Then $\mathcal{H}\mathrm{om}(\mathscr{S}, V)$ has dimension ns; therefore if we fix $\varphi : \mathscr{S} \to V$, the expected dimension of $\mathcal{H}\mathrm{om}_{\{\varphi\}}^d(\mathbf{P}_k^1, V)$ ought to be $n(1 - s) + \langle d, \omega_V^{-1} \rangle$. But obstruction theory in that setting relates the deformation at f to the vector bundle $f^*(TV) \otimes \mathcal{O}(-D)$ therefore a sufficient condition for the dimension of the moduli space $\mathcal{H}\mathrm{om}_{\{\varphi\}}^d(\mathbf{P}_k^1, V)$ at f to be the correct one is $a_n - s \geqslant 0$. In particular a curve is said to be *very free* if $a_n > 0$. Therefore if one wishes to have geometric equidistribution, then one ought to look at the limit as a_n goes to $+\infty$.

One should note that the counter-examples introduced in Sect. 4.4, like the intersection of two quadrics, also show the necessity to go beyond degrees in the geometric setting.

6 Slopes à la Bost

Following the geometric analogue, we need a notion which is the arithmetic translation of the notion of very free curves. This analogue, introduced in [48], is given by Arakelov geometry and is based upon the slopes as they are considered by J.-B. Bost.

6.1 Definition

In this section, we again consider a nice variety V over a number field $\mathbf{K}$.

Slopes of an Adelic Vector Bundle Over Spec(K)

The following definition is a variant of the definition described in Chapter II, Sect. 3.2 of this volume.

Definition 6.1 Let E be a $\mathbf{K}$-vector space of finite dimension n equipped with

- A projective $\mathscr{O}_{\mathbf{K}}$-submodule Λ_E of rank n;
- For any complex place $w \in \mathrm{Val}(\mathbf{K})$, a map

$$\| \cdot \|_w : E_w = E \otimes_{\mathbf{K}} \mathbf{K}_w \longrightarrow \mathbf{R}_{\geqslant 0}$$

such that there exists a positive definite hermitian form ϕ on E_w so that $\|y\|_w = \phi(y, y)$;
- For any real place $w \in \mathrm{Val}(\mathbf{K})$ a euclidean norm

$$\| \cdot \|_w : E_w \longrightarrow \mathbf{R}_{\geqslant 0}.$$

Let F be a vector subspace of E. We equip it with $\Lambda_F = \Lambda \cap F$ and the restrictions of the norms. The *Newton polygon*, which we denote by $\mathscr{P}(E)$ is defined as the convex hull of the set of pairs $(\dim(F), \widehat{\deg}(F))$ where F describes the set of vector subspaces of E.

Remark 6.2 Let us assume that $\mathbf{K} = \mathbf{Q}$. If we consider the subspaces F of dimension 1, then $\widehat{\deg}(F)$ is given as $-\log(\|y_0\|_\infty)$ where y_0 is a generator of $\Lambda \cap F$. Thus we get the points $(1, -\log(\|y\|_\infty))$ where y goes over the primitive elements of the lattice Λ. In particular, there is an upper bound for the possible values of the second coordinate. More generally $\mathscr{P}(E)$ is bounded from above. In Fig. 5, we represented how the points $(\dim(F), \widehat{\deg}(F))$ and the upper part of the convex hull may look like.

Fig. 5 Convex hull

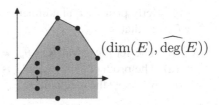

$(\dim(E), \widehat{\deg}(E))$

Construction 6.3 Since the set $\mathscr{P}(E)$ is bounded from above, we may define the function $m_E : [0, n] \to \mathbf{R}$ by

$$m_E(x) = \max\{\, y \in \mathbf{R} \mid (x, y) \in \mathscr{P}(E) \,\}.$$

This function is concave and affine in each interval $[i-1, i]$ for $i \in \{1, \ldots, \dim(E)\}$. The slopes of E are then given as

$$\mu_i(E) = m_E(i) - m_E(i - 1)$$

for $i \in \{1, \ldots, \dim(E)\}$.

Remarks 6.4

(i) By construction, we have the inequalities

$$\mu_1(E) \geqslant \mu_2(E) \geqslant \cdots \geqslant \mu_{\dim(E)}(E).$$

These inequalities might not be strict. Moreover

$$\widehat{\deg}(E) = \sum_{i=1}^{\dim(E)} \mu_i(E).$$

Therefore the *slope* of E, which is defined as $\mu(E) = \frac{\widehat{\deg}(E)}{\dim(E)}$ is the mean of the slopes:

$$\mu(E) = \frac{1}{\dim(E)} \sum_{i=1}^{\dim(E)} \mu_i(E).$$

(ii) The value of $m_E(i)$ may differ from $\max_{\dim(F)=i}(\widehat{\deg}(F))$. However, following E. Gaudron [26, definition 5.18], we may define the successive minima of the arithmetic lattice E as follows: for $i \in \{1, \ldots, \dim(E)\}$, the i-th minima $\lambda_i(E)$ is the infimum of the numbers $\theta \in \mathbf{R}_{>0}$ such that there exists a family of strictly positive real numbers $(\theta_w)_{w \in \mathrm{Val}(\mathbf{K})}$ and a free family $(x_1, \ldots, x_i)$ in E such that

(i) The set $\{\, w \in \mathrm{Val}(\mathbf{K}) \mid \theta_w \neq 1 \,\}$ is finite;
(ii) The product $\prod_{w \in \mathrm{Val}(\mathbf{K})} \theta_w$ is equal to θ;
(iii) We have the inequalities

$$\|x_j\|_w \leqslant \theta_w$$

for $j \in \{1, \ldots, i\}$ and $w \in \mathrm{Val}(\mathbf{K})$.

Then Minkowski's theorem gives an explicit constant $C_{\mathbf{K}}$ such that

$$0 \leqslant \log(\lambda_i(E)) + \mu_i(E) \leqslant C_{\mathbf{K}}$$

for $i \in \{1, \ldots, \dim(E)\}$. Other definitions of successive minima are given in Chapter II, Sect. 3.1 and are similarly related to slopes.

(iii) In this chapter, the slopes are not invariant under field extensions since we did not normalise them by $\frac{1}{[\mathbf{K}:\mathbf{Q}]}$. This conforms to the usual convention for heights in Manin's program, which has been chosen to get a formulation of the expected estimate which does not depend on the degree of the field.

Slopes on Varieties, Freeness

We now apply the constructions of last paragraph to vector bundles on varieties.

Definition 6.5 Let E be a vector bundle on the nice variety V of dimension n. We assume that E is equipped with an adelic norm $(\|\cdot\|_w)_{w \in \mathrm{Val}(\mathbf{K})}$ then for any rational point $P \in V(\mathbf{K})$, the fibre E_P is an adelic vector bundle over $\mathrm{Spec}(\mathbf{K})$ and we may define

$$\mu_i^E(P) = \mu_i(E_P).$$

In particular, if V is equipped with an adelic metric (see Definition 2.5), we may define the *slopes* of a rational point $P \in V(\mathbf{K})$ as

$$\mu_i(P) = \mu_i(T_P V)$$

for $i \in \{1, \ldots, n\}$.

Remarks 6.6

(i) From Remark 6.4 (i), we deduce that for any rational point $P \in V(\mathbf{K})$, we have

$$\mu_n(P) \leqslant \mu_{n-1}(P) \leqslant \cdots \leqslant \mu_1(P)$$

and $\widehat{\deg}(T_P V) = \sum_{k=1}^{n} \mu_i(P)$. But we may interpret this degree $\widehat{\deg}(T_P V) = \widehat{\deg}((\omega_V^{-1})_P)$ as the logarithmic height of P, that is $h(P) = \log(H(P))$, where the height H is defined by the induced metric on the anticanonical line bundle.

(ii) From the previous remark we deduce the inequalities

$$\mu_n(P) \leqslant \frac{h(P)}{n} \leqslant \mu_1(P)$$

for any rational point $P \in V(\mathbf{K})$.

Definition 6.7 The *freeness* of a rational $P \in V(\mathbf{K})$ is defined by

$$l(P) = \begin{cases} n\frac{\mu_n(P)}{h(P)} & \text{if } \mu_n(P) > 0, \\ 0 & \text{otherwise.} \end{cases}$$

Remarks 6.8

(i) By definition the freeness of a point $l(P)$ belongs to the interval $[0, 1]$.

(ii) We have the equality $l(P) = 0$ if and only if the minimal slope $\mu_n(P) \leqslant 0$.

(iii) The equality $l(P) = 1$ occurs if and only if the lattice $T_P V$ is semi-stable, that is $\mu_1(P) = \cdots = \mu_n(P)$ and $h(P) > 0$. In other words this means that $\mu(F) \leqslant \mu(T_P V)$ for any subspace F of $T_P V$. This is, for example, the case if the lattice is the usual lattice $\mathbf{Z}^n$ in $\mathbf{R}^n$ equipped with its standard euclidean structure. Up to scaling, this occurs for a point $(P, \ldots, P)$ on the diagonal of $(\mathbf{P}_{\mathbf{K}}^1)^n$. Another example of a semi-stable lattice in dimension 2 is the classical hexagonal lattice $\mathbf{Z}[j]$ generated by a primitive third root of 1, as shown in Fig. 6. More generally for two dimensional lattices we may consider that Λ is isomorphic to the lattice $a(\mathbf{Z}+\mathbf{Z}\tau) \subset \mathbf{C}$, where a is some positive real number, $\Re(\tau) \in [-1/2, 1/2], |\tau| \geqslant 1$ and $\Im(\tau) > 0$. Then a lattice is semistable if and only if $\Im(\tau) \leqslant 1$, which is drawn in grey on Fig. 7.

(iv) For any rational point on a curve, we have $l(P) = 1$.

(v) For a surface S over $\mathbf{Q}$, an adelic metric define two invariants, namely the height H and a map $S(\mathbf{Q}) \to \mathbf{H}/\mathrm{PSL}_2(\mathbf{Z})$, where $\mathbf{H}$ denotes the Poincaré half-plane $\{z \in \mathbf{C} \mid \Im(z) > 0\}$ which sends a point P to the class of τ_P such

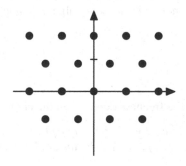

Fig. 6 Hexagonal lattice

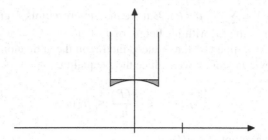

Fig. 7 Semi-stable lattices

that the lattice in $T_P S$ is isomorphic to $a_P(\mathbf{Z} + \mathbf{Z}\tau_P)$ with $a_P \in \mathbf{R}_{>0}$. Then, taking τ_P in the usual fundamental domain, the freeness of P is given by

$$l(P) = \begin{cases} 1 \text{ if } \Im(\tau_P) \leqslant 1 \text{ and } h(P) > 0, \\ 1 - \frac{\log(\Im(\tau_P))}{h(P)} \text{ if } 1 < \Im(\tau_P) < H(P), \\ 0 \text{ otherwise.} \end{cases}$$

Indeed, in that case, we have $h(P) = -2\log(a_p) - \log(\Im(\tau_P))$ and, since $|\tau_P| \geqslant 1$, the first slope is given by

$$\mu_1(P) = \max\left(-\log(a_P), \frac{h(P)}{2}\right)$$

We get that $\mu_1(P) = \mu_2(P)$ if and only if $\Im(\tau_P) \leqslant 1$ and

$$\mu_2(P) = -\log(a_p) - \log(\Im(\tau_P)) = \frac{h(P) - \log(\Im(\tau_P))}{2}$$

otherwise.

(vi) By definition, the freeness $l(P)$ is invariant under field extensions. Thus a condition of the form $l(P) > \varepsilon$ does not depend on the field of definition and makes sense for algebraic points in $V(\overline{\mathbf{K}})$. On the other hand the defining condition for a thin subset, namely $P \in \varphi(X(\mathbf{K}))$ for a morphism φ as in Definition 3.34 does not make sense for algebraic points.

6.2 Properties

Let us first describe how the freeness depends on the choice of the metric.

Proposition 6.9 *Let* $\varphi : E \to F$ *be a morphism of vector bundles and let* $(\| \cdot \|_w)_{w \in \mathrm{Val}(\mathbf{K})}$ *(resp.* $(\| \cdot \|'_w)_{w \in \mathrm{Val}(\mathbf{K})}$*) be an adelic norm on* E *(resp.* F*) then* *there exists a family* $(\lambda_w)_{w \in \mathrm{Val}(\mathbf{K})}$ *such that*

(i) *For any* $w \in \mathrm{Val}(\mathbf{K})$, *any* $P \in V(\mathbf{K}_w)$, *and any* $y \in E_P$, *we have*

$$\|\varphi(y)\|'_w \leqslant \lambda_w \|y\|_w;$$

(ii) *The set* $\{\, w \in \mathrm{Val}(\mathbf{K}) \mid \lambda_w \neq 1 \,\}$ *is finite.*

Proof Let $\mathbf{P}(E)$ be the projective bundle of the lines in E and $E^\times$ be the complement of the zero section in E. Then for any place w of $\mathbf{K}$, we may define a map $f_w : E^\times(\mathbf{K}_w) \to \mathbf{R}_{\geqslant 0}$ by $f_w(y) = \frac{\|\varphi(y)\|'_w}{\|y\|_w}$. This map is constant on the lines and induces a continuous map $\mathbf{P}(E)(\mathbf{K}_w) \to \mathbf{R}_{\geqslant 0}$. Since the space $\mathbf{P}(E)(\mathbf{K}_w)$ is compact, this function is bounded from above by a constant λ_w. Moreover for almost all $w \in \mathrm{Val}(\mathbf{K})$ the norms on E and F are defined by model and the morphism φ is defined over $\mathscr{O}_w$. For such a place w, for any $P \in V(\mathbf{K}_w)$, we get that

$$\varphi(\{\, y \in E_P \mid \|y\|_w \leqslant 1 \,\}) \subset \{\, y \in F_P \mid \|y\|'_w \leqslant 1 \,\},$$

therefore we may take $\lambda_w \leqslant 1$. $\qquad\qquad\qquad\qquad\qquad\qquad\qquad\qquad\square$

Remark 6.10 From this proposition, it follows that, if we consider norms $(\| \cdot \|_w)_{w \in \mathrm{Val}(\mathbf{K})}$ and $(\| \cdot \|'_w)_{w \in \mathrm{Val}(\mathbf{K})}$ on a vector bundle, then the quotient $\frac{\| \cdot \|'_w}{\| \cdot \|_w}$ is bounded from above and from below by a strictly positive constant. Moreover, by definition the norms are equal for almost all places. This implies the existence of a constant C such that, for any rational point $P \in V(\mathbf{K})$ and any subspace F of $T_P V$,

$$|\widehat{\deg}(F) - \widehat{\deg}'(F)| \leqslant C.$$

where $\widehat{\deg}'$ is the degree corresponding to the second norm.

Corollary 6.11 *Let* μ_i *and* μ'_i *be the slopes defined by two different metrics on* V *and let* l *and* l' *be the corresponding freeness, then*

(i) *The difference* $|\mu_i - \mu'_i|$ *is bounded on* $V(\mathbf{K})$;
(ii) *There exists* $C \in \mathbf{R}_{>0}$ *such that*

$$|l(P) - l'(P)| < \frac{C}{h(P)}$$

for any $P \in V(\mathbf{K})$ *such that* $h(P) > 0$.

We now wish to describe a strong link between the geometric and arithmetic settings. Let us first define the freeness in the geometric setting.

Definition 6.12 Let $\varphi : \mathbf{P}^1_{\mathbf{K}} \to V$ be a morphism of varieties. The pull-back of the tangent bundle $\varphi^*(TV)$ is isomorphic to a direct sum $\bigoplus_{i=1}^n \mathscr{O}_{\mathbf{P}^1_{\mathbf{K}}}(a_i)$ with $a_1 \geqslant$

$a_2 \geqslant \cdots \geqslant a_n$. The slopes of φ are the integers $\mu_i(\varphi) = a_i$. We may consider $\deg_{\omega_V^{-1}}(\varphi) = \sum_{i=1}^n \mu_i(\varphi)$ and the *freeness of* φ is defined by

$$l(\varphi) = \begin{cases} \dfrac{na_n}{\deg_{\omega_V^{-1}}(\varphi)} & \text{if } a_n > 0, \\ 0 & \text{otherwise.} \end{cases}$$

Remark 6.13 By construction $l(\varphi) \in [0, 1] \cap \mathbf{Q}$ and $l(\varphi) > 0$ if and only if φ is very free, that is $a_n > 0$.

Proposition 6.14 *Let* $\varphi : \mathbf{P}_{\mathbf{K}}^1 \to V$ *be a non constant morphism of varieties and assume that V is equipped with an adelic metric. Then*

$$l(\varphi(P)) \longrightarrow l(\varphi)$$

as $h_{\mathscr{O}(1)}(P) \to +\infty$.

Proof Let us fix an isomorphism from $\varphi^*(TV)$ to a direct sum $\bigoplus_{i=1}^n \mathscr{O}_{\mathbf{P}_{\mathbf{K}}^1}(a_i)$ with $a_1 \geqslant a_2 \geqslant \cdots \geqslant a_n$. On $\varphi^*(TV)$ we consider the pull-back of the adelic metric on V and we equip the sum $\bigoplus_{i=1}^n \mathscr{O}_{\mathbf{P}_{\mathbf{K}}^1}(a_i)$ with the direct sums of the norms induced by a norm on $\mathscr{O}_{\mathbf{P}_{\mathbf{K}}^1}(1)$. Using the Corollary 6.11, we get that the differences $|\mu_i(\varphi(P)) - a_i h_{\mathscr{O}(1)}(P)|$ is bounded, as well as $|h(\varphi(P)) - \sum_{i=1}^n a_i h_{\mathscr{O}(1)}(P)|$. If $a_n \geqslant 0$, then the sum $\sum_{i=1}^n a_i$ is strictly positive since the morphism is not constant and we get

$$\left| l(\varphi(P)) - \frac{a_n n}{\sum_{i=1}^n a_i} \right| < \frac{C}{h_{\mathscr{O}(1)}(P)}.$$

If $a_n < 0$, then we get that $l(\varphi(P)) = 0$ except for a finite number of $P \in \mathbf{P}_{\mathbf{K}}^1$. □

6.3 Explicit Computations

In the Projective Space

Let us compute the freeness for points of the projective space. We denote by H the usual height on $\mathbf{P}_{\mathbf{K}}^N$ relative to $\omega_{\mathbf{P}_{\mathbf{K}}^N}^{-1} = \mathscr{O}_{\mathbf{P}_{\mathbf{K}}^N}(N+1)$ and write $h = \log \circ H$.

Proposition 6.15 *Let* $P \in \mathbf{P}^n(\mathbf{K})$, *then*

$$l(P) = \frac{n}{n+1} + \min_F \left(\frac{-n\, \widehat{\deg}(F)}{\mathrm{codim}(F) h(P)} \right)$$

where F goes over the subspaces $F \subsetneq \mathbf{K}^{n+1}$ such that $P \in \mathbf{P}(F)$.

Proof Let $D \subset E$ be the line in E corresponding to the projective point P. There is a canonical isomorphism from the tangent space $T_P \mathbf{P}_{\mathbf{K}}^n$ to the quotient $D^\vee \otimes E / D^\vee \otimes D$ where $D^\vee$ is the dual of D. This gives a bijection from the set of subspaces F of E such that $D \subset F \subsetneq E$ to the strict subspaces of $T_P \mathbf{P}_{\mathbf{K}}^n$ which maps the subspace F to the quotient $D^\vee \otimes F / D^\vee \otimes D$. Since $D^\vee \otimes D$ is canonically isomorphic to $\mathbf{K}$, the arithmetic degree of the subspace of $T_P \mathbf{P}_{\mathbf{K}}^n$ is given by

$$\widehat{\deg}(D^\vee \otimes F / D^\vee \otimes D) = \widehat{\deg}(D^\vee \otimes F) - \widehat{\deg}(\mathbf{K})$$

$$= \widehat{\deg}(F) - \dim(F)\,\widehat{\deg}(D).$$

On the other hand, by the description of the tangent space,

$$h(P) = -(n+1)\,\widehat{\deg}(D).$$

We get that the smallest slope is given by

$$\mu_n(P) = -\widehat{\deg}(D) + \min_F \left(\frac{-\widehat{\deg} F}{\operatorname{codim}_E(F)} \right)$$

and the freeness by

$$l(P) = \frac{n}{n+1} + \min_F \left(\frac{-n\,\widehat{\deg}(F)}{\operatorname{codim}_E(F)h(P)} \right). \qquad \square$$

Corollary 6.16 *For any point $P \in \mathbf{P}^n(\mathbf{K})$, we have*

$$l(P) \geqslant \frac{n}{n+1}.$$

Remarks 6.17

(i) If we take a fixed projective subspace F in E, then $l(P)$ converges to $\frac{n}{n+1}$ as $h(P)$ goes to $+\infty$ with $P \in F$.

(ii) One can show that for any $\eta > 0$, there exists a constant $C > 0$ such that, for $B > 1$,

$$\sharp\{ P \in \mathbf{P}^n(\mathbf{K}) \mid H(P) \leqslant B \text{ and } l(P) < 1 - \eta \} < CB^{1-\eta}.$$

Since we have an equivalence

$$\sharp\{ P \in \mathbf{P}^n(\mathbf{K}) \mid H(P) \leqslant B \} \sim C(\mathbf{P}_{\mathbf{K}}^n)B$$

as B goes to infinity, this means that the number of points P with a freeness $l(P) < 1 - \eta$ is in fact asymptotically negligible.

Products of Lines

Despite the previous example, the freeness of points can be very small even on a homogeneous variety. Let us prove that for $(\mathbf{P}_{\mathbf{K}}^1)^n$. We equip $(\mathbf{P}_{\mathbf{K}}^1)^n$ with the product of the adelic metrics. We denote by H the usual height on $\mathbf{P}_{\mathbf{K}}^1$ relative to $\omega_{\mathbf{P}_{\mathbf{K}}^1}^{-1} = \mathscr{O}_{\mathbf{P}_{\mathbf{K}}^1}(2)$ and write $h = \log \circ H$. We shall also use h (resp. H) to denote the logarithmic (resp exponential) height on $(\mathbf{P}_{\mathbf{K}}^1)^n$.

Proposition 6.18 *For any $P = (P_1, \ldots, P_n) \in \mathbf{P}^1(\mathbf{K})^n$, one gets*

$$l(P) = \frac{n \, \min_{1 \leqslant i \leqslant n}(h(P_i))}{\sum_{i=1}^n h(P_i)}.$$

Proof The tangent space $T_P(\mathbf{P}_{\mathbf{K}}^1)^n$ is canonically isomorphic to $\bigoplus T_{P_i} \mathbf{P}_{\mathbf{K}}^1$ and, by construction, this isomorphism is compatible with the norms. Let us choose a permutation $\sigma \in \mathfrak{S}_n$ such that

$$h(P_{\sigma(1)}) \geqslant h(P_{\sigma(2)}) \geqslant \cdots \geqslant h(P_{\sigma(n)}).$$

Then we get that $\mu_i(P) = h(P_{\sigma(i)})$, since the subspace of dimension i with the biggest arithmetic degree is given by $\bigoplus_{j=1}^i T_{P_{\sigma(j)}} \mathbf{P}_{\mathbf{K}}^1$. $\qquad\qquad\square$

Corollary 6.19 *For any $\varepsilon > 0$, there exist a constant C_ε such that*

$$\frac{\sharp\{ P \in \mathbf{P}^1(\mathbf{K})^n \mid H(P) \leqslant B \text{ and } l(P) > \varepsilon \}}{\sharp\{ P \in \mathbf{P}^1(\mathbf{K})^n \mid H(P) \leqslant B \}} \longrightarrow C_\varepsilon$$

as $B \to +\infty$. Moreover $1 - C_\varepsilon = O(\varepsilon)$.

Proof Let us consider the map $h : \mathbf{P}^1(\mathbf{K})^n \to \mathbf{R}_{\geqslant 0}^n$ given by $(P_i)_{1 \leqslant i \leqslant n} \mapsto (h(P_i))_{1 \leqslant i \leqslant n}$ and, for $t = (t_i)_{1 \leqslant i \leqslant n}$, write $|t| = \sum_{i=1}^n t_i$. The height of point P in $\mathbf{P}^1(\mathbf{K})^n$ is given by $h(P) = |h(P)|$. By Proposition 6.18, we only have to estimate the cardinal of the set

$$\left\{ (P_i)_{1 \leqslant i \leqslant n} \in \mathbf{P}^1(\mathbf{K})^n \,\middle|\, \left|\sum_{i=1}^n h(P_i)\right| \leqslant \min\left(\log(B), \frac{n}{\varepsilon} \min_{1 \leqslant i \leqslant n}(h(P_i))\right) \right\}.$$

Let us introduce the compact simplex $\Delta_\varepsilon(B)$ in $\mathbf{R}_{\geqslant 0}^n$ defined by

$$|t| \leqslant \min\left(\log(B), \frac{n}{\varepsilon} \min_{1 \leqslant i \leqslant n}(t_i)\right).$$

Then we may write the above set as

$$\{ P \in \mathbf{P}^1(\mathbf{K})^n \mid h(P) \in \Delta_\varepsilon(B) \}.$$

Using the estimate of S. H. Schanuel [52, theorem 1], we get

$$\sharp\{\, P \in \mathbf{P}^1(\mathbf{K}) \mid H(P) \leqslant B \,\} = C(\mathbf{P}^1_{\mathbf{K}})B + O(B^{1/2}\log(B)),$$

we get that, for real numbers η, δ with $0 < \eta < 1$ and $0 < \delta < 1/2$ and any $t = (t_1, \dots, t_n) \in \mathbf{R}^n_{\geqslant 0}$, we have

$$\sharp\left\{\, P \in \mathbf{P}^1(\mathbf{K})^n \,\middle|\, h(P) \in \prod_{i=1}^{n}[t_i, t_i + \eta] \,\right\}$$

$$= C(\mathbf{P}^1_{\mathbf{K}})^n e^{|t|}(e^\eta - 1)^n + O(e^{|t| - \delta \min_{1 \leqslant i \leqslant n}(t_i)}) \qquad (11)$$

$$= C(\mathbf{P}^1_{\mathbf{K}})^n e^{|t|}\eta^n + O(e^{|t|}\eta^{n+1}) + O(e^{|t| - \delta \min_{1 \leqslant i \leqslant n}(t_i)}).$$

Covering $\Delta_\varepsilon(B)$ with cubes with edges of length η, the number of such cubes meeting the boundary of the simplex is bounded by $O((\log(B)/\eta)^{n-1})$. Therefore comparing sum and integral, we get the following estimate for the cardinal of our set:

$$C(\mathbf{P}^1_{\mathbf{K}})^n \int_{\Delta_\varepsilon(B)} e^{|t|}\mathrm{d}t + O(B(\log(B))^n \eta) + O\left(\left(\frac{\log(B)}{\eta}\right)^n B^{1-\delta\varepsilon/n}\right).$$

We may take $\eta = B^{-\delta\varepsilon/(2n^2)}$ to have a sufficiently small error term. The computation of the integral gives $BP_\varepsilon(\log(B))$ where P_ε is a polynomial of degree $n-1$ and leading coefficient $\frac{1}{(n-1)!} + O(\varepsilon)$. To conclude, we note that $C((\mathbf{P}^1_{\mathbf{K}})^n) = \frac{1}{(n-1)!}C(\mathbf{P}^1_{\mathbf{K}})^n$. $\qquad\square$

Remarks 6.20

(i) The proof shows that the number of points with freeness $< \varepsilon$ is not negligible in this case!

(ii) If we consider as in Sect. 4 the points P in $\mathbf{P}^1(\mathbf{K})^n_{h \in \mathscr{D}_B}$ where $\mathscr{D}_B = \mathscr{D}_1 + \log(B)u$, with $u = (u_i)_{1 \leqslant i \leqslant n}$, then

$$l(P) \longrightarrow \frac{n \min_{1 \leqslant i \leqslant n}(u_i)}{\sum_{i=1}^{n} u_i}$$

as B goes to infinity. Thus, in this case, the set

$$\{\, P \in V(\mathbf{K}) \mid h(P) \in \mathscr{D}_B, \, l(P) < \varepsilon \,\}$$

is empty for B big enough.

6.4   Accumulating Subsets and Freeness

We are now going to show that the freeness gives valuable information about points related to accumulating phenomena.

Rational Curves of Low Degree

Conjecturally the accumulating subsets on projective surfaces are rational curves of low degree. More precisely, the number of points on a rational curve L in a nice variety V for a height given by an adelic metric is equivalent to $C(L) B^{2/\langle L, \omega_V^{-1} \rangle}$. Therefore such a curve would be accumulating if $\langle L, \omega_V^{-1} \rangle < 2$ and could be weakly accumulating if $\langle L, \omega_V^{-1} \rangle = 2$ and the rank of the Picard group of the variety is 1. On a surface S, by the adjunction formula,

$$-2 = \deg(\omega_L) = \langle L, L \rangle + \langle L, \omega_S \rangle.$$

If the rank of the Picard group $\mathrm{Pic}(V)$ is one, any effective divisor is ample since S is projective, in that case $\langle L, L \rangle > 0$, hence $\langle L, \omega_S^{-1} \rangle > 2$ which excludes the last case for a surface. The remaining cases are covered by the following proposition.

Proposition 6.21 *Let V be a nice variety on the number field $\mathbf{K}$, and let L be a rational curve in V such that $\langle L, \omega_V^{-1} \rangle < 2$. Then the set*

$$\{ P \in L(\mathbf{K}) \mid l(P) > 0 \}$$

is finite.

Proof Choose a morphism $\varphi : \mathbf{P}_{\mathbf{K}}^1 \to L$ which is birational and an isomorphism $\varphi^*(TS) \xrightarrow{\sim} \bigoplus_{i=1}^n \mathscr{O}_{\mathbf{P}_{\mathbf{K}}^1}(a_i)$ with $a_1 \geqslant a_2 \geqslant \cdots \geqslant a_n$. Then $\mu_i(\varphi) = a_i$ and $\sum_{i=1}^n \mu_i(\varphi) = \langle L, \omega_V^{-1} \rangle < 2$. We have a natural morphism $T\mathbf{P}_{\mathbf{K}}^1 \to \varphi^*(TV)$ which implies that $a_1 \geqslant 2$. Therefore $a_2 < 0$ and we may apply Proposition 6.14. $\qquad\square$

Remarks 6.22

(i) If we consider only the rational points which satisfy the condition $l(P) > \varepsilon(B)$ for some decreasing function ε with values in $\mathbf{R}_{>0}$, then we exclude all points of L outside a finite set.

(ii) In dimension $\geqslant 3$, if $\langle L, \omega_V^{-1} \rangle = 2$, then we get that the freeness $l(P)$ goes to 0 on L. This applies to the projective lines in cubic volumes or complete intersections of two quadrics in $\mathbf{P}^6$.

Fibrations

We remind the reader that, in the counter-example of Batyrev and Tschinkel [4], the accumulating subset is the reunion of fibers of a fibration. We are now going to explain that the freeness also detects such abnormality.

Proposition 6.23 *Let $\varphi : X \to Y$ be a dominant morphism of nice varieties. Then there exists a constant C such that for any $P \in X(\mathbf{K})$ such that the linear map $T_P\varphi$ is onto,*

$$\mu_{\dim(X)}(P) \leqslant \mu_{\dim(Y)}(\varphi(P)) + C.$$

If, moreover, the logarithmic height of P is strictly positive, we get the inequality:

$$l(P) \leqslant \frac{mh(\varphi(P))}{nh(P)} l(\varphi(P)) + \frac{mC}{h(P)}$$

with $m = \dim(X)$ and $n = \dim(Y)$.

Proof The linear map $T_P\varphi$ induces a dual map $T_P\varphi^\vee : T_{\varphi(P)}Y^\vee \to T_P X^\vee$ which is injective. Using $\|\|\cdot\|\|$ to denote the usual operator norm, we get an inequality

$$\mu_1(T_{\varphi(P)}Y^\vee) \leqslant \mu_1(T_P X^\vee) + \max_{1 \leqslant k \leqslant \dim(Y)} \left(\frac{\log\left(\|\|\bigwedge^k T_P\varphi^\vee\|\|\right)}{k} \right)$$

$$\leqslant \mu_1(T_P X^\vee) + C.$$

We conclude with the duality formula for slopes. $\square$

Corollary 6.24 *Let $Q \in Y(\mathbf{K})$ be a non critical value of φ, then $l(P)$ converges to 0 as $h(P)$ goes to $+\infty$ with P in the fibre $X_Q(\mathbf{K})$.*

Remark 6.25 In particular, this detects bad points in the counter-example of Batyrev and Tschinkel. Of course this result applies to $(\mathbf{P}_{\mathbf{K}}^1)^2$ as well. In fact it is the very property which makes freeness efficient to detect bad points in the counter-example of Batyrev and Tschinkel which implies that the proportion of rational points in $(\mathbf{P}_{\mathbf{K}}^1)^2$ with small freeness is not negligible. Section 7 will show how the freeness reveals subvarieties which are locally accumulating even if they are not globally accumulating.

6.5   Combining Freeness and Heights

To conclude this part, let us suggest a formula which takes into account both the freeness and all the heights.

Definition 6.26 Let $\mathcal{D}_1$ be a compact polyhedron in $\mathrm{Pic}(V)_{\mathbf{R}}^{\vee}$ and let $u \in C_{\mathrm{eff}}^{\circ}(V)^{\vee}$. For any $B > 1$ we define $\mathcal{D}_B = \mathcal{D}_1 + \log(B)u$. Let $\varepsilon \in \mathbf{R}_{>0}$ be small enough, relatively to the distance from u to the boundary of $C_{\mathrm{eff}}(V)^{\vee}$. Then we define

$$V(\mathbf{K})_{h \in \mathcal{D}_B}^{l > \varepsilon} = \{\, P \in V(\mathbf{K}) \mid h(P) \in \mathcal{D}_B, l(P) > \varepsilon \,\}.$$

Instead of using a constant ε, we could also consider a slowly decreasing function in B as in [48]. With these notations, we can present our final problematic:

Question 6.27 *We assume that our nice variety V satisfies the conditions of the Hypothesis 3.27. Do we have an equivalence*

$$\sharp V(\mathbf{K})_{h \in \mathcal{D}_B}^{l > \varepsilon} \sim \beta(V)\nu(\mathcal{D}_1)\omega_V(V(A_{\mathbf{K}})^{\mathrm{Br}})B^{\langle \omega_V^{-1}, u \rangle} \tag{12}$$

as B goes to infinity?

Equidistribution 6.28 *We shall say that free points are equidistributed for h if the measure $\delta_{V(\mathbf{K})_{h \in \mathcal{D}_B}^{l > \varepsilon}}$ converges weakly to μ_V^{Br} as B goes to infinity.*

7 Local Accumulation

The rational points on $\mathbf{P}_{\mathbf{K}}^2$ and $(\mathbf{P}_{\mathbf{K}}^1)^2$ are equidistributed in the sense of naïve equidistribution 3.9. But if one looks at Figs. 1 and 2, we see lines, which are all projective lines for the projective plane and the fibres of the two projections for the product of two projective lines. To interpret these lines, we need to go beyond the global distribution.

7.1 Local Distribution

Let us assume that $\mathbf{K} = \mathbf{Q}$ to simplify the discussion. Instead of looking at the proportion of points in a fixed open subset U in the adelic space, we may look at the rational points of bounded height in a open subset U_B depending on B and ask the very broad question

Question 7.1 *For which families $(U_B)_{B>1}$ of open subsets in $V(A_{\mathbf{Q}})$ can we hope to have*

$$\frac{\sharp U_B \cap V(\mathbf{Q})_{h \in \mathcal{D}_B}}{\sharp V(\mathbf{Q})_{h \in \mathcal{D}_B}} \sim \mu_V^{\mathrm{Br}}(U_B)$$

as B goes to infinity?

A particularly interesting case is the distribution around a rational point. Fix $P_0 \in V(\mathbf{Q})$ and choose a local diffeomorphism $\rho : W \to W'$, where W is an open subset in $V(\mathbf{R})$ and W' is an open subset of $T_{P_0} V_{\mathbf{R}}$, which maps P_0 to 0 and such that the differential at P_0 is the identity map. Then we may try to zoom in on the point P_0 with some power of B. More precisely, let us consider the ball

$$\mathscr{B}(0, R) = \{\, y \in T_{P_0} V_{\mathbf{R}} \mid \|y\|_\infty \leqslant R \,\}.$$

We may then introduce the probability measure on $\mathscr{B}(0, R)$ defined by

$$\delta_{R,B}^\alpha = \frac{1}{\sharp(V(\mathbf{Q})_{H \leqslant B} \cap \rho^{-1}(\mathscr{B}(O, RB^{-\alpha})))} \sum_{\substack{P \in V(\mathbf{Q})_{H \leqslant B} \\ \rho(P) \in \mathscr{B}(0, RB^{-\alpha})}} \delta_{B^\alpha \rho(P)}.$$

Remarks 7.2

(i) Let us assume that P_0 belongs to a Zariski open subset of V on which the rational points of bounded height are equidistributed in the sense of 3.31. For $\alpha = 0$, we get the measure induced on $B(0, R)$ by $\rho_*(\mu_\infty)$.

(ii) Under the same hypothesis, if α is small, corresponding to a small zoom, we may expect that the points are evenly distributed: the measure converges to the probability measure induced by the Lebesgue measure.

(iii) If α is big enough, diophantine approximation tells us that there is no rational point that near to the rational point P_0. In other words, for α big enough the above measure is the Dirac measure at P_0.

We are interested in the critical values of α, that is those for which the asymptotic behaviour of the measure $\delta_{R,B}^\alpha$ changes. In particular, we can consider the smallest value of α for which the measure is not the Dirac measure at P_0, which is the biggest of the critical values. This is directly related to the generalisation of the measures of irrationality introduced by D. McKinnon and M. Roth in [39]. In our context, with a height defined by an adelic metric on V, the archimedean metric defines a distance d_∞ on $V(\mathbf{R})$. Then if W is a constructible subset of V containing P_0, we define in this text $\alpha_W(P_0)$ as the infimum of the set of $\alpha \in \mathbf{R}_{>0}$ such that for any $C \in \mathbf{R}$ the set

$$\left\{ Q \in W(\mathbf{Q}) \,\middle|\, d_\infty(Q, P_0) < \frac{C}{H(Q)^\alpha} \right\}$$

is finite. Since ρ is a diffeomorphism, $\alpha_V(P_0)$ corresponds to the biggest critical value.

Remark 7.3 In this text, we take the inverse of the constant defined by D. McKinnon and M. Roth in their paper (*loc. cit.*), since it better expresses the power appearing in the zoom factor.

In [38], D. McKinnon suggests that there should exist rational curves L in V such that $\alpha_V(P_0) = \alpha_L(P_0)$. In other words the best approximations should come from rational curves. On the other hand D. McKinnon and M. Roth [39, theorem 2.16] give the following formula for $\alpha_L(P_0)$: let $\varphi : \mathbf{P}^1_{\mathbf{K}} \to L$ be a normalisation of the curve L

$$\alpha_L(P_0) = \max_{Q \in \varphi^{-1}(P_0)} \frac{r_Q m_Q}{d}$$

where $d = \deg(\varphi^*(\omega_V^{-1}))$, m_Q is the multiplicity of the branch of L through x corresponding to Q and r_Q corresponds to the approximation of Q by rational points in $\mathbf{P}^1_{\mathbf{Q}}$ and is given by Roth theorem [49]:

$$r_Q = \begin{cases} 0 \text{ if } \kappa(Q) \not\subset \mathbf{R}, \\ 1 \text{ if } \kappa(Q) = \mathbf{Q}, \\ 2 \text{ otherwise.} \end{cases}$$

On the other hand, if we take a sequence of rational points $(Q_n)_{n \in \mathbf{N}}$ on $L(\mathbf{Q})$ which converges to P_0 then $(H(Q_n))_{n \in \mathbf{N}}$ goes to $+\infty$ and therefore, by Proposition 6.14, we have that $(l(Q_n))_{n \in \mathbf{N}}$ converges to $l(\varphi)$. In the case where there exists a branch of degree 1 through P_0, if the deformations of the morphism φ are contained in a strict subvariety, this means that all the tangent vectors in $T_{P_0} V$ can not be obtained by a deformation of φ and thus φ can not be very free. Under these assumptions, we get that $l(\varphi) \leqslant 0$ and therefore $(l(Q_n))_{n \in \mathbf{N}}$ converges to 0. Therefore, if the locally accumulating subvarieties are dominantly covered by rational curves, we may expect that the freeness of the points on these locally accumulating subvarieties tends to 0.

In [29, 30, 31], 黄治中 studies the local distribution of points on various toric surfaces, exhibiting phenomena like local accumulating subvarieties, and locally accumulating thin subsets.

8   Another Description of the Slopes

Construction 8.1 For any vector bundle E of rank r on V, we may define the *frame bundle of* E, denoted by $F(E)$, as the GL_r-torsor of the basis in E: for any extension $\mathbf{L}$ of $\mathbf{K}$ and any point $P \in V(\mathbf{L})$, the fibre of $F(E)$ at P is the set of basis of the fibre E_P. For a line bundle L, the frame bundle $F(L)$ is equal to $L^\times$.

Let us now assume that the vector bundle E is equipped with an adelic norm $(\| \cdot \|_w)_{w \in \text{Val}(\mathbf{K})}$. Then for any place w, any point $P \in V(\mathbf{K}_w)$ and any basis $e = (e_1, \ldots, e_r) \in F(E)_P$ we get an element M_w in $\text{GL}_r(\mathbf{K}_w)/K_w$ where

$$K_w = \begin{cases} \text{GL}_r(\mathscr{O}_w) \text{ if } w \text{ is ultrametric,} \\ O_r(\mathbf{R}) \text{ if } w \text{ is real,} \\ U_r(\mathbf{R}) \text{ if } w \text{ is complex.} \end{cases}$$

which is the class of the matrix of the coordinates of $(e_1, \ldots, e_r)$ in a basis of the $\mathscr{O}_w$ lattice (resp. orthonornal basis) defined by $\| \cdot \|_w$ if w is ultrametric (resp. non-archimedean). We get a map

$$F(E)(A_{\mathbf{K}}) \longrightarrow \text{GL}_r(A_{\mathbf{K}})/K,$$

where K is the compact subgroup $\prod_{w \in \text{Val}(\mathbf{K})} K_w$. Taking the quotient by $\text{GL}_r(\mathbf{K})$ for the rational points we get a map

$$V(\mathbf{K}) \longrightarrow \text{GL}_r(\mathbf{K}) \backslash \text{GL}_r(A_{\mathbf{K}})/K.$$

Let us denote by Q_r the biquotient on the right, we get a map

$$\tau_E : V(\mathbf{K}) \longrightarrow Q_r.$$

The determinant composed with product of the norms gives a morphism of groups from the adelic group $\text{GL}_n(A_{\mathbf{K}})$ to $\mathbf{R}_{>0}$ which is invariant under the action of K on the right and the action of $\text{GL}_n(\mathbf{K})$ on the left, this gives a map $|\det| : Q_r \to \mathbf{R}_{>0}$. The composition $|\det| \circ \tau_E$ coincides with the exponential height H_E defined by E with its adelic norm.

Similarly, since the slopes μ_i^E are defined in terms of the $\mathscr{O}_{\mathbf{K}}$-module defined by the norms at the ultrametric places equipped with the non-archimedean norms, we may factorise the slopes through Q_r, and the freeness of a rational point P may also be computed in terms of $\tau_{TV}(P)$.

Remarks 8.2

(i) In Q_r, we may consider the subset Q_r^1 of points P such that $|\det|(P) = 1$. The determinant map then defines a map $Q_r^1 \to \mathbf{K}^* \backslash \mathbf{G}_m(A_{\mathbf{K}})^1/K_{\mathbf{G}_m}$ where $K_{\mathbf{G}_m}$ is the product over the places w of the maximal compact subgroup in $\mathbf{G}_m(\mathbf{K}_w)$. We get a map $c : Q_r^1 \to \text{Pic}(\mathscr{O}_{\mathbf{K}})$; the composition map $c \circ \tau_E$ maps a rational point P onto the class of the projective $\mathscr{O}_{\mathbf{K}}$-module defined by the ultrametric norms in E_P. As an example, for the projective space $\mathbf{P}_{\mathbf{K}}^n$, with $E = TV$, this maps a point $P = [y_0 : \ldots : y_n]$ with integral homogeneous coordinates to $(n + 1)$ times the class of the ideal $(y_0, \ldots, y_n)$.

(ii) For surfaces, as described in Remark 6.8, the slopes, and thus the freeness, measures the deformation of the lattice or the proximity to the cusp in the modular curve $X(1)$. The above construction generalises this description in higher dimension.

(iii) The frame bundle would enable GL_n descent on varieties for which the lifting to versal torsors is not sufficient. In fact we may extend this and consider bundles giving geometric elements in the Brauer group. This may provide a method to generalise the construction of Salberger [50] in the case the geometric Brauer group is not trivial.

9 Conclusion and Perspectives

In these notes we made a quick survey of the various directions to upgrade the principle of Batyrev and Manin to include the cases of Zariski dense accumulating subsets. Let me summarize these options:

1. Remove accumulating thin subsets. This method has been successful in several cases. However, this notion depends on the ground field and we could imagine situations in which there are infinitely many thin subsets to remove, similar to the situation of K3-surfaces containing infinitely many rational lines which are all accumulating.

2. Consider all heights. This method may apply to fibrations and other cases in which the accumulating subsets come from line bundles. However, as shown by examples of Picard rank one, this is not enough to detect accumulating subsets of higher codimension.

3. As in [48], we could use a height defined by an adelic metric and the freeness. But the freeness condition tends to remove too many points as shown by the product of projective lines. A recent example by W. Sawin [51] indicates this is not enough.

4. Combine all heights and freeness. This combination is inspired by the geometric analogue.

This list is far from exhaustive. In fact, we could consider the slopes given by norms on any vector bundle on our variety which gives a profusion of probably redundant invariants. Arakelov geometry is a very natural tool to attack this question of redundancy and look for a minimal set of slopes controlling the distribution of points.

The freeness, which is in part suggested by the analogy with the geometric setting, is very efficient to detect local adelic deformations which correspond to local or global accumulation. However this invariant is particularly difficult to compute efficiently. Indeed its explicit computation is related to the finding of a non-zero vector of minimal length in a lattice which is known to be computationally difficult. At the time of writing, the following question is still open:[2]

---

[2]A result in that direction was obtained by T. Browning and W. Sawin [13]

Question 9.1 *Let V be a smooth hypersurface of degree d in $\mathbf{P}_{\mathbf{Q}}^N$, with $d \geqslant 3$ and $N > (d-1)2^d$. Is the cardinal of points $x \in V(\mathbf{Q})$ with $l(x) < \varepsilon$ and $H(x) < B$ negligible as B goes to infinity?*

In other words, the author is still lacking methods giving lower bounds for the smallest slope, but again we may hope that the techniques of Arakelov geometry may provide the necessary tools.

Acknowledgements The author was supported by the ANR Grant Gardio 14-CE25-0015

References

1. E. Artin, Über eine neue Art von L-Reihen. Abh. Math. Semin. Univ. Hamburg **3**, 89–108 (1924)
2. V.V. Batyrev, The cone of effective divisors of threefolds, in *Proceedings of the International Conference on Algebra, Part 3 (Novosibirsk, 1989)*. Contemporary Mathematics, vol. 131(Part 3) (American Mathematical Society, Providence, 1992), pp. 337–352
3. V.V. Batyrev, Y.I. Manin, Sur le nombre des points rationnels de hauteur bornée des variétés algébriques. Math. Ann. **286**, 27–43 (1990)
4. V.V. Batyrev, Y. Tschinkel, Rational points on some Fano cubic bundles. C. R. Acad. Sci. Paris Sér. I Math. **323**, 41–46 (1996)
5. V.V. Batyrev, Y. Tschinkel, Tamagawa numbers of polarized algebraic varieties, in *Nombre et répartition de points de hauteur bornée, Astérisque*, vol. 251 (SMF, Paris, 1998), pp. 299–340
6. M. Bilu, *Motivic Euler Products and Motivic Height Zeta Functions* (2018). http://arxiv.org/abs/1802.06836
7. B.J. Birch, Forms in many variables. Proc. Roy. Soc. London **265A**, 245–263 (1962)
8. L.A. Borisov, The class of the affine line is a zero divisor in the Grothendieck ring. J. Algebraic Geometry **27**, 203–209 (2018)
9. N. Bourbaki, *Topologie Algébrique, Chapitre 4* (Springer, Berlin, 2015)
10. D. Bourqui, Produit eulérien motivique et courbes rationnelles sur les variétés toriques. Compos. Math. **145**, 1360–1400 (2009)
11. T. Browning, D.R. Heath-Brown, *Density of Rational Points on a Quadric Bundle in $\mathbf{P}_{\mathbf{Q}}^3 \times \mathbf{P}_{\mathbf{Q}}^3$* (2018). http://arxiv.org/abs/1805.10715 .
12. T. Browning, D. Loughran, Varieties with too many rational points. Math. Zeit. **285**, 1249–1267 (2017)
13. T. Browning, W. Sawin, *Free Rational Points on Smooth Hypersurfaces* (2019), pp. 1–23. http://arxiv.org/abs/1906.08463
14. J.-L. Colliot-Thélène et J.-J. Sansuc, Torseurs sous des groupes de type multiplicatif; applications à l'étude des points rationnels de certaines variétés algébriques. C. R. Acad. Sci. Paris Sér. A **282**, 1113–1116 (1976)
15. J.-L. Colliot-Thélène et J.-J. Sansuc, La descente sur les variétés rationnelles, in *Journées de géométrie algébrique d'Angers* ed. by A. Beauville (Sijthoff and Noordhoff, Alphen aan den Rijn, 1979/1980), pp. 223–237
16. J.-L. Colliot-Thélène et J.-J. Sansuc, La descente sur les variétés rationnelles, II. Duke Math. J. **54**, 375–492 (1987)
17. O. Debarre, *Higher Dimensional Algebraic Geometry* (Universitext, Springer, New York, 2001)
18. R. de la Bretèche, Nombre de points de hauteur bornée sur les surfaces de Del Pezzo de degré 5. Duke Math. J. **113**, 421–464 (2002)

19. R. de la Bretèche, T.D. Browning, E. Peyre, On Manin's conjecture for a family of Châtelet surfaces. Ann. Math. **175**, 297–343 (2012)
20. P. Deligne, La conjecture de Weil I.. Publ. Math. I.H.E.S. **43**, 273–307 (1974)
21. J. Denef, F. Loeser, Germs of arcs on singular algebraic varieties and motivic integration. Invent. Math. **135**, 201–232 (1999)
22. K. Destagnol, La conjecture de Manin sur les surfaces de Châtelet. Acta Arith. **174**, 31–97 (2016)
23. J.S. Ellenberg, A. Venkatesh, C. Westerland, Homological stability for Hurwitz spaces and the Cohen-Lenstra conjecture over function fields. Ann. Math. (2) **183**, 729–786 (2016)
24. J. Franke, Y.I. Manin, Y. Tschinkel, Rational points of bounded height on Fano varieties. Invent. Math. **95**, 421–435 (1989)
25. C. Frei, D. Loughran, E. Sofos, Rational points of bounded height on general conic bundle surfaces. Proc. London Math. Soc. **117**, 407–440 (2018). http://arxiv.org/abs/1609.04330
26. É. Gaudron, Pentes des fibrés vectoriels adéliques sur un corps global. Rend. Semin. Mat. Univ. Padova **119**, 21–95 (2008)
27. A. Grothendieck, Technique de descente et théorèmes d'existence en géométrie algébrique. IV. Les schémas de Hilbert. Séminaire Bourbaki 13-ème année (n° 221) (1960/61)
28. R. Hartshorne, Algebraic geometry, in *Graduate Texts in Mathematical*, vol. 52 (Springer, Berlin, 1977)
29. 黄治中(Z. Huáng), Distribution locale des points rationnels de hauteur bornée sur une surface de del Pezzo de degré 6. Int. J. Number Theory **7**, 1895–1930 (2017)
30. 黄治中(Z. Huáng), Approximation diophantienne et distribution locale sur une surface torique II. Bull. Soc. Math. Fr., To appear (2018)
31. 黄治中(Z. Huáng), Approximation diophantienne et distribution locale sur une surface torique, Acta Arith. **189**, 1–94 (2019)
32. M. Kapranov, *The Elliptic Curve in the S-duality Theory and Eisenstein Series for Kac-Moody Groups* (2001). https://arxiv.org/abs/math/0001005
33. R.P. Langlands, in *On the Functional Equations Satisfied by Eisenstein Series*. Lecture Notes in Mathematical, vol. 544 (Springer, Berlin, 1976)
34. B. Lehmann, S. Tanimoto, Y. Tschinkel, Balanced line bundles on Fano varieties. Journ. Reine und Angew. Math. To appear (2016)
35. J. Leray, *Hyperbolic Differential Equations* (The Institute for Advanced Study, Princeton, 1953)
36. C. Le Rudulier, *Points algébriques de hauteur bornée* (Université de Rennes 1, Rennes, 2014), Ph.D. thesis
37. Y.I. Manin, Le groupe de Brauer-Grothendieck en géométrie diophantienne, in *Actes du congrès international des mathématiciens, Tome 1 (Nice, 1970)* (Gauthiers-Villars, Paris, 1971), pp. 401–411
38. D. McKinnon, A conjecture on rational approximations to rational points. J. Algebraic Geom. **16**, 253–303 (2007)
39. D. McKinnon, M. Roth, Seshadri constants, diophantine approximation and Roth's theorem for arbitrary varieties. Invent. Math. **200**, 513–583 (2015)
40. D.G. Northcott, An inequality in the theory of arithmetic on algebraic varieties. Proc. Cambridge Phil. Soc. **45**, 502–509 (1949)
41. D.G. Northcott, Further inequality in the theory of arithmetic on algebraic varieties. Proc. Cambridge Phil. Soc. **45**, 510–518 (1949)
42. T. Ono, On some arithmetic properties of linear algebraic groups. Ann. of Math. (2) **70**, 266–290 (1959)
43. T. Ono, Arithmetic of algebraic tori. Ann. of Math. (2) **74**, 101–139 (1961)
44. T. Ono, On the Tamagawa number of algebraic tori. Ann. Math. (2) **78**, 47–73 (1963)
45. E. Peyre, Hauteurs et mesures de Tamagawa sur les variétés de Fano. Duke Math. J. **79**, 101–218 (1995)
46. E. Peyre, Torseurs universels et méthode du cercle, in *Rational Points on Algebraic Varieties*. Progress in Mathematical, vol. 199 (Birkhaüser, Basel, 2001), pp. 221–274

47. E. Peyre, Obstructions au principe de Hasse et à l'approximation faible, in *Séminaire Bourbaki 56-ème année* (n° 931) (2003/2004),
48. E. Peyre, Liberté et accumulation. Documenta Math. **22**, 1615–1659 (2017)
49. K.F. Roth, Rational approximations to algebraic numbers. Mathematika **2**, 1–20 (1955). corrigendum ibid. 2, 168 (1955)
50. P. Salberger, *Tamagawa Measures on Universal Torsors and Points of Bounded Height on Fano Varieties*. Nombre et répartition de points de hauteur bornée, Astérisque, vol. 251 (SMF, Paris, 1998), pp. 91–258
51. W. Sawin, *Freeness Alone is Insufficient for Manin-Peyre* (2020). http://arxiv.org/abs/2001. 06078
52. S.H. Schanuel, Heights in Number Fields. Bull. Soc. Math. France **107**, 433–449 (1979)
53. J.-P. Serre, Corps locaux, in *Actualités scientifiques et industrielles*, vol. 1296 (Hermann, Paris, 1968)
54. J.-P. Serre, *Lectures on the Mordell-Weil theorem*. Aspects of Mathematics, vol. E15 (Vieweg, Braunschweig, Wiesbaden, 1989)
55. J.-P. Serre, *Topics in Galois Theory*, 2nd edn. Research Notes in Mathematical, vol. 1 (A K Peters, Wellesley, 2007)
56. A. Weil, *Adèles and Algebraic Groups*. Progress in Mathematics, vol. 23 (Birkhaüser, Boston, 1982)

Chapter VI: On the Determinant Method and Geometric Invariant Theory

Per Salberger

1 Introduction

Let $X \subset \mathbf{P}_{\mathbf{Q}}^N$ be a closed subvariety over $\mathbf{Q}$ and Ξ be the scheme-theoretic closure of X in $\mathbf{P}_{\mathbf{Z}}^N$. Let $B_0, \ldots, B_N$ be positive real numbers and let $X(\mathbf{Q}; B_0, \ldots, B_N)$ be the set of rational points on X which may be represented by an integral $(N + 1)$-tuple $(x_0, \ldots, x_N)$ with $|x_m| \leq B_m$ for $0 \leq m \leq N$. This chapter is concerned with the asymptotic behaviour of $\sharp X(\mathbf{Q}; B_0, \ldots, B_N)$. To apply the p-adic determinant method, one first divides $X(\mathbf{Q}; B_0, \ldots, B_N)$ into congruence classes. Let $X(\mathbf{Q}; B_0, \ldots, B_N; P)$ be the set of all points in $X(\mathbf{Q}; B_0, \ldots, B_N)$ which specialize to a fixed $\mathbf{F}_p$-point P on $X_p = \Xi \times_{\mathbf{Z}} \mathbf{F}_p$. The method consists of constructing an auxiliary hypersurface $Y_P \subset \mathbf{P}_{\mathbf{Q}}^N$ of bounded degree such that $X(\mathbf{Q}; B_0, \ldots, B_N; P) \subset Y_P$ but $X \not\subset Y_P$. The name determinant method comes from the fact that one obtains the hypersurface Y_P by showing that certain determinants of values of monomials in $X(\mathbf{Q}; B_0, \ldots, B_N; P)$ vanish.

The first major result of this type is the following result of Heath-Brown, which he stated somewhat differently (see Theorem 14 in [12]).

Theorem 1.1 (Heath-Brown) *Let*

$$X \subset \mathbf{P}_{\mathbf{Q}}^N = \mathrm{Proj}(\mathbf{Q}[x_0, \ldots, x_N])$$

be a hypersurface defined by a homogeneous polynomial $F(x_0, \ldots, x_N)$ of degree d. Let $B_0, \ldots, B_N \in \mathbf{R}_{\geq 1}$, $r = N - 1$, $V = \prod_{i=0}^{N} B_i$ and T be the maximum of all

P. Salberger (✉)
Mathematical Sciences, Chalmers University of Technology, Göteborg, Sweden

University of Gothenburg, Gothenburg, Sweden
e-mail: salberg@chalmers.se

© The Editor(s) (if applicable) and The Author(s), under exclusive license
to Springer Nature Switzerland AG 2021
E. Peyre, G. Rémond (eds.), *Arakelov Geometry and Diophantine Applications*,
Lecture Notes in Mathematics 2276, https://doi.org/10.1007/978-3-030-57559-5_7

$F_i(B_0, \ldots, B_N)$ *for monomials* F_i *which occur in* F *with non-zero coefficient. Let* $\varepsilon > 0$. *Then there exists for any prime* $p > \left(\frac{V}{T^{1/d}}\right)^{\frac{1}{rd^{1/r}}} V^\varepsilon$ *and any non-singular* $\mathbf{F}_p$*-point* P *on* X_p *a hypersurface* $Y_P \subset \mathbf{P}_{\mathbf{Q}}^N$ *of degree bounded only in terms of* d, N *and* ε *such that* $X(\mathbf{Q}; B_0, \ldots, B_N; P) \subset Y_P$ *and* $X \not\subset Y_P$.

Heath-Brown used this result to prove new estimates for rational points of bounded height on curves and surfaces. It is desirable to generalise Theorem 1.1 to subvarieties of higher codimension. If $B_0 = \cdots = B_N = B$, then $\left(\frac{V}{T^{1/d}}\right)^{\frac{1}{rd^{1/r}}} = B^{\frac{r+1}{rd^{1/r}}}$ if X is a hypersurface of dimension r. It is thus natural to guess that for such boxes Theorem 1.1 should hold for $p > B^{\frac{r+1}{rd^{1/r}}+\epsilon}$ also when X is of arbitrary codimension in $\mathbf{P}_{\mathbf{Q}}^N$ and this turns out to be true. But it is not clear what the generalisation of Theorem 1.1 to varieties of higher codimension should be for boxes with unequal sides. This problem was solved by Broberg [2] in his thesis from 2002 by means of Gröbner basis techniques. His result was further refined and generalised by the author [22] and reinterpreted in terms of Arakelov theory by Huayi Chen [4, 5]. But the generalisation of Theorem 1.1 in these papers were still not as intrinsic as one would like since they depended on the choice of a graded monomial ordering.

The aim of this chapter is to remove the use of graded monomials orderings and give a more conceptual generalisation of Theorem 1.1 based on Chow forms and a result of Mumford (see [17, §2.11]) in geometric invariant theory. Heath-Brown's invariant $\left(\frac{V}{T^{1/d}}\right)^{\frac{1}{rd^{1/r}}}$ has then a natural generalisation to $\exp\left(\frac{w_X(\boldsymbol{b})}{rd}\right)$ for the Chow weight $w_X(\boldsymbol{b})$ of $X \subset \mathbf{P}^N$ of X with respect to $\boldsymbol{b} = (\log(B_0), \ldots, \log(B_N))$ (see Definition 2.2 and Theorem 5.3).

Chow forms have not been used before for Diophantine problems in analytic number theory as far as we know. But they have been used extensively to solve problems about Diophantine approximation and in transcendence theory. Evertse and Ferretti gave in [7] another very different proof of a theorem of Faltings and Wüstholz in which they estimated determinants of values of monomials similar to those studied in [2] and [22].

There are also similarities with papers on Kähler-Einstein metrics. Donaldson's test functions give algebraic stability conditions for the existence of such metrics similar to those introduced by Mumford. We were originally inspired by the paper [19] on generalised Futaki invariants when we first developed the theory of Chow and Hilbert weights for estimating the determinants. We were then not aware of [7] but we have now tried to present this material in a way such that the reader can compare the two determinant methods for counting points and for generalising the linear subspace theorem.

2   Chow Forms and Chow Weights

Let K be a field of characteristic 0 and

$$X \subset \mathbf{P}_K^N = \mathrm{Proj}(K[x_0, \ldots, x_N])$$

be a closed integral subscheme of dimension r and degree d. Let $(\mathbf{P}_K^N)^\vee$ denote the dual projective space where a point $\boldsymbol{h} = (h_0, \ldots, h_N)$ in $(\mathbf{P}_K^N)^\vee$ corresponds to the hyperplane $H \subset \mathbf{P}_K^N$ given by the equation $\sum_{i=0}^N h_i x_i = 0$. There is then an incidence correspondence Σ_X in $X \times ((\mathbf{P}_K^N)^\vee)^{r+1}$ where

$$\Sigma_X = \left\{ (x, H_0, \ldots, H_r) \in X \times \left((\mathbf{P}_K^N)^\vee\right)^{r+1} \; \middle| \; x \in \bigcap_{i=1}^r H_i \right\} \qquad (1)$$

so that Σ_X is a $(\mathbf{P}^{N-1})^{r+1}$-bundle over X. Its projection to $((\mathbf{P}_K^N)^\vee)^{r+1}$ is therefore an irreducible hypersurface in $((\mathbf{P}_K^N)^\vee)^{r+1}$ defined by an irreducible multihomogeneous polynomial:

$$F_X(\boldsymbol{h}_0, \ldots, \boldsymbol{h}_r) = F_X(h_{0,0}, \ldots, h_{0,N}; \ldots; h_{r,0}, \ldots, h_{r,N}), \qquad (2)$$

which is unique up to a constant factor. We will refer to F_X as a *Chow form of* X. It is sometimes named after Cayley as in [14, chapter X] or after Van der Waerden. For good accounts on Chow forms see [21] and [24, §9]. It is proved there that F_X is of multidegree $(d. \ldots, d)$.

More generally, we define Chow forms of effective r-cycles $Z = n_1 X_1 + \cdots + n_r X_r$ on $\mathbf{P}_K^N$ by letting $F_Z = \prod_{i=1}^r F_{X_i}^{n_i}$. This polynomial is unique up to a constant factor and of multidegree $(d, \ldots, d)$ for the degree $d = \sum_{i=1}^r n_i \deg(X_i)$ of Z. It is also a Chow form of the r-cycle Z_L on $\mathbf{P}_L^n$ (cf. [8, §1.7]) for any larger field L.

Definition 2.1 Let $X \subset \mathbf{P}_K^N$ be a closed subscheme with Hilbert polynomial $p(x) = \frac{d}{r!} x^r + O(x^{r-1})$ with $d \geq 1$ and $X_1, \ldots X_l$ be the irreducible r-dimensional components of X. Let m_i be the multiplicity of X along X_i for $i \in \{1, \ldots, l\}$ and $Z = \sum_{i=1}^l m_i X_i$ be the r-cycle associated to X. Then the *Chow divisor of* X is the divisor of multidegree $(d, \ldots, d)$ in $\left((\mathbf{P}_K^N)^\vee\right)^{r+1}$ defined by the Chow form of Z.

Definition 2.2 Let $X \subset \mathbf{P}_K^N$ be a closed subscheme with Hilbert polynomial of degree r and Z be the effective r-cycle on $\mathbf{P}_K^N$ associated to X. Let $e = (e_0, \ldots, e_N) \in \mathbf{R}^{N+1}$, t be an auxiliary variable and F_Z be the Chow form of Z. Write

$$F_Z(t^{e_0} h_{0,0}, \ldots, t^{e_N} h_{0,N}; \ldots; t^{e_0} h_{r,0}, \ldots, t^{e_N} h_{r,N}) = \sum_{i=0}^s t^{\varepsilon_i} F_i \qquad (3)$$

with $F_0, \ldots, F_s \in K[h_{0,0}, \ldots, h_{r,N}]$ and $\varepsilon_0 > \cdots > \varepsilon_s$. Then $w_X(e) = \varepsilon_s$ is called the *Chow weight of X with respect to e* and F_s *the final term of F_Z* with respect to *e*.

This function occurs naturally, when one applies the determinant method to count rational points on varieties. A similar function was used by Evertse and Ferretti [7, §3.3] in their proof of a theorem of Faltings and Wüstholz on Diophantine approximation. But they define their Chow weight $w_X^{EF}(e)$ to be the *highest* exponent ε_0 of t in the expansion in (3). We have thus the relation $w_X(e) = -w_X^{EF}(-e)$ between these two weights.

The weight of a cycle will not change if we extend the base field to a larger field L as F_Z is also a Chow form for the r-cycle Z_L on $\mathbf{P}_K^N$ defined in [8, §1.7]. It is therefore sufficient to consider the case where K is algebraically closed in most proofs of results on Chow weights. It is also clear from the definition that

$$w_X(e) = m_1 w_{Y_1}(e) + \cdots + m_l w_{Y_l}(e) \tag{4}$$

for the r-cycle $Z = m_1 Y_1 + \cdots + m_l Y_l$ associated to X.

The following remark will be used to reduce the study of Chow weights to the case where e is integral.

Remarks 2.3

(i) The function $w_X : \mathbf{R}^{N+1} \to \mathbf{R}$ is continuous and homogeneous with respect to multiplication with positive real numbers in the sense that $w_X(\lambda e) = \lambda w_X(e)$ for $\lambda > 0$.

(ii) One has

$$w_X(e) \leq \max(|e_0|, \ldots, |e_N|) w_X((1, \ldots, 1))$$

$$= \max(|e_0|, \ldots, |e_N|)(r + 1)d$$

for any $e \in \mathbf{R}^{N+1}$.

To any $(N + 1)$-tuple $e \in \mathbf{Z}^{N+1}$, we associate the 1-parameter subgroup $\mu : \mathbf{G}_m \to \mathrm{GL}_{N+1}$ which sends t to the diagonal matrix $(t^{-e_0}, \ldots, t^{-e_N})$. The GL_{N+1}-action on $\mathbf{P}^N$ will then restrict to a $\mathbf{G}_m$-action on $\mathbf{P}^N$, which in its turn will induce a $\mathbf{G}_m$-action on Hilbert scheme of $\mathbf{P}^N$. This $\mathbf{G}_m$-action will play a central role in the study of Chow weights with respect to $e \in \mathbf{Z}^{N+1}$.

We now fix a closed subscheme $X \subset \mathbf{P}_K^N$ and let $\mathbf{H}_{p(x)}$ be the Hilbert K-scheme of all closed subschemes of $\mathbf{P}_K^N$ with the same Hilbert polynomial $p(x) = \frac{d}{r!}x^r + O(x^{r-1})$ as X. We then have for each $e \in \mathbf{Z}^{N+1}$ an action on $\mathbf{H}_{p(x)}$ by μ and a $\mathbf{G}_m$-orbit of the K-point h on $\mathbf{H}_{p(x)}$ representing $X \subset \mathbf{P}_K^N$. This will represent a family $\mathscr{F}_e \subset \mathbf{G}_m \times \mathbf{P}^N$ of closed subschemes of $\mathbf{P}^N$, which if $X \subset \mathbf{P}^N$ is defined by the homogeneous polynomials $F_i(x_0, \ldots, x_N)$ for $1 \le i \le s$, will be the closed subscheme of $\mathbf{G}_m \times \mathbf{P}^N$ defined by

$$F_1(t^{-e_0}x_0, \ldots, t^{-e_N}x_N) = \cdots = F_s(t^{-e_0}x_0, \ldots, t^{-e_N}x_N) = 0.$$

The projection from $\mathscr{F}_e$ to $\mathbf{G}_m = \mathbf{A}^1 \backslash \{0\}$ is a proper flat morphism which extends to a proper morphism $q_e : \mathscr{X}_e \to \mathbf{A}^1$ from the scheme-theoretic closure $\mathscr{X}_e$ of $\mathscr{F}_e$ in $\mathbf{A}^1 \times \mathbf{P}^N$. This extension is also flat (see [11, prop. III.9.8]) and we shall denote the fibre of q_e where $t = 0$ by X_e. It is a subscheme of $\mathbf{P}^N$ with the same Hilbert polynomial as $X \subset \mathbf{P}^N_K$. The main goal of this section is to show that X and X_e have the same Chow weights with respect to $e \in \mathbf{Z}^{N+1}$. This will be useful as it is normally easier to determine the Chow weights of X_e than those of X. To this end, let $\mathrm{Div}^{d,\ldots,d}\left(((\mathbf{P}^N_K)^\vee)^{r+1}\right)$ be the projective space of all effective divisors (or hypersurfaces) of multidegree $(d, \ldots, d)$ in $((\mathbf{P}^N_K)^\vee)^{r+1}$ and

$$\phi : \mathbf{H}_{p(x)} \to \mathrm{Div}^{d,\ldots,d}\left(((\mathbf{P}^N_K)^\vee)^{r+1}\right)$$

be the canonical morphism defined by Mumford in [18, §5.4]. This map sends h to the point in $\mathrm{Div}^{d,\ldots,d}\left(((\mathbf{P}^N_K)^\vee)^{r+1}\right)$ representing the Chow divisor $D \subset ((\mathbf{P}^N_K)^\vee)^{r+1}$ of $X \subset \mathbf{P}^N_K$ (cf. Definition 2.1). There is further a GL_{N+1}-action on $\mathrm{Div}^{d,\ldots,d}\left(((\mathbf{P}^N_K)^\vee)^{r+1}\right)$ which makes ϕ equivariant. This action is induced by the GL_{N+1}-action on $(\mathbf{P}^N_K)^\vee$ dual to the evident GL_{N+1}-action on $\mathbf{P}^N_K$. If we restrict to the 1-parameter subgroup μ, then we obtain a $\mathbf{G}_m$-action on $\mathrm{Div}^{d,\ldots,d}\left(((\mathbf{P}^N_K)^\vee)^{r+1}\right)$ induced by the $\mathbf{G}_m$-action on $(\mathbf{P}^N_K)^\vee$, which sends $(t; (h_0, \ldots, h_N))$ to $(t^{e_0}h_0, \ldots, t^{e_N}h_N)$. The $\mathbf{G}_m$-orbit of $\phi(h)$ will therefore represent the family $\mathscr{G}_e \subset \mathbf{G}_m \times ((\mathbf{P}^N_K)^\vee)^{r+1}$ of divisors of degree $(d, \ldots, d)$ in $((\mathbf{P}^N_K)^\vee)^{r+1}$ defined by the equation

$$F_Z(t^{e_0}h_{0,0}, \ldots, t^{e_N}h_{0,N}; \ldots; t^{e_0}h_{r,0}, \ldots, t^{e_N}h_{r,N}) = 0 \qquad (5)$$

for a Chow form F_Z of the r-cycle Z associated to $X \subset \mathbf{P}^N_K$. The projection from $\mathscr{G}_e$ to $\mathbf{G}_m = \mathbf{A}^1 \backslash \{0\}$ is a proper flat morphism which extends to a proper morphism $r_e : \mathscr{D}_e \to \mathbf{A}^1$ from the scheme-theoretic closure $\mathscr{D}_e$ of $\mathscr{G}_e$ in $\mathbf{A}^1 \times ((\mathbf{P}^N_K)^\vee)^{r+1}$. By Hartshorne [11, prop. III.9.8], the morphism r_e is flat as its restriction to $\mathscr{G}_e$ is flat over $\mathbf{G}_m$. This is the family of effective divisors of degree $(d, \ldots, d)$ in $((\mathbf{P}^N_K)^\vee)^{r+1}$ represented by the $\mathbf{G}_m$-equivariant morphism

$$\phi \circ b : \mathbf{A}^1 \longrightarrow \mathrm{Div}^{d,\ldots,d}\left(((\mathbf{P}^N_K)^\vee)^{r+1}\right)$$

for the morphism $b : \mathbf{A}^1 \to \mathbf{H}_{p(x)}$ representing $q_e : \mathscr{X}_e \to \mathbf{A}^1$.

Proposition 2.4 *Let* $X \subset \mathbf{P}_K^N$ *be a closed subscheme with Hilbert polynomial of degree r and* F_Z *be a Chow form of the r-cycle Z associated to X. Let* $e = (e_0, \ldots, e_N)$ *be an* $(N + 1)$-*tuple of integers and* $X_e \subset \mathbf{P}_K^N$ *be the closed fibre of* $q_e : \mathscr{X}_e \to \mathbf{A}^1$ *over the point where* $t = 0$. *Then the final term of* F_Z *is a Chow form of the r-cycle associated to* X_e.

Proof Let

$$F_Z(t^{e_0}h_{0,0}, \ldots, t^{e_N}h_{0,N}; \ldots; t^{e_0}h_{r,0}, \ldots, t^{e_N}h_{r,N}) = t^{\epsilon_0}F_0 + \cdots + t^{\epsilon_s}F_s$$

with $\epsilon_0 > \cdots > \epsilon_s$. Then $\mathscr{D}_e$ is the closed subscheme of $\mathbf{A}^1 \times ((\mathbf{P}^N)^\vee)^{r+1}$ defined by

$$t^{\epsilon_0 - \epsilon_s}F_0 + t^{\epsilon_1 - \epsilon_s}F_1 + \cdots + F_s = 0.$$

The closed fibre of $r_e : \mathscr{D}_e \to \mathbf{A}^1$ with $t = 0$ is thus the hypersurface of $((\mathbf{P}^N)^\vee)^{r+1}$ defined by the final term of F_Z. But it is also the divisor in $((\mathbf{P}_K^N)^\vee)^{r+1}$ parameterized by $\phi(b(0))$ thereby proving the assertion. $\qquad\square$

Corollary 2.5 *Let* $X \subset \mathbf{P}_K^N$, $e = (e_0, \ldots, e_N) \in \mathbf{Z}^{N+1}$ *and* $X_e \subset \mathbf{P}_K^N$ *be as in the last proposition. Then X and* X_e *have the same Chow weights with respect to* e.

Proof It follows from the expansion

$$F_Z(t^{e_0}h_{0,0}, \ldots, t^{e_N}h_{0,N}; \ldots; t^{e_0}h_{r,0}, \ldots, t^{e_N}h_{r,N}) = t^{\epsilon_0}F_0 + \cdots + t^{\epsilon_s}F_s$$

with $\epsilon_0 > \cdots > \epsilon_s$ that

$$F_s(t^{e_0}h_{0,0}, \ldots, t^{e_N}h_{0,N}; \ldots; t^{e_0}h_{r,0}, \ldots, t^{e_N}h_{r,N}) = t^{\epsilon_s}F_s.$$

The assertion is therefore a direct consequence of the proposition and of the definition of Chow weights in 2.2. $\qquad\square$

To compute the Chow weight of X_e, we now give a more concrete interpretation of $X_e \subset \mathbf{P}_K^N$.

Definition 2.6 Let $H(x_0, \ldots, x_N)$ be a homogeneous polynomial in $K[x_0, \ldots, x_N]$ which is not the zero polynomial, $e = (e_0, \ldots, e_N)$ be a real $(N + 1)$-tuple and t be an auxiliary variable. Write

$$H(t^{-e_0}x_0, \ldots, t^{-e_N}x_N) = t^{\epsilon_0}H_0 + \cdots + t^{\epsilon_s}H_s \tag{6}$$

with $H_0, \ldots, H_s \in K[x_0, \ldots, x_N]$ and $\epsilon_0 > \cdots > \epsilon_s$. Then we call H_s the *e-final term of H* and $-\epsilon_s$ *the weight of H with respect to* e. We denote this weight by $w(e, H)$.

Lemma 2.7 *Let* $X \subset \mathbf{P}_K^N$, $e = (e_0, \ldots, e_N) \in \mathbf{Z}^{N+1}$ *and* $X_e \subset \mathbf{P}_K^N$ *be as in Proposition 2.4. Let* $I \subset K[x_0, \ldots, x_N]$ *be the saturated homogeneous ideal corresponding to* $X \subset \mathbf{P}_K^N$. *Then the* e-*final terms of the homogeneous polynomials in* I *generate a homogeneous ideal* $I_e \subset K[x_0, \ldots, x_N]$, *which defines* $X_e \subset \mathbf{P}_K^N$.

Proof Set $R = K[t, t^{-1}]$. Then $\mathscr{F}_e$ is the closed subscheme of $\mathbf{P}_{K[t,t^{-1}]}^N$ defined by all polynomials $H(t^{-e_0} x_0, \ldots, t^{-e_N} x_N)$ in $R[x_0, \ldots, x_N]$ derived from homogeneous polynomials H in I. But then $\mathscr{X}_e$ must be the closed subscheme of $\mathbf{P}_{K[t]}^N$ defined by all $t^{w(e;H)} H(t^{-e_0} x_0, \ldots, t^{-e_N} x_N)$ in $K[t][x_0, \ldots, x_N]$ for all such H. This proves the assertion as $t^{w(e;H)} H(t^{-e_0} x_0, \ldots, t^{-e_N} x_N)$ reduces to the e-final term of H modulo the ideal in $K[t][x_0, \ldots, x_N]$ generated by t. $\qquad \square$

This lemma will be particularly useful in the proof of Theorem 3.8 where I is generated by homogeneous polynomials of degree at most D and $e = (e_0, \ldots, e_N) \in \mathbf{Z}^{N+1}$ is chosen to be outside all hyperplanes defined by the equations $k_0 x_0 + \ldots k_N x_N = 0$ with integer coefficients in $[-D, D]$. Then $X_e \subset \mathbf{P}_K^N$ will be defined by monomials in $(x_0, \ldots, x_N)$. We shall use the following notation.

Notation 2.8 Let $J = \{j_1, \ldots, j_{N-r}\} \subset \{0, \ldots, N\}$ be a subset of $N - r$ integers. Then

(a) Π_J is the coordinate r-plane defined by the equations $x_j = 0$ for all $j \in J$.
(b) If F is a monomial in $(x_0, \ldots, x_N)$ then F_J is the monomial in $x_{j_1}, \ldots, x_{j_{N-r}}$ obtained by putting $x_i = 1$ in F for all $i \in \{i_0, \ldots, i_r\} = \{0, \ldots, N\} \backslash J$.
(c) If $I \subset K[x_0, \ldots, x_N]$ is an ideal generated by monomials $F_1, \ldots, F_s$, then we will write I_J for the ideal generated by the monomials $(F_1)_J, \ldots, (F_s)_J$.

The following results will be used to determine the Chow form and Chow weight of a monomial scheme.

Proposition 2.9 *Let* $I \subset K[x_0, \ldots, x_N]$ *be an ideal generated by monomials in* $(x_0, \ldots, x_N)$ *with Hilbert polynomial of degree* r *and* $X \subset \mathbf{P}_K^N$ *be the closed subscheme defined by* I. *Then the following holds:*

(a) *The* r-*dimensional components are all coordinate* r-*planes and the multiplicity* m_J *of* X *along such an* r-*plane* Π_J *is the same as the multiplicity of the scheme* X_J *defined by* I_J *along* Π_J.
(b) *The intersection* $\mathbf{I}$ *of the ideals* I_J *in (a) defines a scheme* $\mathbf{X} \subset \mathbf{P}_K^N$ *of dimension* r *such that the* r-*cycles associated to* X *and* $\mathbf{X}$ *coincide. In particular,* $w_X(e) = w_{\mathbf{X}}(e)$ *for any* $e \in \mathbf{R}^{N+1}$.

Proof If I is generated by the monomials $F_1, \ldots, F_s$ and Y is an irreducible component of X, then there are linear factors $L_1, \ldots, L_s$ of $F_1, \ldots, F_s$ which vanish on Y. The components of X must thus be coordinate planes on X. To prove the other assertions, we use that $I \subset \mathbf{I} \subset I_J$ and $I_{P(J)} = (I_J)_{P(J)}$ in the ring $K[x_0, \ldots, x_N]_{P(J)}$ for the ideals $P(J) = (x_{j_1}, \ldots, x_{j_{N-r}})$ of $K[x_0, \ldots, x_N]$ defining the r-planes Π_J on X. $\qquad \square$

For a more combinatorial interpretation of the multiplicities see [26, Prop. 3.4]. The following result about Chow forms is well-known (see [7, (2.7)]).

Proposition 2.10 *Let* $J = \{j_1, \ldots, j_{N-r}\} \subset \{0, \ldots, N\}$ *be a subset of* $N - r$ *integers and* $\Pi_J \subset \mathbf{P}_K^N$ *be the coordinate* r-*plane defined in Notation 2.8. Let* $\{i_0, \ldots, i_r\} = \{0, \ldots, N\} \backslash J$ *and* $\boldsymbol{h}$ *be the* $(r + 1) \times (r + 1)$-*matrix where the* l-*th row* $\boldsymbol{h}_l$ *is given by* $(h_{l,i_0}, \ldots, h_{l,i_r})$. *Then the determinant of* $\boldsymbol{h}$ *is a Chow form* $F_J(\boldsymbol{h}_0; \ldots; \boldsymbol{h}_r)$ *of* $\Pi_J \subset \mathbf{P}_K^N$.

Corollary 2.11 *The Chow weight of* Π_J *with respect to* $\boldsymbol{e} \in \mathbf{R}^{N+1}$ *is* $e_{i_0} + \cdots + e_{i_r}$.

We end this section by determining the Chow weights for hypersurfaces. We will thereby write $x^{\boldsymbol{m}} = x_0^{m_0} \ldots x_N^{m_N}$ for $(N + 1)$-tuples $\boldsymbol{m} = (m_0, \ldots, m_N)$ of non-negative integers.

Proposition 2.12 *Let* $X \subset \mathbf{P}_K^N$ *be the hypersurface defined by the homogeneous polynomial*

$$F(x_0, \ldots, x_N) = \sum_{\boldsymbol{m} \in M} a_{\boldsymbol{m}} x^{\boldsymbol{m}} \in K[x_0, \ldots, x_N]$$

of degree d *with* $a_{\boldsymbol{m}} \neq 0$ *for all* $\boldsymbol{m} \in M \subset \mathbf{Z}_{\geq 0}^{N+1}$. *Let* $\boldsymbol{e} \in \mathbf{R}^{N+1}$. *Then,*

$$w_X(\boldsymbol{e}) = \min_{\boldsymbol{m} \in M} \left(\sum_{i=0}^N m_i \sum_{\substack{0 \leq k \leq N \\ k \neq i}} e_k \right) = d \left(\sum_{k=0}^N e_k \right) - \max_{\boldsymbol{m} \in M} \left(\sum_{i=0}^N m_i e_i \right).$$

Proof This follows from the well known formula for Chow forms of hypersurfaces just as in the proof of the formula for $w_X^{\mathrm{EF}}(\boldsymbol{e}) = -w_X(-\boldsymbol{e})$ in [7, 3.22]. $\qquad \square$

3 Hilbert Polynomials and Hilbert Weights

In Definition 2.6, we have defined the weight $w(\boldsymbol{e}, H)$ of a homogeneous polynomial $H \in K[x_0, \ldots, x_N]$ with respect to $\boldsymbol{e}$. For $x^{\boldsymbol{\alpha}} = \prod_{j=0}^N x_j^{\alpha_j}$ with $\boldsymbol{\alpha} = (\alpha_0, \ldots, \alpha_N) \in \mathbf{Z}_{\geq 0}^{N+1}$ we have in particular

$$w(\boldsymbol{e}, x^{\boldsymbol{\alpha}}) = \alpha_0 e_0 + \cdots + \alpha_N e_N. \tag{7}$$

We shall more generally for a finite set T of such monomials $x^{\boldsymbol{\alpha}}$ write $w(\boldsymbol{e}, T)$ for the sum of all $w(\boldsymbol{e}, x^{\boldsymbol{\alpha}})$ with $x^{\boldsymbol{\alpha}} \in T$.

We now define Hilbert weights of subschemes of $\mathbf{P}_K^N$.

Definition 3.1 Let K be a field and $S = \bigoplus_{k \geq 0} S_k$ be the graded quotient ring $K[x_0, \ldots, x_N]/I$ of a homogeneous ideal I in $K[x_0, \ldots, x_N]$. Then the k-th Hilbert weight of I with respect to $e = (e_0, \ldots, e_N) \in \mathbf{R}^{N+1}$ is given by

$$w_I(e; k) = \min w(e; T)$$

where T runs over all sets of monomials x^{α} of degree k which give a basis of S_k. If X is a closed subscheme of $\mathbf{P}_K^N = \operatorname{Proj}(K[x_0, \ldots, x_N])$, then we let $w_X(e; k) = w_I(e; k)$ for the saturated homogeneous ideal I of $K[x_0, \ldots, x_N]$ corresponding to X (cf. p. 125 in [11]).

Remarks 3.2

(i) We have just as for Chow weights that $w_I(e, k)$ is continuous with respect to $e \in \mathbf{R}^{N+1}$ and that $w_I(\lambda e; k) = \lambda w_I(e; k)$ for $\lambda > 0$.

(ii) $w_I((1, \ldots, 1); k) = kh_I(k)$ for the Hilbert function $h_I(k) = \dim(S_k)$ of I.

(iii) One has

$$|w_I(e; k)| \leq w_I((|e_0|, \ldots, |e_N|); k) \leq \max(|e_0|, \ldots, |e_N|)kh_I(k).$$

Example 3.3 If $I \subset K[x_0, \ldots, x_N]$ is generated by monomials $x^{\alpha_1}, \ldots, x^{\alpha_r}$ where $\alpha_1, \ldots, \alpha_r \in \mathbf{Z}_{\geq 0}^{N+1}$, then there is only one monomial basis for each S_k. It is given by the monomials x^{α} of degree k where $\alpha - \alpha_i \notin \mathbf{Z}_{\geq 0}^{N+1}$ for $i \in \{1, \ldots, r\}$. Therefore, $w_I(e; k)$ is **R**-linear with respect to e for monomial ideals I. If $\bar{I}$ is another monomial ideal containing I, then $w_I(e; k) - w_{\bar{I}}(e; k) = w(e; T)$ for the set T of monomials of degree k in $\bar{I} \setminus I$. In particular,

$$|w_I(e; k) - w_{\bar{I}}(e; k)| \leq \max(|e_0|, \ldots, |e_N|)k(h_I(k) - h_{\bar{I}}(k)). \tag{8}$$

The following lemma is a slight reformulation of lemma 1 in [2].

Lemma 3.4 *Let $I \subset K[x_0, \ldots, x_N]$ be a monomial ideal $(x^{\alpha_1}, \ldots, x^{\alpha_r})$ generated by monomials of degree at most D and P_I be the Hilbert polynomial of I. Let $e = (e_0, \ldots, e_N) \in \mathbf{R}^{N+1}$. Then the following holds.*

(a) *There exists a positive integer k_0 depending solely on D and N such that $w_I(e, k)$ is given by a polynomial function $Q_{I,e}(k)$ for $k \geq k_0$.*

(b) *If $e = (1, \ldots, 1)$, then $Q_{I,e}(k) = kP_I(k)$.*

(c) *The polynomial $Q_{I,e}$ is of degree at most $1 + \deg(P_I)$ and there exists a positive constant C depending solely on D and N such that the absolute values of the coefficients of $Q_{I,e}$ are bounded by $C \max(|e_0|, \ldots, |e_N|)$.*

Proof It suffices by the **R**-linearity of $w_I(e; k)$ to handle the case where e has a single non-zero entry e_i with $e_i = 1$. Then $w_I(e; k)$ is the function $\sigma_i(k)$ studied by Broberg in (*loc. cit.*). He gives an explicit formula for $\sigma_i(k)$ from which (a) follows. To prove (b), use Remark 3.2 (ii) and to prove (c), use Remark 3.2 (iii) and the fact that there are only finitely many ideals in $K[x_0, \ldots, x_N]$ generated by monomials of degree at most D. $\square$

We now relate the leading coefficient of $Q_{I,e}$ to the Chow weight of I with respect to e.

Lemma 3.5 *Let $I \subset K[x_0, \ldots, x_N]$ be an ideal with Hilbert polynomial of degree r, which is generated by monomials of degree at most D. Let $X \subset \mathbf{P}^N_K$ be the subscheme defined by I and $e = (e_0, \ldots, e_N) \in \mathbf{R}^{N+1}$. Then the polynomial $Q_{I,e}$ is of the form*

$$Q_{I,e}(k) = w_X(e) \frac{k^{r+1}}{(r+1)!} + O(\max(|e_0|, \ldots, |e_N|)k^r),$$

where the implicit constant only depends on D and N.

Proof Let I, I_J and X be as in Proposition 2.9. Then X and X are r-dimensional subschemes of the same degree such that $P_I(k) - P_{\overline{I}}(k) = O(k^{r-1})$ (cf.[11, §I.7]). We also have by Lemma 3.4 that the implicit constant depends only on D and N and hence that

$$|Q_{I,e}(k) - Q_{\overline{I},e}(k)| = O_{D,N}(\max(|e_0|, \ldots, |e_N|)k^r)$$

by (8). This means that is enough to prove the lemma for $\overline{I}$ as $w_X(e) = w_{\overline{X}}(e)$ (see Proposition 2.9). By Example 3.3, and the inclusion-exclusion principle, we obtain the following identity:

$$w_{I,e}(k) = \sum w_{I_J,e}(k) - \sum w_{\langle I_J, I_{J'} \rangle, e}(k) + \sum w_{\langle I_J, I_{J'}, I_{J''} \rangle, e}(k) \ldots \qquad (9)$$

where in the first sum J runs over all $(N-r)$-subsets J of $\{0, \ldots, N\}$ with $\Pi_J \subset X$ and where in the second sum $\{J, J'\}$ runs over all pairs of such $(N-r)$-subsets. The ideals $\langle I_J, I_{J'} \rangle$ in the second sum and the ideals in the following sums will all define schemes of dimension less than r. The total contribution from these sums is thus by Lemma 3.4 (a) and (c) of order $O(\max(|e_0|, \ldots, |e_N|)k^r)$ with an implicit constant depending only on D and N. Since $w_X(e) = \sum w_{X_J}(e)$ by (4), it therefore suffices to prove the lemma for the ideals I_J defined in Notation 2.8.

If, say, $J = \{r+1, \ldots, N\}$, then $w_{X_J}(e) = m_J(e_0 + \cdots + e_r)$ by (4) and Corollary 2.11. Let $\mathbf{e}_i \in \mathbf{Z}^{N+1}$ be the $(N+1)$-tuple with a single non-zero entry $e_i = 1$. Then $w_{I_J, \mathbf{e}_i}(k)$ will not depend on i for $i \notin J$ as I_J is generated by monomials in $(x_{r+1}, \ldots, x_N)$. As $w_{I_J,e}(k)$ is linear with respect to e, we have thus by Remark 3.2 (ii) the following identity for any $e = (e_0, \ldots, e_N) \in \mathbf{R}^{N+1}$.

$$w_{I_J}(k) = \frac{1}{r+1}(e_0 + \cdots + e_r)kh_{I_J}(k)$$

$$+ \sum_{i=r+1}^{N} (e_i - \frac{1}{r+1}(e_0 + \cdots + e_r))w_{I_J, \mathbf{e}_i}(k).$$

But one of the monomials $(F_1)_J, \ldots, (F_r)_J$ which generate I_J must be a power of x_j for each $j \in J$, since otherwise X_J would have a coordinate s-plane with $s > r$ as component. The x_j-degree of a monomial outside I_J is therefore at most $D - 1$ and $w_{I_J,e_i}(k) \le h_{I_J}(k)(D - 1) = O_{D,N}(k^r)$ for $i \in \{r + 1, \ldots, N\}$. Hence

$$Q_{I_J,e}(k) = \frac{1}{r+1}(e_0 + \cdots + e_r)k P_{I_J}(k) + O_{D,N}(\max(|e_0|, \ldots, |e_N|)k^r)$$

with implicit constant depending only on D and N. But $\deg X_J = m_J \deg(\Pi_j) = m_J$. Therefore, $P_{I_J}(k) = m_J k^r / r! + O_{D,N}(k^{r-1})$ and we are done. $\square$

Lemma 3.6 *Let $k \in \mathbf{N}$ and $I \subset K[x_0, \ldots, x_N]$ be a homogeneous ideal generated by homogeneous polynomials $F_1, \ldots, F_r$ and $I_e \subset K[x_0, \ldots, x_N]$ be the ideal generated by the e-final terms of $F_1, \ldots, F_r$ for some $e \in \mathbf{R}^{N+1}$. Then the following holds*

(a) *If T is a monomial basis of $K[x_0, \ldots, x_N]_k / I_k$ such that $w(e, T)$ is minimal then T is also a basis of $K[x_0, \ldots, x_N]_k / I_{e,k}$;*

(b) *If $T = \{x^{m_1}, \ldots, x^{m_l}\}$ is a monomial basis of $K[x_0, \ldots, x_N]_k / I_{e,k}$, then it is also a basis of $K[x_0, \ldots, x_N]_k / I_k$.*

Proof This follows from the arguments in the proofs of lemma 6.2 and lemma 6.3 (iii) in [7] apart from obvious modifications. $\square$

We shall in fact only use Lemma 3.6 when the final terms of $F_1, \ldots, F_r$ are monomials and the result is then well known from the theory of Gröbner basis's (cf. [2, §2]).

The following result is well-known (cf.[10, §1.11]) and an important step in the construction of Hilbert schemes.

Lemma 3.7 *Let $I \subset K[x_0, \ldots, x_N]$ be a saturated homogeneous ideal with Hilbert polynomial $P_I(k) = dk^r/r! + O(k^{r-1})$ with $d \ge 1$. Then I is generated by homogeneous polynomials of degree at most D for some D bounded in terms of N and $P_I(t)$.*

Proof See section 4.3 in [25] for a proof. One can also recover the result from the existence of Hilbert schemes (see [22, §1.3]). $\square$

Theorem 3.8 *Let $X \subset \mathbf{P}_K^N$ be a closed subscheme with Hilbert polynomial $P_X(k) = dk^r/r! + O(k^{r-1})$ with $d \ge 1$ and let $e = (e_0, \ldots, e_N) \in \mathbf{R}^{N+1}$. Then the following holds.*

(a) *There exists an integer k_0 depending only on N and $P_X(t)$ such that $w_X(e; k)$ is given by the value of a polynomial $Q_{X,e}(t) \in \mathbf{R}[t]$ of degree at most $r + 1$ for each $k \ge k_0$.*

(b) *$Q_{X,e}(k) = w_X(e)k^{r+1}/(r+1)! + O(\max(|e_0|, \ldots, |e_N|)k^r)$ where the implicit constant only depends on D and N.*

Proof

(a) Let Δ be the forward difference operator which sends the function $f : \mathbf{Z} \to \mathbf{R}$ to the function $\Delta f : \mathbf{Z} \to \mathbf{R}$ where $\Delta f(k) = f(k+1) - f(k)$. The assertion is then equivalent to the assertion that $\Delta^{r+2} w_X(e,k) = 0$ for $k \geq k_0$. It suffices by a continuity argument to prove this for the dense subset Ω of all $e \in \mathbf{Q}^{N+1}$ such that $m_0 e_0 + \cdots + m_N e_N \neq 0$ for all $(m_0, \ldots, m_N) \in (\mathbf{Z} \cap [-D, D])^{N+1}$ different from $(0, \ldots, 0)$. By Remark 3.2 (i), we may also assume that $e \in \mathbf{Z}^{N+1}$ by replacing e by some multiple of e. Now let D be as in Lemma 3.7. Then the saturated ideal I defining X is generated by homogeneous polynomials of degree at most D and their e-final terms will, for $e \in \Omega$, be monomials generating I_e (see Definition 2.6). We thus have by Lemma 3.6 and Definition 3.1 that $w_I(e; k) = w_{I_e}(e; k)$. It is therefore enough to show a) for $X_e \subset \mathbf{P}_K^N$. But the result then follows from Lemma 3.4 (a) and (c).

(b) It is again enough to prove the assertion for $e \in \Omega$ by a continuity argument and we may further assume that $e \in \mathbf{Z}^{N+1}$ by Remarks 2.3 and 3.2 (i). We then have by Lemma 3.6 that $Q_{X,e} = Q_{X_e,e}$ and by Corollary 2.5 that $w_X(e) = w_{X_e}(e)$. We therefore obtain the desired result from Lemma 3.5 $\square$

Remark 3.9 Hilbert weights were first introduced by Gieseker [9]. Theorem 3.8 (b) is related to proposition 2.11 of Mumford [17]. But it is not clear if his somewhat sketchy arguments are sufficient to prove the more general and precise version that we need.

There are similarities with the paper by Evertse and Ferretti on Diophantine approximation [7]. Their method is also based on Gröbner basis techniques and a reduction to monomial ideals. But we were not able to extract a proof of our version of Mumford's proposition 2.11 from their paper.

The $\mathbf{G}_m$-equivariant degenerations from X to X_e have been used by Donaldson [6] to define his test functions for the existence of Kähler-Einstein metrics. Let us also in this context mention the preprint of Paul and Tian [19]. The theory in section 2 of that paper is related to the methods that we use to give upper estimates for the determinants.

4   Estimates of Some Determinants

An important new tool in the study of counting functions of rational points of bounded height is the determinant method. It was initiated by Bombieri and Pila [1] who used it to give uniform bounds for affine plane curves. Then, Heath-Brown gave a p-adic version of the method and developed it to a powerful tool for projective hypersurfaces (see [12], th. 14). Finally Broberg [2] and the author [22] generalised the p-adic method to arbitrary projective varieties. The goal of the method in the projective case is to construct auxiliary forms of low degree containing all the rational points of bounded height that one would like to count. The existence of these auxiliary forms follows from the vanishing of certain determinants. We shall in this section estimate the absolute values of these determinants. We first relate the determinant to the Hilbert weights introduced in Definition 3.1.

Lemma 4.1 *Let* $X \subset \mathbf{P}_{\mathbf{Q}}^N$ *be a closed subscheme and* I *be the saturated homogeneous ideal of* $\mathbf{Q}[x_0, \ldots, x_N]$ *which defines* X. *Let* $(B_0, \ldots, B_N) \in \mathbf{R}_{\geq 1}^{N+1}$ *and* $\boldsymbol{\xi}_l = (\xi_{l,0}, \ldots, \xi_{l,N})$ *for* $1 \leq l \leq s$ *be* $(N+1)$-*tuples of integers representing rational points on* X *with* $|\xi_{l,m}| \leq B_m$ *for all* l *and all* $m \in \{0, \ldots, N\}$. *Suppose that* $k \in \mathbf{N}$ *is chosen such that* $s \leq h_I(k)$. *Then there exist monomials* $F_1, \ldots, F_s$ *of degree* k, *which are linearly independent in* $\mathbf{Q}[x_0, \ldots, x_N]_k / I_k$ *such that*

$$|\det(F_j(\boldsymbol{\xi}_l))| \leq s^{s/2} \exp(w_X(\boldsymbol{e}; k))$$

for $\boldsymbol{e} = (\log(B_0), \ldots, \log(B_N)) \in \mathbf{R}^{N+1}$.

Proof Let M be the $(s \times s)$-matrix with entries $F_j(\boldsymbol{\xi}_l)/F_j(B_0, \ldots, B_N)$. Then $\det(M) \leq s^{s/2}$ by Hadamard's inequality as

$$|F_j(\boldsymbol{\xi}_l)/F_j(B_0, \ldots, B_N)| \leq 1$$

for all j and l. Hence

$$|\det(F_j(\boldsymbol{\xi}_l))| \leq s^{s/2} \prod_{i=1}^{s} F_i(B_0, \ldots, B_N)$$

$$= s^{s/2} \prod_{i=1}^{s} \exp(w(\boldsymbol{e}; F_i)) = s^{s/2} \exp\left(\sum_{i=1}^{s} w(\boldsymbol{e}; F_i)\right)$$

for any monomials $F_1, \ldots, F_s \in \mathbf{Q}[x_0, \ldots, x_N]_k$. To complete the proof, choose a monomial basis T of $\mathbf{Q}[x_0, \ldots, x_N]_k / I_k$ with $w(\boldsymbol{e}; T) = w_X(\boldsymbol{e}; k)$. Then $\sum_{i=1}^{s} w(\boldsymbol{e}; F_i) \leq w_X(\boldsymbol{e}; k)$ for any subset $\{F_1, \ldots, F_s\}$ of T and we are done. $\square$

We now make a closer analysis of the Hilbert weight in the upper bound. The following result is a corollary of Theorem 3.8.

Lemma 4.2 *Let* $X \subset \mathbf{P}_K^N$ *be a closed geometrically reduced equidimensional subscheme of dimension* r *and degree* d *and let* I *be the saturated homogeneous ideal of* $\mathbf{Q}[x_0, \ldots, x_N]$ *which defines* X. *Let* $(B_0, \ldots, B_N) \in \mathbf{R}_{\geq 1}^{N+1}$, *and* $\boldsymbol{\xi}_l = (\xi_{l,0}, \ldots, \xi_{l,N})$ *for* $l \in \{1, \ldots, s\}$ *be* $(N+1)$-*tuples of integers representing rational points on* X *such that* $|\xi_{l,m}| \leq B_m$ *for all* l *and all* $m \in \{0, \ldots, N\}$, *Then there exist monomials* $F_1, \ldots, F_s$ *of the same degree* $k = (r!/d)^{1/r} s^{1/r} + O_{d,N}(1)$ *which are linearly independent modulo* I *and a constant* $c \geq 0$ *depending only on* d *and* N *such that*

$$|\det(F_j(\boldsymbol{\xi}_l))| \leq s^{s/2} \exp\left(w_X(\boldsymbol{b}) \frac{1}{(r+1)d} \left(\frac{r!}{d}\right)^{1/r} s^{1+1/r}\right) \max_i(B_i)^{cs}$$

for $b = (b_0, \ldots, b_N) = (\log(B_0), \ldots, \log(B_N)) \in \mathbf{R}^{N+1}$.

Proof We first recall that there are only finitely many possibilities for the Hilbert function $h_I : \mathbf{Z}_+ \to \mathbf{Z}$ for closed geometrically reduced equidimensional subschemes $X \subset \mathbf{P}_K^N$ of degree d (see [15, cor. 6.11] or [20, th. 3.4]). We have therefore that $h_I(k) = dk^r/r! + O_{d,N}(k^{r-1})$ and that $s = dk^r/r! + O_{d,N}(k^{r-1})$ for the smallest positive integer k with $s \leq h_I(k)$. In particular, $k = (r!/d)^{1/r}s^{1/r} + O_{d,N}(1)$ and $\frac{k^{r+1}}{(r+1)!} = \frac{1}{(r+1)d}\left(\frac{r!}{d}\right)^{1/r}s^{1+1/r} + O_{d,N}(s)$. By Theorem 3.8 and Remark 2.3 (ii), we obtain therefore

$$w_I(b; k) = w_X(b)\frac{1}{(r+1)d}\left(\frac{r!}{d}\right)^{1/r}s^{1+1/r} + O_{d,N}\left(\max_{0 \leq i \leq N}(|b_i|)s\right). \tag{10}$$

The assertion now follows from Lemma 4.1 and (10).   □

Remark 4.3 This lemma plays a similar role as lemma 1.11 in [22] and Lemma 4.2 is essentially a reformulation of that result. But the use of Chow weights is new and gives a cleaner statement than in (*op. cit.*), where instead of Chow weights we used certain invariants of Broberg [2] based on a choice of a graded monomial ordering.

5   The Determinant Method

We shall in this section generalise the p-adic determinant method for hypersurfaces of Heath-Brown [12] to arbitrary subvarieties X of $\mathbf{P}_\mathbf{Q}^n$. This was already done by Broberg [2] in his thesis with some further improvements in [22]. But these generalisations depended on the choice of graded monomial orderings. We shall here remove the use of graded monomial orderings and give a more conceptual formulation of the p-adic determinant method by means of Chow forms and Chow weights.

The following notation will be used in the sequel.

Notation 5.1 Let $X \subset \mathbf{P}_\mathbf{Q}^N = \mathrm{Proj}(\mathbf{Q}[x_0, \ldots, x_N])$ be a closed subscheme over $\mathbf{Q}$ and let Ξ be the scheme-theoretic closure of X in $\mathbf{P}_\mathbf{Z}^N = \mathrm{Proj}(\mathbf{Z}[x_0, \ldots, x_N])$. Let $p_1, \ldots, p_t$ be primes and P_i be an $\mathbf{F}_{p_i}$-point on $X_{p_i} = \Xi \times_\mathbf{Z} \mathbf{F}_{p_i}$ for each $i \in \{1, \ldots, t\}$.

(a) $X(\mathbf{Q}; B_0, \ldots, B_N)$ is the set of rational points on X which may be represented by an integral $(n + 1)$-tuple $(x_0, \ldots, x_N)$ with $|x_m| \leq B_m$ for $m \in \{0, \ldots, N\}$. If $B_0 = \cdots = B_N = B$, then we denote this set by $X(\mathbf{Q}, B)$.

(b) $X(\mathbf{Q}; B_0, \ldots, B_N; P_1, \ldots, P_t)$ is defined as the subset of points in $X(\mathbf{Q}; B_0, \ldots, B_N)$ which specialise to P_i on X_{p_i} for each $i \in \{1, \ldots, t\}$. If $B_0 = \cdots = B_N = B$, then we write $X(\mathbf{Q}; B; P_1, \ldots, P_t)$ for this set.

The following result is a special case of main lemma 2.5 in [22].

Lemma 5.2 *Let $X \subset \mathbf{P}_{\mathbf{Q}}^N$ be a closed subscheme of dimension r and degree d as in Lemma 4.2. Let p be a prime and let Ξ be the scheme-theoretic closure of X in $\mathbf{P}_{\mathbf{Z}}^N$. Let P be a non-singular $\mathbf{F}_p$-point on $X_p = \Xi \times_{\mathbf{Z}} \mathbf{F}_p$ and $\boldsymbol{\xi}_1, \ldots, \boldsymbol{\xi}_s$ be primitive $(N+1)$-tuples of integers representing $\mathbf{Z}$-points on Ξ with reduction P. Let $F_1, \ldots, F_s$ be forms in $(x_0, \ldots, x_N)$ with integer coefficients and $\det(F_j(\boldsymbol{\xi}_l))$ be the determinant of the $s \times s$-matrix $(F_j(\boldsymbol{\xi}_l))$. Then there exists a non-negative integer $M = r!^{1/r}(r/(r+1))s^{1+1/r} + O_{d,N}(s)$ such that $p^M \mid \det(F_j(\boldsymbol{\xi}_l))$.*

The main idea of the p-adic determinant method is now to choose p sufficiently big so that the factor of $\det(F_j(\boldsymbol{\xi}_l))$ in Lemma 5.2 is bigger than the upper bound of $\det(F_j(\boldsymbol{\xi}_l))$ in Lemma 4.2. We will then have that $\det(F_j(\boldsymbol{\xi}_l)) = 0$ and that some non-trivial linear combination of $F_1, \ldots, F_s$ will vanish on $X(\mathbf{Q}; B_0, \ldots, B_N)$.

Theorem 5.3 *Let $X \subset \mathbf{P}_{\mathbf{Q}}^N$ be a closed geometrically reduced equidimensional subscheme of dimension r and degree d. Let $(B_0, \ldots, B_N) \in \mathbf{R}_{\geq 1}^{N+1}$, $\boldsymbol{b} = (\log(B_0), \ldots, \log(B_N))$ and $p_1, \ldots, p_t$ be a sequence of different primes satisfying*

$$(p_1 \ldots p_t)^{d^{1/r}} \geq \exp(w_X(\boldsymbol{b})/rd) \max_i(B_i^\epsilon). \tag{11}$$

for some $\epsilon > 0$. Finally, let P_i be a non-singular point on X_{p_i} for each $i \in \{1, \ldots, t\}$.

Then there exists a homogeneous polynomial $G(x_0, \ldots, x_N)$ of degree bounded in terms of d, N and ϵ, which vanishes on $X(\mathbf{Q}; B; P_1, \ldots, P_t)$ but not at all generic points of X.

Proof Let $\boldsymbol{\xi}_1, \ldots, \boldsymbol{\xi}_s$ be primitive $(N+1)$-tuples of integers representing the rational points on $X(\mathbf{Q}; B; P_1, \ldots, P_t)$. Let $k \geq 1$ be the smallest integer such that $s \leq h(k)$. Then by Lemma 4.2, there are forms $F_1, \ldots, F_s$ of the same degree $k = (r!/d)^{1/r}s^{1/r} + O_{d,N}(1)$ such that no non-trivial combination of $F_1, \ldots, F_s$ vanishes everywhere on X and such that

$$\log |\det(F_j(\boldsymbol{\xi}_l))|$$

$$\leq s\left(\log(s^{\frac{1}{2}}) + \frac{w_X(\boldsymbol{b})}{(r+1)d}\left(\frac{r!s}{d}\right)^{\frac{1}{r}} + c\max_i(\log(B_i))\right) \tag{12}$$

for some constant $c \geq 0$ depending only on d and N.

By the p-adic estimates in Lemma 5.2, there exists also a positive factor f of $|\det(F_j(\boldsymbol{\xi}_l))|$ such that

$$\log(f) = (r!)^{1/r}\frac{r}{r+1}s(s^{1/r} - \eta)\log(q) \tag{13}$$

for $q = p_1 \ldots p_t$ and some constant $\eta \geq 0$ depending only on d and N.

By (12) and (13), we will thus have that $\det(F_j(\boldsymbol{\xi}_l)) = 0$ if q is large enough so that

$$\log(s^{1/2}) + w_X(\boldsymbol{b})\frac{1}{(r+1)d}\left(\frac{r!}{d}\right)^{1/r} s^{1/r} + c\max_i(\log(B_i))$$

$$< (r!)^{1/r}\frac{r}{r+1}(s^{1/r} - \eta)\log(q). \qquad (14)$$

Suppose now that $\epsilon > 0$ is given. Then there exists a positive integer s_0 depending solely on d, n and ϵ such that (14) holds for all $s \geq s_0$ when $q^{d^{1/r}} \geq \exp(w_X(\boldsymbol{b})/rd)\max_i(B_i^\epsilon)$. We thus have that $\det(F_j(\boldsymbol{\xi}_l)) = 0$ so that there exists a non-trivial linear combination of $F_1, \ldots, F_s$ which vanishes at $X(\mathbf{Q}; \boldsymbol{B}; P_1, \ldots, P_t)$ but not at all the generic points of X.   □

Remark 5.4 One can extend Theorem 5.3 to the case where P_i have arbitrary fixed multiplicities μ_i on X_{p_i} for $i \in \{1, \ldots, t\}$. One then obtains the same conclusion from the following hypothesis:

$$p_1^{(d/\mu_1)^{1/r}} \cdots p_t^{(d/\mu_t)^{1/r}} \geq \exp(w_X(\boldsymbol{b})/rd)\max_i(B_i^\epsilon). \qquad (15)$$

The only difference in the proof is that one has to use the more general lemma 2.5 in [22] instead of Lemma 5.2. Theorem 5.3 should be seen as a more intrinsic reformulation of Theorem 3.2 in [22], which makes use of Broberg's invariants $(a_0, \ldots, a_N)$ in [2]. These invariants depend apart from $X \subset \mathbf{P}^N$ on the choice of a graded monomial ordering. But it can be shown that $(B_0^{a_0} \cdots B_N^{a_N})^{(r+1)/r} = \exp(w_X(\boldsymbol{b})/rd)$ for an optimal choice of a graded monomial ordering so that the two theorems are actually equivalent. If X is a hypersurface, then it follows from Proposition 2.12 that $\exp(w_X(\boldsymbol{b})/rd) = (V/T^{1/d})^{1/r}$. Theorem 5.3 is therefore then equivalent to Theorem 1.1.

Most applications of Theorem 5.3 concern the case where $B_0 = \cdots = B_N = B$. We have then by Remark 2.3 (ii) that $\exp(w_X(\boldsymbol{b})/rd) = B^{(r+1)/r}$. There are applications of Theorem 5.3 for such boxes in [2] and [22] and in other papers by the author. Another important case is when $B_0 = 1$ and $B_1 = \ldots = B_N = B$. It can then be shown that $\exp(w_X(\boldsymbol{b})/rd) \leq B$ if the hyperplane with $x_0 = 0$ intersects X properly (cf. [22, lemma 1.12]). There are important applications to integral points on affine varieties in this case (see [22, 3] and [23]). There are applications of Theorem 1.1 for boxes given by more complicated $(N + 1)$-tuples $(B_0, \ldots, B_N)$ in papers of Heath-Brown. He uses, for example, results like Theorem 1.1 to study class numbers of quadratic number fields in [13].

Remark 5.5 There is also a real-analytic determinant method first used by Bombieri and Pila [1], which was extended to projective varieties in arbitrary codimension by Marmon [16]. He used thereby graded monomial orderings as in [2] and [22] to formulate his main result. One can use Lemma 4.2 to reformulate his result in terms of Chow weights instead just as for the p-adic version in Theorem 5.3.

References

1. E. Bombieri, J. Pila, The number of integral points on arcs and ovals. Duke Math. J. **59**, 337–357 (1989)
2. N. Broberg, A note on a paper by R. Heath-Brown: "the density of rational points on curves and surfaces". J. Reine Angew. Math. **571**, 159–178 (2004)
3. T.D. Browning, R. Heath-Brown, P. Salberger, Counting rational points on algebraic varieties. Duke Math. J. **132**, 545–578 (2006)
4. H. Chen, Explicit uniform estimation of rational points I. Estimation of heights. J. Reine Angew. Math. **668**, 59–88 (2012)
5. H. Chen, Explicit uniform estimation of rational points II. Hypersurface coverings. J. Reine Angew. Math. **668**, 89–108 (2012)
6. S.K. Donaldson, Scalar curvature and stability of toric varieties. J. Differ. Geom. **62**, 289–349 (2002)
7. J.-H. Evertse, R. Ferretti, Diophantine inequalities on projective varieties. Int. Math. Res. Notices **25**, 1295–1330 (2002)
8. W. Fulton, *Intersection Theory*. Ergebnisse der Mathematik und ihrer Grenzgebiete 3, vol. 2 (Springer, Berlin, 1984)
9. D. Gieseker, Global moduli for surfaces of general type. Invent. Math. **43**, 233–282 (1977)
10. J. Harris, I. Morrison, *Moduli of Curves*. Graduate Texts in Mathematics, vol. 187 (Springer, New York, 1998)
11. R. Hartshorne, *Algebraic Geometry*. Graduate Texts in Mathematics, vol. 52 (Springer, New-York, 1977)
12. R. Heath-Brown, The density of rational points on curves and surfaces. Ann. Math. **155**, 553–595 (2002)
13. R. Heath-Brown, Imaginary quadratic fields with class group exponent 5. Forum Math. **20**, 275–283 (2008)
14. W.V.D. Hodge, D. Pedoe, *Methods of Algebraic Geometry*, vol. 2 (Cambridge University Press, Cambridge, 1953)
15. S.J. Kleiman, Les théorèmes de finitude pour le foncteur de Picard, in *Théorie des Intersections et Théorème de Riemann-Roch*, Séminaire Géométrie Algébrique du Bois Marie 1966/67, SGA 6. Lecture Notes in Mathematic, vol. 225 (Springer, Berlin, 1971), pp. 616–666
16. O. Marmon, A generalization of the Bombieri-Pila determinant method. J. Math. Sci. **171**, 736–744 (2010)
17. D. Mumford, Stability of projective varieties. Enseign. Math. **23**, 39–110 (1977)
18. D. Mumford, J. Fogarty, *Geometric Invariant Theory*. Ergebnisse der Mathematik und ihrer Grenzgebiete 2, vol. 34 (Springer, Berlin, 1966)
19. S. Paul, G. Tian, CM stability and the generalized Futaki invariant I (2006). http://arxiv.org/abs/math/0605278
20. M.E. Rossi, N.V. Trung, G. Valla, Castelnuovo-mumford regularity and finiteness of Hilbert functions, in *Commutative Algebra*. Lecture Notes in Pure and Applied Mathematics, vol. 244 (Chapman Hall/CRC, Boca Raton, 2006), pp. 193–209
21. D. Rydh, Chow varieties, Master's Thesis, Royal Institute of Technology, Stockholm, 2003
22. P. Salberger, On the density of rational and integral points on algebraic varieties. J. Reine Angew. Math. **606**, 123–147 (2007)
23. P. Salberger, Uniform bounds for rational points on cubic hypersurfaces, in *Arithmetic and Geometry*. London Mathematical Society Lecture Note Series, vol. 420 (Cambridge University Press, Cambridge, 2015), pp. 401–421
24. P. Samuel, *Méthodes d'algèbre abstraite en géométrie algébrique* (Springer, Berlin, 1955)

25. E. Sernesi, *Deformations of Algebraic Schemes*. Grundlehren der Mathematischen Wissenschaften, vol. 334 (Springer, Berlin, 2007)
26. B. Sturmfels, Sparse Elimination Theory, in *Computational Algebraic Geometry amd Commutative Algebra*, ed. by D. Eisenbud, L. Robbiano (Cambridge University Press, Cambridge, 1993), p. 264

Chapter VII: Arakelov Geometry, Heights, Equidistribution, and the Bogomolov Conjecture

Antoine Chambert-Loir

To the memory of Lucien Szpiro

1 Introduction

Fermat's method of infinite descent studies the solutions to diophantine equations by constructing, from a given solution of a diophantine equation, a smaller solution, and ultimately deriving a contradiction. In order to formalize the intuitive notion of "size" of an algebraic solution of a diophantine equation, Northcott [35] and Weil [43] have introduced the notion of *height* of an algebraic point of an algebraic variety defined over a number field and established their basic functorial properties, using the decomposition theorem of Weil [42]. The *height machine* is now an important tool in modern diophantine geometry.

The advent of arithmetic intersection theory with Arakelov [3] and, above all, its extension in any dimension by Gillet and Soulé [28] ("Arakelov geometry") has led Faltings [22] to extend the concept further by introducing the height of a subvariety, defined in pure analogy with its degree, replacing classical intersection theory with arithmetic intersection theory. This point of view has been developed in great depth by Bost et al. [10] and Zhang [47].

Although I shall not use it in these notes, I also mention the alternative viewpoint of Philippon [36] who defines the height of a subvariety as the height of the coefficients-vector of its "Chow form".

During the preparation of this paper, the author's research was partially supported by ANR-13-BS01-0006 (Valcomo) and by ANR-15-CE40-0008 (Défigéo).

A. Chambert-Loir (✉)
Université de Paris and Sorbonne Université, CNRS, IMJ-PRG, Paris, France
e-mail: antoine.chambert-loir@u-paris.fr

© The Editor(s) (if applicable) and The Author(s), under exclusive license to Springer Nature Switzerland AG 2021
E. Peyre, G. Rémond (eds.), *Arakelov Geometry and Diophantine Applications*,
Lecture Notes in Mathematics 2276, https://doi.org/10.1007/978-3-030-57559-5_8

The viewpoint of adelic metrics introduced in Zhang [48] is strengthened by the introduction of Berkovich spaces in this context, based on Gubler [30], and leading to the definition by Chambert-Loir [13] of measures at all places analogous to product of Chern forms at the archimedean place.

We describe these notions in the first sections of the text. We then present the equidistribution theorem of Szpiro et al. [40] and its extension by Yuan [44]. Finally, we use these ideas to explain the proof of Bogomolov's conjecture, following Ullmo [41] and Zhang [49]. I refer readers to [1, 34, 39] for alternative expositions of the material presented in these notes.

I thank the editors of this volume for their invitation to lecture on this topic in Grenoble, and the referees for their patient comments that helped me improve this survey.

I started my graduate education under the guidance of Lucien Szpiro. He introduced me to heights, abelian varieties and Arakelov geometry almost 30 years ago. As a sad coincidence, he passed away a few weeks before I was to make the final modifications to this text, at a time when a viral pandemy locked down half of humanity. I dedicate this survey to his memory.

2 Arithmetic Intersection Numbers

2.1 Let $\mathscr{X}$ be a proper flat scheme over $\mathbf{Z}$. For every integer $d \geq 0$, let $Z_d(\mathscr{X})$ be the group of d-cycles on $\mathscr{X}$: it is the free abelian group generated by integral closed subschemes of dimension d.

Remark 2.2 Let $f : \mathscr{X} \to \mathrm{Spec}(\mathbf{Z})$ be the structural morphism. By assumption, f is proper so that the image of an integral closed subscheme Z of $\mathscr{X}$ is again an integral closed subscheme of $\mathrm{Spec}(\mathbf{Z})$. There are thus two cases:

1. Either $f(Z) = \mathrm{Spec}(\mathbf{Z})$, in which case we say that Z is horizontal;
2. Or $f(Z) = \{(p)\}$ for some prime number p, in which case we say that Z is vertical.

2.3 The set $\mathscr{X}(\mathbf{C})$ of complex points of $\mathscr{X}$ has a natural structure of a complex analytic space, smooth if and only if $\mathscr{X}_{\mathbf{Q}}$ is regular. This gives rise to the notions of continuous, resp. smooth, resp. holomorphic function on $\mathscr{X}(\mathbf{C})$: by definition, this is a function which, for every local embedding of an open subset U of $\mathscr{X}(\mathbf{C})$ into $\mathbf{C}^n$, extends to a continuous, resp. smooth, resp. holomorphic function around the image of U.

Let $\mathscr{L}$ be a line bundle on $\mathscr{X}$. A *hermitian metric* on $\mathscr{L}$ is the datum, for every open subset U of $\mathscr{X}(\mathbf{C})$ and every section $s \in \Gamma(U, \mathscr{L})$ of a continuous function $\|s\| : U \to \mathbf{R}_+$, subject to the following conditions:

1. For every open subset V of U, one has $\|s|_V\| = \|s\| |_V$;
2. For every holomorphic function $f \in \mathscr{O}_{\mathscr{X}}(U)$, one has $\|fs\| = |f| \|s\|$;
3. If s does not vanish on U, then the function $\|s\|$ is strictly positive and smooth.

A *hermitian line bundle* $\overline{\mathcal{L}}$ on $\mathcal{X}$ is a line bundle $\mathcal{L}$ endowed with a hermitian metric.

With respect to the tensor product of underlying line bundles and the tensor product of hermitian metrics, the set of isomorphism classes of hermitian line bundles on $\mathcal{X}$ is an abelian group, denoted by $\widehat{\mathrm{Pic}}(\mathcal{X})$. This group fits within an exact sequence of abelian groups:

$$\Gamma(\mathcal{X}, \mathcal{O}_{\mathcal{X}}^{\times}) \to \mathscr{C}^{\infty}(\mathcal{X}(\mathbf{C}), \mathbf{R}) \to \widehat{\mathrm{Pic}}(\mathcal{X}) \to \mathrm{Pic}(\mathcal{X}) \to 0, \qquad (1)$$

where the first map is $f \mapsto \log|f|$, the second associates with $\phi \in \mathscr{C}^{\infty}(\mathcal{X}(\mathbf{C}), \mathbf{R})$ the trivial line bundle $\mathcal{O}_{\mathcal{X}}$ endowed with the hermitian metric for which $\log\|1\|^{-1} = \phi$, and the last one forgets the metric.

2.4 The starting point of our lectures will be the following theorem that asserts existence and uniqueness of "arithmetic intersection degrees" of cycles associated with hermitian line bundles. It fits naturally within the arithmetic intersection theory of Gillet and Soulé [28], we refer to the foundational article by Bost et al. [10] for such an approach; see also Faltings [23] for a direct construction, as well as to the notes of Soulé in this volume.

Theorem 2.5 *Let* $n = \dim(\mathcal{X})$ *and let* $\overline{\mathcal{L}}_1, \ldots, \overline{\mathcal{L}}_n$ *be hermitian line bundles on* $\mathcal{X}$. *There exists a unique family of linear maps:*

$$\widehat{\deg}\left(\hat{c}_1(\overline{\mathcal{L}}_1) \cdots \hat{c}_1(\overline{\mathcal{L}}_d) \mid \cdot\right) : Z_d(\mathcal{X}) \to \mathbf{R},$$

for $d \in \{0, \ldots, n\}$ *satisfying the following properties:*

1. *For every integer* $d \in \{1, \ldots, n\}$, *every integral closed subscheme* Z *of* $\mathcal{X}$ *such that* $\dim(Z) = d$, *every integer* $m \neq 0$ *and every regular meromorphic[1] section* s *of* $\mathcal{L}_d^m|_Z$, *one has*

$$m\,\widehat{\deg}\left(\hat{c}_1(\overline{\mathcal{L}}_1) \cdots \hat{c}_1(\overline{\mathcal{L}}_d) \mid Z\right)$$
$$= \widehat{\deg}\left(\hat{c}_1(\overline{\mathcal{L}}_1) \cdots \hat{c}_1(\overline{\mathcal{L}}_{d-1}) \mid \mathrm{div}(s)\right)$$
$$+ \int_{Z(\mathbf{C})} \log\|s\|^{-1} c_1(\overline{\mathcal{L}}_1) \cdots c_1(\overline{\mathcal{L}}_{d-1}). \qquad (2)$$

2. *For every closed point* z *of* $\mathcal{X}$, *viewed as a integral closed subscheme of dimension* $d = 0$, *one has*

$$\widehat{\deg}(z) = \log(\mathrm{Card}(\kappa(z))), \qquad (3)$$

---

[1] That is, defined over a dense open subscheme of Z.

where $\kappa(z)$ is the residue field of z.[2]

Moreover, these maps are multilinear and symmetric in the hermitian line bundles $\overline{\mathscr{L}_1}, \ldots, \overline{\mathscr{L}_n}$ and only depend on their isomorphism classes in $\widehat{\mathrm{Pic}}(\mathscr{X})$.

Remark 2.6 This theorem should be put in correspondence with the analogous geometric result for classical intersection numbers. Let F be a field and let X be a proper scheme over F, let $n = \dim(X)$ and let $L_1, \ldots, L_n$ be line bundles over X. The degree $\deg(c_1(L_1) \ldots c_1(L_d) \mid Z)$ of a d-cycle Z in X is characterized by the relations:

1. It is linear in Z;
2. If $d = 0$ and Z is a closed point z whose residue field $\kappa(Z)$ is a finite extension of F, then $\deg(Z) = [\kappa(Z) : F]$;
3. If $d \geq 1$ and Z is an integral closed subscheme of X of dimension d, m a non-zero integer, s a regular meromorphic section of $L_d^m|_Z$, then

$$m \deg(c_1(L_1) \ldots c_1(L_d) \mid Z) = \deg(c_1(L_1) \ldots c_1(L_{d-1}) \mid \mathrm{div}(s)). \qquad (4)$$

The additional integral that appears in the arithmetic degree takes into account the fact that $\mathrm{Spec}(\mathbf{Z})$ does not behave as a proper variety.

Example 2.7 Assume that Z is vertical and lies over a maximal ideal (p) of $\mathrm{Spec}(\mathbf{Z})$. Then Z is a proper scheme over $\mathbf{F}_p$ and it follows from the inductive definition and the analogous formula in classical intersection theory that

$$\widehat{\deg}\left(\hat{c}_1(\overline{\mathscr{L}_1}) \cdots \hat{c}_1(\overline{\mathscr{L}_d}) \mid Z\right)$$
$$= \deg\left(c_1(\mathscr{L}_1|_{\mathscr{X}_{\mathbf{F}_p}}) \cdots c_1(\mathscr{L}_d|_{\mathscr{X}_{\mathbf{F}_p}}) \mid Z\right) \log(p).$$

Example 2.8 Assume that $d = 1$ and that Z is horizontal, so that Z is the Zariski-closure in $\mathscr{X}$ of a closed point $z \in \mathscr{X}_{\mathbf{Q}}$. Let $F = \kappa(z)$ and let $\mathfrak{o}_F$ be its ring of integers; by properness of $\mathscr{X}$, the canonical morphism $\mathrm{Spec}(F) \to \mathscr{X}$ with image z extends to a morphism $\varepsilon_z : \mathrm{Spec}(\mathfrak{o}_F) \to \mathscr{X}$, whose image is Z. The formula

$$\widehat{\deg}\left(\hat{c}_1(\overline{\mathscr{L}}) \mid Z\right) = \widehat{\deg}(\varepsilon_z^* \overline{\mathscr{L}})$$

expresses the arithmetic intersection number as the *arithmetic degree* of the hermitian line bundle $\varepsilon_z^* \overline{\mathscr{L}}$ over $\mathrm{Spec}(\mathfrak{o}_F)$.

Proposition 2.9 *Let $f : \mathscr{X}' \to \mathscr{X}$ be a generically finite morphism of proper flat schemes over $\mathbf{Z}$, let Z be an integral closed subscheme of $\mathscr{X}'$ and let $d = \dim(Z)$.*

---

[2]It is indeed a finite field.

1. If $\dim(f(Z)) < d$, *then*

$$\left(\hat{c}_1(f^*\overline{\mathscr{L}_1}) \cdots \hat{c}_1(f^*\overline{\mathscr{L}_d}) \mid Z\right) = 0;$$

2. Otherwise, $\dim(f(Z)) = d$ *and*

$$\left(\hat{c}_1(f^*\overline{\mathscr{L}_1}) \cdots \hat{c}_1(f^*\overline{\mathscr{L}_d}) \mid Z\right) = \left(\hat{c}_1(\overline{\mathscr{L}_1}) \cdots \hat{c}_1(\overline{\mathscr{L}_d}) \mid f_*(Z)\right),$$

where $f_*(Z) = [\kappa(Z) : \kappa(f(Z))]f(Z)$ *is a* d-*cycle on* $\mathscr{X}$.

This is the "projection formula" for height. It can be proved by induction using the inductive definition of the heights and the change of variables formula.

Remark 2.10 Let $n = \dim(\mathscr{X})$ and assume that $\mathscr{X}$ is regular. As the notation suggests rightly, the arithmetic intersection theory of Gillet and Soulé [28] allows another definition of the real number $\widehat{\deg}\left(\hat{c}_1(\overline{\mathscr{L}_1}) \cdots \hat{c}_1(\overline{\mathscr{L}_n}) \mid \mathscr{X}\right)$ as the arithmetic degree of the zero-dimensional arithmetic cycle $\hat{c}_1(\overline{\mathscr{L}_1}) \cdots \hat{c}_1(\overline{\mathscr{L}_n}) \in \widehat{\mathrm{CH}}_0(\mathscr{X})$.

In fact, while the theory of Gillet and Soulé [28] imposes regularity conditions on $\mathscr{X}$, the definition of arithmetic product of classes of the form $\hat{c}_1(\overline{\mathscr{L}})$ requires less stringent conditions; in particular, the regularity of the generic fiber $\mathscr{X}_{\mathbf{Q}}$ is enough. See Faltings [23] for such an approach, as well as Soulé's notes in this volume. More generally, for every birational morphism $f : Z' \to Z$ such that $Z'_{\mathbf{Q}}$ is regular, one has

$$\widehat{\deg}\left(\hat{c}_1(\overline{\mathscr{L}_1}) \cdots \hat{c}_1(\overline{\mathscr{L}_d}) \mid Z\right) = \widehat{\deg}\left(\hat{c}_1(f^*\overline{\mathscr{L}_1}) \cdots \hat{c}_1(f^*\overline{\mathscr{L}_d}) \mid Z'\right).$$

3 The Height of a Variety

3.1 Let X be a proper $\mathbf{Q}$-scheme and let L be a line bundle on X. The important case is when the line bundle L is ample, an assumption which will often be implicit below; in that case, the pair (X, L) is called a polarized variety.

3.2 Let $\mathscr{X}$ be a proper flat scheme over $\mathbf{Z}$ and let $\overline{\mathscr{L}}$ be a hermitian line bundle on $\mathscr{X}$ such that $\mathscr{X}_{\mathbf{Q}} = X$ and $\mathscr{L}_{\mathbf{Q}} = L$. Let Z be a closed integral subscheme of X and let $d = \dim(Z)$. Let $\mathscr{Z}$ be the Zariski-closure of Z in $\mathscr{X}$; it is an integral closed subscheme of $\mathscr{X}$ and $\dim(\mathscr{Z}) = d + 1$.

Definition 3.3 The *degree* and the *height* of Z relative to $\overline{\mathscr{L}}$ are defined by the formulas (provided $\deg_{\mathscr{L}}(Z) \neq 0$).

$$\deg_{\mathscr{L}}(Z) = \deg(c_1(L)^d \mid Z) \tag{5}$$

$$h_{\overline{\mathscr{L}}}(Z) = \widehat{\deg}\left(\hat{c}_1(\overline{\mathscr{L}}^{d+1}) \mid \mathscr{Z}\right) / ((d+1)\deg_{\mathscr{L}}(Z)). \tag{6}$$

Note that the degree $\deg_{\mathscr{A}}(Z)$ is computed on X, hence only depends on L. Moreover, the condition that $\deg_{\mathscr{A}}(Z) \neq 0$ is satisfied (for every Z) when L is ample on X.

Proposition 3.4 *Let* $f \colon \mathscr{X}' \to \mathscr{X}$ *be a generically finite morphism of proper flat schemes over* $\mathbf{Z}$, *let* Z *be a closed integral subscheme of* $\mathscr{X}'_{\mathbf{Q}}$ *and let* $d = \dim(Z)$. *Assume that* L *is ample on* X *and that* $\dim(f(Z)) = d$. *Then* $\deg_{f_* \mathscr{A}}(Z) > 0$ *and*

$$h_{f_* \overline{\mathscr{L}}}(Z) = h_{\overline{\mathscr{L}}}(f(Z)).$$

Proof This follows readily from Proposition 2.9 and its analogue for geometric degrees. Indeed, when one compares formula (6) for Z and for $f(Z)$, both the numerator and the denominator get multiplied by $[\kappa(Z) : \kappa(f(Z))]$. □

Example 3.5 For every $x \in X(\overline{\mathbf{Q}})$, let $[x]$ denote its Zariski closure in X. The function $X(\overline{\mathbf{Q}}) \to \mathbf{R}$ given by $x \mapsto h_{\overline{\mathscr{L}}}([x])$ is a height function relative to the line bundle $\mathscr{L}_{\mathbf{Q}}$ on X in the sense of Weil. Note that while the height functions defined by Weil which are associated with a line bundle L on X are only defined modulo the space of bounded functions on $X(\overline{\mathbf{Q}})$, the techniques of Arakelov geometry show that specifying a model $\mathscr{X}$ of X and a hermitian line bundle $\overline{\mathscr{L}}$ on $\mathscr{X}$ such that $\mathscr{L}_{\mathbf{Q}}$ is isomorphic to L allows to obtain a well-defined representative of this class, for which precise theorems, such as the equidistribution theorem below, can be proved.

Example 3.6 Let us assume that X is an abelian variety over a number field F, with everywhere good reduction, and let $\mathscr{X}$ be an $\mathfrak{o}_F$-abelian scheme such that $\mathscr{X}_F = X$. Let o be the origin of X and let $\varepsilon_o \colon \operatorname{Spec}(\mathfrak{o}_F) \to \mathscr{X}$ be the corresponding section. Let L be a line bundle on X with a trivialisation ℓ of $L|_o$. There exists a unique line bundle $\mathscr{L}$ on $\mathscr{X}$ such that $\mathscr{L}_F = L$ and such that the given trivialisation of $L|_o$ extends to a trivialisation of $\varepsilon_o^* \mathscr{L}$. Moreover, for every embedding $\sigma \colon F \hookrightarrow \mathbf{C}$ the theory of Riemann forms on complex tori endows L_σ with a canonical metric $\|\cdot\|_\sigma$ whose curvature form $c_1(L_\sigma, \|\cdot\|_\sigma)$ is invariant by translation and such that $\|\ell\|_\sigma = 1$; this is in fact the unique metric possessing these two properties. We let $\overline{\mathscr{L}}$ be the hermitian line bundle on $\mathscr{X}$ so defined.

The associated height function will be denoted by $\widehat{h}_L$: it extends the *Néron–Tate height* from $X(\overline{\mathbf{Q}})$ to all integral closed subschemes.

Assume that L is even, that is $[-1]^* L \simeq L$. Then $[n]^* L \simeq L^{n^2}$ for every integer $n \geq 1$, and this isomorphism extends to an isomorphism of hermitian line bundles $[n]^* \overline{\mathscr{L}} \simeq \overline{\mathscr{L}}^{n^2}$. Consequently, for every integral closed subscheme Z of X, one has the following relation

$$\widehat{h}_L([n](Z)) = n^2 \widehat{h}_L(Z). \tag{7}$$

Assume otherwise that L is odd, that is $[-1]^* L \simeq L^{-1}$. Then $[n]^* L \simeq L^n$ for every integer $n \geq 1$; similarly, this isomorphism extends to an isomorphism of hermitian line bundles $[n]^*\overline{\mathscr{L}} \simeq \overline{\mathscr{L}}^n$. Consequently, for every integral closed subscheme Z of X, one has the following relation

$$\widehat{h}_L([n](Z)) = n\widehat{h}_L(Z). \tag{8}$$

Proposition 3.7 *Let $\mathscr{X}'$ be a proper flat scheme over $\mathbf{Z}$ such that $\mathscr{X}'_{\mathbf{Q}} = X$; let $\overline{\mathscr{L}'}$ be a hermitian line bundle on $\mathscr{X}'$ such that $\mathscr{L}'_{\mathbf{Q}} = L$. Assume that L is ample. Then there exists a real number c such that*

$$|h_{\overline{\mathscr{L}}}(Z) - h_{\overline{\mathscr{L}'}}(Z)| \le c$$

for every integral closed subscheme Z of X.

Proof One proves in fact the existence of a real number c such that

$$|\widehat{\deg}(\widehat{c}_1(\overline{\mathscr{L}})^{d+1} \mid \mathscr{Z}) - \widehat{\deg}(\widehat{c}_1(\overline{\mathscr{L}'})^{d+1} \mid \mathscr{Z}')| \le c \deg(c_1(L)^d \mid Z)$$

for every integral d-dimensional subvariety Z of X, where $\mathscr{Z}$ and $\mathscr{Z}'$ are the Zariski closures of Z in $\mathscr{X}$ and $\mathscr{X}'$ respectively. Considering a model $\mathscr{X}''$ that dominates $\mathscr{X}$ and $\mathscr{X}'$ (for example, the Zariski closure in $\mathscr{X} \times_{\mathbf{Z}} \mathscr{X}'$ of the diagonal), we may assume that $\mathscr{X} = \mathscr{X}'$, hence $\mathscr{Z} = \mathscr{Z}'$. A further reduction, that we omit here, allows us to assume that $\mathscr{L}$ is a nef line bundle, and that its hermitian metric is semipositive, and similarly for $\mathscr{L}'$.

By multilinearity, the left hand side that we wish to bound from above is the absolute value of

$$\sum_{i=0}^{d} \widehat{\deg}(\widehat{c}_1(\overline{\mathscr{L}} \otimes \overline{\mathscr{L}'}^{-1})\widehat{c}_1(\overline{\mathscr{L}'})^i \widehat{c}_1(\overline{\mathscr{L}})^{d-i} \mid \mathscr{Z}).$$

Then we view the section 1 of $\mathscr{L}'_{\mathbf{Q}} \otimes \mathscr{L}^{-1}_{\mathbf{Q}}$ as a meromorphic section s of $\mathscr{L}' \otimes \mathscr{L}^{-1}$. Note that its divisor is purely vertical, and its hermitian norm $\|s\|$ is a non-vanishing continuous function on $X(\mathbf{C})$. The definition of the arithmetic intersection numbers then leads us to estimate algebraic intersection numbers

$$\deg(c_1(\mathscr{L}')^i c_1(\mathscr{L})^{d-i} \mid \operatorname{div}(s|_{\mathscr{X}}))$$

and an integral

$$\int_{Z(\mathbf{C})} \log(\|s\|^{-1}) c_1(\overline{\mathscr{L}'})^i c_1(\overline{\mathscr{L}})^{d-i}.$$

By positivity of the curvatures forms $c_1(\overline{\mathscr{L}})$ and $c_1(\overline{\mathscr{L}'})$, the latter integral is bounded from above by

$$\left\| \log(\|s\|)^{-1} \right\|_\infty \int_{Z(\mathbf{C})} c_1(\overline{\mathscr{L}'})^i c_1(\overline{\mathscr{L}})^{d-i} = \left\| \log(\|s\|)^{-1} \right\|_\infty \deg(c_1(L)^d \mid Z).$$

The algebraic terms can be bounded as well. Observe that there exists an integer $n \geq 1$ such that ns extends to a global section of $\mathscr{L}' \otimes \mathscr{L}^{-1}$, and ns^{-1} extends to a global section of its inverse $(\mathscr{L}')^{-1} \otimes \mathscr{L}$. (This is the ultrametric counterpart to the fact that the section s has non-vanishing norm on $X(\mathbf{C})$.) Consequently, $\operatorname{div}(ns \mid \mathscr{Z})$ and $\operatorname{div}(ns^{-1} \mid \mathscr{Z})$ are both effective, so that

$$-\sum_p v_p(n)[\mathscr{Z}_{\mathbf{F}_p}] \leq \operatorname{div}(s \mid \mathscr{Z}) \leq \sum_p v_p(n)[\mathscr{Z}_{\mathbf{F}_p}].$$

This inequality of cycles is preserved after taking intersections, so that

$$\deg(c_1(\mathscr{L}')^i c_1(\mathscr{L})^{d-i} \mid \operatorname{div}(s \mid \mathscr{Z})_p)$$
$$\leq v_p(n) \deg(c_1(\mathscr{L}')^i c_1(\mathscr{L})^{d-i} \mid [\mathscr{Z}_{\mathbf{F}_p}])$$
$$= v_p(n) \deg(c_1(L)^d \mid Z),$$

where $\operatorname{div}(s \mid \mathscr{Z})_p$ is the part of $\operatorname{div}(s \mid \mathscr{Z})$ that lies above the maximal ideal (p) of $\operatorname{Spec}(\mathbf{Z})$. There is a similar lower bound.

Adding all these contributions, this proves the proposition. We refer to Bost et al. [10], §3.2.2, for more details. $\qquad\qquad\qquad\qquad\qquad\qquad\qquad\qquad\qquad\qquad\qquad\qquad\qquad$ □

Proposition 3.8 *Let us assume that L is ample. For every real number B, the set of integral closed subschemes Z of X such that $\deg_L(Z) \leq B$ and $h_{\overline{\mathscr{L}}}(Z) \leq B$ is finite.*

The case of closed points is Northcott's theorem, and the general case is Theorem 3.2.5 of Bost et al. [10]. The principle of its proof goes by reducing to the case where $X = \mathbf{P}^N$ and $\overline{\mathscr{L}} = \overline{\mathscr{O}}(1)$, and comparing the height $h_{\overline{\mathscr{L}}}(Z)$ of a closed integral subscheme Z with the height of its Chow form. (That paper also provides a more elementary proof, relying on the fact that a finite set of sections of powers of $\mathscr{O}(1)$ are sufficient to compute by induction the height of any closed integral subscheme of $\mathbf{P}^N$ of given degree.)

4 Adelic Metrics

4.1 Let $S = \{2, 3, \ldots, \infty\}$ be the set of places of $\mathbf{Q}$.

Each prime number p is identified with the p-adic absolute value on $\mathbf{Q}$, normalized by $|p|_p = 1/p$; these places are said to be finite. We denote by $\mathbf{Q}_p$ the completion of $\mathbf{Q}$ for this p-adic absolute value and fix an algebraic closure $\overline{\mathbf{Q}_p}$ of $\mathbf{Q}_p$. The p-adic absolute value extends uniquely to $\overline{\mathbf{Q}_p}$; the corresponding completion is denoted by $\mathbf{C}_p$: this is an algebraically closed complete valued field.

The archimedean place is represented by the symbol ∞, and is identified with the usual absolute value on $\mathbf{Q}$; it is also called the infinite place. For symmetry of notation, we may write $\mathbf{Q}_\infty = \mathbf{R}$ and $\mathbf{C}_\infty = \mathbf{C}$, the usual fields of real and complex numbers.

The notion of adelic metrics that we now describe is close in spirit to the one exposed by Peyre in this volume. However, Peyre is only interested in rational points and it suffices for him to consider local fields such as $\mathbf{Q}_p$, and "naïve p-adic analytic varieties". We will however need to consider algebraic points. The topological spaces associated with varieties over $\mathbf{C}_p$ are not well behaved enough, and we will have to use a rigid analytic notion of p-adic varieties, specifically the one introduced by Berkovich.

4.2 Let X be a proper scheme over $\mathbf{Q}$. Let $v \in S$ be a place of $\mathbf{Q}$.

Assume $v = \infty$. Then we set $X_\infty^{\mathrm{an}} = X(\mathbf{C}_\infty)/F_\infty$, the set of complex points of X modulo the action of complex conjugation F_∞.

Assume now that $v = p$ is a finite place. Then we set X_p^{an} to be the analytic space associated by Berkovich [5] to the $\mathbf{Q}_p$-scheme $X_p = X_{\mathbf{Q}_p}$. As a set, X_p^{an} is the quotient of the class of all pairs (K, x), where K is a complete valued field containing $\mathbf{Q}_p$ and $x \in X(K)$, modulo the identification $(K, x) \sim (K', x')$ if there exists a common valued field extension K'' of K and K' such that x and x' induce the same point of $X(K'')$. Its topology is the coarsest topology such that for every open subscheme U of X and every $f \in \mathcal{O}_X(U)$, the subset U_p^{an} (consisting of classes $[K, x]$ such that $x \in U(K)$) is open in X_p^{an}, and the function $[K, x] \mapsto |f(x)|$ from U_p^{an} to $\mathbf{R}_+$ is continuous.

The space X_p^{an} is a compact metrizable topological space, locally contractible (in particular locally arcwise connected). There is a canonical continuous map $X(\mathbf{C}_p) \to X_p^{\mathrm{an}}$; it identifies the (totally discontinuous) topological space $X(\mathbf{C}_p)/\mathrm{Gal}(\mathbf{C}_p/\mathbf{Q}_p)$ with a dense subset of X_p^{an}. It is endowed with a sheaf in local rings $\mathcal{O}_{X_p^{\mathrm{an}}}$; for every open subset U of X_p^{an}, every holomorphic function $f \in \mathcal{O}_{X_p^{\mathrm{an}}}(U)$ admits an absolute value $|f|: U \to \mathbf{R}_+$.

We gather all places together and consider the topological space $X_{\mathrm{ad}} = \coprod_{v \in S} X_v^{\mathrm{an}}$, coproduct of the family $(X_v^{\mathrm{an}})_{v \in S}$. By construction, a function ϕ on X_{ad} consists in a family $(\phi_v)_{v \in S}$, where ϕ_v is a function on X_v^{an}, for every $v \in S$.

4.3 Let L be a line bundle on X; it induces a line bundle L_v^{an} on X_v^{an} for every place v.

A continuous v-adic metric on L_v^{an} is the datum, for every open subset U of X_v^{an} and every section s of L_v^{an} on U, of a continuous function $\|s\| \colon U \to \mathbf{R}_+$, subject to the requirements:

1. For every open subset V of U, one has $\|s|_V\| = \|s\| |_V$;
2. For every holomorphic function $f \in \mathscr{O}_{X_v^{\mathrm{an}}}(U)$, one has $\|fs\| = |f| \|s\|$.
3. If s does not vanish, then $\|s\|$ does not vanish as well.

If L and M are line bundles on X equipped with v-adic metrics, then L^{-1} and $L \otimes M$ admit natural v-adic metrics, and the canonical isomorphism $L^{-1} \otimes L \simeq \mathscr{O}_X$ is an isometry.

The trivial line bundle $\mathscr{O}_X$ admits a canonical v-adic metric for which $\|f\| = |f|$ for every local section of $\mathscr{O}_X$. More generally, for every v-adic metric $\|\cdot\|$ on $\mathscr{O}_X$, $\phi = \log\|1\|^{-1}$ is a continuous function on X_v^{an}, and any v-adic metric on $\mathscr{O}_X$ is of this form. The v-adically metrized line bundle associated with ϕ is denoted by $\mathscr{O}_X(\phi)$.

If $\overline{L}$ is a line bundle endowed with an v-adic metric and $\phi \in \mathscr{C}(X_v^{\mathrm{an}}, \mathbf{R})$, we denote by $\overline{L}(\phi)$ the v-adically metrized line bundle $\overline{L} \otimes \mathscr{O}_X(\phi)$. Explicitly, its v-adic metric is that of $\overline{L}$ multiplied by $e^{-\phi}$.

Example 4.4 Let $\mathscr{X}$ be a proper flat scheme over $\mathbf{Z}$ such that $\mathscr{X}_{\mathbf{Q}} = X$, let d be a positive integer and let $\mathscr{L}$ be a line bundle on $\mathscr{X}$ such that $\mathscr{L}_{\mathbf{Q}} = L^d$. Let us show that this datum endows L with an p-adic metric, for every finite place $p \in S$.

Let thus fix a prime number p. There exists a canonical *specialization map*, $X_p^{\mathrm{an}} \to \mathscr{X} \otimes_{\mathbf{Z}} \mathbf{F}_p$; it is anticontinuous (the inverse image of an open subset is closed). For every open subset $\mathscr{U} \subset \mathscr{X} \otimes_{\mathbf{Z}} \mathbf{F}_p$, let $]\mathscr{U}[$ be the preimage of $\mathscr{U}$.

There exists a unique continuous metric on L_p^{an} such that for every open subscheme $\mathscr{U}$ of $\mathscr{X} \otimes_{\mathbf{Z}} \mathbf{Z}_p$ and every basis ℓ of $\mathscr{L}$ on $\mathscr{U}$, one has $\|\ell\| \equiv 1$ on $]\mathscr{U} \otimes \mathbf{F}_p[$. Explicitly, if s is a section of L_p^{an} on an open subset U of $]\mathscr{U} \otimes \mathbf{F}_p[$, there exists a holomorphic function $f \in \mathscr{O}_{X_p^{\mathrm{an}}}(U)$ such that $s^d = f\ell$ and $\|s\| = |f|^{1/d}$ on U.

Such p-adic metrics are called *algebraic*.

4.5 An adelic metric on L is the datum, for every place $v \in S$, of a v-adic metric on the line bundle L_v^{an} on X_v^{an}, subject to the additional requirement that there exists a model $(\mathscr{X}, \mathscr{L})$ of (X, L) inducing the given p-adic metric for all but finitely many primes p.

If L and M are line bundles on X equipped with adelic metrics, then L^{-1} and $L \otimes M$ admit natural adelic metrics, and the canonical isomorphism $L^{-1} \otimes L \simeq \mathscr{O}_X$ is an isometry.

The trivial line bundle $\mathscr{O}_X$ admits a canonical adelic metric for which $\|f\| = |f|$ for every local section of $\mathscr{O}_X$. More generally, for every adelic metric $\|\cdot\|$ on $\mathscr{O}_X$, and every place $v \in S$, then $\phi_v = \log\|1\|_v^{-1}$ is a continuous function on X_v^{an}, and is identically zero for all but finitely many places v; in other words, the function $\phi = (\phi_v) \in \mathscr{C}(X_{\mathrm{ad}}, \mathbf{R})$ has compact support. The adelically metrized line bundle associated with ϕ is denoted by $\mathcal{O}_X(\phi)$. Conversely, any adelic metric on $\mathcal{O}_X$ is of this form.

If $\overline{L}$ is a line bundle endowed with an adelic metric and $\phi \in \mathscr{C}_c(X_{ad}, \mathbf{R})$, we denote by $\overline{L}(\phi)$ the adelically metrized line bundle $\overline{L} \otimes \mathcal{O}_X(\phi)$. Explicitly, for every place v, its v-adic metric is that of $\overline{L}$ multiplied by $e^{-\phi_v}$.

Remark 4.6 Let $(\mathscr{X}, \mathscr{L})$ and $(\mathscr{X}', \mathscr{L}')$ be two models of the polarized variety (X, L). Since X is finitely presented, there exists a dense open subscheme U of $\mathrm{Spec}(\mathbf{Z})$ such that the isomorphism $\mathscr{X}_\mathbf{Q} = X = \mathscr{X}'_\mathbf{Q}$ extends to an isomorphism $\mathscr{X}_U \simeq \mathscr{X}'_U$. Then, up to shrinking U, we may assume that the isomorphism $\mathscr{L}_\mathbf{Q} = L = \mathscr{L}'_\mathbf{Q}$ extends to an isomorphism $\mathscr{L}_U \simeq \mathscr{L}'_U$. In particular, for every prime number p such that $(p) \in U$, the p-adic norms on L induced by $\mathscr{L}$ and $\mathscr{L}'$ coincide.

4.7 Let $\overline{\mathrm{Pic}}(X_{ad})$ be the abelian group of isomorphism classes of line bundles on X endowed with adelic metrics. It fits within an exact sequence

$$\Gamma(X, \mathcal{O}_X^\times) \to \mathscr{C}_c(X_{ad}, \mathbf{R}) \to \overline{\mathrm{Pic}}(X_{ad}) \to \mathrm{Pic}(X) \to 0. \qquad (9)$$

The morphism on the left is given by $u \mapsto (\log|u|_v^{-1})_{v \in S}$. It is injective up to torsion, as a consequence of Kronecker's theorem: if $|u|_v = 1$ for every place v, then there exists $m \geq 1$ such that $u^m = 1$. Its image is the kernel of the morphism $\mathscr{C}(X_{ad}, \mathbf{R}) \to \overline{\mathrm{Pic}}(X)$; indeed, an isometry $\mathcal{O}_X(\phi) \to \mathcal{O}_X(\psi)$ is given by an element $u \in \Gamma(X, \mathcal{O}_X^\times)$ such that $\psi_v + \log|u|_v^{-1} = \phi_v$, for every place $v \in S$.

We denote by $\hat{c}_1(\overline{L})$ the isomorphism class in $\overline{\mathrm{Pic}}(X_{ad})$ of an adelically metrized line bundle $\overline{L}$ on X.

Remark 4.8 Let D be an effective Cartier divisor on X and let $\mathcal{O}_X(D)$ be the corresponding line bundle; let s_D be its canonical section. Assume that $\mathcal{O}_X(D)$ is endowed with an adelic metric.

Let $v \in S$ be a place of $\mathbf{Q}$. The function $g_{D,v} = \log\|s_D\|_v^{-1}$ is a continuous function on $X_v^{an} \backslash |D|$, and is called a v-adic Green function for D. For every open subscheme U of X and any equation f of D on U, $g_{D,v} + \log|f|_v$ extends to a continuous function on U_v^{an}. Conversely, this property characterizes v-adic Green functions for D.

The family $g_D = (g_{D,v})$ is called an adelic Green function for D.

Lemma 4.9 (Chambert-Loir and Thuillier [15], prop. 2.2) *Let $\mathscr{X}$ be a proper flat integral scheme over $\mathbf{Z}$, let $\overline{\mathscr{L}}$ be a hermitian line bundle on $\mathscr{X}$. Let $X = \mathscr{X}_\mathbf{Q}$ and let $L = \mathscr{L}_\mathbf{Q}$, endowed with the algebraic adelic metric associated with $(\mathscr{X}, \overline{\mathscr{L}})$. Assume that $\mathscr{X}$ is integrally closed in its generic fiber (for example, that it is normal).*

Then the canonical map $\Gamma(\mathscr{X}, \mathscr{L}) \to \Gamma(X, L)$ is injective and its image is the set of sections s such that $\|s\|_v \leq 1$ for every finite place $v \in S$.

Equivalently, effective Cartier divisors on $\mathscr{X}$ correspond to v-adic Green functions which are positive at all finite places v.

Proof Injectivity follows from the fact that $\mathscr{X}$ is flat, so that X is schematically dense in $\mathscr{X}$. Surjectivity is a generalization of the fact that an integrally closed domain is the intersection of its prime ideals of height 1.   □

4.10 Let $\|\cdot\|$ and $\|\cdot\|'$ be two adelic metrics on L. The ratio of these metrics is a metric on the trivial line bundle, hence is of the form $\mathscr{O}_X(\phi)$, for some function $\phi = (\phi_v) \in \mathscr{C}_c(X_{\mathrm{ad}}, \mathbf{R})$. For all but finitely many places, ϕ_v is identically 0. For every place v, we let

$$\delta_v(\|\cdot\|, \|\cdot\|') = \|\phi_v\|_\infty = \sup_{x \in X_v^{\mathrm{an}}} \left| \log \frac{\|\cdot\|'}{\|\cdot\|}(x) \right|.$$

Since X_v^{an} is compact, this is a positive real number; it is equal to 0 for all but finitely many places v.

We then define the distance between the two given adelic metrics by

$$\delta(\|\cdot\|, \|\cdot\|') = \sum_{v \in S} \delta_v(\|\cdot\|', \|\cdot\|).$$

The set of adelic metrics on a given line bundle L is a real affine space, its underlying vector space is the subspace $\mathscr{C}_c(X_{\mathrm{ad}}, \mathbf{R})$ of $\mathscr{C}(X_{\mathrm{ad}}, \mathbf{R}) = \prod_v \mathscr{C}(X_v^{\mathrm{an}}, \mathbf{R})$ consisting of families (ϕ_v) such that $\phi_v \equiv 0$ for all but finitely many places $v \in S$.

The space $\mathscr{C}_c(X_{\mathrm{ad}}, \mathbf{R})$ is the union of the subspaces $\mathscr{C}_U(X_{\mathrm{ad}}, \mathbf{R})$ of functions with (compact) support above a given finite set U of places of S. We thus endow it with its natural inductive limit topology.

Example 4.11 (Algebraic Dynamics) We expose here the point of view of Zhang [48] regarding the canonical heights of Call and Silverman [11] in the context of algebraic dynamics. Let X be a proper $\mathbf{Q}$-scheme, let $f : X \to X$ be a morphism, let L be a line bundle on X such that $f^*L \simeq L^q$, for some integer $q \geq 2$. We fix such an isomorphism ε. The claim is that *there exists a unique adelic metric on L for which the isomorphism ε is an isometry.*

Let us first fix a place v and prove that there is a unique v-adic metric on L for which ε is an isometry. To that aim, let us consider, for any v-adic metric $\|\cdot\|$ on L, the induced v-adic metric on f^*L and transfer it to L^q via ε. This furnishes a v-adic metric $\|\cdot\|^f$ on L such that ε is an isometry from $(L, f^*\|\cdot\|)$ to $(L, \|\cdot\|^f)^q$, and it is the unique v-adic metric on L satisfying this property. Within the real affine space of v-adic metrics on L, normed by the distance δ_v, and complete, the self-map $\|\cdot\| \mapsto \|\cdot\|^f$ is contracting with Lipschitz constant $1/q$. Consequently, the claim follows from Banach's fixed point theorem.

We also note that there exists a dense open subscheme U of $\mathrm{Spec}(\mathbf{Z})$, a model $(\mathscr{X}, \mathscr{L})$ of (X, L) over U such that $f : X \to X$ extends to a morphism $\phi : \mathscr{X} \to \mathscr{X}$ and the isomorphism $\varepsilon : f^*L \simeq L^q$ extends to an isomorphism $\phi^*\mathscr{L} \simeq \mathscr{L}^q$, still denoted by ε. This implies that for every finite place p above U, the canonical v-adic metric is induced by the model $(\mathscr{X}, \mathscr{L})$.

Consequently, the family $(\|\cdot\|_v)$ of v-adic metrics on L for which ε is an isometry is an adelic metric.

5 Arithmetic Ampleness

Definition 5.1 Let $\mathscr{X}$ be a proper scheme over $\mathbf{Z}$ and let $\overline{\mathscr{L}}$ be a hermitian line bundle on $\mathscr{X}$. One says that $\overline{\mathscr{L}}$ is relatively *semipositive* if:

1. For every vertical integral curve C on $\mathscr{X}$, one has $\deg_{\mathscr{L}}(C) \geq 0$;
2. For every holomorphic map $f : \mathbf{D} \to \mathscr{X}(\mathbf{C})$, the curvature of $f^*\overline{\mathscr{L}}$ is semipositive.

If $\overline{\mathscr{L}}$ is relatively semipositive, then $\mathscr{L}_{\mathbf{Q}}$ is nef. (The degree of a curve Z in $X_{\mathbf{Q}}$ with respect to $L_{\mathbf{Q}}$ of the integral over $Z(\mathbf{C})$ of the curvature of $\overline{\mathscr{L}}$, hence is positive.)

Example 5.2 Let us consider the tautological line bundle $\mathscr{O}(1)$ on the projective space $\mathbf{P}_{\mathbf{Z}}^N$. Its local sections correspond to homogeneous rational functions of degree 1 in indeterminates $T_0, \ldots, T_N$. It is endowed with a natural hermitian metric such that, if f is such a rational function, giving rise to the section s_f, and if $x = [x_0 : \ldots : x_N] \in \mathbf{P}^N(\mathbf{C})$, one has the formula

$$\|s_f\|(x) = \frac{|f(x_0, \ldots, x_N)|}{\left(|x_0|^2 + \cdots + |x_N|^2\right)^{1/2}}.$$

(By homogeneity of f, the right hand side does not depend on the choice of the system of homogeneous coordinates for x.) The corresponding hermitian line bundle $\overline{\mathscr{O}(1)}$ is relatively semipositive. It is in fact the main source of relatively semipositive hermitian line bundles, in the following way.

Let $\mathscr{X}$ be a proper scheme over $\mathbf{Z}$ and let $\overline{\mathscr{L}}$ be a hermitian line bundle on $\mathscr{X}$. One says that $\overline{\mathscr{L}}$ is relatively *ample* if there exists an embedding $\phi : \mathscr{X} \hookrightarrow \mathbf{P}_{\mathbf{Z}}^N$, a metric with strictly positive curvature on $\mathscr{O}_{\mathbf{P}^N}(1)$ and an integer $d > 1$ such that $\overline{\mathscr{L}}^d \simeq \phi^*\overline{\mathscr{O}_{\mathbf{P}^N}(1)}$.

Proposition 5.3 *Let X be a proper scheme over $\mathbf{Q}$ and let $L_0, \ldots, L_d$ be line bundles on X.*

Let $\mathscr{X}, \mathscr{X}'$ be proper flat schemes over $\mathbf{Z}$ such that $X = \mathscr{X}_{\mathbf{Q}} = \mathscr{X}'_{\mathbf{Q}}$, let $\overline{\mathscr{L}_0}, \ldots, \overline{\mathscr{L}_d}$ (respectively $\overline{\mathscr{L}'_0}, \ldots, \overline{\mathscr{L}'_d}$) be semipositive hermitian line bundles on $\mathscr{X}$ (respectively $\mathscr{X}'$) such that $\mathscr{L}_{j,\mathbf{Q}} = \mathscr{L}'_{j,\mathbf{Q}} = L_j$. We write $\overline{L_0}, \ldots, \overline{L_d}$ (respectively $\overline{L'_0}, \ldots, \overline{L'_d}$) for the corresponding adelically metrized line bundles on X.

Then for every closed subscheme Z of $\mathcal{X}$, one has

$$\left| \widehat{\deg} \left(\hat{c}_1(\overline{\mathcal{L}'_0}) \dots \hat{c}_1(\overline{\mathcal{L}'_d}) \mid Z \right) - \widehat{\deg} \left(\hat{c}_1(\overline{\mathcal{L}_0}) \dots \hat{c}_1(\overline{\mathcal{L}_d}) \mid Z \right) \right|$$

$$\leq \sum_{j=0}^{d} \delta(\overline{L}_j, \overline{L}'_j) \deg \left(c_1(L_0) \dots \widehat{c_1(L_j)} \dots c_1(L_d) \mid Z \right),$$

where the factor $c_1(L_j)$ is omitted in the jth term.

Proof Using the projection formula, we first reduce to the case where $\mathcal{X} = \mathcal{X}'$ is normal. We then write

$$\widehat{\deg} \left(\hat{c}_1(\overline{\mathcal{L}'_0}) \dots \hat{c}_1(\overline{\mathcal{L}'_d}) \mid Z \right) - \widehat{\deg} \left(\hat{c}_1(\overline{\mathcal{L}_0}) \dots \hat{c}_1(\overline{\mathcal{L}_d}) \mid Z \right)$$

$$= \sum_{j=0}^{d} \widehat{\deg} \left(\hat{c}_1(\overline{\mathcal{L}'_0}) \dots \hat{c}_1(\overline{\mathcal{L}'_{j-1}})(\hat{c}_1(\overline{\mathcal{L}'_j}) - \hat{c}_1(\overline{\mathcal{L}_j}))\hat{c}_1(\overline{\mathcal{L}_{j+1}}) \dots \hat{c}_1(\overline{\mathcal{L}_d}) \mid Z \right)$$

and bound the jth term as follows. Let s_j be the regular meromorphic section of $\mathcal{O}_X = \mathcal{L}'_j \otimes (\mathcal{L}_j)^{-1}$ corresponding to 1. By definition, one has

$$\widehat{\deg} \left(\hat{c}_1(\overline{\mathcal{L}'_0}) \dots \hat{c}_1(\overline{\mathcal{L}'_{j-1}})(\hat{c}_1(\overline{\mathcal{L}'_j}) - \hat{c}_1(\overline{\mathcal{L}_j}))\hat{c}_1(\overline{\mathcal{L}_{j+1}}) \dots \hat{c}_1(\overline{\mathcal{L}_d}) \mid Z \right)$$

$$= \widehat{\deg} \left(\hat{c}_1(\overline{\mathcal{L}'_0}) \dots \hat{c}_1(\overline{\mathcal{L}'_{j-1}})\hat{c}_1(\overline{\mathcal{L}_{j+1}}) \dots \hat{c}_1(\overline{\mathcal{L}_d})(\hat{c}_1(\overline{\mathcal{L}'_j}) - \hat{c}_1(\overline{\mathcal{L}_j})) \mid Z \right)$$

$$= \widehat{\deg} \left(\hat{c}_1(\overline{\mathcal{L}'_0}) \dots \hat{c}_1(\overline{\mathcal{L}'_{j-1}})\hat{c}_1(\overline{\mathcal{L}_{j+1}}) \dots \hat{c}_1(\overline{\mathcal{L}_d}) \mid \mathrm{div}(s_j|_Z) \right)$$

$$+ \int_{Z(\mathbb{C})} \log\|s_j\|^{-1} c_1(\overline{\mathcal{L}'_0}) \dots c_1(\overline{\mathcal{L}'_{j-1}})c_1(\overline{\mathcal{L}_{j+1}}) \dots c_1(\overline{\mathcal{L}_d}).$$

Moreover, all components of $\mathrm{div}(s_j|_Z)$ are vertical. For every $j \in \{0, \dots, d\}$ and every $v \in S$, let $\delta_{j,v} = \delta_v(\overline{L}_j, \overline{L}'_j)$ (this is zero for all but finitely many places v). Using the fact that $|\log\|s_j\|_v| \leq \delta_v(\overline{L}_j, \overline{L}'_j)$ for every place $v \in S$, the normality assumption on $\mathcal{X}$ implies that

$$\mathrm{div}(s_j|_Z) \leq \sum_{p \in S \setminus \{\infty\}} \delta_{j,p}(\log p)^{-1}[Z \otimes \mathbf{F}_p].$$

Since the line bundles $\mathcal{L}_k$ and $\mathcal{L}'_k$ are semipositive, this implies the bound

$$\widehat{\deg} \left(\hat{c}_1(\overline{\mathcal{L}'_0}) \dots \hat{c}_1(\overline{\mathcal{L}'_{j-1}})\hat{c}_1(\overline{\mathcal{L}_{j+1}}) \dots \hat{c}_1(\overline{\mathcal{L}_d}) \mid \mathrm{div}(s_j|_Z) \right)$$

$$= \sum_p \deg \left(c_1(\mathcal{L}'_0) \dots c_1(\mathcal{L}'_{j-1})c_1(\mathcal{L}_{j+1}) \dots c_1(\mathcal{L}_d) \mid \mathrm{div}(s_j|_Z)_p \right) \log p$$

$$\leq \sum_p \delta_{j,p} \deg \left(c_1(\mathscr{L}_0')\ldots c_1(\mathscr{L}_{j-1}')c_1(\mathscr{L}_{j+1})\ldots c_1(\mathscr{L}_d) \mid [Z \otimes \mathbf{F}_p]\right)$$

$$\leq \left(\sum_p \delta_{j,p}\right) \deg \left(c_1(L_0)\ldots c_1(L_{j-1})c_1(L_{j+1})\ldots c_1(L_d) \mid Z\right).$$

Similarly, the curvature forms $c_1(\overline{L_k})$ and $c_1(\overline{L_k'})$ are semipositive, so that the upper bound $\log\|s_j\|^{-1} \leq \delta_{j,\infty}$ implies

$$\int_{Z(\mathbf{C})} \log\|s_j\|^{-1}c_1(\overline{\mathscr{L}_0'})\ldots c_1(\overline{\mathscr{L}_{j-1}'})c_1(\overline{\mathscr{L}_{j+1}})\ldots c_1(\overline{\mathscr{L}_d})$$

$$\leq \delta_{j,\infty} \int_{Z(\mathbf{C})} c_1(\overline{\mathscr{L}_0'})\ldots c_1(\overline{\mathscr{L}_{j-1}'})c_1(\overline{\mathscr{L}_{j+1}})\ldots c_1(\overline{\mathscr{L}_d})$$

$$\leq \delta_{j,\infty} \deg \left(c_1(L_0)\ldots c_1(L_{j-1})c_1(L_{j+1})\ldots c_1(L_d) \mid Z\right).$$

Adding these contributions, we get one of the desired upper bound, and the other follows by symmetry. □

Definition 5.4 An adelic metric on a line bundle L on X is said to be *semipositive* if it is a limit of a sequence of semipositive algebraic adelic metrics on L.

Let $\overline{\mathrm{Pic}}^+(X)$ be the set of all isomorphism classes of line bundles endowed with a semipositive metric. We endow it with the topology for which the set of adelic metrics on a given line bundle is open and closed in $\overline{\mathrm{Pic}}^+(X)$ and endowed with the topology induced by the distance of adelic metrics.

It is a submonoid of $\overline{\mathrm{Pic}}(X)$; moreover, its image in $\mathrm{Pic}(X)$ consists of (isomorphism classes of) nef line bundles on X. I thank the referee for pointing out an example [21, 1.7] of a nef line bundle on a complex projective variety admitting no smooth semipositive metric, as well as for the numerous weaknesses of exposition he/she pointed out.

Corollary 5.5 *Let Z be an integral closed subscheme of X, let $d = \dim(Z)$. The arithmetic degree map extends uniquely to a continuous function $\overline{\mathrm{Pic}}^+(X)^{d+1} \to \mathbf{R}$. This extension is multilinear and symmetric.*

Proof This follows from Proposition 5.3 and from the classical extension theorem of uniformly continuous maps. □

Definition 5.6 Let X be a projective $\mathbf{Q}$-scheme and let L be a line bundle on X. An adelic metric on L is said to be *admissible* if there exists two line bundles endowed with semipositive adelic metrics, $\overline{M_1}$ and $\overline{M_2}$, such that $\overline{L} \simeq \overline{M_1} \otimes \overline{M_2}^{-1}$.

More generally, we say that a v-adic metric on L is admissible if it is the v-adic component of an admissible adelic metric on L. The set of all admissible adelically metrized line bundles on X is denoted by $\overline{\mathrm{Pic}}^{\mathrm{adm}}(X)$; it is the subgroup generated by $\overline{\mathrm{Pic}}^+(X)$.

By construction, the arithmetic intersection product extends by linearity to $\overline{\text{Pic}}^{\,\text{adm}}(X)$. We use the notation $\widehat{\deg}(\hat{c}_1(\overline{L}_0)\cdots\hat{c}_1(\overline{L}_d)\mid Z)$ for the arithmetic degree of a d-dimensional integral closed subscheme Z of X with respect to admissible adelically metrized line bundles $\overline{L}_0,\ldots,\overline{L}_d$.

This gives rise to a natural notion of height parallel to that given in Definition 3.3.

Example 5.7 Let us retain the context and notation of Example 4.11. Let us moreover assume that L is ample and let us prove that the canonical adelic metric on L is semipositive.

We make the observation that if $\|\cdot\|$ is an algebraic adelic metric on L induced by a relatively semipositive hermitian line bundle $\overline{\mathscr{L}}$ on a proper flat model $\mathscr{X}$ of X, then the metric $\|\cdot\|^f$ is again relatively semipositive. Indeed, the normalization of $\mathscr{X}$ in the morphism $f\colon X \to X$ furnishes a proper flat scheme $\mathscr{X}'$ over $\mathbf{Z}$ such that $\mathscr{X}'_{\mathbf{Q}} = X$ and a morphism $\phi\colon \mathscr{X}' \to \mathscr{X}$ that extends f. Then $\phi^*\overline{\mathscr{L}}$ is a relatively semipositive hermitian line bundle on $\mathscr{X}'$, model of L^d, which induces the algebraic adelic metric $\|\cdot\|^f$ on L.

Starting from a given algebraic adelic metric induced by a relatively semipositive model (for example, a relatively ample one), the proof of Picard's theorem invoked in Example 4.11 proves that the sequence of adelic metrics obtained by the iteration of the operator $\|\cdot\| \mapsto \|\cdot\|^f$ converges to the unique fixed point. Since this iteration preserves algebraic adelic metrics induced by a relatively semipositive model, the canonical adelic metric on L is semipositive, as claimed.

For a generalization of this construction, see theorem 4.9 of Yuan and Zhang [46].

6 Measures

Definition 6.1 Let X be a projective $\mathbf{Q}$-scheme. A function $\phi \in \mathscr{C}(X_{\text{ad}}, \mathbf{R})$ is said to be admissible if the adelically metrized line bundle $\mathscr{O}_X(\phi)$ is admissible.

The set $\mathscr{C}_{\text{adm}}(X_{\text{ad}}, \mathbf{R})$ of admissible functions (ϕ_v) is a real vector subspace of $\mathscr{C}_{\text{c}}(X_{\text{ad}}, \mathbf{R})$. One has an exact sequence

$$\Gamma(X, \mathscr{O}_X^\times) \to \mathscr{C}_{\text{adm}}(X_{\text{ad}}, \mathbf{R}) \to \overline{\text{Pic}}^{\,\text{adm}}(X) \to \text{Pic}(X) \to 0 \qquad (10)$$

analogous to (9).

More generally, we say that a function $\phi_v \in \mathscr{C}(X_v^{\text{an}}, \mathbf{R})$ is admissible if it is the v-adic component of an admissible function $\phi = (\phi_v)$. This defines a real vector subspace $\mathscr{C}_{\text{adm}}(X_v^{\text{an}}, \mathbf{R})$ of $\mathscr{C}(X_v^{\text{an}}, \mathbf{R})$.

Proposition 6.2 (Gubler [30, theorem 7.12]) *For every place $v \in S$, the subspace $\mathscr{C}_{\text{adm}}(X_v^{\text{an}}, \mathbf{R})$ is dense in $\mathscr{C}(X_v^{\text{an}}, \mathbf{R})$.*

The space $\mathscr{C}_{\text{adm}}(X_{\text{ad}}, \mathbf{R})$ of admissible functions is dense in $\mathscr{C}_{\text{c}}(X_{\text{ad}}, \mathbf{R})$.

Proof Observe that X_v^{an} is a compact topological space. By corollary 7.7 and lemma 7.8 of Gubler [30], the subspace of $\mathscr{C}_{adm}(X_v^{an}, \mathbf{R})$ corresponding to algebraic v-adic metrics on L separates points and is stable under sup and inf. The first part of the proposition thus follows from Stone's density theorem.

The second part follows from the first one and a straightforward argument. □

Theorem 6.3 *Let Z be an integral closed subscheme of X, let $d = \dim(Z)$, let $\overline{L_1}, \ldots, \overline{L_d}$ be admissible adelically metrized line bundles on X.*

1. *There exists a unique measure $c_1(\overline{L_1}) \ldots c_1(\overline{L_d})\delta_Z$ on X_{ad} such that*

$$\int_{X_{ad}} \phi_0 c_1(\overline{L_1}) \ldots c_1(\overline{L_d})\delta_Z = \widehat{\deg}(\hat{c}_1(\mathscr{O}_X(\phi_0))\hat{c}_1(\overline{L_1}) \ldots \hat{c}_1(\overline{L_d}) \mid Z)$$

   for every compactly supported admissible function ϕ_0 on X_{ad}.
2. *This measure is supported on Z_{ad}; for every place v of S, the mass of its restriction to X_v^{an} is equal to*

$$\int_{X_v^{an}} c_1(\overline{L_1}) \ldots c_1(\overline{L_d})\delta_Z = \deg(c_1(L_1) \cdots c_1(L_d) \mid Z).$$

   If $\overline{L_1}, \ldots, \overline{L_d}$ are semipositive, then this measure is positive.
3. *The induced map from $\overline{\mathrm{Pic}}^{adm}(X)^d$ to the space $\mathscr{M}(X_{ad})$ of measures on X_{ad} is d-linear and symmetric.*
4. *Every admissible function is integrable for this measure.*

Proof Let us first assume that $\overline{L_1}, \ldots, \overline{L_d}$ are semipositive. It then follows from the definition of the arithmetic intersection degrees that the map

$$\phi_0 \mapsto \widehat{\deg}(\hat{c}_1(\mathscr{O}_X(\phi_0))\hat{c}_1(\overline{L_1}) \ldots \hat{c}_1(\overline{L_d}) \mid Z)$$

is a positive linear form on $\mathscr{C}_{adm}(X_{ad}, \mathbf{R})$. By the density theorem, it extends uniquely to a positive linear form on $\mathscr{C}_c(X_{ad}, \mathbf{R})$, which then corresponds to an inner regular, locally finite, positive Borel measure on X_{ad}.

The rest of the theorem follows from this. □

Remark 6.4

1. At archimedean places, the construction of the measure $c_1(\overline{L_1}) \ldots c_1(\overline{L_d})\delta_Z$ shows that it coincides with the measure defined by Bedford and Taylor [4] and Demailly [20].
2. At finite places, it has been first given in Chambert-Loir [13]. By approximation, the definition of the measure in the case of a general semipositive p-adic metric is then deduced from the case of algebraic metrics, given by a model $(\mathscr{X}, \mathscr{L})$. In this case, the measure $c_1(\overline{L_1}) \ldots c_1(\overline{L_d})\delta_Z$ on X_p^{an} has finite support. Let us describe it when $Z = X$ and the model $\mathscr{X}$ (the general case follows). For each component $\mathscr{Y}$ of $\mathscr{X} \otimes \mathbf{F}_p$, there exists a unique point $y \in X_p^{an}$ whose specialization

is the generic point of $\mathscr{Y}$. The contribution of the point y to the measure is then equal to

$$m_{\mathscr{Y}} \deg(c_1(\mathscr{L}_1) \ldots c_1(\mathscr{L}_d) \mid \mathscr{Y}),$$

where $m_{\mathscr{Y}}$ is the multiplicity of $\mathscr{Y}$ in the special fiber, that is, the length of the ideal (p) at the generic point of $\mathscr{Y}$.

Example 6.5 Let X be an abelian variety of dimension d over a number field F. Let $\overline{L}$ be an ample line bundle equipped with a canonical adelic metric; let us then describe the measure $c_1(\overline{L})^d$ on X_v^{an}, for every place $v \in S$. For simplicity, we assume that $F = \mathbf{Q}$.

1. First assume $v = \infty$. Then X_∞^{an} is the quotient, under complex conjugation, of the complex torus $X(\mathbf{C})$, and the canonical measure on X_∞^{an} is the direct image of the unique Haar measure on $X(\mathbf{C})$ with total mass $\deg(c_1(L)^d \mid X)$.
2. The situation is more interesting in the case of a finite place p.

    If X has good reduction at p, that is, if it extends to an abelian scheme $\mathscr{X}$ over $\mathbf{Z}_p$, then the canonical measure is supported at the unique point of X_p^{an} whose specialization is the generic point of $\mathscr{X} \otimes \mathbf{F}_p$.

    Let us assume, on the contrary, that X has (split) totally degenerate reduction. In this case, the uniformization theory of abelian varieties shows that X_p^{an} is the quotient of a torus $(\mathbf{G}_m^d)^{\mathrm{an}}$ by a lattice Λ. The definition of $(\mathbf{G}_m^d)^{\mathrm{an}}$ shows that this analytic space contains a canonical d-dimensional real vector space V, and V/Λ is a real d-dimensional torus $S(X_p^{\mathrm{an}})$ contained in X_p^{an}, sometimes called its skeleton. Gubler [31] has shown that the measure $c_1(\overline{L})^d$ on X_p^{an} coincides with the Haar measure on $S(X_p^{\mathrm{an}})$ with total mass $\deg(c_1(L)^d \mid X)$.

    The general case is a combination of these two cases, see Gubler [32].

Remark 6.6 At finite places, the theory described in this section defines measures $c_1(\overline{L}_1) \ldots c_1(\overline{L}_d) \delta_Z$ without defining the individual components $c_1(\overline{L}_1), \ldots, c_1(\overline{L}_d), \delta_Z$.

In Chambert-Loir and Ducros [14], we propose a theory of real differential forms and currents on Berkovich analytic spaces that allows a more satisfactory analogy with the theory on complex spaces. In particular, we provide an analogue of the Poincaré–Lelong equation, and semipositive metrized line bundles possess a curvature current $c_1(\overline{L})$ (curvature form in the "smooth" case) whose product can be defined and coincides with the measures of the form $c_1(\overline{L}_1) \ldots c_1(\overline{L}_d) \delta_Z$ that we have discussed.

7 Volumes

7.1 Let X be a proper $\mathbf{Q}$-scheme and let $\overline{L}$ be a line bundle endowed with an adelic metric.

The Riemann-Roch space $H^0(X, L)$ is a finite dimensional $\mathbf{Q}$-vector space. For every place $v \in S$, we endow it with a v-adic semi-norm:

$$\|s\|_v = \sup_{x \in X_v^{an}} \|s(x)\|$$

for $s \in H^0(X, L)$. If X is reduced, then this is a norm; let then B_v be its unit ball.

In the case where the metric on L is algebraic, given by a model $\mathscr{L}$ on a *normal* model $\mathscr{X}$ of X, the real vector space $H^0(X, L)_{\mathbf{R}}$ admits a natural lattice $H^0(\mathscr{X}, \mathscr{L})$ and the norm $\|\cdot\|_\infty$ at the archimedean place. The real space $H^0(X, L)_{\mathbf{R}}$ can then be endowed with the unique Haar measure for which B_∞ has volume 1, and we can then consider the covolume of the lattice $H^0(\mathscr{X}, \mathscr{L})$. Minkowski's first theorem will then eventually provide non-trivial elements of $H^0(\mathscr{X}, \mathscr{L})$ of controlled norm at the archimedean place.

The following construction, due to Bombieri and Vaaler [9], provides the analogue in the adelic setting.

Let $\mathbf{A}$ be the ring of adeles of $\mathbf{Q}$ and let μ be a Haar measure on $H^0(X, L) \otimes \mathbf{A}$. Then $\prod_{v \in S} B_v$ has finite strictly positive volume in $H^0(X, L)_{\mathbf{A}}$, and one defines

$$\chi(X, \overline{L}) = -\log\left(\frac{\mu(H^0(X, L) \otimes \mathbf{A}/H^0(X, L))}{\mu(\prod_v B_v)}\right). \tag{11}$$

This does not depend on the choice of the Haar measure μ.

One also defines

$$\widehat{H}^0(X, \overline{L}) = \{s \in H^0(X, L) \,;\, \|s\|_v \leq 1 \text{ for all } v \in S\}. \tag{12}$$

This is a finite set. We then let

$$\widehat{h}^0(X, \overline{L}) = \log\left(\mathrm{Card}(\widehat{H}^0(X, \overline{L}))\right). \tag{13}$$

The following inequality is an adelic analogue of Blichfeldt's theorem, itself a close companion to the adelic analogue of Minkowski's first theorem proved by Bombieri and Vaaler [9]. We refer to corollaire 2.12 of Gaudron [24] for details.

Lemma 7.2 *One has*

$$\chi(X, \overline{L}) \leq \widehat{h}^0(X, \overline{L}) + h^0(X, L)\log(2).$$

7.3 The volume and the χ-volume of $\overline{L}$ are defined by the formulas:

$$\widehat{\mathrm{Vol}}(X, \overline{L}) = \varlimsup_{n \to \infty} \frac{\widehat{h}^0(X, \overline{L}^n)}{n^{d+1}/(d+1)!} \tag{14}$$

$$\widehat{\mathrm{Vol}}_\chi(X, \overline{L}) = \lim_{n \to \infty} \frac{\chi(X, \overline{L}^n)}{n^{d+1}/(d+1)!}. \tag{15}$$

One thus has the inequality

$$\widehat{\mathrm{Vol}}_\chi(X, \overline{L}) \le \widehat{\mathrm{Vol}}(X, \overline{L}). \tag{16}$$

In fact, it has been independently shown by Yuan [45] and Chen [16] that the volume is in fact a limit.

The relation between volumes and heights follows from the following result.

Lemma 7.4 *Assume that L is ample. Then, for every real number t such that*

$$t < \frac{\widehat{\mathrm{Vol}}_\chi(X, \overline{L})}{(d+1)\,\mathrm{Vol}(X, L)},$$

the set of closed points $x \in X$ such that $h_{\overline{L}}(x) \le t$ is not dense for the Zariski topology.

Proof Consider the adelically metrized line bundle $\overline{L}(-t)$, whose metric at the archimedean place has been multiplied by e^t. It follows from the definition of the χ-volume that

$$\widehat{\mathrm{Vol}}_\chi(X, \overline{L}(-t)) = \widehat{\mathrm{Vol}}_\chi(X, \overline{L}) - (d+1)t\,\mathrm{Vol}(X, L).$$

Indeed, for every finite place p, changing $\overline{L}$ to $\overline{L}(-t)$ does not modify the balls B_p in $H^0(X, \overline{L}^n) \otimes_{\mathbf{Q}} \mathbf{Q}_p$, while it dilates it by the ratio e^{-nt} at the archimedean place, so that its volume is multiplied by $e^{-nt\,\dim(H^0(X,L^n))}$.

Consequently,

$$\widehat{\mathrm{Vol}}(X, \overline{L}(-t)) \ge \widehat{\mathrm{Vol}}_\chi(X, \overline{L}(-t)) \ge \widehat{\mathrm{Vol}}_\chi(X, \overline{L}) - (d+1)t\,\mathrm{Vol}(X, L) > 0.$$

In particular, there exists an integer $n \ge 1$ and a nonzero section $s \in H^0(X, \overline{L}^n)$ such that $\|s\|_p \le 1$ for all finite places p, and $\|s\|_\infty \le e^{-nt}$. Let now $x \in X$ be a closed point that is not contained in $|\mathrm{div}(s)|$; one then has

$$h_{\overline{L}}(x) = \sum_{v \in S} \int_{X_v^{\mathrm{an}}} \log\|s\|_v^{-1/n} \delta_v(x) \ge t,$$

whence the lemma.   □

Theorem 7.5 *Assume that $\overline{L}$ is semipositive. Then one has*

$$\widehat{\mathrm{Vol}}(X, \overline{L}) = \widehat{\mathrm{Vol}}_\chi(X, \overline{L}) = \widehat{\deg}\left(\hat{c}_1(\overline{L})^{d+1} \mid X\right). \tag{17}$$

This is the arithmetic Hilbert–Samuel formula, due to Gillet and Soulé [27], Bismut and Vasserot [7] when X is smooth and the adelic metric of $\overline{L}$ is algebraic. Abbes and Bouche [2] later gave an alternative proof. In the given generality, the formula is a theorem of Zhang [47, 48].

Theorem 7.6

1. *The function* $\overline{L} \mapsto \widehat{\mathrm{Vol}}(X, \overline{L})$ *extends uniquely to a continuous function on the real vector space* $\widehat{\mathrm{Pic}}(X) \otimes_{\mathbf{Q}} \mathbf{R}$.
2. *If* $\widehat{\mathrm{Vol}}(X, \overline{L}) > 0$, *then* $\widehat{\mathrm{Vol}}$ *is differentiable at* $\overline{L}$.
3. *If* $\overline{L}$ *is semipositive, then* $\widehat{\mathrm{Vol}}$ *and* $\widehat{\mathrm{Vol}}_\chi$ *are differentiable at* $\overline{L}$, *with differential*

$$\overline{M} \mapsto (d+1)\,\widehat{\deg}(\hat{c}_1(\overline{L})^d \hat{c}_1(\overline{M}) \mid X).$$

This theorem is proved by Chen [17] as a consequence of results of Yuan [44, 45]. It essentially reduces from the preceding one in the case $\overline{L}$ is defined by an ample line bundle on a model of X, and its metric has strictly positive curvature. Reaching the "boundary" of the cone of semipositive admissible metrized line bundles was the main result of Yuan [44] who proved that for every admissible metrized line bundle $\overline{M}$ and every large enough integer t, one has

$$t^{-(d+1)}\,\widehat{\mathrm{Vol}}_\chi(X, \overline{L}^t \otimes \overline{M})$$

$$\geq \widehat{\mathrm{Vol}}_\chi(X, \overline{L}) + \frac{1}{t}(d+1)\,\widehat{\deg}(\hat{c}_1(\overline{L})^d \hat{c}_1(\overline{M}) \mid X) + o(1/t).$$

It is this inequality, an arithmetic analogue of an inequality of Siu, will be crucial for the applications to equidistribution in the next section.

8 Equidistribution

The main result of this section is the equidistribution theorem 8.4. It has been first proved in the case $v = \mathbf{C}$ by Szpiro et al. [40], under the assumption that the given archimedean metric is smooth and has a strictly positive curvature form, and the general case is due to Yuan [44]. However, our presentation derives it from a seemingly more general result, Lemma 8.2, whose proof, anyway, closely follows their methods. Note that for the application to Bogomolov's conjecture in Sect. 9, the initial theorem of Szpiro et al. [40] is sufficient.

Definition 8.1 Let X be a proper $\mathbf{Q}$-scheme, let $\overline{L}$ be an ample line bundle on X endowed with an admissible adelically metric. Let (x_n) be a sequence of closed points of X.

1. One says that (x_n) is *generic* if for every strict closed subscheme Z of X, the set of all $n \in \mathbf{N}$ such that $x_n \in Z$ is finite; in other words, this sequence converges to the generic point of X.
2. One says that (x_n) is *small* (relative to $\overline{L}$) if

$$h_{\overline{L}}(x_n) \to h_{\overline{L}}(X).$$

Lemma 8.2 *Let X be a proper $\mathbf{Q}$-scheme, let $d = \dim(X)$, let $\overline{L}$ be a semipositive adelically metrized line bundle on X such that L is ample. Let (x_n) be a generic sequence of closed points of X which is small relative to $\overline{L}$. For every line bundle $\overline{M}$ on X endowed with an admissible adelic metric, one has*

$$\lim_{n \to \infty} h_{\overline{M}}(x_n) = \frac{\widehat{\deg}(\hat{c}_1(\overline{L})^d \hat{c}_1(\overline{M}) \mid X)}{\deg_L(X)} - dh_{\overline{L}}(X) \frac{\deg(c_1(L)^{d-1} c_1(M) \mid X)}{\deg_L(X)}.$$

Proof Since L is ample, $L^t \otimes M$ is ample for every large enough integer t, and the classical Hilbert-Samuel formula implies that

$$\frac{1}{t^d} \operatorname{Vol}(X, L^t \otimes M) = \deg(c_1(L)^d \mid X)$$

$$+ dt^{-1} \deg(c_1(L)^{d-1} c_1(M) \mid X) + \mathrm{O}(t^{-2})$$

when $t \to \infty$. Since $\overline{L}$ is semipositive and L is ample, the main inequality of Yuan [44] implies that

$$\frac{1}{t^{d+1}} \widehat{\operatorname{Vol}}_\chi(X, \overline{L}^t \otimes \overline{M}) \geq \widehat{\deg}(\hat{c}_1(\overline{L})^{d+1} \mid X)$$

$$+ (d+1)t^{-1} \widehat{\deg}(\hat{c}_1(\overline{L})^d \hat{c}_1(\overline{M}) \mid X) + \mathrm{o}(t^{-1}).$$

Consequently, when $t \to \infty$, one has

$$\frac{\widehat{\operatorname{Vol}}_\chi(X, \overline{L}^t \otimes \overline{M})}{\operatorname{Vol}(X, L^t \otimes M)} \geq t \frac{\widehat{\deg}(\hat{c}_1(\overline{L})^{d+1} \mid X)}{\deg(c_1(L)^d \mid X)}$$

$$+ (d+1) \frac{\widehat{\deg}(\hat{c}_1(\overline{L})^d \hat{c}_1(\overline{M}) \mid X)}{\deg(c_1(L)^d \mid X)}$$

$$- d \frac{\widehat{\deg}(\hat{c}_1(\overline{L})^{d+1} \mid X)}{\deg(c_1(L)^d \mid X)} \frac{\deg(c_1(L)^{d-1} c_1(M) \mid X)}{\deg(c_1(L)^d \mid X)} + \mathrm{o}(1).$$

The sequence (x_n) is generic, hence Lemma 7.4 furnishes the inequality:

$$\varliminf_n h_{\overline{L}^t \otimes \overline{M}}(x_n) \geq \frac{\widehat{\mathrm{Vol}}_\chi(X, \overline{L}^t \otimes M)}{(d+1)\,\mathrm{Vol}(X, L^t \otimes M)}.$$

We observe that

$$\varliminf_n h_{\overline{L}^t \otimes \overline{M}}(x_n) = t \lim h_{\overline{L}}(x_n) + \varliminf_n h_{\overline{M}}(x_n),$$

so that, when $t \to \infty$, we have

$$\varliminf_n h_{\overline{M}}(x_n) \geq \frac{\widehat{\deg}(\hat{c}_1(\overline{L})^d \hat{c}_1(\overline{M}) \mid X)}{\deg(c_1(L)^d \mid X)}$$

$$- \frac{d}{d+1} \frac{\widehat{\deg}(\hat{c}_1(\overline{L})^{d+1} \mid X)}{\deg(c_1(L)^d \mid X)} \frac{\deg(c_1(L)^{d-1} c_1(M) \mid X)}{\deg(c_1(L)^d \mid X)}.$$

Applying this inequality for $\overline{M}^{-1}$ shows that $\varlimsup_n h_{\overline{M}}(x_n)$ is bounded above by its right hand side. The lemma follows. $\qquad\qquad\square$

8.3 Let X be a proper $\mathbf{Q}$-scheme. Let $v \in S$ be a place of $\mathbf{Q}$.

Let $x \in X$ be a closed point. Let $F = \kappa(x)$; this is a finite extension of $\mathbf{Q}$, and there are exactly $[F : \mathbf{Q}]$ geometric points on $X(\mathbf{C}_v)$ whose image is x, permuted by the Galois group $\mathrm{Gal}(\mathbf{C}_v/\mathbf{Q}_v)$. We consider the corresponding "probability measure" in $X(\mathbf{C}_v)$, giving mass $1/[F : \mathbf{Q}]$ to each of these geometric points, and let $\delta_v(x)$ be its image under the natural map $X(\mathbf{C}_p) \to X_v^{\mathrm{an}}$.

By construction, $\delta_v(x)$ is a probability measure on X_v^{an} with finite support, a point of X_v^{an} being counted proportionaly to the number of its liftings to a geometric point supported by x.

The space of measures on X_v^{an} is the dual of the space $\mathscr{C}(X_v^{\mathrm{an}}, \mathbf{R})$; it is endowed with the topology of pointwise convergence ("weak-$*$ topology").

Theorem 8.4 *Let X be a proper $\mathbf{Q}$-scheme, let $d = \dim(X)$, let $\overline{L}$ be a semipositive adelically metrized line bundle on X such that L is ample. Let (x_n) be a generic sequence of closed points of X which is small relative to $\overline{L}$. Then for each place $v \in S$, the sequence of measures $(\delta_v(x_n))$ on X_v^{an} converges to the measure $c_1(\overline{L})^d / \deg(c_1(L)^d \mid X)$.*

Proof Let $\mu_{\overline{L}}$ denote the probability measure $c_1(\overline{L})^d / \deg_L(X)$ on X_v^{an} and let $f \in \mathscr{C}(X_v^{\mathrm{an}}, \mathbf{R})$ be an admissible function, extended by zero to an element of $\mathscr{C}_{\mathrm{adm}}(X_{\mathrm{ad}}, \mathbf{R})$. We apply Lemma 8.2 to the metrized line bundle $\overline{M} = \overline{\mathscr{O}_X(f)}$ whose underlying line bundle on X is trivial. For every closed point $x \in X$, one has

$$h_{\overline{M}}(x) = \int_{X_v^{\mathrm{an}}} f \delta_v(x).$$

Moreover,

$$\widehat{\deg}(\hat{c}_1(\overline{L})^d \hat{c}_1(\overline{M}) \mid X) = \int_{X_v^{an}} f c_1(\overline{L})^d.$$

It thus follows from Lemma 8.2 that

$$\lim_{n \to \infty} \int_{X_v^{an}} f \delta_v(x_n) = \frac{\widehat{\deg}(\hat{c}_1(\overline{L})^d \hat{c}_1(\overline{M}) \mid X)}{\deg(c_1(L)^d \mid X)}$$

$$= \frac{1}{\deg(c_1(L)^d \mid X)} \int_{X_v^{an}} f c_1(\overline{L})^d.$$

The case of an arbitrary continuous function on X_v^{an} follows by density. $\qquad \square$

9 The Bogomolov Conjecture

9.1 Let X be an abelian variety over a number field F, that is, a projective connected algebraic variety over F which is endowed with an additional structure of an algebriac group. For every integer n, we write $[n]$ for the multiplication by n-morphism on X.

Let us first explain how the theory of canonical adelic metrics allows to extend the Néron–Tate height to arbitrary integral closed subschemes. For alternative and independent presentations, see Philippon [36], Gubler [29], and Bost et al. [10].

Let L be a line bundle on X trivialized at the origin.

If L is even ($[-1]^*L \simeq L$), the theorem of the cube implies that there exists, for every integer n, a unique isomorphism $[n]^*L \simeq L^{n^2}$ which is compatible with the trivialization at the origin. If $n \geq 2$, then by Example 4.11, it admits a unique adelic metric for which this isomorphism is an isometry; these isomorphisms are then all isometries.

Similarly, if L is odd ($[-1]^*L \simeq L^{-1}$), then it admits a unique adelic metric for which the canonical isomorphisms $[n]^*L \simeq L^n$ are isometries, for all integer n.

In general, one can write $L^2 \simeq (L \otimes [-1]^*L) \otimes (L \otimes [-1]^*L^{-1})$, as the sum of an even and an odd line bundle, and this endows L with an adelic metric. This adelic metric is called the *canonical adelic metric* on L (compatible with the given trivialization at the origin).

If L is ample and even, then the canonical adelic metric on L is semipositive. This implies that the canonical adelic metric of an arbitrary even line bundle is admissible.

Assume that L is odd. Fix an even ample line bundle M. Up to extending the scalars, there exists a point $a \in X(F)$ such that $L \simeq \tau_a^*M \otimes M^{-1}$, where τ_a is the translation by a on X. Then there exists a unique isomorphism $L \simeq \tau_a^*M \otimes M^{-1} \otimes M_a^{-1}$ which is compatible with the trivialization at the origin, and this gives rise to an isometry $\overline{L} \simeq \tau_a^* \overline{M} \otimes \overline{M}^{-1} \otimes \overline{M}_a^{-1}$. In particular, the adelic metric of $\overline{M}$ is admissible. In fact, it follows from a construction of Künnemann that it is even semipositive, see Chambert-Loir [12].

9.2 In particular, let us consider an ample even line bundle L on X endowed with a canonical adelic metric. This furnishes a *height*

$$h_{\overline{L}}(Z) = \frac{\widehat{\deg}(\hat{c}_1(\overline{L})^{d+1} \mid Z)}{(d+1) \deg(c_1(L)^d \mid Z)},$$

for every integral closed subscheme Z of X, where $d = \dim(Z)$.
   In fact, if $(\mathscr{X}, \overline{\mathscr{L}})$ is any model of (X, L), one has

$$h_{\overline{L}}(Z) = \lim_{n \to +\infty} n^{-2} h_{\overline{\mathscr{L}}}([n](Z)),$$

which shows the relation of the point of view of adelic metrics with Tate's definition of the Néron–Tate height, initially defined on closed points. This formula also implies that $h_{\overline{L}}$ is positive.
   More generally, if Z is an integral closed subscheme of $X_{\overline{F}}$, we let $h_{\overline{L}}(Z) = h_{\overline{L}}([Z])$, where $[Z]$ is its Zariski-closure in X (more precisely, the smallest closed subscheme of X such that $[Z]_{\overline{F}}$ contains Z).

Lemma 9.3 *The induced height function $h_{\overline{L}}: X(\overline{F}) \to \mathbf{R}$ is a positive quadratic form. It induces a positive definite quadratic form on $X(\overline{F}) \otimes \mathbf{R}$. In particular, a point $\in X(\overline{F})$ satisfies $h_{\overline{L}}(x) = 0$ if and only if x is a torsion point.*

Proof For $I \subset \{1, 2, 3\}$, let $p_I: X^3 \to X$ be the morphism given by $p_I(x_1, x_2, x_3) = \sum_{i \in I} x_i$. The cube theorem asserts that the line bundle

$$\mathscr{D}_3(L) = \bigotimes_{\emptyset \neq I \subset \{1,2,3\}} (p_I^* L)^{(-1)^{\mathrm{Card}(I)-1}}$$

on X^3 is trivial, and admits a canonical trivialisation. The adelic metric of $\overline{L}$ endows it with an adelic metric which satisfies $[2]^* \mathscr{D}_3(\overline{L}) \simeq \mathscr{D}_3(\overline{L})^4$, hence is the trivial metric. This implies the following relation on heights:

$$h_{\overline{L}}(x+y+z) - h_{\overline{L}}(y+z) - h_{\overline{L}}(x+z) - h_{\overline{L}}(x+y) + h_{\overline{L}}(x) + h_{\overline{L}}(y) + h_{\overline{L}}(z) \equiv 0$$

on $X(\overline{F})^3$. Consequently,

$$(x, y) \mapsto h_{\overline{L}}(x+y) - h_{\overline{L}}(x) - h_{\overline{L}}(y)$$

is a symmetric bilinear form on $X(\overline{F})$. Using furthermore that $h_{\overline{L}}(-x) = h_{\overline{L}}(x)$ for all x, we deduce that $h_{\overline{L}}$ is a quadratic form on $X(\overline{F})$.

Since L is ample, $h_{\overline{L}}$ is bounded from below. The formula $h_{\overline{L}}(x) = h_{\overline{L}}(2x)/4$ then implies that $h_{\overline{L}}$ is positive. By what precedes, it induces a positive quadratic form on $X(\overline{F})_{\mathbf{R}}$.

Let us prove that it is in fact positive definite. By definition, it suffices that its restriction to the subspace generated by finitely many points $x_1, \ldots, x_m \in X(\overline{F})$ is positive definite. Let E be a finite extension of F such that $x_1, \ldots, x_m \in X(E)$. On the other hand, Northcott's theorem implies that for every real number t, the set of $(a_1, \ldots, a_m) \in \mathbf{Z}^m$ such that $h_{\overline{L}}(a_1 x_1 + \cdots + a_m x_m) \leq t$ is finite. One deduces from that the asserted positive definiteness. $\square$

Definition 9.4 A *torsion subvariety* of $X_{\overline{F}}$ is a subvariety of the form $a + Y$, where $a \in X(\overline{F})$ is a torsion point and Y is an abelian subvariety of $X_{\overline{F}}$.

Theorem 9.5

(a) *Let Z be an integral closed subscheme of $X_{\overline{F}}$. One has $h_{\overline{L}}(Z) = 0$ if and only if Z is a torsion subvariety of $X_{\overline{F}}$.*

(b) *Let Z be an integral closed subscheme of $X_{\overline{F}}$ which is not a torsion subvariety. There exists a strictly positive real number δ such that the set*

$$\{x \in Z(\overline{F})\,;\ h_{\overline{L}}(x) \leq \delta\}$$

is not Zariski-dense in $Z_{\overline{F}}$.

Assertion (a) has been independently conjectured by Philippon [36, 37] and Zhang [48]. Assertion (b) has been conjectured by Bogomolov [8] in the particular case where Z is a curve of genus $g \geq 2$ embedded in its jacobian variety; for this reason, it is called the "generalized Bogomolov conjecture". The equivalence of (a) and (b) is a theorem of Zhang [48]. In fact, the implication (a)$\Rightarrow$(b) already follows from Theorem 7.5 and Lemma 7.4. Assume indeed that (a) holds and let Z be an integral closed subscheme of X which is not a torsion subvariety; by (a), one has $h_{\overline{L}}(Z) > 0$, and one may take for δ any real number such that $0 < \delta < h_{\overline{L}}(Z)$.

Theorem 9.5 has been proved by Zhang [49], following a breakthrough of Ullmo [41] who treated the case of a curve embedded in its jacobian; their proof makes use of the equidistribution theorem. Soon after, David and Philippon [18] gave an alternative proof; when Z is not a translate of an abelian subvariety, their proof provides a strictly positive lower bound for $h_{\overline{L}}(Z)$ (in (a)) as well as an explicit real number δ (in (b)) which only depends on the dimension and the degree of Z with respect to L.

As a corollary of Theorem 9.5, one obtains a new proof of the Manin–Mumford conjecture in characteristic zero, initially proved by Raynaud [38].

Corollary 9.6 *Let X be an abelian variety over an algebraically closed field of characteristic zero, let Z be an integral closed subscheme of X which is not a torsion subvariety. Then the set of torsion points of X which are contained in Z is not Zariski-dense in Z.*

Proof A specialization argument reduces to the case where X is defined over a number field F. In this case, the torsion points of X are defined over $\overline{F}$ and are characterized by the vanishing of their Néron–Tate height relative to an(y) ample line bundle L on X. It is thus clear that the corollary follows from Theorem 9.5, (b).

□

9.7 The proof of Theorem 9.5, (b), begins with the observation that the statement does not depend on the choice of the ample line L on X. More precisely, if $\overline{M}$ is another symmetric ample line bundle on X endowed with a canonical metric, then there exists an integer $a \geq 1$ such that $\overline{L}^a \otimes \overline{M}^{-1}$ is ample, as well as $\overline{M}^a \otimes \overline{L}^{-1}$. Consequently, $h_{\overline{L}} \geq a^{-1} h_{\overline{M}}$ and $h_{\overline{M}} \geq a^{-1} h_{\overline{L}}$. From these two inequalities, one deduces readily that the statement holds for $\overline{L}$ if and only if it holds for $\overline{M}$.

For a similar reason, if $f \colon X' \to X$ is an isogeny of abelian varieties, then the statements for X and X' are equivalent. Let indeed Z be an integral closed subvariety of $X_{\overline{F}}$ and let Z' be an irreducible component of $f^{-1}(Z)$. Then Z is a torsion subvariety of $X_{\overline{F}}$ if and only if Z' is a torsion subvariety of $X'_{\overline{F}}$. On the other hand, the relation $h_{f*\overline{L}}(x) = h_{\overline{L}}(f(x))$ shows that $h_{f*\overline{L}}$ has a strictly positive lower bound on Z' outside of a strict closed subset E' if and only if $h_{\overline{L}}$ has a strictly positive lower bound on Z outside of the strict closed subset $f(E')$.

9.8 Building on that observation, one reduces the proof of the theorem to the case where the stabilizer of Z is trivial.

Let indeed X'' be the neutral component of this stabilizer and let $X' = X/X''$; this is an abelian variety. By Poincaré's complete reducibility theorem, there exists an isogeny $f \colon X' \times X'' \to X$. This reduces us to the case where $X = X' \times X''$ and $Z = Z' \times X''$, for some integral closed subscheme Z' of $X'_{\overline{F}}$. We may also assume that $\overline{L} = \overline{L}' \boxtimes \overline{L}''$. It then clear that the statement for (X', Z') implies the desired statement for (X, Z).

Lemma 9.9 *Assume that* $\dim(Z) > 0$ *and that its stabilizer is trivial. Then, for every large enough integer* $m \geq 1$, *the morphism*

$$ f \colon Z^m \to X_{\overline{F}}^{m-1}, \qquad (x_1, \ldots, x_m) \mapsto (x_2 - x_1, \ldots, x_m - x_{m-1}) $$

is birational onto its image but not finite.

Proof Let m be an integer and let $x = (x_1, \ldots, x_m)$ be an $\overline{F}$-point of Z^m. Then a point $y = (y_1, \ldots, y_m) \in Z(\overline{F})^m$ belongs to the same fiber as x if and only if $y_2 - y_1 = x_2 - x_1, \ldots$, that is, if and only if, $y_1 - x_1 = y_2 - x_2 = \cdots = y_m - x_m$. This identifies $f^{-1}(f(x))$ with the intersection $(Z - x_1) \cap \ldots \cdots \cap (Z - x_m)$ of translates of Z. If m is large enough and $x_1, \ldots, x_m$ are well chosen in Z, then this intersection is equal to stabilizer of Z in $X_{\overline{F}}$, hence is reduced to a point. In that case, the morphism f has a fiber reduced to a point, hence it is generically injective.

On the other hand, the preimage of the origin $(o, \ldots, o)$ contains the diagonal of Z^m, which has strictly positive dimension by hypothesis.   □

9.10 For the proof of Theorem 9.5, (b), we now argue by contradiction and assume the existence of a generic sequence (x_n) in $Z(\overline{F})$ such that $h_{\overline{L}}(x_n) \to 0$.

Having reduced, as explained above, to the case where the stabilizer of Z is trivial, we consider an integer $m \geq 1$ such that the morphism $f \colon Z^m \to X_{\overline{F}}^{m-1}$ is birational onto its image, but not finite.

Since the set of strict closed subschemes of Z is countable, one can construct a generic sequence (y_n) in Z^m where y_n is of the form $(x_{i_1}, \ldots, x_{i_m})$. One has $h_{\overline{L}}(y_n) \to 0$, where, by abuse of language, we write $h_{\overline{L}}$ for the height on X^m induced by the adelically metrized line bundle $\overline{L} \boxtimes \cdots \boxtimes \overline{L}$ on X^m. This implies that $h_{\overline{L}}(Z) = 0$, hence the sequence (y_n) is *small*.

For every integer n, let $z_n = f(y_n)$. By continuity of a morphism of schemes, the sequence (z_n) is generic in $f(Z^m)$. Moreover, we deduce from the quadratic character of the Néron–Tate height $h_{\overline{L}}$ that $h_{\overline{L}}(z_n) \to 0$. In particular, $h_{\overline{L}}(f(Z^m)) = 0$, and the sequence (z_n) is small.

Fix an archimedean place σ of F. Applied to the sequences (y_n) and (z_n), the equidistribution theorem 8.4 implies the following convergences:

$$\lim_{n \to \infty} \delta_\sigma(y_n) \propto c_1(\overline{L} \boxtimes \cdots \boxtimes \overline{L})^{m \dim(Z)} \delta_{Z^m}$$

$$\lim_{n \to \infty} \delta_\sigma(z_n) \propto c_1(\overline{L} \boxtimes \cdots \boxtimes \overline{L})^{\dim(f(Z^m))} \delta_{f(Z^m)},$$

where, by $\propto$, I mean that both sides are proportional. (The proportionality ratio is the degree of Z^m, *resp.* of $f(Z^m)$, with respect to the indicated measure.) Since $f(y_n) = z_n$, we conclude that the measures

$$f_*(c_1(\overline{L} \boxtimes \cdots \boxtimes \overline{L})^{m \dim(Z)} \delta_{Z^m}) \quad \text{and} \quad c_1(\overline{L} \boxtimes \cdots \boxtimes \overline{L})^{\dim(f(Z^m))} \delta_{f(Z^m)}$$

on $f(Z^m)$ are proportional.

Recall that the archimedean metric of $\overline{L}$ has the property that it is smooth and that its curvature form $c_1(\overline{L})$ is a smooth strictly positive $(1, 1)$-form on $X_\sigma(\mathbf{C})$. Consequently, on a dense smooth open subscheme of $f(Z^m)$ above which f is an isomorphism, both measures are given by differential forms, which thus coincide there. We can pull back them to Z^m by f and obtain a proportionality of differential forms

$$c_1(\overline{L} \boxtimes \cdots \boxtimes \overline{L})^{m \dim(Z)} \propto f^* c_1(\overline{L} \boxtimes \cdots \boxtimes \overline{L})^{m \dim(Z)}$$

on $Z_\sigma(\mathbf{C})^m$. At this point, the contradiction appears: the differential form on the left is strictly positive at every point, while the one on the right vanishes at every point of $Z_\sigma^m(\mathbf{C})$ at which f is not smooth.

This concludes the proof of Theorem 9.5.

Remark 9.11 The statement of 9.5 can be asked in more general contexts that allow for canonical heights. The case of toric varieties has been proved by Zhang [47], while in that case the equidistribution result is first due to Bilu [6]. The case of semiabelian varieties is due to David and Philippon [19], by generalization of their proof for abelian varieties; I had proved in Chambert-Loir [12] the equidistribution result for almost-split semi-abelian varieties, and the general case has just been announced by Kühne [33].

The general setting of algebraic dynamics (X, f) is unclear. For a polarized dynamical system as in 4.11, the obvious and natural generalization proposed in Zhang [48] asserts that subvarieties of height zero are exactly those whose forward orbit is finite. However, Ghioca and Tucker have shown that it does not hold; see Ghioca et al. [26] for a possible rectification. The case of dominant endomorphisms of $(\mathbf{P}^1)^n$ is a recent theorem of Ghioca et al. [25].

References

1. A. Abbes, Hauteurs et discrétude. Séminaire Bourbaki, 1996/1997. Astérisque **245**, 141–166 (1997). Exp. 825
2. A. Abbes, T. Bouche, Théorème de Hilbert–Samuel "arithmétique". Ann. Inst. Fourier (Grenoble) **45**, 375–401 (1995)
3. S.J. Arakelov, An intersection theory for divisors on an arithmetic surface. Izv. Akad. Nauk SSSR Ser. Mat. **38**, 1179–1192 (1974)
4. E. Bedford, B. Taylor, A new capacity for plurisubharmonic functions. Acta Math. **149**, 1–40 (1982)
5. V.G. Berkovich, *Spectral Theory and Analytic Geometry Over Non-Archimedean Fields.* Mathematical Surveys and Monographs, vol. 33 (American Mathematical Society, Providence, 1990)
6. Yu. Bilu, Limit distribution of small points on algebraic tori. Duke Math. J. **89**, 465–476 (1997)
7. J.-M. Bismut, É. Vasserot, The asymptotics of the Ray–Singer analytic torsion associated with high powers of a positive line bundle. Comm. Math. Phys. **125**, 355–367 (1989)
8. F.A. Bogomolov, Points of finite order on abelian varieties. Izv. Akad. Nauk. SSSR Ser. Mat. **44**, 782–804, 973 (1980)
9. E. Bombieri, J. Vaaler, On Siegel's lemma. Invent. Math. **73**, 11–32 (1983)
10. J.-B. Bost, H. Gillet, C. Soulé, Heights of projective varieties and positive Green forms. J. Am. Math. Soc. **7**, 903–1027 (1994)
11. G. Call, J. Silverman, Canonical heights on varieties with morphisms. Compos. Math. **89**, 163–205 (1993)
12. A. Chambert-Loir, Géométrie d'Arakelov et hauteurs canoniques sur des variétés semi-abéliennes. Math. Ann. **314**, 381–401 (1999)
13. A. Chambert-Loir, Mesures et équidistribution sur des espaces de Berkovich. J. Reine Angew. Math. **595**, 215–235 (2006). Math.NT/0304023
14. A. Chambert-Loir, A. Ducros, Formes différentielles réelles et courants sur les espaces de Berkovich (2012). ArXiv:1204.6277, p. 124, revision in progress
15. A. Chambert-Loir, A. Thuillier, Mesures de Mahler et équidistribution logarithmique. Ann. Inst. Fourier (Grenoble) **59**, 977–1014 (2009)
16. H. Chen, Arithmetic Fujita approximation. Ann. Sci. Éc. Norm. Supér. (4) **43**, 555–578 (2010)
17. H. Chen, Differentiability of the arithmetic volume function. J. Lond. Math. Soc. (2) **84**, 365–384 (2011)
18. S. David, P. Philippon, Minorations des hauteurs normalisées des sous-variétés de variétés abéliennes. *International Conference on Discrete Mathematics and Number Theory.* Contemporary Mathematics, vol. 210 (Tiruchirapelli, 1998), pp. 333–364

19. S. David, P. Philippon, Minorations des hauteurs normalisées des sous-variétés de variétés abéliennes. II. Comment. Math. Helv. **77**, 639–700 (2002)
20. J.-P. Demailly, Mesures de Monge-Ampère et caractérisation géométrique des variétés algébriques affines. Mém. Soc. Math. France **19**, 124 (1985)
21. J.-P. Demailly, T. Peternell, M. Schneider, Compact complex manifolds with numerically effective tangent bundles. J. Algebraic Geom. **3**, 295–345 (1994)
22. G. Faltings, Diophantine approximation on abelian varieties. Ann. Math. (2) **133**, 549–576 (1991)
23. G. Faltings, *Lectures on the Arithmetic Riemann–Roch Theorem*. Annals of Mathematics Studies, vol. 127 (Princeton University Press, Princeton, 1992). Notes taken by Shouwu Zhang
24. É. Gaudron, Géométrie des nombres adélique et lemmes de Siegel généralisés. Manuscripta Math. **130**, 159–182 (2009)
25. D. Ghioca, K.D. Nguyen, H. Ye, The dynamical Manin–Mumford conjecture and the dynamical Bogomolov conjecture for endomorphisms of $(\mathbf{P}^1)^n$ (2017). ArXiv:1705.04873
26. D. Ghioca, T.J. Tucker, S. Zhang, Towards a dynamical Manin–Mumford conjecture. Internat. Math. Res. Not. **22**, 5109–5122 (2011)
27. H. Gillet, C. Soulé, Amplitude arithmétique. C. R. Acad. Sci. Paris Sér. I Math. **307**, 887–890 (1988)
28. H. Gillet, C. Soulé, Arithmetic intersection theory. Publ. Math. Inst. Hautes Études Sci. **72**, 94–174 (1990)
29. W. Gubler, Höhentheorie. Math. Ann. **298**, 427–455 (1994). With an appendix by Jürg Kramer
30. W. Gubler, Local heights of subvarieties over non-archimedean fields. J. Reine Angew. Math. **498**, 61–113 (1998)
31. W. Gubler, The Bogomolov conjecture for totally degenerate abelian varieties. Invent. Math. **169**, 377–400 (2007)
32. W. Gubler, Non-archimedean canonical measures on abelian varieties. Compos. Math. **146**, 683–730 (2010)
33. L. Kühne, Points of small height on semiabelian varieties (2018). arXiv:1808.00855
34. A. Moriwaki, *Arakelov Geometry*. Translations of Mathematical Monographs, vol. 244 (American Mathematical Society, Providence, 2014). Translated from the 2008 Japanese original
35. D.G. Northcott, Periodic points on an algebraic variety. Ann. Math. **51**,167–177 (1950)
36. P. Philippon, Sur des hauteurs alternatives, I. Math. Ann. **289**, 255–283 (1991)
37. P. Philippon, Sur des hauteurs alternatives, III. J. Math. Pures Appl. **74**, 345–365 (1995)
38. M. Raynaud, Sous-variétés d'une variété abélienne et points de torsion, in *Arithmetic and Geometry. Papers dedicated to I.R. Shafarevich*, ed. by M. Artin, J. Tate. Progress in Mathematics, vol. 35 (Birkhäuser, Basel, 1983), pp. 327–352
39. C. Soulé, D. Abramovich, J.-F. Burnol, J. Kramer, *Lectures on Arakelov Geometry*. Cambridge Studies in Advanced Mathematics, vol. 33 (Cambridge University Press, Cambridge, 1992)
40. L. Szpiro, E. Ullmo, S.-W. Zhang, Équidistribution des petits points. Invent. Math. **127**, 337–348 (1997)
41. E. Ullmo, Positivité et discrétion des points algébriques des courbes. Ann. Math. **147**, 167–179 (1998)
42. A. Weil, L'arithmétique sur les courbes algébriques. Acta Math. **52**, 281–315 (1929)
43. A. Weil, Arithmetic on algebraic varieties. Ann. Math. (2) **53**, 412–444 (1951)
44. X. Yuan, Big line bundles on arithmetic varieties. Invent. Math. **173**, 603–649 (2008)
45. X. Yuan, On volumes of arithmetic line bundles. Compos. Math. **145**, 1447–1464 (2009)
46. X. Yuan, S.-W. Zhang, The arithmetic Hodge index theorem for adelic line bundles. Math. Ann. **367**, 1123–1171 (2017)
47. S.-W. Zhang, Positive line bundles on arithmetic varieties. J. Am. Math. Soc. **8**, 187–221 (1995)
48. S.-W. Zhang, Small points and adelic metrics. J. Algebraic Geometry **4**, 281–300 (1995)
49. S.-W. Zhang, Equidistribution of small points on abelian varieties. Ann. Math. **147**, 159–165 (1998)

Chapter VIII : Autour du théorème de Fekete-Szegő

Pascal Autissier

1 Introduction

Soit $\mathscr{K}$ un compact de $\mathbf{C}$. En théorie du potentiel, on associe à $\mathscr{K}$ sa capacité $\gamma(\mathscr{K})$, qui est un réel positif. Ce nombre mesure en quelque sorte la « taille » de $\mathscr{K}$ et est défini de la manière suivante :

D'abord, pour toute mesure de probabilité ν sur $\mathscr{K}$, on considère son *énergie* définie par

$$I(\nu) = -\int_{\mathscr{K}}\int_{\mathscr{K}} \ln|x - y| \mathrm{d}\nu(y)\mathrm{d}\nu(x) \quad ;$$

c'est un élément de $\mathbf{R} \cup \{+\infty\}$.

Définition 1.1 On pose $V(\mathscr{K}) = \inf\{I(\nu) \; ; \; \nu \text{ mesure de probabilité sur } \mathscr{K}\}$. La *capacité* de $\mathscr{K}$ est le réel $\gamma(\mathscr{K}) = e^{-V(\mathscr{K})}$.

Exemples 1.2 La capacité d'un cercle est égale à son rayon. La capacité d'un compact dénombrable est nulle.

Fekete [3] et Fekete-Szegő [4] ont découvert que la capacité de $\mathscr{K}$, bien que de nature analytique, est en fait reliée à la présence d'entiers algébriques dont tous les conjugués sont « près » de $\mathscr{K}$:

Lorsque U est une partie de $\mathbf{C}$, on note ici $\mathscr{Y}(U)$ l'ensemble des entiers algébriques α tels que α et tous ses conjugués (par Galois) soient dans U.

P. Autissier (✉)
I.M.B., Université de Bordeaux, Talence cedex, France
e-mail: pascal.autissier@math.u-bordeaux.fr

© The Editor(s) (if applicable) and The Author(s), under exclusive license to Springer Nature Switzerland AG 2021
E. Peyre, G. Rémond (eds.), *Arakelov Geometry and Diophantine Applications*, Lecture Notes in Mathematics 2276, https://doi.org/10.1007/978-3-030-57559-5_9

Théorème 1.3 (Fekete 1923) *Soit $\mathcal{K}$ un compact de $\mathbf{C}$ symétrique par rapport à l'axe réel. On suppose $\gamma(\mathcal{K}) < 1$. Il existe un ouvert U de $\mathbf{C}$ contenant $\mathcal{K}$ tel que $\mathscr{Y}(U)$ soit fini. En particulier $\mathscr{Y}(\mathcal{K})$ est fini.*

Théorème 1.4 (Fekete-Szegő 1955) *Soit $\mathcal{K}$ un compact de $\mathbf{C}$ symétrique par rapport à l'axe réel. On suppose $\gamma(\mathcal{K}) \geqslant 1$. Pour tout ouvert U de $\mathbf{C}$ contenant $\mathcal{K}$, $\mathscr{Y}(U)$ est infini.*

Remarques 1.5

(i) Si U est une partie de $\mathbf{C}$, alors $U^* = \{z \in U \mid \bar{z} \in U\}$ est symétrique par rapport à l'axe réel, et on a $\mathscr{Y}(U^*) = \mathscr{Y}(U)$. L'hypothèse de symétrie est donc naturelle et on peut toujours s'y ramener.

(ii) Sous les hypothèses du théorème 1.4, $\mathscr{Y}(\mathcal{K})$ peut être vide : si $\mathcal{K}$ est un cercle de centre 0 et de rayon transcendant > 1, alors $\gamma(\mathcal{K}) > 1$ mais $\mathcal{K}$ ne contient aucun nombre algébrique.

Après un rappel de théorie du potentiel, on prouvera le théorème 1.3 via la théorie d'Arakelov sur $\mathbf{P}_{\mathbf{Z}}^1$ (*cf.* paragraphe 4). On donnera ensuite une démonstration « classique » du théorème 1.4.

La théorie du potentiel se généralise en fait aux surfaces de Riemann compactes. On expliquera au paragraphe 8 comment ceci permet d'étendre les énoncés précédents au cas des surfaces arithmétiques.

Enfin, dans le cas critique $\gamma(\mathcal{K}) = 1$, on montrera un théorème d'équidistribution dû à Bilu [2] et Rumely [8].

Je remercie le rapporteur pour ses suggestions qui m'ont permis d'améliorer la présentation de ce texte.

2 Théorie du potentiel sur C

Pour tout $p \in \mathbf{P}^1(\mathbf{C}) = \mathbf{C} \cup \{\infty\}$, on désigne par δ_p la masse de Dirac en p.

Soit $\mathcal{K}$ un compact de $\mathbf{C}$. Lorsque ν est une mesure de probabilité sur $\mathcal{K}$, on pose

$$\forall x \in \mathbf{C} \quad u_\nu(x) = -\int_{\mathcal{K}} \ln|x - y| d\nu(y) \,.$$

On a les propriétés suivantes (*cf* chapitre 3 de [6]).

Proposition 2.1 *Soit ν une mesure de probabilité sur $\mathcal{K}$. La fonction u_ν est harmonique sur $\mathbf{C} \setminus \mathcal{K}$, sur-harmonique sur $\mathbf{C}$. Au voisinage de ∞, on a $u_\nu(x) = -\ln|x| + O\left(\frac{1}{|x|}\right)$. De plus, on a $\frac{i}{\pi}\partial\bar{\partial}u_\nu = \delta_\infty - \nu$ au sens des courants sur $\mathbf{P}^1(\mathbf{C})$.*

Si $\gamma(\mathcal{K}) > 0$ (*i.e.* si $V(\mathcal{K}) \neq +\infty$), il existe une unique mesure de probabilité $\mu_{\mathcal{K}}$ réalisant l'infimum dans $V(\mathcal{K})$; elle est appelée la *mesure à l'équilibre* de $\mathcal{K}$. La fonction $u_{\mu_{\mathcal{K}}}$ est alors notée $u_{\mathcal{K}}$ et est appelée la *fonction potentiel* de $\mathcal{K}$.

Exemple 2.2 Si $\mathcal{K}$ est le cercle de centre a et de rayon $r > 0$, alors $u_{\mathcal{K}}(z) = -\ln \max(|z - a|, r)$ pour tout $z \in \mathbf{C}$, et $\mu_{\mathcal{K}}$ est la mesure de Haar sur $\mathcal{K}$.

Proposition 2.3 (Frostman) *On désigne par $\mathcal{D}$ la composante connexe de $\mathbf{P}^1(\mathbf{C}) \setminus \mathcal{K}$ contenant ∞. Supposons $\gamma(\mathcal{K}) > 0$. Alors :*

(a) *Sur $\mathbf{C} \setminus \overline{\mathcal{D}}$, $u_{\mathcal{K}}$ est constante égale à $V(\mathcal{K})$;*
(b) *Sur la frontière $\partial \mathcal{D}$, on a $u_{\mathcal{K}} \leqslant V(\mathcal{K})$ avec inégalité stricte seulement sur une réunion dénombrable de compacts de capacité nulle ;*
(c) *Sur $\mathcal{D}$, on a en tout point $u_{\mathcal{K}} < V(\mathcal{K})$.*

Remarque 2.4 Les compacts $\mathcal{K}$ et $\partial \mathcal{D}$ ont même capacité, même mesure à l'équilibre et même fonction potentiel.

Proposition 2.5 *Soit $x \in \partial \mathcal{D}$. Notons C la composante connexe de $\mathcal{K}$ contenant x. Si $C \neq \{x\}$, alors $\gamma(\mathcal{K}) > 0$, $u_{\mathcal{K}}$ est continue en x et $u_{\mathcal{K}}(x) = V(\mathcal{K})$.*

Définition 2.6 On dit que $\mathcal{K}$ est *à bord continu* lorsque $\mathcal{K}$ est la réunion non vide de parties connexes non réduites à un point.

En résumé, si $\mathcal{K}$ est à bord continu, alors $u_{\mathcal{K}}$ est continue sur $\mathbf{C}$ et constante (égale à $V(\mathcal{K})$) sur $\mathcal{K}$.

3 Lien avec l'intersection arithmétique

Si $\widehat{\mathcal{L}}$ est un fibré en droites hermitien (continu) sur $\mathbf{P}_{\mathbf{Z}}^1$ et D un diviseur sur $\mathbf{P}_{\mathbf{Z}}^1$, on désigne par $\mathrm{h}_{\widehat{\mathcal{L}}}(D)$ la hauteur d'Arakelov de D relativement à $\widehat{\mathcal{L}}$ (*cf* chapitre I). Lorsque $\widehat{\mathcal{L}}$ et $\widehat{\mathcal{M}}$ sont deux fibrés en droites hermitiens admissibles sur $\mathbf{P}_{\mathbf{Z}}^1$, on note $\left(\widehat{\mathcal{L}}.\widehat{\mathcal{M}}\right)$ leur nombre d'intersection arithmétique (*cf* définition 5.6 et théorème 2.5 du chapitre VII).

Soit $\mathcal{K}$ un compact de $\mathbf{C}$ invariant par conjugaison complexe. On suppose $\mathcal{K}$ à bord continu. On pose $X = \mathbf{P}_{\mathbf{Z}}^1$ et on munit $\mathcal{L} = \mathcal{O}_X([\infty])$ de la métrique continue $\| \ \|$ vérifiant

$$\forall x \in \mathbf{C} \quad \|1(x)\| = \exp[u_{\mathcal{K}}(x) - V(\mathcal{K})], \quad (*)$$

où 1 désigne la section globale de $\mathcal{L}$ définie par le diviseur $[\infty]$.

Proposition 3.1 *Posons $\widehat{\mathcal{L}} = (\mathcal{L}, \| \ \|)$. La courbure de $\widehat{\mathcal{L}}$ est égale à $\mu_{\mathcal{K}}$; en particulier $\widehat{\mathcal{L}}$ est admissible. Et on a la formule $\left(\widehat{\mathcal{L}}.\widehat{\mathcal{L}}\right) = V(\mathcal{K})$.*

Démonstration On trouve la courbure de $\widehat{\mathcal{L}}$ en appliquant la formule de Poincaré-Lelong et la Proposition 2.1.

Montrons la formule. Munissons $\mathcal{L}$ de la métrique $\|\ \|'$ telle que $\|1(x)\|' = \min\left(\dfrac{1}{|x|}, 1\right)$ pour tout $x \in \mathbf{C}$, et posons $\widehat{\mathcal{L}}' = (\mathcal{L}, \|\ \|')$. On vérifie aisément que $\mathrm{h}_{\widehat{\mathcal{L}}'}([\infty]) = 0$. Par ailleurs, la fonction $\varphi = \ln \dfrac{\|\ \|'}{\|\ \|}$ est continue sur $\mathbf{P}^1(\mathbf{C})$ et vaut $V(\mathcal{K})$ en ∞. D'après la propriété des hauteurs d'Arakelov (cf théorème 2.5 du chapitre VII), on a

$$\left\langle \widehat{\mathcal{L}}.\widehat{\mathcal{L}} \right\rangle = \mathrm{h}_{\widehat{\mathcal{L}}}([\infty]) - \int_{\mathbf{P}^1(\mathbf{C})} [u_{\mathcal{K}} - V(\mathcal{K})] \mathrm{d}\mu_{\mathcal{K}} = \mathrm{h}_{\widehat{\mathcal{L}}}([\infty]) \,,$$

puisque la mesure $\mu_{\mathcal{K}}$ est à support dans $\mathcal{K}$ et la fonction $u_{\mathcal{K}}$ vaut $V(\mathcal{K})$ sur $\mathcal{K}$ (propositions 2.3 et 2.5). En outre, on a $\mathrm{h}_{\widehat{\mathcal{L}}}([\infty]) = \mathrm{h}_{\widehat{\mathcal{L}}'}([\infty]) + \varphi(\infty) = V(\mathcal{K})$. D'où le résultat. □

4 Théorème de Fekete

Théorème 4.1 *Soit $\mathcal{K}$ un compact de $\mathbf{C}$ invariant par conjugaison complexe. On suppose $\gamma(\mathcal{K}) < 1$. Soit R un réel $> \gamma(\mathcal{K})$. Il existe $P \in \mathbf{Z}[X]$ de degré $d \geqslant 1$ tel que $\forall z \in \mathcal{K} \, |P(z)| < R^{d/2}$.*

Démonstration On peut bien sûr se placer dans le cas $R \leqslant 1$. Quitte à agrandir le compact $\mathcal{K}$ (on le recouvre par un nombre fini de petites boules fermées), on peut le supposer à bord continu. Considérons $X = \mathbf{P}^1_{\mathbf{Z}}$ et munissons $\mathcal{L} = \mathcal{O}_X([\infty])$ de la métrique $\|\ \|$ vérifiant ($*$). Le théorème de Hilbert-Samuel arithmétique (cf chapitre II) permet d'estimer le quotient V_n du covolume du réseau $\Gamma(X, \mathcal{L}^{\otimes n})$ par le volume de la boule $\left\{ s \in \Gamma(X, \mathcal{L}^{\otimes n})_{\mathbf{R}} \mid \max_{X(\mathbf{C})} \|s\| \leqslant 1 \right\}$:

$$\ln V_n = -\left\langle \widehat{\mathcal{L}}.\widehat{\mathcal{L}} \right\rangle \frac{n^2}{2} + o(n^2) \,.$$

En posant $\varepsilon = \dfrac{1}{2} \ln \dfrac{R}{\gamma(\mathcal{K})}$, de sorte que $\varepsilon > 0$, le théorème de Minkowski ([5] page 40) fournit un entier $n \geqslant 1$ et une section $s \in \Gamma(X, \mathcal{L}^{\otimes n}) \smallsetminus \{0\}$ tels que

$$\max_{X(\mathbf{C})} \|s\| \leqslant 2 V_n^{1/(n+1)} < \exp\left(-\left\langle \widehat{\mathcal{L}}.\widehat{\mathcal{L}} \right\rangle \frac{n}{2} + \varepsilon n \right) = R^{n/2} \,.$$

La section s définit un polynôme $P \in \mathbf{Z}[X]$ non nul de degré $d \leqslant n$. Alors pour tout $z \in \mathcal{K}$, on a $|P(z)| = \|s(z)\| < R^{n/2} \leqslant R^{d/2}$ (ce qui implique $d \geqslant 1$). □

On en déduit le théorème de Fekete de la manière suivante.

Démonstration du théorème 1.3 D'après le théorème 4.1, il existe $P \in \mathbf{Z}[X]$ non nul tel que $|P(z)| < 1$ pour tout $z \in \mathcal{K}$. Considérons l'ouvert $U = \{z \in \mathbf{C} \mid |P(z)| < 1\}$, qui contient donc $\mathcal{K}$. Soit $\alpha \in \mathcal{U}(U)$; notons $\alpha_1, \cdots, \alpha_n$ ses conjugués. La norme $P(\alpha_1) \cdots P(\alpha_n)$ de $P(\alpha)$ est un entier de valeur absolue < 1, donc est nulle. D'où $P(\alpha) = 0$. On en conclut que $\mathcal{U}(U)$ est contenu dans l'ensemble fini des racines de P. □

Remarquons que l'inégalité du théorème 4.1 est en fait presque optimale :

Proposition 4.2 *Soit n un entier $\geqslant 1$; désignons par $\mathscr{C}$ le cercle de centre $\dfrac{1}{n}$ et de rayon $\dfrac{1}{n^2}$. Pour tout $P \in \mathbf{Z}[X]$ non nul de degré d, on a $\max\limits_{\mathscr{C}} |P| \geqslant \gamma(\mathscr{C})^{d/2}$.*

Démonstration Le polynôme $nX - 1$ est irréductible dans $\mathbf{Z}[X]$. On peut donc factoriser P sous la forme $P = (nX - 1)^k Q$ avec $k \in \{0, \cdots, d\}$ et $Q \in \mathbf{Z}[X]$ de degré $d - k$ vérifiant $Q\left(\dfrac{1}{n}\right) \neq 0$. On a alors

$$\max_{\mathscr{C}} |P| = \frac{1}{n^k} \max_{\mathscr{C}} |Q| \geqslant \frac{1}{n^k}\left|Q\left(\frac{1}{n}\right)\right| \qquad \text{(principe du maximum)}$$
$$\geqslant \frac{1}{n^k} \frac{1}{n^{d-k}} = \gamma(\mathscr{C})^{d/2} . \qquad \square$$

5 Théorème de Fekete-Szegő

On aura besoin du résultat suivant de théorie du potentiel :

Proposition 5.1 *Soit $\mathcal{K}$ un compact non vide de $\mathbf{C}$. Soit U un ouvert de $\mathbf{C}$ contenant $\mathcal{K}$. Il existe $P \in \mathbf{C}[X]$ unitaire de degré $k \geqslant 1$ tel que $|P(z)| > \gamma(\mathcal{K})^k$ pour tout $z \in \mathbf{C} \smallsetminus U$.*

Si $\mathcal{K}$ est invariant par conjugaison complexe, alors on peut imposer à P d'être à coefficients réels.

Démonstration La première partie de l'énoncé se déduit aisément du théorème 5.5.8 de [6].

Montrons la deuxième partie : on suppose donc $\mathcal{K}$ invariant par conjugaison complexe. Quitte à réduire l'ouvert U, on peut le supposer symétrique par rapport à l'axe réel. D'après la première partie, il existe $P_0 \in \mathbf{C}$ unitaire de degré $k' \geqslant 1$ tel que $|P_0(z)| > \gamma(\mathcal{K})^{k'}$ pour tout $z \in \mathbf{C} \smallsetminus U$. Considérons le polynôme $P \in \mathbf{R}[X]$ caractérisé par la propriété $P(z) = P_0(z)\overline{P_0(\overline{z})}$ pour tout $z \in \mathbf{C}$. Alors P est unitaire de degré $k = 2k'$. Et pour tout $z \in \mathbf{C} \smallsetminus U$, on a $|P(z)| = |P_0(z)||P_0(\overline{z})| > \gamma(\mathcal{K})^{2k'} = \gamma(\mathcal{K})^k$. □

Théorème 5.2 *Soit $\mathcal{K}$ un compact de $\mathbf{C}$ invariant par conjugaison complexe. On suppose $\gamma(\mathcal{K}) \geqslant 1$. Soit U un ouvert de $\mathbf{C}$ contenant $\mathcal{K}$. Il existe $P \in \mathbf{Z}[X]$ unitaire de degré $d \geqslant 1$ tel que $|P(z)| > \gamma(\mathcal{K})^d$ pour tout $z \in \mathbf{C} \smallsetminus U$.*

Démonstration La proposition 5.1 fournit un $P_0 \in \mathbf{R}[X]$ unitaire de degré $k \geqslant 1$ tel que $|P_0(z)| > \gamma(\mathscr{K})^k$ pour tout $z \in \mathbf{C} \smallsetminus U$. Par compacité de $\mathbf{P}^1(\mathbf{C}) \smallsetminus U$, on a $\inf\limits_{\mathbf{C} \smallsetminus U} |P_0| > \gamma(\mathscr{K})^k$. On approche P_0 par un $P_1 \in \mathbf{Q}[X]$ unitaire de degré k de telle sorte qu'il existe un réel $R > \gamma(\mathscr{K})^k$ vérifiant $|P_1(z)| \geqslant R$ pour tout $z \in \mathbf{C} \smallsetminus U$.

Pour tout entier $n \geqslant -1$, désignons par $\mathbf{Q}[X]_n$ l'espace des polynômes à coefficients rationnels de degré $\leqslant n$. Posons $T = \max\limits_{z \in \mathbf{C} \smallsetminus U} \sum\limits_{b=0}^{k-1} \dfrac{|z|^b}{|P_1(z)|}$. Observant que $R > 1$, on choisit un entier $v \geqslant 1$ tel que $R^v \geqslant \dfrac{RT}{R-1}$.

L'idée est maintenant de trouver un $n \geqslant v$ et un $P \in \mathbf{Z}[X]$ vérifiant

$$P_1^n - P \in \sum_{a=0}^{n-v-1} \sum_{b=0}^{k-1} \left[-\frac{1}{2}, \frac{1}{2}\right] P_1^a X^b \, .$$

Le polynôme P_1 s'écrit $P_1 = X^k + \dfrac{1}{q} Q$ avec $Q \in \mathbf{Z}[X]$ de degré $\leqslant k-1$ et q entier $\geqslant 1$. Pour tout entier $n \geqslant 1$, on a $P_1^n = \sum\limits_{j=0}^{n} \dfrac{\mathrm{C}_n^j}{q^j} X^{(n-j)k} Q^j$.

On choisit un entier n multiple de $(vk)! q^{vk}$ et tel que $\left(\dfrac{R}{\gamma(\mathscr{K})^k}\right)^n > 2$. Pour tout $j \in \{1, \cdots, vk\}$, $\dfrac{\mathrm{C}_n^j}{q^j}$ est alors un entier. En posant

$$F = \sum_{j=0}^{vk} \frac{\mathrm{C}_n^j}{q^j} X^{(n-j)k} Q^j \quad \text{et} \quad G = \sum_{j=vk+1}^{n} \frac{\mathrm{C}_n^j}{q^j} X^{(n-j)k} Q^j \, ,$$

on a donc $P_1^n = F + G$ avec $F \in \mathbf{Z}[X]$ unitaire de degré $d = nk$ et $G \in \mathbf{Q}[X]_{d-vk-1}$.

On construit 3 suites (c_j), (G_j), (H_j) par récurrence descendante sur j :

On commence par poser $G_{d-vk-1} = G \in \mathbf{Q}[X]_{d-vk-1}$. Soit $j \in \{0, \cdots, d-vk-1\}$. Écrivons j sous la forme $j = ak+b$ avec $0 \leqslant a \leqslant n-v-1$ et $0 \leqslant b \leqslant k-1$. Le coefficient de degré j de G_j s'écrit $c_j + \delta_j$ avec c_j entier et $|\delta_j| \leqslant \dfrac{1}{2}$. On pose alors $H_j = \delta_j P_1^a X^b$ et $G_{j-1} = G_j - c_j X^j - H_j$, de sorte que $H_j \in \mathbf{Q}[X]_j$ et $G_{j-1} \in \mathbf{Q}[X]_{j-1}$. Remarquons que $|H_j(z)| \leqslant \dfrac{1}{2} |z|^b |P_1(z)|^a$ pour tout $z \in \mathbf{C}$.

On pose finalement $P = F + \sum\limits_{j=0}^{d-vk-1} c_j X^j$; c'est un polynôme à coefficients entiers, unitaire de degré d.

Par construction, on obtient $G_{-1} = 0$ puis $G = G_{d-vk-1} = \displaystyle\sum_{j=0}^{d-vk-1} (c_j X^j + H_j)$.

Il en découle la relation $P_1^n - P = \displaystyle\sum_{j=0}^{d-vk-1} H_j$. Soit $z \in \mathbf{C} \smallsetminus U$. Alors on a

$$
\begin{aligned}
|P_1(z)^n - P(z)| &\leqslant \frac{1}{2}\Big(\sum_{b=0}^{k-1} |z|^b\Big)\Big(\sum_{a=0}^{n-v-1} |P_1(z)|^a\Big) \\
&\leqslant \frac{T}{2}|P_1(z)| \sum_{a=0}^{n-v-1} \frac{|P_1(z)|^{n-1}}{R^{n-1-a}} \leqslant \frac{T|P_1(z)|^n}{2R^{v-1}(R-1)} \leqslant \frac{1}{2}|P_1(z)|^n .
\end{aligned}
$$

On en déduit que $|P(z)| \geqslant \dfrac{1}{2}|P_1(z)|^n \geqslant \dfrac{R^n}{2} > \gamma(\mathscr{K})^d$. $\qquad\qquad\square$

Prouvons maintenant le théorème de Fekete-Szegő.

Démonstration du théorème 1.4 D'après le théorème 5.2, il existe $P \in \mathbf{Z}[X]$ unitaire de degré $d \geqslant 1$ tel que $|P(z)| > 1$ pour tout $z \in \mathbf{C} \smallsetminus U$. Soit n un entier $\geqslant 1$. On note Φ_n le n-ième polynôme cyclotomique et on choisit une racine α_n du polynôme unitaire $\Phi_n \circ P$. Pour tout conjugué β de α_n, $P(\beta)$ est une racine de l'unité, donc de module 1 ; en particulier $\beta \in U$. On en déduit que α_n est un élément de $\mathscr{U}(U)$. En outre $P(\alpha_n)$ est d'ordre n dans le groupe $\mathbf{C}^*$. Les $(\alpha_n)_{n\geqslant 1}$ sont donc distincts deux à deux. $\qquad\qquad\square$

6   Théorie du potentiel sur les courbes

Soit M une surface de Riemann compacte connexe, munie d'une forme volume μ de masse totale 1.

Soit p un point de M. Rappelons que la *fonction de Green* pour p est l'unique fonction g_p de classe C^∞ sur $M \smallsetminus \{p\}$ telle que :

(i) Dans une carte (U, z) contenant p, on ait $g_p = -\ln|z - z(p)| + \varphi$, où φ est C^∞ sur U ;

(ii) Sur $M \smallsetminus \{p\}$, on ait $\dfrac{i}{\pi}\partial\bar{\partial}g_p = \mu$;

(iii) On ait $\displaystyle\int_M g_p\mu = 0$.

Pour tout $(x, y) \in M^2$ tel que $x \neq y$, on pose $g(x, y) = g_x(y)$. La fonction g ainsi définie est de classe C^∞ sur $M \times M \smallsetminus \Delta$, où Δ désigne la diagonale de $M \times M$. Et on a $g(x, y) = g(y, x)$ pour tout $(x, y) \in M^2 \smallsetminus \Delta$.

Pour tout $(p, x, y) \in M^3$ tel que $y \neq p$, on pose $[x, y]_p = \exp[g(x, p) + g(y, p) - g(x, y)]$. Ceci mesure la « pseudo-distance » entre x et y (c'est l'analogue de la distance induite par la valeur absolue sur $\mathbf{C}$; le point p joue le rôle de ∞).

Fixons maintenant un compact $\mathcal{K}$ de M et un point p de $M \smallsetminus \mathcal{K}$. Lorsque ν est une mesure de probabilité sur $\mathcal{K}$, on pose

$$\forall x \in M \quad u_\nu(x) = - \int_{\mathcal{K}} \ln[x, y]_p \mathrm{d}\nu(y) \,.$$

On a les propriétés suivantes (*cf* chapitre 3 de [7]).

Proposition 6.1 *Soit ν une mesure de probabilité sur $\mathcal{K}$. La fonction u_ν est harmonique sur $M \smallsetminus (\mathcal{K} \cup \{p\})$, sur-harmonique sur $M \smallsetminus \{p\}$. Au voisinage de p, on a $u_\nu = -g_p + \varphi$ avec φ de classe C^∞ au voisinage de p et nulle en p. De plus, on a $\dfrac{i}{\pi} \partial \bar{\partial} u_\nu = \delta_p - \nu$ au sens des courants sur M.*

On définit ensuite l'*énergie* de ν :

$$I(\nu) = \int_{\mathcal{K}} u_\nu \mathrm{d}\nu = - \int_{\mathcal{K}} \int_{\mathcal{K}} \ln[x, y]_p \mathrm{d}\nu(y) \mathrm{d}\nu(x) \,.$$

Définition 6.2 On pose $V_p(\mathcal{K}) = \inf\{I(\nu) \; ; \; \nu \text{ mesure de probabilité sur } \mathcal{K}\}$. La *capacité* de $\mathcal{K}$ par rapport à p est le réel $\gamma_p(\mathcal{K}) = e^{-V_p(\mathcal{K})}$.

Si $\gamma_p(\mathcal{K}) > 0$, il existe une unique mesure de probabilité $\mu_{\mathcal{K},p}$ réalisant l'infimum dans $V_p(\mathcal{K})$; elle est appelée la *mesure à l'équilibre* de $\mathcal{K}$ par rapport à p. On pose alors $g_{\mathcal{K},p}(x) = V_p(\mathcal{K}) - u_{\mu_{\mathcal{K},p}}(x)$ pour tout $x \in M \smallsetminus \{p\}$. La fonction $g_{\mathcal{K},p}$ et la mesure $\mu_{\mathcal{K},p}$ ne dépendent pas du choix de μ.

Proposition 6.3 *On note $\mathcal{D}_p$ la composante connexe de $M \smallsetminus \mathcal{K}$ contenant p. Supposons $\gamma_p(\mathcal{K}) > 0$. Alors :*

(a) *Sur $M \smallsetminus \overline{\mathcal{D}_p}$, $g_{\mathcal{K},p}$ est nulle ;*
(b) *Sur $\partial \mathcal{D}_p$, on a $g_{\mathcal{K},p} \geqslant 0$ avec inégalité stricte seulement sur une réunion dénombrable de compacts de capacité nulle ;*
(c) *Sur $\mathcal{D}_p$, on a en tout point $g_{\mathcal{K},p} > 0$.*

Proposition 6.4 *Soit $x \in \partial \mathcal{D}_p$. Désignons par C la composante connexe de $\mathcal{K}$ contenant x. Si $C \neq \{x\}$, alors $\gamma_p(\mathcal{K}) > 0$, $g_{\mathcal{K},p}$ est continue en x et $g_{\mathcal{K},p}(x) = 0$.*

Définition 6.5 On dit que $\mathcal{K}$ est *à bord continu* lorsque $\mathcal{K}$ est la réunion non vide de parties connexes non réduites à un point.

Plus généralement, pour toute partie finie Z de $M \smallsetminus \mathcal{K}$, on posera

$$g_{\mathcal{K},Z} = \sum_{p \in Z} g_{\mathcal{K},p} \,, \quad \mu_{\mathcal{K},Z} = \sum_{p \in Z} \mu_{\mathcal{K},p} \quad \text{et} \quad \delta_Z = \sum_{p \in Z} \delta_p \,.$$

7 Points entiers

Soit K un corps de nombres. On note G_K l'ensemble des plongements $\sigma : K \hookrightarrow \mathbf{C}$. Soit X une surface arithmétique sur O_K, *i.e.* un schéma intègre, régulier, projectif et plat sur O_K, tel que la fibre générique X_K soit une courbe géométriquement irréductible sur K.

Remarque 7.1 Soit Y un fermé intègre de X. On a deux possibilités :

- Y est plat et surjectif sur $B = \mathrm{Spec}(O_K)$; Y est alors dit *horizontal* ;
- Y est au-dessus d'un point fermé de B ; Y est alors dit *vertical*.

Définition 7.2 Soit V un ouvert de X. Un *point entier* sur V est un fermé E intègre horizontal (de X) de dimension 1 contenu dans V. E est alors affine : $E = \mathrm{Spec}(A)$, avec un ordre A d'un corps de nombres $k(E)$.

Exemple 7.3 L'ensemble des points entiers sur $\mathbf{P}^1_{\mathbf{Z}} \smallsetminus [\infty]$ est en bijection naturelle avec l'ensemble des entiers algébriques modulo l'action du groupe de Galois absolu de $\mathbf{Q}$.

Définition 7.4 Soient $\widehat{\mathscr{L}}$ un fibré en droites hermitien (continu) sur X et E un point entier sur X. La *hauteur normalisée* de E relativement à $\widehat{\mathscr{L}}$ est le réel

$$\mathrm{h}'_{\widehat{\mathscr{L}}}(E) = \frac{\mathrm{h}_{\widehat{\mathscr{L}}}(E)}{[k(E) : K]} \ .$$

Rappelons qu'à tout diviseur d'Arakelov continu $\widehat{D} = (D, g_0)$ sur X, on associe le fibré en droites $\mathscr{O}_X(D)$ muni de la métrique continue $\| \ \|$ telle que $\|1_D(x)\| = e^{-g_0(x)}$ pour tout $x \in X(\mathbf{C}) \smallsetminus |D_{\mathbf{C}}|$, où 1_D désigne la section rationnelle de $\mathscr{O}_X(D)$ définie par le diviseur D. On note $\mathscr{O}_X(D, g_0)$ ce fibré en droites hermitien (continu) sur X.

Lorsque $\widehat{D}_1$ et $\widehat{D}_2$ sont deux diviseurs d'Arakelov admissibles sur X, on désignera par $\langle\widehat{D}_1.\widehat{D}_2\rangle$ leur nombre d'intersection arithmétique (*cf* chapitre VII).

8 Théorèmes de Rumely

Soient K un corps de nombres et X une surface arithmétique sur O_K. On se donne un fermé Z de X purement de dimension 1 tel que $Z(\mathbf{C})$ ne soit pas vide, et un compact $\mathscr{K}$ de $X(\mathbf{C})$ invariant par conjugaison complexe. On suppose que pour tout $\sigma \in G_K$, le compact $\mathscr{K}_\sigma = X_\sigma(\mathbf{C}) \cap \mathscr{K}$ est à bord continu et est disjoint de $Z_\sigma(\mathbf{C})$. On suppose aussi chaque $X_\sigma(\mathbf{C})$ munie d'une forme volume de masse totale 1. On utilise les notations du paragraphe 6.

Notons $Z_1, \cdots, Z_r$ les composantes irréductibles de Z, avec $Z_1, \cdots, Z_q$ horizontales et $Z_{q+1}, \cdots, Z_r$ verticales. Pour tout $i \in \{1, \cdots, q\}$, on désigne par $g_{\mathcal{K},i}$ la fonction sur $X(\mathbf{C})$ égale à $g_{\mathcal{K}_\sigma, z_{i_\sigma}}$ sur chaque $X_\sigma(\mathbf{C})$ et on pose $\widehat{Z}_i = (Z_i, g_{\mathcal{K},i})$. Pour tout $i \in \{q+1, \cdots, r\}$, on pose $\widehat{Z}_i = (Z_i, 0)$.

On pose finalement $a_{ij} = \langle \widehat{Z}_i.\widehat{Z}_j \rangle$ pour tout $(i, j) \in \{1, \cdots, r\}^2$.

Lorsque U est une partie de $X(\mathbf{C})$, on note ici $\mathcal{Y}(U)$ l'ensemble des points entiers E sur $X \smallsetminus Z$ tels que $E(\mathbf{C}) \subset U$. Rumely ([7] théorème 6.3.1) a établi la généralisation suivante du théorème de Fekete :

Théorème 8.1 (Rumely 1989) *On suppose qu'il existe* $(\lambda_1, \cdots, \lambda_r) \in \mathbf{R}_+^r$ *vérifiant* $\displaystyle\sum_{i=1}^r \sum_{j=1}^r \lambda_i a_{ij} \lambda_j > 0$. *Il existe un ouvert U de $X(\mathbf{C})$ contenant $\mathcal{K}$ tel que* $\mathcal{Y}(U)$ *soit fini.*

Démonstration Par densité, on peut supposer que $(\lambda_1, \cdots, \lambda_r) \in \mathbf{Q}_+^{*r}$. Choisissons un entier $m \geqslant 1$ tel que $m\lambda_i$ soit entier pour tout $i \in \{1, \cdots, r\}$. Posons

$$D = \sum_{i=1}^r m\lambda_i Z_i \quad \text{et} \quad g_0 = \sum_{i=1}^q m\lambda_i g_{\mathcal{K},i} .$$

On pose aussi $\widehat{\mathscr{L}} = \widehat{\mathscr{O}}_X(D, g_0)$ et $\varepsilon = \dfrac{\langle \widehat{\mathscr{L}}.\widehat{\mathscr{L}} \rangle}{4 \deg(\mathscr{L}_K)}$. Remarquons que $\varepsilon > 0$ par hypothèse.

On considère l'ouvert $U = \left\{ x \in X(\mathbf{C}) \ \middle|\ g_0(x) < \dfrac{\varepsilon}{[K : \mathbf{Q}]} \right\}$, qui contient donc $\mathcal{K}$. Soit E un point entier sur $X \smallsetminus Z$ tel que $E(\mathbf{C}) \subset U$. Par définition des hauteurs d'Arakelov, on a

$$\mathrm{h}_{\widehat{\mathscr{L}}}(E) = \langle D.E \rangle_{\text{fini}} + \sum_{p \in E(\mathbf{C})} g_0(p) = \sum_{p \in E(\mathbf{C})} g_0(p) < \varepsilon[k(E) : K] ,$$

où $\langle D.E \rangle_{\text{fini}}$ désigne le nombre d'intersection de D et E au-dessus des places finies. On vient ainsi de vérifier que $\mathrm{h}'_{\widehat{\mathscr{L}}}(E) < \varepsilon$ pour tout $E \in \mathcal{Y}(U)$. Or d'après le corollaire de Hilbert-Samuel arithmétique (*cf* lemme 7.4 du chapitre VII), X n'admet qu'un nombre fini de points entiers E tels que

$$\mathrm{h}'_{\widehat{\mathscr{L}}}(E) < \frac{\langle \widehat{\mathscr{L}}.\widehat{\mathscr{L}} \rangle}{2 \deg(\mathscr{L}_K)} - \varepsilon = \varepsilon .$$

En particulier $\mathcal{Y}(U)$ est fini. $\qquad\qquad\qquad\qquad\qquad\qquad\qquad\qquad\qquad\qquad\qquad\square$

Rumely a également montré l'extension suivante du théorème de Fekete-Szegő (*cf* théorème 6.3.2 de [7]) :

Théorème 8.2 *On suppose que* $\displaystyle\sum_{i=1}^{r}\sum_{j=1}^{r}\lambda_i a_{ij}\lambda_j < 0$ *pour tout* $(\lambda_1,\cdots,\lambda_r) \in$ $\mathbf{R}_{+}^{r} \smallsetminus \{0\}$. *Pour tout ouvert* U *de* $X(\mathbf{C})$ *contenant* $\mathscr{K}$, $\mathscr{Y}(U)$ *est infini.*

9   Équidistribution dans le cas critique

Lorsque α est un nombre algébrique, on note ici d_α son degré et $O(\alpha)$ l'ensemble des conjugués de α.

Rumely [8], généralisant un théorème de Bilu [2], a obtenu le résultat d'équiré-partition suivant :

Théorème 9.1 (Bilu, Rumely) *Soit* $\mathscr{K}$ *un compact de* $\mathbf{C}$ *symétrique par rapport à l'axe réel. On suppose* $\mathscr{K}$ *à bord continu et* $\gamma(\mathscr{K}) = 1$. *Soit* $(\alpha_n)_{n\geqslant 1}$ *une suite d'entiers algébriques distincts deux à deux vérifiant : pour tout ouvert* U *de* $\mathbf{C}$ *contenant* $\mathscr{K}$, *il existe* n_0 *tel que* $O(\alpha_n) \subset U$ *pour tout* $n \geqslant n_0$. *Alors la suite de mesures* $\left(\dfrac{1}{d_{\alpha_n}}\delta_{O(\alpha_n)}\right)_{n\geqslant 1}$ *converge faiblement vers* $\mu_{\mathscr{K}}$, *ce qui signifie que pour toute* $f : \mathbf{P}^1(\mathbf{C}) \to \mathbf{R}$ *continue, on a*

$$\lim_{n\to+\infty}\frac{1}{d_{\alpha_n}}\sum_{\beta\in O(\alpha_n)} f(\beta) = \int_{\mathbf{C}} f \, d\mu_{\mathscr{K}}.$$

Démonstration On pose $X = \mathbf{P}_{\mathbf{Z}}^1$ et $\widehat{\mathscr{L}} = \widehat{\mathscr{O}}_X([\infty], -u_{\mathscr{K}})$. D'après la proposition 3.1, la courbure de $\widehat{\mathscr{L}}$ vaut $\mu_{\mathscr{K}}$, et on a $\left(\widehat{\mathscr{L}}.\widehat{\mathscr{L}}\right) = 0$. Pour tout $n \geqslant 1$, notons E_n l'adhérence de Zariski de α_n dans X, de sorte que E_n est un point entier sur $X \smallsetminus [\infty]$.

Soit ε un réel > 0. Appliquons l'hypothèse à l'ouvert $U = \{x \in X(\mathbf{C}) \mid u_{\mathscr{K}}(x) > -\varepsilon\}$: il existe n_0 tel que $O(\alpha_n) \subset U$ pour tout $n \geqslant n_0$. On a alors, pour tout $n \geqslant n_0$:

$$0 \leqslant \mathrm{h}_{\widehat{\mathscr{L}}}(E_n) = \big\langle [\infty].E_n\big\rangle_{\text{fini}} - \sum_{\beta\in E_n(\mathbf{C})} u_{\mathscr{K}}(\beta) = -\sum_{\beta\in O(\alpha_n)} u_{\mathscr{K}}(\beta) < \varepsilon[k(E_n):\mathbf{Q}].$$

On vient ainsi de montrer que $\mathrm{h}'_{\widehat{\mathscr{L}}}(E_n)$ converge vers 0 lorsque n tend vers $+\infty$. On conclut par le théorème d'équidistribution (*cf* théorème 8.4 du chapitre VII). $\qquad\square$

On peut en fait étendre cet énoncé au cas des surfaces arithmétiques (*cf* proposition 4.7.1 de [1]) :

On reprend les notations du paragraphe 8 concernant K, X, Z, $\mathscr{K}$, les $\widehat{Z}_i$, les a_{ij}. On note de plus $\mu_{\mathscr{K},i}$ la mesure sur $X(\mathbf{C})$ égale, sur chaque $X_\sigma(\mathbf{C})$, à $\mu_{\mathscr{K}_\sigma, Z_{i\sigma}}$.

Proposition 9.2 *On suppose qu'il existe* $(\lambda_1, \cdots, \lambda_r) \in \mathbf{R}_+^r$ *tel que*

$$\sum_{i=1}^r \sum_{j=1}^r \lambda_i a_{ij} \lambda_j = 0 \quad et\ que \quad d = \sum_{i=1}^q \lambda_i [k(Z_i) : K] > 0 .$$

Soit $(E_n)_{n \geqslant 1}$ *une suite de points entiers sur* $X \setminus Z$ *distincts deux à deux vérifiant : pour tout ouvert* U *de* $X(\mathbf{C})$ *contenant* $\mathscr{K}$, *il existe* n_0 *tel que* $E_n(\mathbf{C}) \subset U$ *pour tout* $n \geqslant n_0$. *Posons* $e_n = [k(E_n) : K]$ *pour tout* $n \geqslant 1$. *Alors la suite de mesures* $\left(\frac{1}{e_n} \delta_{E_n(\mathbf{C})} \right)_{n \geqslant 1}$ *converge faiblement vers* $\frac{1}{d} \sum_{i=1}^q \lambda_i \mu_{\mathscr{K},i}$.

Observons que la mesure $\frac{1}{d} \sum_{i=1}^q \lambda_i \mu_{\mathscr{K},i}$ ne dépend pas du choix de $(\lambda_1, \cdots, \lambda_r)$.

Références

1. P. Autissier, Points entiers sur les surfaces arithmétiques. J. Reine Angew. Math. **531**, 201–235 (2001)
2. Y. Bilu, Limit distribution of small points on algebraic tori. Duke Math. J. **89**, 465–476 (1997)
3. M. Fekete, Über die Verteilung der Wurzeln bei gewissen algebraischen Gleichungen mit ganzzahligen Koeffizienten. Math. Z. **17**, 228–249 (1923)
4. M. Fekete, G. Szegő, On algebraic equations with integral coefficients whose roots belong to a given point set. Math. Z. **63**, 158–172 (1955)
5. P.M. Gruber, C.G. Lekkerkerker, *Geometry of Numbers* 2nd edn. vol. 37. North-Holland Mathematical Library (1987)
6. T. Ransford, *Potential Theory in the Complex Plane*. LMS Student Texts, vol. 28 (1995)
7. R. Rumely, *Capacity Theory on Algebraic Curves*. Lecture Notes in Mathematics, vol. 1378 (1989)
8. R. Rumely, On Bilu's equidistribution theorem. Contemp. Math. **237**, 159–166 (1999)

Chapter IX: Some Problems of Arithmetic Origin in Rational Dynamics

Romain Dujardin

1 Introduction

In the recent years a number of classical ideas and problems in arithmetics have been transposed to the setting of rational dynamics in one and several variables. A main source of motivation in these developments is the analogy between torsion points on an Abelian variety and (pre-)periodic points of rational maps. This is actually more than an analogy since torsion points on an Abelian variety A are precisely the preperiodic points of the endomorphism of A induced by multiplication by 2. Thus, problems about the distribution or structure of torsion points can be translated to dynamical problems. This analogy also applies to spaces of such objects: in this way elliptic curves with complex multiplication would correspond to post-critically finite rational maps. Again one may ask whether results about the distribution of these "special points" do reflect each other. This point of view was in particular put forward by J. Silverman (see [39, 40] for a detailed presentation and references).

Our goal is to present a few recent results belonging to this line of research. More precisely we will concentrate on some results in which potential theory and arithmetic equidistribution, as presented in this volume by P. Autissier and A. Chambert-Loir (see [2, 11]), play a key role. This includes:

- an arithmetic proof of the equidistribution of periodic orbits towards the equilibrium measure as well as some consequences (Sect. 7);
- the equidistribution of post-critically finite mappings in the parameter space of degree 2 (Sect. 8) and degree $d \geqslant 3$ polynomials (Sect. 9);

R. Dujardin (✉)
Sorbonne Université, CNRS, Laboratoire de Probabilités, Statistiques et Modélisations (LPSM, UMR 8001), Paris, France
e-mail: romain.dujardin@upmc.fr

© The Editor(s) (if applicable) and The Author(s), under exclusive license to Springer Nature Switzerland AG 2021
E. Peyre, G. Rémond (eds.), *Arakelov Geometry and Diophantine Applications*, Lecture Notes in Mathematics 2276, https://doi.org/10.1007/978-3-030-57559-5_10

– the classification of special curves in the space of cubic polynomials (Sect. 10).

A large part of these results is based on the work of Baker–DeMarco [3, 4] and Favre–Gauthier [21, 22].

These notes are based on a series of lectures given by the author in a summer school in Grenoble in June 2017, which were intended for an audience with minimal knowledge in complex analysis and dynamical systems. The style is deliberately informal and favors reading flow against precision, in order to arrive rather quickly at some recent advanced topics. In particular the proofs are mostly sketched, with an emphasis on the dynamical parts of the arguments. The material in the first part is standard and covered with much greater detail in classical textbooks: see e.g. Milnor [35] or Carleson–Gamelin [10] for holomorphic dynamics, and Silverman [39] for the arithmetic side. Silverman's lecture notes [40] and DeMarco's 2018 ICM address [14] contain similar but more advanced material.

PART IX.A: Basic Holomorphic and Arithmetic Dynamics on P^1

2 A Few Useful Geometric Tools

2.1 Uniformization

Theorem (Uniformization Theorem) *Every simply connected Riemann surface is biholomorphic to the open unit disk* $\mathbf{D}$, *the complex plane* $\mathbf{C}$ *or the Riemann sphere* $\mathbf{P}^1(\mathbf{C})$.

See [16] for a beautiful and thorough treatment of this result and of its historical context. A Riemann surface S is called *hyperbolic* (resp. *parabolic*) if its universal cover is the unit disk (resp. the complex plane). Note that in this terminology, an elliptic curve $E \simeq \mathbf{C}/\Lambda$ is parabolic. In a sense, "generic" Riemann surfaces are hyperbolic, however, interesting complex dynamics occurs only on parabolic Riemann surfaces or on $\mathbf{P}^1(\mathbf{C})$.

Theorem *If a, b, c are distinct points on* $\mathbf{P}^1(\mathbf{C})$, *then* $\mathbf{P}^1(\mathbf{C})\setminus\{a, b, c\}$ *is hyperbolic.*

Proof Since the Möbius group acts transitively on triples of points, we may assume that $\{a, b, c\} = \{0, 1, \infty\}$. Fix a base point $\star \in \mathbf{C}\setminus\{0, 1\}$, then the fundamental group $\pi_1(\mathbf{C}\setminus\{0, 1\}, \star)$ is free on two generators. Let $S \to \mathbf{C}\setminus\{0, 1\}$ be a universal cover. Since $\mathbf{C}\setminus\{0, 1\}$ is non-compact, then S is biholomorphic to $\mathbf{D}$ or $\mathbf{C}$. The deck transformation group is a group of automorphisms of S acting discretely and isomorphic to the free group on two generators. The affine group $\mathrm{Aut}(\mathbf{C})$ has no free subgroups (since for instance its commutator subgroup is Abelian) so necessarily $S \simeq \mathbf{D}$.   □

Corollary *If Ω is an open subset of $\mathbf{P}^1(\mathbf{C})$ whose complement contains at least 3 points, then Ω is hyperbolic.*

Recall that a family $\mathscr{F}$ of meromorphic functions on some open set Ω is said *normal* if it is equicontinuous w.r.t. the spherical metric, or equivalently, if it is relatively compact in the compact-open topology. Concretely, if (f_n) is a sequence in a normal family $\mathscr{F}$, then there exists a subsequence converging locally uniformly on Ω to a meromorphic function.

Theorem (Montel) *If Ω is an open subset of $\mathbf{P}^1(\mathbf{C})$ and $\mathscr{F}$ is a family of meromorphic functions in Ω avoiding 3 values, then $\mathscr{F}$ is normal.*

Proof By post-composing with a Möbitz transformation we may assume that $\mathscr{F}$ avoids $\{0, 1, \infty\}$. Pick a disk $D \subset \Omega$ and a sequence $(f_n) \in \mathscr{F}^{\mathbf{N}}$. It is enough to show that $(f_n|_D)$ is normal. A first possibility is that $(f_n|_D)$ diverges uniformly in $\mathbf{C} \backslash \{0, 1\}$, that is converges to $\{0, 1, \infty\}$, and in this case we are done. Otherwise there exists $t_0 \in D$, $z_0 \in \Omega$ and a subsequence (n_j) such that $f_{n_j}(t_0) \to z_0$. Then if $\psi : \mathbf{D} \to \mathbf{C} \backslash \{0, 1\}$ is a universal cover such that $\psi(0) = z_0$, lifting (f_{n_j}) under ψ yields a sequence $\hat{f}_{n_j} : D \to \mathbf{D}$ such that $\hat{f}_{n_j}(t_0) \to 0$. The Cauchy estimates imply that $(\hat{f}_{n_j})$ is locally uniformly Lipschitz in D so by the Ascoli–Arzela theorem, extracting further if necessary, $(\hat{f}_{n_j})$ converges to some $\hat{f} : D \to \mathbf{D}$. Finally, f_{n_j} converges to $\psi \circ \hat{f}$ and we are done. $\qquad\square$

2.2 The Hyperbolic Metric

First recall the **Schwarz-Lemma**: *if $f : \mathbf{D} \to \mathbf{D}$ is a holomorphic map such that $f(0) = 0$, then $|f'(0)| \leqslant 1$, with equality if and only if f is an automorphism, which must then be a rotation.*

More generally, any automorphism of $\mathbf{D}$ is of the form $f(z) = e^{i\theta} \frac{z-\alpha}{1-\bar{\alpha}z}$, for some $\theta \in \mathbf{R}$ and $|\alpha| < 1$, and the inequality in the Schwarz Lemma can then be propagated as follows: let ρ be the Riemannian metric on $\mathbf{D}$ defined by the formula $\rho(z) = \frac{2|dz|}{1-|z|^2}$, i.e. if v is a tangent vector at z, $v \in T_z\mathbf{D} \simeq \mathbf{C}$, then $\|v\|_\rho = \frac{2|v|}{1-|z|^2}$. This metric is referred to as the *hyperbolic (or Poincaré) metric* and the Riemannian manifold $(\mathbf{D}, \rho)$ is the *hyperbolic disk*.

Theorem (Schwarz–Pick Lemma) *Any holomorphic map $f : \mathbf{D} \to \mathbf{D}$ is a weak contraction for the hyperbolic metric. It is a strict contraction unless f is an automorphism.*

If now S is any hyperbolic Riemann surface, since the deck transformation group of the universal cover $\mathbf{D} \to S$ acts by isometries, we can push ρ to a well-defined *hyperbolic metric* on S. As before any holomorphic map $f : S \to S'$ between hyperbolic Riemann surfaces is a weak contraction, and it is a strict contraction unless f lifts to an isometry between their universal covers.

3 Review of Rational Dynamics on $\mathbf{P}^1(\mathbf{C})$

Let $f : \mathbf{P}^1(\mathbf{C}) \to \mathbf{P}^1(\mathbf{C})$ be a rational map of degree d, that is f can be written in homogeneous coordinates as $[P(z, w) : Q(z, w)]$, where P and Q are homogeneous polynomials of degree d without common factors. Equivalently, $f(z) = \frac{P(z)}{Q(z)}$ in some affine chart. It is an elementary fact that any holomorphic self-map on $\mathbf{P}^1(\mathbf{C})$ is rational. In particular the group of automorphisms of $\mathbf{P}^1(\mathbf{C})$ is the Möbius group $\mathrm{PGL}(2, \mathbf{C})$.

We consider f as a dynamical system, that is we wish to understand the asymptotic behavior of the iterates $f \circ \cdots \circ f =: f^n$. General references for the results of this section include [35, 10]. Throughout these notes, we make the standing assumption that $d \geqslant 2$.

3.1 Fatou–Julia Dichotomy

The *Fatou set* $F(f)$ is the set of points $z \in \mathbf{P}^1(\mathbf{C})$ such that there exists a neighborhood $N \ni z$ on which the sequence of iterates $(f^n|_N)$ is normal. It is open by definition. A typical situation occurring in the Fatou set is that of an attracting orbit: that is an invariant finite set A such that every nearby point converges under iteration to A. The locus of non-normality or *Julia set* $J(f)$ is the complement of the Fatou set: $J(f) = F(f)^{\complement}$. When there is no danger of confusion we feel free to drop the dependence on f.

The Fatou and Julia sets are invariant ($f(X) \subset X$, where $X = F(f)$ or $J(f)$) and even totally invariant ($f^{-1}(X) = X$). Since $d \geqslant 2$ it easily follows that $J(f)$ is nonempty. On the other hand the Fatou set may be empty.

If $z \in J(f)$ and N is a neighborhood of z, it follows from Montel's theorem that $\bigcup_{n \geqslant 0} f^n(N)$ avoids at most 2 points. Define

$$E_z = \bigcap_{N \ni z} \left(\mathbf{P}^1(\mathbf{C}) \setminus \bigcup_{n \geqslant 0} f^n(N) \right).$$

Proposition 3.1 *The set* $E := E_z$ *is independent of* $z \in J(f)$. *Its cardinality is at most 2 and it is the maximal totally invariant finite subset. Furthermore:*

– *If* $\#E = 1$ *then* f *is conjugate in* $\mathrm{PGL}(2, \mathbf{C})$ *to a polynomial.*
– *If* $\#E = 2$ *then* f *is conjugate in* $\mathrm{PGL}(2, \mathbf{C})$ *to* $z \mapsto z^{\pm d}$.

The set $E(f)$ is called the *exceptional set* of f. Note that it is always an attracting periodic orbit, in particular it is contained in the Fatou set.

Proof It is immediate that $f^{-1}(E_z) \subset E_z$ and by Montel's Theorem $\#E_z \leqslant 2$. If $\#E_z = 1$, then we may conjugate so that $E_z = \{\infty\}$ and since $f^{-1}(\infty) = \infty$ it follows that f is a polynomial. If $\#E_z = 2$ we conjugate so that $E_z = \{0, \infty\}$. If both points are fixed then it is easy to see that $f(z) = z^d$, and in the last case that $f(z) = z^{-d}$. The independence with respect to z follows easily. $\square$

The following is an immediate consequence of the definition of E.

Corollary 3.2 *For every $z \notin E$, $\overline{\bigcup_{n \geqslant 0} f^{-n}(z)} \supset J$.*

Let us also note the following consequence of Montel's Theorem.

Corollary 3.3 *If the Julia set J has non-empty interior, then $J = \mathbf{P}^1(\mathbf{C})$.*

3.2 What Does $J(f)$ Look Like?

The Julia set is closed, invariant, and infinite (apply Corollary 3.2 to $z \in J$). It can be shown that J is perfect (i.e. has no isolated points).

A first possibility is that J **is the whole sphere** $\mathbf{P}^1(\mathbf{C})$. It is a deep result by M. Rees [37] that this occurs with positive probability if a rational map is chosen at random in the space of all rational maps of a given degree. Note that this never happens for polynomials because ∞ is an attracting point.

Important explicit examples of rational maps with $J = \mathbf{P}^1$ are *Lattès mappings*, which are rational maps coming from multiplication on an elliptic curve. In a nutshell: let E be an elliptic curve, viewed as a torus $\mathbf{C}/\Lambda$, where Λ is a lattice, and let $m : \mathbf{C} \to \mathbf{C}$ be a $\mathbf{C}$-linear map such that $m(\Lambda) \subset \Lambda$. Then m commutes with the involution $z \mapsto -z$, so it descends to a self-map of the quotient Riemann surface $E/(z \sim (-z))$ which turns out to be $\mathbf{P}^1(\mathbf{C})$. The calculations can be worked out explicitly for the doubling map using elementary properties of the Weierstraß $\wp$-function, and one finds for instance that $f(z) = \frac{(z^2+1)^2}{4z(z^2-1)}$ is a Lattès example (see e.g. [36] for this and more about Lattès mappings).

It can also happen that J **is a smooth curve** (say of class C^1). A classical theorem of Fatou then asserts that J must be contained in a circle (this includes lines in $\mathbf{C}$), and more precisely:

- either it is a circle: this happens for $z \mapsto z^{\pm d}$ but also for some Blaschke products;
- or it is an interval in a circle: this happens for interval maps such as the Chebychev polynomial $z \mapsto z^2 - 2$ for which $J = [-2, 2]$.

Otherwise J **is a self-similar "fractal" set** with often complicated topological structure (see Fig. 1).

Fig. 1 Gallery of quadratic Julia sets $J(f_c)$ with $f_c(z) = z^2 + c$. Left: $z^2 + (-0.5 + 0.556i)$, middle: $z^2 + 0.2531$, right: $z^2 + i$

3.3 Periodic Points

A point $z \in \mathbf{P}^1(\mathbf{C})$ is *periodic* if there exists n such that $f^n(z) = z$. The period of z is the minimal such positive n, And a fixed point is a point of period 1. Elementary algebra shows that f^n admits $d^n + 1$ fixed points counting multiplicities. If z_0 has exact period n, the *multiplier* of z_0 is the complex number $(f^n)'(z_0)$ (which does not depend on the chosen Riemannian metric on the sphere). It determines a lot of the local dynamics of f^n near z_0. There are three main cases:

- *attracting:* $\left|(f^n)'(z_0)\right| < 1$: then $z_0 \in F$ and for z near z_0, $f^{nk}(z) \to z_0$ as $k \to \infty$;
- *repelling:* $\left|(f^n)'(z_0)\right| > 1$: then $z_0 \in J$ since $\left|(f^{nk})'(z_0)\right| \to \infty$ as $k \to \infty$;
- *neutral:* $\left|(f^n)'(z_0)\right| = 1$: then z_0 belongs to either F or J.

The neutral case subdivides further into *rationally neutral* (or *parabolic*) when $(f^n)'(z_0)$ is a root of unity and *irrationally neutral* otherwise. Parabolic points belong to the Julia set. The dynamical classification of irrationally neutral points, which essentially boils down to the question whether they belong to the Fatou or Julia set, is quite delicate (and actually still not complete) and we will not need it.

Theorem 3.4 (Fatou, Julia) *Repelling periodic points are dense in $J(f)$.*

This is an explanation for the local self-similarity of J. In particular we see that if z_0 is repelling with a non-real multiplier, then J has a "spiralling structure" at z_0, in particular it cannot be smooth. A delicate result by Eremenko and Van Strien [20] asserts conversely that if all periodic points multipliers are real, then J is contained in a circle.

Idea of proof The first observation is that if g is a holomorphic self-map of an open topological disk Δ such that $\overline{g(\Delta)} \subset \Delta$, then by contraction of the Poincaré metric, g must have an attracting fixed point. If now $z_0 \in J$ is arbitrary we know from Montel's Theorem and the classification of exceptional points that for any neighborhood $N \ni z_0$, $\bigcup_{n \geq 0} f^n(N)$ eventually covers J, hence z_0. With some more work, it can be ensured that there exists a small open disk D close to z_0 and an integer n such that f^n is univalent on D and $\overline{D} \subset f^n(D)$. Then the result follows from the initial observation.   □

In particular a rational map admits infinitely many repelling points. It turns out that conversely the number of non-repelling periodic points is finite. The first step is the following basic result.

Theorem 3.5 (Fatou) *Every attracting periodic orbit attracts a critical point.*

By this mean that for every attracting periodic orbit A there exists a critical point c such that $f^n(c)$ tends to A as $n \to \infty$. This is a basic instance of a general heuristic principle: the dynamics is determined by the behavior of critical points.

Proof Let z_0 be an attracting point of period n. For expositional ease we replace f by f^n so assume $n = 1$. Let

$$\mathscr{B} = \left\{ z \in \mathbf{P}^1, \ f^k(z) \xrightarrow[k \to \infty]{} z_0 \right\}$$

be the *basin* of z_0 and $\mathscr{B}_0$ be the *immediate basin*, that is the connected component of z_0 in $\mathscr{B}$. Then f maps $\mathscr{B}_0$ into itself and $\mathscr{B}_0$ is hyperbolic since $\mathscr{B}_0 \cap J = \emptyset$. If $f|_{\mathscr{B}_0}$ had no critical points, then $f|_{\mathscr{B}_0}$ would be a covering, hence a local isometry for the hyperbolic metric. This contradicts the fact that z_0 is attracting.   □

Since f has $2d - 2$ critical points we infer:

Corollary 3.6 *A rational map of degree d admits at most $2d - 2$ attracting periodic orbits.*

Theorem 3.7 (Fatou, Shishikura) *A rational map admits only finitely many non-repelling periodic orbits.*

Idea of Proof The bound $6d - 6$ was first obtained by Fatou by a beautiful perturbative argument. First, the previous theorem (and its corollary) can be easily extended to the case of periodic orbits with multiplier equal to 1 (these are attracting in a certain direction). So there are at most $2d - 2$ periodic points with multiplier 1. Let now N be the number of neutral periodic points with multiplier different from 1. Fatou shows that under a generic perturbation of f, at least half of them become attracting. Hence $N/2 \leqslant 2d - 2$ and the result follows.

The sharp bound $2d - 2$ for the total number of non-repelling cycles was obtained by Shishikura.   □

3.4 Fatou Dynamics

The dynamics in the Fatou set can be understood completely. Albeit natural, this is far from obvious: one can easily imagine an open set U on which the iterates form a normal family, yet the asymptotic behavior of $(f^n|_U)$ is complicated to analyze.

Such a phenomenon actually happens in transcendental or higher dimensional dynamics, but not in our setting: the key point is the following deep and celebrated "non-wandering domain theorem".

Theorem 3.8 (Sullivan) *Every component U of the Fatou set is ultimately periodic, i.e. there exist integers $l > k$ such that $f^l(U) = f^k(U)$.*

It remains to classify periodic components.

Theorem 3.9 *Every component U of period k of the Fatou set is*

- *either an attraction domain: as $n \to \infty$, (f^{nk}) converges locally uniformly on U to a periodic cycle (attracting or parabolic);*
- *or a rotation domain: U is biholomorphic to a disk or an annulus and $f^k|_U$ is holomorphically conjugate to an irrational rotation.*

4   Equilibrium Measure

4.1   Definition and Main Properties

In many arithmetic applications the most important dynamical object is the *equilibrium measure*. The following theorem summarizes its construction and main properties.

Theorem 4.1 (Brolin, Lyubich, Freire–Lopes–Mañé) *Let f be a rational map on $\mathbf{P}^1(\mathbf{C})$ of degree $d \geqslant 2$. Then for any $z \notin E(f)$, the sequence of probability measures*

$$\mu_{n,z} = \frac{1}{d^n} \sum_{w \in f^{-n}(z)} \delta_w$$

(where pre-images are counted with their multiplicity) converges weakly to a Borel probability measure $\mu = \mu_f$ on $\mathbf{P}^1(\mathbf{C})$ enjoying the following properties:

  (i) μ *is invariant, that is $f_*\mu = \mu$, and has the "constant Jacobian" property $f^*\mu = d\mu$;*
 (ii) μ *is ergodic;*
(iii) $\mathrm{Supp}(\mu) = J(f)$;
 (iv) μ *is repelling, that is for μ-a.e. z, $\displaystyle\lim_{n\to\infty} \frac{1}{n} \log \left\| df_z^n \right\| \geqslant \frac{\log d}{2}$, where the norm of the differential is computed with respect to any smooth Riemannian metric on $\mathbf{P}^1(\mathbf{C})$;*
  (v) μ *describes the asymptotic distribution of repelling periodic orbits, that is, if RPer_n denotes the set of repelling periodic points of exact period n, then $\frac{1}{d^n} \sum_{z \in \mathrm{RPer}_n} \delta_z$ converges to μ as $n \to \infty$.*

Recall that if μ is a probability measure, its image $f_*\mu$ under f is defined by $f_*\mu(A) = \mu(f^{-1}(A))$ for any Borel set A. The pull-back $f^*\mu$ is conveniently defined by its action on continuous functions: $\langle f^*\mu, \varphi \rangle = \langle \mu, f_*\varphi \rangle$, where $f_*\varphi$ is defined by $f_*\varphi(x) = \sum_{y \in f^{-1}(x)} \varphi(y)$. Recall also that ergodicity means that any measurable invariant subset has measure 0 or 1. A stronger form of ergodicity actually holds: μ is *exact*, that is, if A is any measurable subset of positive measure, then $\mu(f^n(A)) \to 1$ as $n \to \infty$.

The proof of this theorem is too long to be explained in these notes (see [26] for a detailed treatment). Let us only discuss the convergence statement.

Proof of the Convergence of the $\mu_{n,z}$ Consider the Fubini-Study $(1, 1)$ form ω associated to the spherical metric on $\mathbf{P}^1$. It expresses in coordinates as $\omega = \frac{i}{\pi} \partial\bar{\partial} \log \|\sigma\|^2$ where $\sigma : \mathbf{P}^1 \to \mathbf{C}^2 \backslash \{0\}$ is any local section of the canonical projection $\mathbf{C}^2 \backslash \{0\} \to \mathbf{P}^1$ and $\|\cdot\|$ is the standard Hermitian norm on $\mathbf{C}^2$, i.e $\|(z, w)\|^2 = |z|^2 + |w|^2$. Then

$$f^*\omega - d\omega = \frac{i}{\pi} \partial\bar{\partial} \log \frac{\|(P(z, w), Q(z, w))\|^2}{\|(z, w)\|^{2d}} =: \frac{i}{\pi} \partial\bar{\partial} g_0,$$

where by homogeneity of the polynomials P and Q, g_0 is a globally well-defined smooth function on $\mathbf{P}^1$. Thus we infer that

$$\frac{1}{d^n}(f^n)^*\omega - \omega = \sum_{k=0}^{n-1} \left(\frac{1}{d^{k+1}}(f^{k+1})^*\omega - \frac{1}{d^k}(f^k)^*\omega \right)$$

$$= \sum_{k=0}^{n-1} \frac{1}{d^{k+1}}(f^k)^*(f^*\omega - d\omega)$$

$$= \frac{i}{\pi} \partial\bar{\partial} \left(\sum_{k=0}^{n-1} \frac{1}{d^{k+1}} g_0 \circ f^k \right) =: \frac{i}{\pi} \partial\bar{\partial} g_k.$$

One readily sees that the sequence of functions g_k converges uniformly to a continuous function g_∞, therefore basic distributional calculus implies that

$$\frac{1}{d^n}(f^n)^*\omega \xrightarrow[n \to \infty]{} \omega + \frac{i}{\pi} \partial\bar{\partial} g_\infty.$$

Being a limit in the sense of distributions of a sequence of probability measures, this last term can be identified to a positive measure μ by declaring that

$$\langle \mu, \varphi \rangle = \int_{\mathbf{P}^1} \varphi\, \omega + \frac{i}{\pi} \int g_\infty \partial\bar{\partial} \varphi$$

for any smooth function φ. By definition this measure is the equilibrium measure μ_f.

Now for any $z \in \mathbf{P}^1$, identifying (1,1) forms and signed measures as above we write

$$\frac{1}{d^n} \sum_{w \in f^{-n}(z)} \delta_w = \frac{1}{d^n}(f^n)^* \delta_z = \frac{1}{d^n}(f^n)^* \omega + \frac{1}{d^n}(f^n)^*(\delta_z - \omega).$$

There exists a L^1 function on $\mathbf{P}^1$ such that in the sense of distributions $\frac{i}{\pi}\partial\bar{\partial}g_z = \delta_z - \omega$, so that

$$\frac{1}{d^n}(f^n)^*(\delta_z - \omega) = \frac{i}{\pi}\partial\bar{\partial}\left(\frac{1}{d^n}g_z \circ f^n\right)$$

and to establish the convergence of the $\mu_{n,z}$ the problem is to show that for $z \notin E$, $d^{-n}g_z \circ f^n$ converges to 0 in L^1. It is rather easy to prove that this convergence holds for a.e. z with respect to Lebesgue measure. Indeed, suppose that the g_z are chosen with some uniformity, for instance by assuming $\sup_{\mathbf{P}^1} g_z = 0$. In this case the average $\int g_z \mathrm{Leb}(dz)$ is a bounded function g (such that $\frac{i}{\pi}\partial\bar{\partial}g = \mathrm{Leb}$) and the result essentially follows from the Borel-Cantelli Lemma: indeed, on average, $d^{-n}g_z \circ f^n$ is bounded by C^{st}/d^n.

Proving the convergence for every $z \notin E$ requires a finer analysis, see [26] for details.   □

4.2   The Case of Polynomials

If f is a polynomial, rather than the Fubini-Study measure, we can use the Dirac mass at the totally invariant point ∞, to give another formulation of these results. Indeed, if ν is any probability measure (say with compact support) in $\mathbf{C}$, write $\nu - \delta_\infty = \frac{i}{\pi}\partial\bar{\partial}g$ as before. In $\mathbf{C}$, this rewrites as $\nu = \Delta g$, where[1] g is a subharmonic function with logarithmic growth at infinity: $g(z) = \log|z| + c + o(1)$. If ν_{S^1} be the normalized Lebesgue measure on the unit circle, then $\nu_{S^1} = \Delta\left(\log^+|z|\right)$ (where $\log^+ t = \max(\log t, 0)$), so

$$\frac{1}{d^n}(f^n)^*(\nu_{S^1}) = \frac{1}{d^n}\Delta\left(\log^+\left|f^n(z)\right|\right).$$

---

[1] For convenience we have swallowed the normalization constant in Δ, so that Δ is $1/4\pi$ times the ordinary Laplacian.

An argument similar to that of the proof of Theorem 4.1 yields the following:

Proposition 4.2 *If f is a polynomial of degree d, the sequence of functions $d^{-n} \log^+ |f^n|$ converges locally uniformly to a subharmonic function $G : \mathbf{C} \to \mathbf{R}$ satisfying $G \circ f = dG$.*

The function $G = G_f$ is by definition the *dynamical Green function* of f. Introduce the *filled Julia set*

$$K(f) = \{z \in \mathbf{C}, \ (f^n(z)) \text{ is bounded in } \mathbf{C}\}$$
$$= \{z \in \mathbf{C}, \ (f^n(z)) \text{ does not tend to } \infty\}.$$

It is easy to show that $J(f) = \partial K(f)$.

Proposition 4.3 *The dynamical Green function has the following properties:*

(i) G_f is continuous, non-negative and subharmonic in $\mathbf{C}$;
(ii) $\{G_f = 0\} = K(f)$;
(iii) $\Delta G_f = \mu_f$ is the equilibrium measure (in particular $\mathrm{Supp}(\Delta G_f) = \partial \{G_f = 0\}$);
(iv) if f is a monic polynomial, then $G_f(z) = \log |z| + o(1)$ as $z \to \infty$.

Properties *(i)*–*(iii)* show that the dynamically defined function G_f coincides with the Green function of $K(f)$ from classical potential theory. Property *(iv)* implies that if f is monic, then $K(f)$ is of capacity 1, which is important in view of its number-theoretic properties (see the lectures by P. Autissier in this volume [2]).

Another useful consequence is that the Green function G_f is completely determined by the compact set $K(f)$ (or by its boundary $J(f)$). In particular if f and g are polynomials of degree at least 2 then

$$J(f) = J(g) \Leftrightarrow \mu_f = \mu_g. \tag{1}$$

The same is not true for rational maps: $J(f)$ does not determine μ_f in general. For instance there are many rational maps f such that $J = \mathbf{P}^1(\mathbf{C})$, however Zdunik [42] proved that μ_f is absolutely continuous with respect to the Lebesgue measure on $\mathbf{P}^1(\mathbf{C})$ (in this case automatically $J(f) = \mathbf{P}^1(\mathbf{C})$) if and only if f is a Lattés example.

5 Non-archimedean Dynamical Green Function

Dynamics of rational functions over non-Archimedean valued fields has been developing rapidly during the past 20 years, greatly inspired by the analogy with holomorphic dynamics. Most of the results of the previous section have analogues in the non-Archimedean setting. We will not dwell upon the details in these notes, and rather refer the interested reader to [39, 6]. We shall content ourselves with recalling the vocabulary of valued fields and heights, and the construction of the dynamical Green function.

5.1 Vocabulary of Valued Fields

To fix notation and terminology let us first recall a few standard notions and facts.

Definition 5.1 An *absolute value* $|\cdot|$ on a field K is a function $K \to \mathbf{R}^+$ such that

- for every $\alpha \in K$, $|\alpha| \geqslant 0$ and $|\alpha| = 0$ iff $\alpha = 0$;
- for all α, β, $|\alpha\beta| = |\alpha| \cdot |\beta|$;
- for all α, β, $|\alpha + \beta| \leqslant |\alpha| + |\beta|$.

If in addition $|\cdot|$ satisfies the ultrametric triangle inequality $|\alpha + \beta| \leqslant \max(|\alpha|, |\beta|)$, then it is said non-Archimedean.

Besides the usual modulus on $\mathbf{C}$, a basic example is the p-adic norm on $\mathbf{Q}$, defined by $|\alpha|_p = p^{-v_p(\alpha)}$ where v_p is the p-adic valuation: $v_p(a/b) = \ell$ where $a/b = p^\ell(a'/b')$ and p does not divide a' nor b'. On any field we can define the trivial absolute value by $|0| = 0$ and $|x| = 1$ if $x \neq 0$.

If $(K, |\cdot|)$ is a non-Archimedean valued field, we define the *spherical metric* on $\mathbf{P}^1(K)$ by

$$\rho(p_1, p_2) = \frac{|x_1 y_2 - x_2 y_1|}{\max(|x_1|, |y_1|) \max(|x_2|, |y_2|)},$$

where $p_1 = [x_1 : y_1]$ and $p_2 = [x_2 : y_2]$. From the ultrametric inequality we infer that the ρ-diameter of $\mathbf{P}^1(K)$ equals 1.

Definition 5.2 Two absolute values $|\cdot|_1$ and $|\cdot|_2$ on K are said equivalent if there exists a positive real number r such that $|\cdot|_1 = |\cdot|_2^r$.

An equivalence class of absolute values on K is called a *place*. The set of places of K is denoted by $\mathscr{M}_K$.

Theorem 5.3 (Ostrowski) *A set of representatives of the set of places of* $\mathbf{Q}$ *is given by*

- *the trivial absolute value;*
- *the usual Archimedean absolute value* $|\cdot|_\infty$;
- *the set of p-adic absolute values* $|\cdot|_p$.

Decomposition into prime factors yields the *product formula*

$$\forall x \in \mathbf{Q}^*, \quad \prod_{p \in \mathscr{P} \cup \{\infty\}} |x|_p = 1,$$

where $\mathscr{P}$ denotes the set of prime numbers.

For number fields, places can also be described. First one considers $\mathbf{Q}_p$ the completion of $\mathbf{Q}$ relative to the distance induced by $|\cdot|_p$. The p-adic absolute value extends to an algebraic closure of $\mathbf{Q}_p$, so we may consider $\mathbf{C}_p$ the completion of this algebraic closure. (For $p = \infty$ of course $\mathbf{C}_p = \mathbf{C}$.)

A number field K admits a number of embeddings $\sigma : K \hookrightarrow \mathbf{C}_p$, which define by restriction a family of absolute values $a \mapsto |\sigma(a)|_p$ on K. Distinct embeddings may induce the same absolute value. For any non-trivial absolute value $|\cdot|_v$ on K there exists $p_v \in \mathscr{P} \cup \{\infty\}$ such that $|\cdot|_v \, |_{\mathbf{Q}}$ is equivalent to $|\cdot|_{p_v}$ and ε_v distinct embeddings $\sigma : K \hookrightarrow \mathbf{C}_{p_v}$ such that the restriction of $|\cdot|_{\mathbf{C}_{p_v}}$ to $\sigma(K)$ induces $|\cdot|_v$ up to equivalence. Thus for each place of K we can define an absolute value $|\cdot|_v$ on K that is induced by $|\cdot|_{\mathbf{C}_{p_v}}$ by embedding K in $\mathbf{C}_{p_v}$ and the following product formula holds

$$\forall x \in K^*, \quad \prod_{v \in \mathcal{M}_K} |x|_v^{\varepsilon_v} = \prod_{p \in \mathscr{P} \cup \{\infty\}} \prod_{\sigma:K \hookrightarrow \mathbf{C}_p} |\sigma(x)|_p = 1. \tag{2}$$

A similar formalism holds for certain non-algebraic extensions of $\mathbf{Q}$, giving rise to a general notion of *product formula field*.

5.2   Dynamical Green Function

Let $(K, |\cdot|)$ be a complete valued field. If f is a rational map on $\mathbf{P}^1(K)$ we can define a dynamical Green function as before, by working with homogeneous lifts to $\mathbf{A}^2(K)$. For $(x, y) \in \Lambda^2(K)$ we put $\|(x, y)\| = \max(|x|, |y|)$, and fix a lift $F : \mathbf{A}^2(K) \to \mathbf{A}^2(K)$ of f.

Theorem 5.4 *For every $(x, y) \in \mathbf{A}^2(K)$, the limit*

$$G_F(x, y) := \lim_{n \to \infty} \frac{1}{d^n} \log \| F^n(x, y) \|$$

exists.

The function G_F is continuous on $\mathbf{A}^2(K)$ (relative to the norm topology) and satisfies

- $G_F \circ F = d G_F$;
- $G_F(x, y) = \log \|(x, y)\| + O(1)$ *at infinity;*
- $G_F(\lambda x, \lambda y) = \log |\lambda| + G_F(x, y)$.

Note that the function G_F depends on the chosen lift, more precisely $G_{\alpha F} = \frac{1}{d-1}\log|\alpha| + G_F$. The proof of the theorem is similar to that of the Archimedean case. The key point is that there exists positive numbers c_1 and c_2 such that the inequality

$$c_1 \leqslant \frac{\|F(x,y)\|}{\|(x,y)\|^d} \leqslant c_2 \tag{3}$$

holds. The upper bound in (3) is obvious, and the lower bound is a consequence of the Nullstellensatz.

What is more delicate is to make sense of the measure theoretic part, i.e. to give a meaning to ΔG. For this one has to introduce the formalism of Berkovich spaces, which is well-suited to measure theory (see [39, 6] for details).

If f is a polynomial, instead of lifting to $\mathbf{A}^2$, we can directly consider the Green function on K as in the complex case. More precisely: if f is a polynomial of degree d, the sequence of functions $d^{-n}\log^+|f^n|$ converges locally uniformly to a function G_f on K satisfying $G_f \circ f = dG_f$.

6 Logarithmic Height

6.1 Definition and Basic Properties

We refer to [12, 39] for details and references on the material in this section. For $x \in \mathbf{P}^1(\mathbf{Q})$ we can write $x = [a : b]$ with $(a,b) \in \mathbf{Z}^2$ and $\gcd(a,b) = 1$, we define the *naive height* $h(x) = h_{\text{naive}}(x) = \log\max(|a|_\infty, |b|_\infty)$. A fundamental (obvious) property is that for every $M \geqslant 0$

$$\left\{ x \in \mathbf{P}^1(\mathbf{Q}), h(x) \leqslant M \right\}$$

is finite.

We now explain how to extend h to $\mathbf{P}^1(\overline{\mathbf{Q}})$ using the formalism of *places*. The key observation (left as an exercise to the reader) is that the product formula implies that

$$h(x) = \sum_{p \in \mathscr{P} \cup \{\infty\}} \log^+|x|_p \ .$$

This formula can be generalized to number fields, giving rise to a notion of *logarithmic height* in that setting. Let K be a number field and $x = [x_0, x_1] \in \mathbf{P}^1(K)$. Then with notation as above we define

$$h(x) = \frac{1}{[K:\mathbf{Q}]} \sum_{p \in \mathscr{P} \cup \{\infty\}} \sum_{\sigma:K \hookrightarrow \mathbf{C}_p} \log\|(\sigma(x_0), \sigma(x_1))\|_p$$

$$= \frac{1}{[K : \mathbf{Q}]} \sum_{v \in \mathcal{M}_K} \varepsilon_v \log \| (\sigma(x_0), \sigma(x_1)) \|_v \, ,$$

where as before $\| (x_0, x_1) \|_v = \max(|x_0|_v \, , |x_1|_v)$.Note that by the product formula, $h(x)$ does not depend on the lift $(x_0, x_1) \in \mathbf{A}^2$. By construction, h is invariant under the action of the Galois group $\mathrm{Gal}(K : \mathbf{Q})$. Furthermore, the normalization in the definition of h was chosen in such a way that $h(x)$ does not depend on the choice of the number field containing x, therefore h is a well-defined function on $\mathbf{P}^1(\overline{\mathbf{Q}})$ invariant under the absolute Galois group of $\mathbf{Q}$.

If $x \in K$ we can write $x = [x : 1]$ and we recover

$$\log \| (x, 1) \|_p = \log \max(|x_0|_p \, , |x_1|_p) = \log^+ |x|_p \, .$$

Theorem 6.1 (Northcott) *For all $M > 0$ and $D \leqslant 1$, the set*

$$\left\{ x \subset \mathbf{P}^1(\overline{\mathbf{Q}}), h(x) \leqslant M \text{ and } [\mathbf{Q}(x) : \mathbf{Q}] \leqslant D \right\}$$

is finite.

Corollary 6.2 *For $x \in \overline{\mathbf{Q}}^*$, $h(x) = 0$ if and only if x is a root of unity.*

Proof Indeed, note that by construction $h(x^N) = Nh(x)$ and also $h(1) = 0$. In particular if x is a root of unity, then $h(x) = 0$. Conversely if $x \neq 0$ and $h(x) = 0$ then by Northcott's Theorem, $\{x^N, N \geqslant 1\}$ is finite so there exists $M < N$ such that $x^M = x^N$ and we are done. $\square$

6.2 Action Under Rational Maps

Let now $f : \mathbf{P}^1(\overline{\mathbf{Q}}) \to \mathbf{P}^1(\overline{\mathbf{Q}})$ be a rational map defined on $\overline{\mathbf{Q}}$, of degree d.

Proposition 6.3 (Northcott) *There exists a constant C depending only on f such that*

$$\forall x \in \mathbf{P}^1(\overline{\mathbf{Q}}), \ |h(f(x)) - dh(x)| \leqslant C.$$

In other words, h is almost multiplicative. The proof follows from the decomposition of h as a sum of local contributions together with the Nullstellensatz as in (3) (note that $h(f(x)) = h(x)$ for all but finitely many places).

We are now in position to define the *canonical height* associated to f.

Theorem 6.4 (Call–Silverman) *There exists a unique function* $h_f : \mathbf{P}^1(\overline{\mathbf{Q}}) \to \mathbf{R}_+$ *such that:*

- $h - h_f$ *is bounded;*
- $h_f \circ f = d\, h_f.$

Note that the naive height h is the canonical height associated to $x \mapsto x^d$.

Proof This is similar to the construction of the dynamical Green function. For every $n \geqslant 0$, let $h_n = d^{-n} h \circ f^n$. By Proposition 6.3 we have that $|h_{n+1} - h_n| \leqslant C/d^n$, so the sequence h_n converges pointwise. We define h_f to be its limit, and the announced properties are obvious. $\square$

Proposition 6.5 *The canonical height h_f enjoys the following properties:*

- $h_f \geqslant 0;$
- *for $x \in K^*$, $h_f(x) = 0$ if and only if x is preperiodic, i.e. there exists $k < l$ such that $f^k(x) = f^l(x)$.*
- *the set $\left\{ x \in \mathbf{P}^1(\overline{\mathbf{Q}}), h(x) \leqslant M \text{ and } [\mathbf{Q}(x) : \mathbf{Q}] \leqslant D \right\}$ is finite.*

Proof The first item follows from the positivity of h and the formula $h_f(x) = \lim_n d^{-n} h(f^n(x))$. The second one is proved exactly as Corollary 6.2, and the last one follows from the estimate $h_f - h = O(1)$. $\square$

It is often useful to know that the logarithmic height expresses as a sum of local contributions coming from the places of K. For the canonical height, the existence of such a decomposition does not clearly follow from the Call–Silverman definition. On the other hand the construction of the dynamical Green function from the previous paragraph implies that a similar description holds for h_f.

Proposition 6.6 *Let f be a rational map on $\mathbf{P}^1(\overline{\mathbf{Q}})$ and F be a homogeneous lift of f on $\mathbf{A}^2(\overline{\mathbf{Q}})$. Then for every $x \in \mathbf{P}^1(\overline{\mathbf{Q}})$, we have that*

$$h_f(x) = \frac{1}{[K : \mathbf{Q}]} \sum_{v \in \mathcal{M}_K} \varepsilon_v G_{F,v}(x_0, x_1), \text{ where } x = [x_0 : x_1]$$

where K is any number field such that $x \in \mathbf{P}^1(K)$, $G_{F,v}$ is the dynamical Green function associated to $(K, |\cdot|_v)$ and ε_v is as in (2).

Proof Let $\kappa(x_0, x_1)$ denote the right hand side of the formula. The product formula together with the identities $G_F(\lambda x, \lambda y) = \log |\lambda| + G_F(x, y)$ and $G_{\alpha F} = G_F + \frac{1}{d-1} \log |\alpha|$ imply that κ is independent of the lifts F of f and $[x_0 : x_1]$ of x. In particular κ is a function of x only.

The invariance relations for the local Green functions imply that $\kappa \circ f = d\, \kappa$. Since furthermore $G_{F,v}(x_0, x_1) = \log \|(x_0, x_1)\|_v$ for all but finitely many places, we infer from the second property of Theorem 5.4 that $\kappa - h = O(1)$. Then the result follows from the uniqueness assertion in Theorem 6.4. $\square$

7 Consequences of Arithmetic Equidistribution

7.1 Equidistribution of Preperiodic Points

It turns out that the Call–Silverman canonical height satisfies the assumptions of Yuan's equidistribution theorem for points of small height (see [11, Thm. 7.4]). From this for rational maps defined over number fields we can obtain refined versions of previously known equidistribution theorems.

Theorem 7.1 *Let* $f \in \overline{\mathbf{Q}}(X)$ *be a rational map of degree* $d \geqslant 2$ *and* (x_n) *be any infinite sequence of preperiodic points of* f. *Then in* $\mathbf{P}^1(\mathbf{C})$ *we have that*

$$\frac{1}{[\mathbf{Q}(x_n) : \mathbf{Q}]} \sum_{y \in G(x_n)} \delta_y \xrightarrow[n \to \infty]{} \mu_f, \tag{4}$$

where $G(x_n)$ *denotes the set of Galois conjugates of* x_n.

The reason of the appearance of μ_f in (4) is to be found in the description of the canonical height given in Proposition 6.6. When $G(x_n)$ is the set of all periodic points of a given period, this result is a direct consequence of the last item of Theorem 4.1. Indeed, by Theorem 3.7, f admits at most finitely many non-repelling points. Likewise, we may also obtain in the same way an arithmetic proof of the convergence of the measures $\mu_{n,z}$ of Theorem 4.1.

7.2 Rigidity

As an immediate consequence of Theorem 7.1 we obtain the following rigidity statement.

Corollary 7.2 *If* f *and* g *are rational maps defined over* $\overline{\mathbf{Q}}$ *of degree at least 2 are such that* f *and* g *have infinitely many common preperiodic points, then* $\mu_f = \mu_g$ *(so in particular* $J(f) = J(g)$*).*

Rational maps with the same equilibrium measure were classified by Levin and Przytycki [31]. The classification is a bit difficult to state precisely, but the idea is that if f and g have the same equilibrium measure and do not belong to a small list of well understood exceptional examples (monomial maps, Chebychev polynomials, and Lattès maps) then some iterates f^n and g^m are related by a certain correspondence. If f and g are polynomials of the same degree, then the answer becomes simple: $\mu_f = \mu_g$ if and only if there exists a linear transformation L in $\mathbf{C}$ such that $g = L \circ f$ and $L(J) = J$, where $J = J(f) = J(g)$. A consequence of this classification is that for non-exceptional f and g, if $\mu_f = \mu_g$ then f and g have the same sets of preperiodic points (see [31, Rmk 2]).

Note also that Yuan's equidistribution theorem says more: a meaning can be given to the convergence (4) at finite places, and indeed Theorem 7.1 holds at all places. It then follows from the description of the canonical height given in Proposition 6.6 that if f and g are as in Corollary 7.2, then $h_f = h_g$, in particular the sets of preperiodic points of f and g coincide. Denote by Preper(f) the set of preperiodic points of f.

Theorem 7.3 *If f and g are rational maps of degree at least 2, defined over **C**. Consider the following conditions:*

(1) # Preper(f) $\cap$ Preper(g) $= \infty$;

(2) Preper(f) $=$ Preper(g);

(3) $\mu_f = \mu_g$.

Then (1) is equivalent to (2) and both imply (3). If f and g are not exceptional then the converse implication holds as well.

Proof This statement was already fully justified when f and g have their coefficients in $\overline{\mathbf{Q}}$. Note also that the implication (3) $\Rightarrow$ (2) proven in [31] and mentioned above holds for arbitrary non-exceptional rational maps. The equivalence between (1) and (2) for maps with complex coefficients was established by Baker and DeMarco [3] (see also [41, Thm. 1.3]), using some equidistribution results over arbitrary valued fields as well as specialization arguments. Finally, the implication (2) $\Rightarrow$ (3) in this general setting was obtained by Yuan and Zhang [41, Thm. 1.5].
□

PART IX.B: Parameter Space Questions

The general setting in this second part is the following: let $(f_\lambda)_{\lambda \in \Lambda}$ be a holomorphic family of rational maps of degree $d \geqslant 2$ on $\mathbf{P}^1(\mathbf{C})$, that is for every λ, $f_\lambda(z) = \frac{P_\lambda(z)}{Q_\lambda(z)}$ and P_λ and Q_λ depend holomorphically on λ and have no common factors for every λ. Here the parameter space Λ is an arbitrary complex manifold (which may be the space of all rational maps of degree d). From the dynamical point of view we can define a natural dichotomy $\Lambda = \text{Stab} \cup \text{Bif}$ of the parameter space into an open *stability locus* Stab, where the dynamics is in a sense locally constant, and its complement the *bifurcation locus* Bif. Our purpose is to show how arithmetic ideas can give interesting information on these parameter spaces.

8 The Quadratic Family

The most basic example of this parameter dichotomy is given by the family of quadratic polynomials. Any degree 2 polynomial is affinely conjugate to a (unique) polynomial of the form $z \mapsto z^2 + c$. so dynamically speaking the family of quadratic polynomials *is* $\{f_c : z \mapsto z^2 + c, c \in \mathbb{C}\}$. It was studied in great depth since the beginning of the 1980's, starting with the classic monograph by Douady and Hubbard [17].

Let $f_c(z) = z^2 + c$ as above. A first observation is that if $|z| > \sqrt{|c|} + 3$ then

$$\left|z^2 + c\right| \geqslant |z|^2 - |c| \geqslant 2|z| + 1 \tag{5}$$

so $f_c^n(z) \to \infty$. Hence we deduce that the filled Julia set $K_c = K(f_c)$ is contained in $D(0, \sqrt{|c|} + 3)$. Recall that the Julia set of f_c is $J_c = \partial K_c$. Note that the critical point is 0.

8.1   Connectivity of J

Fix R large enough so that $f^{-1}(D(0, R)) \Subset D(0, R)$ $(R = (|c|+1)^2 + 1$ is enough) then by the maximum principle, $f^{-1}(D(0, R))$ is a union of simply connected open sets (topological disks). Since $f : f^{-1}(D(0, R)) \to D(0, R)$ is proper it is a branched covering of degree 2 so the topology of $f^{-1}(D(0, R))$ can be determined from the Riemann–Hurwitz formula. There are two possible cases:

- Either $f(0) \notin D(0, R)$. Then $f|_{f^{-1}(D(0,R))}$ is a covering and $f^{-1}(D(0, R)) = U_1 \cup U_2$ is the union of two topological disks on which $f|_{U_i}$ is a biholomorphism. Let $g_i = (f|_{U_i})^{-1} : D(0, R) \to U_1$, which is a contraction for the Poincaré metric in $D(0, R)$. It follows that

$$K_c = \bigcap_{n \geqslant 1} \bigcup_{(\varepsilon_i) \in \{1,2\}^n} g_{\varepsilon_n} \circ \cdots \circ g_{\varepsilon_1}(D(0, R))$$

   is a Cantor set.
- Or $f(0) \notin D(0, R)$. Then $f^{-1}(D(0, R))$ is a topological disk and $f : f^{-1}(D(0, R)) \to D(0, R)$ is a branched covering of degree 2.

More generally, start with $D(0, R)$ and pull back under f until $0 \notin f^{-n}((D(0, R))$. Again there are two possibilities:

- Either this never happens. Then K_c is a nested intersection of topological disks, so it is a connected compact set which may or may not have interior.
- Or this happens for some $n \geqslant 1$. Then the above reasoning shows that $K_c = J_c$ is a Cantor set.

Summarizing the above discussion, we get the following alternative:

- either 0 escapes under iteration and $K_c = J_c$ is a Cantor set;
- or 0 does not escape under iteration and K_c is connected (and so is J_c).

We define the *Mandelbrot set* M to be the set of $c \in \mathbf{C}$ such that K_c is connected. From the previous alternative we see that the complement of M is an open subset of the plane. Furthermore it is easily seen from (5) that 0 escapes when $|c| > 4$, so M is a compact set in $\mathbf{C}$. The Mandelbrot set is full (that is, its complement has bounded component) and has non-empty interior: indeed if the critical point is attracted by a periodic cycle for $c = c_0$, then this behavior persists for c close to c_0 and K_c is connected in this case.

The following proposition is an easy consequence of the previous discussion.

Proposition 8.1 *c belongs to ∂M if and only if for every open neighborhood $N \ni c$, $(c \mapsto f_c^n(0))_{n \geqslant 1}$ is not a normal family on N.*

8.2   Aside: Active and Passive Critical Points

We will generalize the previous observation to turn it into a quite versatile concept. Let $(f_\lambda)_{\lambda \in \Lambda}$ be a holomorphic family of rational maps as before. In such a family a critical point for f_{λ_0} cannot always be followed holomorphically due to ramification problems (think of $f_\lambda(z) = z^3 + \lambda z$), however this can always be arranged by passing to a branched cover of Λ (e.g. by replacing $z^3 + \lambda z$ by $z^3 + \mu^2 z$). We say that a holomorphically moving critical point $c(\lambda)$ is *marked*.

Definition 8.2 Let $(f_\lambda, c(\lambda))_{\lambda \in \Lambda}$ be a holomorphic family of rational maps with a marked critical point. We say that c is *passive* on some open set Ω if the sequence of meromorphic mappings $(\lambda \mapsto f_\lambda^n(c(\lambda)))_{n \geqslant 0}$ is normal in Ω. Likewise we say that c is *passive* at λ_0 if it is passive in some neighborhood of λ_0. Otherwise c is said *active* at λ_0.

This is an important concept in the study of the stability/bifurcation theory of rational maps, according to the principle "the dynamics is governed by that of critical points". The terminology is due to McMullen. The next proposition is a kind of parameter analogue of the density of periodic points in the Julia set.

Proposition 8.3 *Let $(f_\lambda, c(\lambda))_{\lambda \in \Lambda}$ be a holomorphic family of rational maps of degree $d \geqslant 2$ with a marked critical point. If c is active at λ_0 there exists an infinite sequence (λ_n) converging to λ_0 such that $c(\lambda_n)$ is preperiodic for f_{λ_n}. More precisely we can arrange so that*

- *either $c(\lambda_n)$ falls under iteration onto a repelling periodic point;*
- *or $c(\lambda_n)$ is periodic.*

Proof Fix a repelling cycle of length at least 3 for f_{λ_0}. By the implicit function theorem, this cycle persists and stays repelling in some neighborhood of λ_0. More precisely if we fix 3 distinct points $\alpha_i(\lambda_0)$, $i = 1, 2, 3$ in this cycle, there exists an open neighborhood $N(\lambda_0)$ and holomorphic maps $\alpha_i : N(\lambda_0) \to \mathbf{P}^1$ such that $\alpha_i(\lambda)$ are holomorphic continuations of $\alpha_i(\lambda_0)$ as periodic points. Since the family $(f_\lambda^n(c(\lambda)))_{n \geqslant 1}$ is not normal in $N(\lambda_0)$ by Montel's theorem, there exists

$\lambda_1 \in N(\lambda_0)$, $n \geqslant 1$ and $i \in \{1, 2, 3\}$ such that $f_{\lambda_1}^n (c(\lambda_1)) = \alpha_i(\lambda_1)$. Thus the first assertion is proved, and the other one is similar (see [29] or [19] for details). □

The previous result may be interpreted by saying that an active critical point is a source of bifurcations. Indeed, given any holomorphic family $(f_\lambda)_{\lambda \in \Lambda}$ of rational maps, Λ can be written as the union of an open *stability locus* Stab where the dynamics is (essentially) locally constant on $\mathbf{P}^1$ and its complement, the *bifurcation locus* Bif where the dynamics changes drastically. When all critical points are marked, the bifurcation locus is exactly the union of the activity loci of the critical points. It is a fact that in any holomorphic family, the bifurcation locus is a fractal set with rich topological structure.

Theorem 8.4 (Shishikura [38], Tan Lei [28], McMullen [34]) *If $(f_\lambda)_{\lambda \in \Lambda}$ is any holomorphic family of rational maps with non-empty bifurcation locus, the Hausdorff dimension of* Bif *is equal to that of Λ.*

Recall that the Hausdorff dimension of a metric space is a non-negative number which somehow measures the scaling behavior of the metric. For a manifold, it coincides with its topological dimension, but for a fractal set it is typically not an integer. In the quadratic family, the bifurcation locus is the boundary of the Mandelbrot set, and Shishikura [38] proved that $\text{HDim}(\partial M) = 2$. Therefore ∂M is a nowhere dense subset which in a sense locally fills the plane. It was then shown in [28, 34] that this topological complexity can be transferred to any holomorphic family, resulting in the above theorem.

8.3 Post-Critically Finite Parameters in the Quadratic Family

A rational map is said *post-critically finite* if its critical set has a finite orbit, that is, every critical point is periodic or preperiodic. The post-critically finite parameters in the quadratic family are the solutions of the countable family of polynomial equations $f_c^k(0) = f_c^l(0)$, with $k > l \geqslant 0$, and form a countable set of "special points" in parameter space.

As a consequence of Proposition 8.3 in the quadratic family we get

Corollary 8.5 *∂M is contained in the closure of the set of post-critically finite parameters.*

Post-critically finite quadratic polynomials can be of two different types:

− either the critical point 0 is periodic. in this case c lies in the interior of M. Indeed the attracting periodic orbit persists in some neighborhood of c, thus persistently attracts the critical point and K_c remains connected;
− or 0 is strictly preperiodic. Then it can be shown that it must fall on a repelling cycle, so it is active and $c \in \partial M$.

The previous corollary can be strengthened to an equidistribution statement.

Theorem 8.6 *Post-critically finite parameters are asymptotically equidistributed.*

There are several ways of formalizing this. For a pair of integers $k > l \geqslant 0$, denote by $\mathrm{PerCrit}(k, l)$ the (0-dimensional) variety defined by the polynomial equation $f_c^k(0) = f_c^l(0)$, and by $[\mathrm{PerCrit}(k, l)]$ the sum of point masses at the corresponding points, counting multiplicities.

Then the precise statement of the theorem is that there exists a probability measure μ_M on the Mandelbrot set such that if $0 \leqslant k(n) < n$ is an arbitrary sequence, then $2^{-n}[\mathrm{PerCrit}(n, k(n))] \to \mu_M$ as $n \to \infty$. Originally proved by Levin [30] (see also [32] for $k(n) = 0$), this result was generalized by several authors (see e.g. [19, 5]). Quantitative estimates on the speed of convergence are also available [23, 24].

We present an approach to this result based on arithmetic equidistribution, along the lines of [5, 23].

Recall the dynamical Green function

$$G_{f_c}(z) = \lim_{n \to \infty} \frac{1}{2^n} \log^+ |f_c^n(z)|,$$

which is a non-negative continuous and plurisubharmonic function of $(c, z) \in \mathbf{C}^2$. Put $G_M(c) = G_{f_c}(c) = 2G_{f_c}(0)$. This function is easily shown to have the following properties:

- G_M is non-negative, continuous and subharmonic on $\mathbf{C}$;
- $G_M(c) = \log |c| + o(1)$ when $c \to \infty$;
- $\{G_M = 0\}$ is the Mandelbrot set;
- G_M is harmonic on $\{G_M > 0\} = M^{\complement}$.

Therefore G_M is the (potential-theoretic) Green function of the Mandelbrot set and $\mu_M := \Delta G_M$ is the harmonic measure of M.

To apply arithmetic equidistribution theory, we need to understand what happens at the non-Archimedean places. For a prime number p, let

$$M_p = \left\{ c \in \mathbf{C}_p, \ (f_c^n(0))_{n \geqslant 0} \text{ is bounded in } \mathbf{C}_p \right\}.$$

Proposition 8.7

(i) For every $p \in \mathscr{P}$, M_p is the closed unit ball of $\mathbf{C}_p$.
(ii) For every $p \in \mathscr{P}$, for every $c \in \mathbf{C}_p$,

$$G_{M_p}(c) = G_{f_c}(c) = \lim_{n \to \infty} 2^{-n} \log^+ |f_c^n(c)|_p = \log^+ |c|_p.$$

(iii) The associated height function $h_{\mathbf{M}}$ defined for $c \in \overline{\mathbf{Q}}$ by

$$h_{\mathbf{M}}(c) = \frac{1}{[\mathbf{Q}(c) : \mathbf{Q}]} \sum_{p \in \mathscr{P} \cup \{\infty\}} \sum_{\sigma : \mathbf{Q}(c) \hookrightarrow \mathbf{C}_p} G_{M_p}(\sigma(c))$$

satisfies

$$\left\{ c \in \overline{\mathbf{Q}}, \, h_{\mathbf{M}}(c) = 0 \right\} = \bigcup_{0 \leqslant k < n} \mathrm{PerCrit}(k, n).$$

The collection $\mathbf{M}$ of the sets M_p for $p \in \mathscr{P} \cup \{\infty\}$ is called the *adelic Mandelbrot set*, and $h_{\mathbf{M}}$ will be referred to as the *parameter height function* associated to the critical point 0.

Proof Using the ultrametric property, we see that $|c|_p \leqslant 1$ implies that $|f_c(c)|_p = \left| c^2 + c \right|_p \leqslant 1$. Conversely $|c|_p > 1$ implies $\left| c^2 + c \right|_p = |c|_p^2$ hence $f_c^n(c) \to \infty$. This proves *(i)* and *(ii)*.

For the last assertion, we observe that the canonical height of f_c is given by

$$h_{f_c}(z) = \frac{1}{[\mathbf{Q}(z) : \mathbf{Q}]} \sum_{p \in \mathscr{P} \cup \{\infty\}} \sum_{\sigma : \mathbf{Q}(z) \hookrightarrow \mathbf{C}_p} G_{f_c, p}(\sigma(z)).$$

Indeed it is clear from this formula that $h_{f_c}(z) = 2 h_{f_c}(z)$ and $h_{f_c} - h_{\mathrm{naive}} = O(1)$ so the result follows from the Call–Silverman Theorem 6.4. Recall that $h_{f_c}(z) = 0$ if and only if z is preperiodic. Assertion *(iii)* follows from these properties by simply plugging c into the formulas. □

An *adelic set* (this terminology is due to Rumely) $\mathbf{E} = \left\{ E_p, p \in \mathscr{P} \cup \infty \right\}$ is a collection of sets $E_p \subset \mathbf{C}_p$ such that

- E_∞ is a full compact set in $\mathbf{C}$;
- for every $p \in \mathscr{P}$, E_p is closed and bounded in $\mathbf{C}_p$, and E_p is the closed unit ball for all but finitely many p;
- for every $p \in \mathscr{P} \cup \infty$, E_p admits a Green function g_p that is continuous on $\mathbf{C}_p$ and satisfies $E_p = \left\{ g_p = 0 \right\}$, $g_p(z) = \log^+ |z|_p - c_p + o(1)$ when $z \to \infty$, and g_p is "harmonic"[2] outside E_p.

The *capacity* of an adelic $\mathbf{E}$ is defined to be $\gamma(\mathbf{E}) = \prod_{p \in \mathscr{P} \cup \infty} e^{c_p}$. We will typically assume that $\gamma(\mathbf{E}) = 1$. Under these assumptions one defines a height function from the local Green functions g_p exactly as before

$$h_{\mathbf{E}}(z) = \frac{1}{[\mathbf{Q}(z) : \mathbf{Q}]} \sum_{p \in \mathscr{P} \cup \{\infty\}} \sum_{\sigma : \mathbf{Q}(z) \hookrightarrow \mathbf{C}_p} g_p(\sigma(z)),$$

---

[2]We do not define precisely what "harmonic" means in the p-adic context: roughly speaking it means that locally $g = \log |h|$ for some non-vanishing analytic function (see [6, Chap. 7] for details on this notion). Note that for the adelic Mandelbrot set we have $g_p = \log^+ |\cdot|_p$ at non-Archimedean places.

and we have the following equidistribution theorem (see [2, 11] for details and references).

Theorem 8.8 (Bilu, Rumely) *Let* E *be an adelic set such that* $\gamma(E) = 1$, *and* $(x_n) \in \mathbf{C}^\mathbf{N}$ *be a sequence of points with disjoint Galois orbits* X_n *and such that* $h(x_n)$ *tends to 0. Then the sequence of equidistributed probability measures on the sets* X_n *converges weakly in* $\mathbf{C}$ *to the potential-theoretic equilibrium measure of* E_∞.

This convergence also holds at finite places, provided one is able to make sense of measure theory in this context. Theorem 8.6 follows immediately.

9 Higher Degree Polynomials and Equidistribution

In this and the next section we investigate the asymptotic distribution of special points in spaces of higher degree polynomials. The situation is far less understood for spaces of rational functions. Polynomials of the form $P(z) = \sum_{k=0}^d a_k z^k$ can be obviously parameterized by their coefficients so that the space $\mathscr{P}_d$ of polynomials of degree d is $\mathbf{C}^* \times \mathbf{C}^d$. By an affine conjugacy we can arrange that $a_d = 1$ and $a_{d-1} = 0$ (monic and centered polynomials).

Assume now that P and Q are monic, centered and conjugate by the affine transformation $z \mapsto az + b$, that is, $P(z) = a^{-1}Q(az + b) - ba^{-1}$. Since P and Q are monic we infer that $a^{d-1} = 1$ and from the centering we get $b = 0$. It follows that the space $\mathscr{M}\mathscr{P}_d$ of polynomials of degree d modulo affine conjugacy is naturally isomorphic to $\mathbf{C}^{d-1}/\langle\zeta\rangle$, where $\zeta = e^{2i\pi/(d-1)}$. In practice it is easier to work on its $(d-1)$-covering by $\mathbf{C}^{d-1}$.

9.1 Special Points

The special points in $\mathscr{M}\mathscr{P}_d$ are the post-critically finite maps. They are dynamically natural since classical results of Thurston, Douady–Hubbard and others (see e.g. [18, 27]) show that the geography of $\mathscr{M}\mathscr{P}_d$ is somehow organized around them.

Proposition 9.1

(i) *The set* PCF *of post-critically finite polynomials is countable and Zariski-dense in* $\mathscr{M}\mathscr{P}_d$.
(ii) PCF *is relatively compact in the usual topology.*

Proof For $d = 2$ this follows from the properties of the Mandelbrot set so it is enough to deal with $d \geqslant 3$. To prove the result it is useful to work in the space $\mathscr{M}\mathscr{P}_c^{\mathrm{cm}}$ of critically marked polynomials modulo affine conjugacy, which is a branched covering of $\mathscr{M}\mathscr{P}_d$. Again this space is singular so we work on the following parameterization. For $(c, a) = (c_1, \ldots, c_{d-2}, a) \in \mathbf{C}^{d-1}$ we consider the polynomial $P_{c,a}$ with critical points at $(c_0 = 0, , c_1, \ldots, c_{d-2})$ and such that $P(0) = a^d$, that is $P_{c,a}$ is the primitive of $z \prod_{i=1}^{d-2}(z - c_i)$ such that $P(0) = a^d$. This defines a map $\mathbf{C}^{d-1} \to \mathcal{M}\mathscr{P}_c^{\mathrm{cm}}$ which is a branched cover of degree $d(d-1)$.

Let $G_{P_{c,a}}$ be the dynamical Green function of $P_{c,a}$, and define

$$G(c, a) = \max \left\{ G_{P_{c,a}}(c_i), \ i = 0, \ldots, d - 2 \right\}.$$

The asymptotic behavior of G is well understood.

Theorem 9.2 (Branner–Hubbard [8]) *As $(c, a) \to \infty$ in $\mathbf{C}^{d-1}$ we have that*

$$G(c, a) = \log^+ \max \{|a|, |c_i|, i = 1, \ldots, d - 2\} + O(1).$$

The seemingly curious normalization $P(0) = a^d$ was motivated by this neat expansion. See [19] for a proof using these coordinates, based on explicit asymptotic expansions of the $P_{(c,a)}(c_i)$. In particular $G(c, a) \to \infty$ as $(c, a) \to \infty$ and it follows that the connectedness locus

$$\mathscr{C} := \left\{ (c, a), \ K(P_{c,a}) \text{ is connected} \right\} = \{(c, a), \ G(c, a) = 0\}$$

is compact. Since post-critically finite parameters belong to $\mathscr{C}$, this proves the second assertion of Proposition 9.1. For the first one, note that the set of post-critically finite parameters is defined by countably many algebraic equations in $\mathbf{C}^d$, and since each component is bounded in $\mathbf{C}^{d-1}$ it must be a point.

The Zariski density of post-critically finite polynomials is a direct consequence of the equidistribution results of [7, 19], based on pluripotential theory. Here we present a simpler argument due to Baker and DeMarco [4]. This requires the following result, whose proof will be skipped.

Theorem 9.3 (McMullen [33], Dujardin–Favre [19])
Let $(f_\lambda, c(\lambda))_{\lambda \in \Lambda}$ be a holomorphic family of rational maps of degree d with a marked critical point, parameterized by a quasiprojective variety Λ. If c is passive along Λ, then:

- *either the family is isotrivial, that is the f_λ are conjugate by Möbius transformations*
- *or c is persistently preperiodic, that is there exists $m < n$ such that $f_\lambda^m(c(\lambda)) \equiv f_\lambda^n(c(\lambda))$.*

Let now S be any proper algebraic subvariety of $\mathcal{M}\mathscr{P}_d^{\mathrm{cm}} =: X$, we want to show that there exists a post-critically finite parameter in $X \setminus S$. Consider the first marked critical point c_0 on $\Lambda = X \setminus S$. Since the family $(P_{c,a})$ is not isotrivial on Λ, by the previous theorem, either it is persistently preperiodic or it must be active somewhere and by perturbation we can make it preperiodic by Proposition 8.3. In any case we can find $\lambda_0 \in \Lambda$ and $m_0 < n_0$ such that $f_{\lambda_0}^{m_0}(c_0(\lambda_0)) \equiv f_{\lambda_0}^{m_0}(c_0(\lambda_0))$. Now define

Λ_1 to be the subvariety of codimension $\leqslant 1$ of $X \setminus S$ where this equation is satisfied. It is quasiprojective and c_0 is persistently preperiodic on it. We now consider the behaviour of c_1 on Λ_1 and continue inductively to get a nested sequence of quasiprojective varieties $\Lambda_k \subset X$ on which $c_0, \ldots, c_k$ are persistently preperiodic. Since the dimension drops by at most 1 at each step and $\dim(X) = d - 1$, we can continue until $k = d - 2$ and we finally find the desired parameter. $\square$

9.2   Equidistribution of Special Points

Pluripotential theory (see e.g. [15, Chap. III]) allows to give a meaning to the exterior product $\left(\frac{i}{\pi} \partial \bar{\partial} G \right)^{\wedge (d-1)}$. This defines a probability measure with compact support in $\mathscr{MP}_d^{\mathrm{cm}}$ which will be referred to as the *bifurcation measure*, denoted by μ_{bif}.

The next theorem asserts that post-critically finite parameters are asymptotically equidistributed. For $0 \leqslant i \leqslant d - 2$ and $m < n$ define the subvariety $\mathrm{Per}_{c_i}(m, n)$ to be the closure of the set of parameters at which $f^k(c_i)$ is periodic exactly for $k \geqslant m$, with exact period $n - m$.

Theorem 9.4 (Favre–Gauthier [21]) *Consider a $(2d - 2)$-tuple of sequences of integers $((n_{k,0}, m_{k,0}), \ldots, (n_{k,d-2}, m_{k,d-2}))_{k \geqslant 0}$ such that:*

- *either the $m_{k,i}$ are equal to 0 and for fixed k the $n_{k,i}$ are distinct and $\min_i(n_{k,i})$ tends to ∞ with k;*
- *or for every (k, i), $n_{k,i} > m_{k,i} > 0$ and $\min_i(n_{k,i} - m_{k,i}) \to \infty$ when $k \to \infty$.*

Then, letting $Z_k = \mathrm{Per}_{c_0}(m_{k,0}, n_{k,0}) \cap \cdots \cap \mathrm{Per}_{c_{d-2}}(m_{k,d-2}, n_{k,d-2})$, the sequence of probability measures uniformly distributed on Z_k converges to the bifurcation measure as $k \to \infty$.

This puts forward the bifurcation measure as the natural analogue in higher degree of the harmonic measure of the Mandelbrot set. It was first defined and studied in [7, 19]. It follows from its pluripotential-theoretic construction that μ_{bif} carries no mass on analytic sets. In particular this gives another argument for the Zariski density of post-critically finite parameters .

The result is a consequence of arithmetic equidistribution. Using the function $G(c, a)$ and its adelic analogues, we can define as before a parameter height function on $\mathscr{MP}_d^{\mathrm{cm}}(\overline{\mathbf{Q}})$ satisfying $h(c, a) = 0$ iff $P_{c,a}$ is critically finite. Then Yuan's equidistribution theorem for points of small height (see [11]) applies in this situation –this requires some non-trivial work on understanding the properties of G at infinity. Specialized to our setting it takes the following form.

Theorem 9.5 (Yuan) *Let $Z_k \subset \mathcal{MP}_d^{cm}(\overline{\mathbf{Q}})$ be a sequence of Galois invariant subsets such that:*

(i) $h(Z_k) = \dfrac{1}{\#Z_k} \displaystyle\sum_{x \in Z_k} h(x) \xrightarrow[k \to \infty]{} 0;$

(ii) For every algebraic hypersurface H over $\mathbf{Q}$, $H \cap Z_k$ is empty for large enough k.

Then $\frac{1}{\#Z_k} \sum_{x \in Z_k} \delta_x$ converges to the bifurcation measure μ_{bif} as $k \to \infty$.

Actually in the application to post-critically finite maps the genericity assumption *(ii)* is not satisfied. Indeed we shall see in the next section that there are "special subvarieties" containing infinitely many post-critically finite parameters. Fortunately the following variant is true.

Corollary 9.6 *In Theorem 9.5, if (ii) is replaced by the weaker condition:*

(ii') For every algebraic hypersurface H over $\mathbf{Q}$, $\displaystyle\lim_{k \to \infty} \dfrac{\#H \cap Z_k}{\#Z_k} = 0,$

then the same conclusion holds.

Proof of the Corollary This is based on a diagonal extraction argument. First, enumerate all hypersurfaces defined over $\mathbf{Q}$ to form a sequence $(H_q)_{q \geq 0}$. Fix $\varepsilon > 0$. For $q = 0$ we have that $\frac{\#H_0 \cap Z_k}{\#Z_k} \to 0$ as $k \to \infty$ so if for $k \geq k_0$ we remove from Z_k the (Galois invariant) set of points belonging to H_0 to get a subset $Z_k^{(0)}$ such that

$$Z_k^{(0)} \cap H_0 = \emptyset \text{ and } \frac{\#Z_k^{(0)}}{\#Z_k} \geq 1 - \frac{\varepsilon}{4}.$$

For $k \leq k_0$ we put $Z_k^{(0)} = Z_k$.

Now for H_1 we do the same. There exists $k_1 > k_0$ and for $k \geq k_1$ a Galois invariant subset $Z_k^{(1)}$ extracted from $Z_k^{(0)}$ such that

$$Z_k^{(1)} \cap H_1 = \emptyset \text{ and } \frac{\#Z_k^{(1)}}{\#Z_k^{(0)}} \geq 1 - \frac{\varepsilon}{8}.$$

For $k < k_0$ we set $Z_k^{(1)} = Z_k$ and for $k_0 \leq k < k_1$ we set $Z_k^{(1)} = Z_k^{(0)}$ Note that for $k \geq k_0$, $Z_k^{(1)} \cap H_0 = \emptyset$.

Continuing inductively this procedure, for every $q \geq 0$ we get a sequence of subsets $(Z_k^{(q)})_{k \geq 0}$ with $Z_k^{(q)} \subset Z_k$ such that

$$\#Z_k^{(q)} \geq \prod_{j=0}^{q} \left(1 - \frac{\varepsilon}{2^{j+2}}\right) \#Z_k \geq (1 - \varepsilon)\#Z_k$$

and Z_k is disjoint from $H_0, \ldots, H_q$ for $k \geqslant k_q$. Finally we define $Z_k^{(\infty)} = \bigcap_q Z_k^{(q)}$, which satisfies that for every q and every $k \geqslant k_q$, $Z_k^{(\infty)}$ is disjoint from $H_0, \ldots, H_q$ and for every $k \geqslant 0$, $\#Z_k^{(\infty)} \geqslant (1-\varepsilon)\#Z_k$. Therefore $Z_k^{(\infty)}$ satisfies the assumptions of Yuan's Theorem so

$$\mu_k^{(\infty)} := \frac{1}{\#Z_k^{(\infty)}} \sum_{x \in Z_k^{(\infty)}} \delta_x \xrightarrow[k \to \infty]{} \left(\frac{i}{\pi} \partial \bar{\partial} G\right)^{\wedge (d-1)}.$$

Finally if we let μ_k be the uniform measure on Z_k, we have that for every continuous function φ with compact support,

$$\left|\mu_k^{(\infty)}(\varphi) - \mu_k(\varphi)\right| \leqslant 2\varepsilon \|\varphi\|_{L^\infty}$$

and we conclude that μ_k converges to the bifurcation measure as well. This finishes the proof.   □

Proof of Theorem 9.4 We treat the first set of assumptions $m_{k,i} = 0$, and denote put $\mathrm{Per}_{c_i}(n_{k,i}) = \mathrm{Per}_{c_i}(0, n_{k,i})$. It is convenient to assume that the $n_{k,i}$ are prime numbers, which simplifies the issues about prime periods. We have to check that the hypotheses Corollary 9.6 are satisfied. First X_k is defined by $(d-1)$ equations over $\mathbf{Q}$ so it is certainly Galois invariant, and it is a set of post-critically finite parameters so its parameter height vanishes. So the point is to check condition *(ii')*. For every $0 \leqslant i \leqslant d-2$ the variety $\mathrm{Per}_{c_i}(n_{k,i})$ is defined by the equation $P_{c,a}^{n_{k,i}}(c_i) - c_i = 0$ which is of degree $d^{n_{k,i}}$. Recall from Proposition 9.1 that Z_k is of dimension 0 and relatively compact in $\mathscr{MP}_d^{\mathrm{cm}}$. Furthermore the analysis leading to Theorem 9.2 shows that these hypersurfaces do not intersect at infinity so by Bézout's theorem the cardinality of Z_k equals $d^{n_{k,0}+\cdots+n_{k,d-1}}$ *counting multiplicities*. Actually multiplicities do not account, due to the following deep result, which is based on dynamical and Teichmüller-theoretic techniques.

Theorem 9.7 (Buff–Epstein [9]) *For every $x \in Z_k$, the varieties* $\mathrm{Per}_{c_0}(n_{k,0}), \ldots,$ $\mathrm{Per}_{c_{d-2}}(n_{k,d-2})$ *are smooth and transverse at x.*

Finally, for every hypersurface H we need to bound $\#H \cap Z_k$. Let $x \in H \cap Z_k$ and assume that x is a regular point of H. A first possibility is that locally (and thus also globally) $H \equiv \mathrm{Per}_{c_i}(n_{k,i})$ for some i. Since $n_{k,i} \to \infty$ this situation can happen only for finitely many k so considering large enough k we may assume that H is distinct from the $\mathrm{Per}_{c_i}(n_{k,i})$. By the transversality Theorem 9.7, H must be transverse to $(d-2)$ of the $\mathrm{Per}_{c_i}(n_{k,i})$ at x (this is the incomplete basis theorem!).

So we can bound $\#H \cap Z_k$ by applying Bézout's theorem to all possible intersections of H with $(d-2)$ of the $\text{Per}_{c_i}(n_{k,i})$, that is

$\#\text{Reg}(H) \cap Z_k$

$$\leqslant \sum_{i=0}^{d-2} \#H \cap \text{Per}_{c_0}(n_{k,0}) \cap \cdots \cap \widehat{\text{Per}_{c_i}(n_{k,i})} \cap \cdots \cap \text{Per}_{c_{d-2}}(n_{k,d-2})$$

$$\leqslant \sum_{i=0}^{d-2} \deg(H) \cdot d^{\left(\sum_{j=0}^{d-2} n_{k,j}\right) - n_{k,i}} = o\left(d^{\sum_{j=0}^{d-2} n_{k,j}}\right) = o\,(\#Z_k).$$

To deal with the singular part of H, we write $\text{Sing}(H) = \text{Reg}(\text{Sing}(H)) \cup \text{Sing}(\text{Sing}(H))$, and using the above argument in codimension 2, by we get a similar estimate for $\#\text{Reg}(\text{Sing}(H)) \cap Z_k$. Repeating inductively this idea we finally conclude that $\#H \cap Z_k = o\,(\#Z_k)$, as required. □

10 Special Subvarieties

10.1 Prologue

There are a number of situations in algebraic geometry where the following happens: an algebraic variety X is given containing countably many "special subvarieties" (possibly of dimension 0). Assume that a subvariety Y admits a Zariski-dense subset of special points, then must it be special, too? Two famous instances of this problem are:

- torsion points on Abelian varieties (and the Manin–Mumford conjecture);
- CM points on Shimura varieties (and the André–Oort conjecture).

The Manin–Mumford conjecture admits a dynamical analogue which will not be discussed in these notes. We will consider an analogue of the André–Oort conjecture which has been put forward by Baker and DeMarco [4]. Without entering into the details of what "André–Oort" refers exactly to let us just mention one positive result which motivates the general conjecture. Let us identify the space of pairs of elliptic curves with $\mathbf{C}^2$ via the j-invariant.

Theorem 10.1 (André [1]) *Let $Y \subset \mathbf{C}^2$ be an algebraic curve containing infinitely many points both coordinates of which are "singular moduli", that is, j-invariants of CM elliptic curves. Then Y is special in the sense that Y is either a vertical or a horizontal line, or a modular curve $X_0(N)$.*

Recall that $X_0(N)$ is the irreducible algebraic curve in $\mathbf{C}^2$ uniquely defined by the property that $(E, E') \in X_0(N)$ if there exists a cyclic isogeny $E \to E'$ of degree

N (this property is actually symmetric in E and E'). Likewise, it is characterized by the property that for every $\tau \in \mathbf{H}$, $(j(N\tau), j(\tau)) \in X_0(N)$.

10.2 Classification of Special Curves

In view of the above considerations it is natural to attempt to classify special subvarieties, that is subvarieties Λ of $\mathscr{MP}_d$ (or $\mathscr{M}_d$) with a Zariski dense subset of post-critically finite parameters. Examples are easy to find: assume that the k critical points $c_0, \ldots, c_{k-1}$ critical points are "dynamically related" on a subvariety Λ of codimension $k - 1$. This happens for instance if they satisfy a relation of the form $P^{k_i}(c_i) = P^{l_i}(c_0)$ for some integers k_i, l_i (say Λ is the subvariety cut out by $k - 1$ such equations, thus $\dim(\Lambda) = d - k$). Then $c_0, c_{k+1}, \ldots, c_{d-2}$ are $(d - 1 - k) + 1 = d - k$ "independent" critical points on Λ and arguing as in Proposition 9.1 shows that post-critically finite maps are Zariski dense on Λ.

A "dynamical André–Oort conjecture" was proposed by Baker and DeMarco [4] which says precisely that a subvariety Λ of dimension q in the moduli space of rational maps of degree d with marked critical points $\mathscr{M}_d^{\mathrm{cm}}$ is special if and only if at most q critical points are "dynamically independent" on Λ (the precise notion of dynamical dependence is slightly delicate to formalize).

Some partial results towards this conjecture are known, including a complete proof for the space of cubic polynomials.[3] Recall that $\mathscr{MP}_3^{\mathrm{cm}}$ is parameterized by $(c, a) \in \mathbf{C}^2$.

Theorem 10.2 (Favre–Gauthier [22], Ghioca–Ye [25]) *An irreducible curve C in the space $\mathscr{MP}_3^{\mathrm{cm}}$ is special if and only if one of the of the following holds:*

- *one of the two critical points is persistently preperiodic along C;*
- *There is a persistent collision between the critical orbits, that is, there exists $(k, l) \in \mathbf{N}^2$ such that $P_{c,a}^k(c_0) = P_{c,a}^l(c_1)$ on C;*
- *C is the curve of cubic polynomials $P_{c,a}$ commuting with $Q_c : z \mapsto -c + z$ (which is given by an explicit equation).*

The proof is too long to be described in these notes, let us just say a few words on how arithmetic equidistribution (again!) enters into play. For $i = 0, 1$, we define the function G_i by $G_i(c, a) = G_{c,a}(c_i)$, which again admit adelic versions. Since C admits infinitely many post-critically finite parameters, Yuan's theorem applies to show that they are equidistributed inside C. Now there are two parameter height functions on C, one associated to c_0 and the other one associated to c_1. Since post-critically finite parameters are of height 0 relative to both functions, we infer that the limiting measure must be proportional to both $\Delta(G_0|_C)$ and $\Delta(G_1|_C)$,

---

[3]Note added in April 2020: Favre and Gauthier have recently obtained a classification of special curves in spaces of polynomials of arbitrary degree.

thus $\Delta(G_0|_C) = \alpha \Delta(G_1|_C)$ for some $\alpha > 0$. This defines a first dynamical relation between c_0 and c_1, which after some work (which involves in particular equidistribution at finite places), is promoted to an analytic relation between the c_i, and finally to the desired dynamical relation.

We give a more detailed argument in a particular case of Theorem 10.2, which had previously been obtained by Baker and DeMarco [4]. We let $\mathrm{Per}_n(\kappa)$ be the algebraic curve in $\mathscr{MP}_3^{\mathrm{cm}}$ defined by the property that polynomials in $\mathrm{Per}_n(\kappa)$ admit a periodic point of exact period n and multiplier κ.

Theorem 10.3 (Baker–DeMarco [4]) *The curve* $\mathrm{Per}_1(\kappa)$ *in* $\mathscr{MP}_3^{\mathrm{cm}}$ *is special if and only if* $\kappa = 0$.

The same result actually holds for $\mathrm{Per}_n(\kappa)$ for every $n \geqslant 1$ by Favre and Gauthier [22, Thm B].

Proof The "if" implication is easy: $\mathrm{Per}_1(0)$ is the set of cubic polynomials where some critical point is fixed. Consider an irreducible component of $\mathrm{Per}_1(0)$, then one critical point, say c_0 is fixed. We claim that c_1 is not passive along that component. Indeed otherwise by Theorem 9.3 it would be persistently preperiodic and we would get a curve of post-critically finite parameters. So there exists a parameter $\lambda_0 \in \mathrm{Per}_1(0)$ at which c_1 is active, hence by Proposition 8.3 we get an infinite sequence $\lambda_n \to \lambda_0$ for which $c_1(\lambda_n)$ is preperiodic. In particular $\mathrm{Per}_1(0)$ contains infinitely many post-critically finite parameters and we are done.

Before starting the proof of the direct implication, observe that if $0 < |\kappa| < 1$ by Theorem 3.5 a critical point must be attracted by the attracting fixed point so $\mathrm{Per}_1(\kappa)$ is not special. A related argument also applies for $|\kappa| = 1$, so the result is only interesting when $|\kappa| > 1$. The proof will be divided in several steps. We fix $\kappa \neq 0$.

Step 1: adapted parameterization. We first change coordinates in order to find a parameterization of C that is convenient for the calculations to come. First we can conjugate by a translation so that the fixed point is 0. The general form of cubic polynomials with a fixed point at 0 of multiplier κ is $\kappa z + az^2 + bz^3$ with $b \neq 0$. Now by a homothety we can adjust $b = 1$ to get $\kappa z + az^2 + z^3$. In this form the critical points are not marked. It turns out that a convenient parameterization of C is given by $(f_s)_{s \in \mathbb{C}^*}$ defined by

$$f_s(z) = \kappa \left(z - \frac{1}{2} \left(s + \frac{1}{s} \right) z^2 + \frac{1}{3} z^3 \right),$$

whose critical points are s and $1/s$. Denote $c^+(s) = s$ and $c^-(s) = s^{-1}$. The parameterization has 2-fold symmetry $s \leftrightarrow s^{-1}$.

Step 2: Green function estimates. We consider as usual the values of the dynamical Green function at critical points and define $G^+(s) = G_{f_s}(s)$ and $G^-(s) = G_{f_s}(s^{-1})$. An elementary calculation shows that $s \mapsto f_s^n(s)$ is a polynomial in s, with the following leading coefficient

$$f_s^n(s) = \frac{\kappa}{3}\left(\frac{\kappa}{3}\right)^3 \cdots \left(\frac{\kappa}{3}\right)^{3^{n-2}} \left(-\frac{\kappa}{6}\right)^{3^{n-1}} s^{3^n} + O\left(s^{3^{n-1}}\right)$$

$$= \left(\frac{\kappa}{3}\right)^{\frac{1}{6}(3^{n-2}-1)} \left(-\frac{\kappa}{6}\right)^{3^{n-1}} s^{3^n} + O\left(s^{3^{n-1}}\right).$$

It follows that

$$G^+(s) = G_{f_s}(s) = \lim_{n \to \infty} \log^+ \left|f_s^n(s)\right| \tag{6}$$

$$= \log|s| + \log\left|\frac{\kappa}{6}\right|^{1/3} + \log\left|\frac{\kappa}{3}\right|^{1/6} + O(1)$$

when $s \to \infty$ (I am cheating here because the coefficient in $O(s^{3^{n-1}})$ is not uniform in n). On the other hand, since $f_s^n(s)$ is a polynomial, for $|s| \leqslant 1$ we have $\left|f_s^n(s)\right| \leqslant \left|f_1^n(1)\right|$ so $G^+(s) \leqslant G^+(1)$ and it follows that G^+ is bounded near $s = 0$. By symmetry G^- is bounded near ∞ and tends to $+\infty$ when $s \to 0$.

Step 3: bifurcations. We now claim that c^+ and c^- are not passive along C. Indeed as before otherwise they would be persistently preperiodic, contradicting the fact that G^+ and G^- are unbounded. So we can define two bifurcation measures $\mu^+ = \Delta G^+$ and $\mu^- = \Delta G^-$ in $\mathbf{C}^*$. Note that μ^+ is a probability measure (because the coefficient of the log in (6) equals 1) and its support is bounded in $\mathbf{C}$ (i.e. does not contain ∞) because the dynamical Green function is harmonic when it is positive. A similar description holds for μ^-, whose support is away from 0.

We saw in Sect. 9 that the bifurcation locus is the union of the activity loci of critical points. It follows that it is contained in $\mathrm{Supp}(\mu^+) \cup \mathrm{Supp}(\mu^-)$ –this is actually an equality by a theorem of DeMarco [13].

Step 4: equidistribution argument and conclusion. Assume now that $\mathrm{Per}_1(\kappa)$ contains infinitely many post-critically finite parameters. Note that the existence of a post-critically finite map in $\mathrm{Per}_1(\kappa)$ implies that $\kappa \in \overline{\mathbf{Q}}$ so $\mathrm{Per}_1(\kappa)$ is defined over some number field K. It can be shown that the functions $G^\pm$ (in their adelic version) satisfy the assumptions of the Yuan's arithmetic equidistribution theorem. Thus if $(s_k)_{k \geqslant 0}$ is any infinite sequence such that $c^+(s_k)$ is preperiodic, the uniform measures on the $\mathrm{Gal}(\overline{K}/K)$ conjugates of s_k equidistribute towards μ^+. Applying this fact along a sequence of post-critically finite parameters we conclude that $\mu^+ = \mu^-$. We want to derive a contradiction from this equality.

We have that $\mathrm{Supp}(\mu^+) = \mathrm{Supp}(\mu^-)$ so this set must be compact in $\mathbf{C}^*$. The function $G^+ - G^-$ is harmonic on $\mathbf{C}^*$ with $G^+(s) - G^-(s) = \log|s| + O(1)$ at $+\infty$ and 0. Therefore $G^+ - G^- - \log|\cdot|$ is harmonic and bounded on $\mathbf{C}^*$ so it is constant, and we conclude that $G^+(s) - G^-(s) = \log|s| + C$ for some C. Now recall that the bifurcation locus Bif is contained in $\text{Supp}(\mu^+) \cup \text{Supp}(\mu^-) = \text{Supp}(\mu^+)$, so Bif $\subset \{G^+ = 0\}$. Likewise Bif $\subset \{G^- = 0\}$ so Bif is contained in $\{G^+ - G^- = 0\}$ which is a circle. On the other hand it follows from Theorem 8.4 that for the family $(f_s)_{s \in \mathbb{C}^*}$, the Hausdorff dimension of Bif is equal to 2. This contradiction finishes the proof. $\qquad\square$

Acknowledgements Research partially supported by ANR project LAMBDA, ANR-13-BS01-0002 and a grant from the Institut Universitaire de France.

References

1. Y. André, Finitude des couples d'invariants modulaires singuliers sur une courbe algébrique plane non modulaire. J. Reine Angew. Math. **505**, 203–208 (1998)
2. P. Autissier, Autour du théorème de Fekete-Szegö, in *Arakelov Geometry and Diophantine Applications* (Springer, Cham, 2020)
3. M. Baker, L. De Marco, Preperiodic points and unlikely intersections. Duke Math. J. **159**, 1–29 (2011)
4. M. Baker, L. De Marco, Special curves and postcritically finite polynomials. Forum Math. Pi **1**, 35 (2013)
5. M.H. Baker, L.-C. Hsia, Canonical heights, transfinite diameters, and polynomial dynamics. J. Reine Angew. Math. **585**, 61–92 (2005)
6. M. Baker, R. Rumely, Potential theory and dynamics on the Berkovich projective line, in *Mathematical Surveys and Monographs*, vol. 159 (American Mathematical Society, Providence, 2010)
7. G. Bassanelli, F. Berteloot, Bifurcation currents in holomorphic dynamics on $\mathbb{P}^k$. J. Reine Angew. Math. **608**, 201–235 (2007)
8. B. Branner, J.H. Hubbard, The iteration of cubic polynomials. I. The global topology of parameter space. Acta Math. **160**, 143–206 (1988)
9. X. Buff, A. Epstein, Bifurcation measure and postcritically finite rational maps, in *Complex Dynamics* (A K Peters, Wellesley, 2009), pp. 491–512
10. L. Carleson, T.W. Gamelin, *Complex Dynamics*. Universitext: Tracts in Mathematics (Springer, New York, 1993)
11. A. Chambert-Loir, Arakelov geometry, heights, equidistribution, and the Bogomolov conjecture, in *Arakelov Geometry and Diophantine Applications* (Springer, Cham, 2020)
12. A. Chambert-Loir, Théorèmes d'équidistribution pour les systèmes dynamiques d'origine arithmétique, in *Quelques aspects des systèmes dynamiques polynomiaux*. Panor. Synthèses, vol. 30 (Society of Mathematics, Paris, 2010), pp. 203–294
13. L. De Marco, Dynamics of rational maps: a current on the bifurcation locus. Math. Res. Lett. **8**, 57–66 (2001)
14. L. De Marco, Critical orbits and arithmetic equidistribution, in *Proceedings of the International Congress of Mathematicians* (2018)
15. J.-P. Demailly, Complex analytic and differential geometry. Disponible à l'adresse https://www-fourier.ujf-grenoble.fr/~demailly/manuscripts/agbook.pdf
16. H.P. de Saint-Gervais, *Uniformisation des surfaces de Riemann. Retour sur un théorème centenaire* (ENS Éditions, Lyon, 2010)
17. A. Douady, J.H. Hubbard, *Étude dynamique des polynômes complexes, parties I et II* (Publications Mathématiques d'Orsay, Paris, 1984/1985)

18. A. Douady, J.H. Hubbard, A proof of Thurston's topological characterization of rational functions. Acta Math. **171**, 263–297 (1993)
19. R. Dujardin, C. Favre, Distribution of rational maps with a preperiodic critical point. Am. J. Math. **130**, 979–1032 (2008)
20. A. Eremenko, S. Van Strien, Rational maps with real multipliers. Trans. Am. Math. Soc. **363**, 6453–6463 (2011)
21. C. Favre, T. Gauthier, Distribution of postcritically finite polynomials. Israel J. Math. **209**, 235–292 (2015)
22. C. Favre, T. Gauthier, Classification of special curves in the space of cubic polynomials. Int. Math. Res. Not. **2018**, 362–411 (2018)
23. C. Favre, J. Rivera-Letelier, équidistribution quantitative des points de petite hauteur sur la droite projective. Math. Ann. **335**, 311–361 (2006)
24. T. Gauthier, G. Vigny, Distribution of postcritically finite polynomials II: speed of convergence. J. Mod. Dyn. **11**, 57–98 (2017)
25. D. Ghioca, H. Ye, A dynamical variant of the André-Oort conjecture. Int. Math. Res. Not. **8**, 2447–2480 (2018)
26. V. Guedj, Propriétés ergodiques des applications rationnelles, in *Quelques aspects des systèmes dynamiques polynomiaux*. Panor. Synthèses, vol. 30 (Society of Mathematics, Paris, 2010), pp. 97–202
27. J.H. Hubbard, *Teichmüller Theory and Applications to Geometry, Topology, and Dynamics. Volume 2: Surface Homeomorphisms and Rational Functions* (Matrix Editions, Ithaca, 2016)
28. T. Lei, Hausdorff dimension of subsets of the parameter space for families of rational maps. (A generalization of Shishikura's result). Nonlinearity **11**, 233–246 (1998)
29. G.M. Levin, Irregular values of the parameter of a family of polynomial mappings. Uspekhi Mat. Nauk **36**, 219–220 (1981)
30. G.M. Levin, On the theory of iterations of polynomial families in the complex plane. Teor. Funktsiĭ Funktsional. Anal. i Prilozhen. **51**, 94–106 (1989)
31. G. Levin, F. Przytycki, When do two rational functions have the same Julia set? Proc. Am. Math. Soc. **125**, 2179–2190 (1997)
32. C.T. McMullen, The motion of the maximal measure of a polynomial, Harvard University, Premilinary Notes (1985)
33. C.T. McMullen, Families of rational maps and iterative root-finding algorithms. Ann. Math. **125**, 467–493 (1987)
34. C.T. McMullen, The Mandelbrot set is universal, in *The Mandelbrot set, Theme and Variations*. London Mathematical Society Lecture Note Series, vol. 274, pp. 1–17 (Cambridge University Press, Cambridge, 2000)
35. J. Milnor, *Dynamics in One Complex Variable*. Annals of Mathematics Studies, vol. 160, 3rd edn. (Princeton University Press, Princeton, 2006)
36. J. Milnor, On Lattès maps, in *Dynamics on the Riemann Sphere*, pp. 9–43 (European Mathematical Society, Zürich, 2006)
37. M. Rees, Positive measure sets of ergodic rational maps. Ann. Sci. École Norm. Sup. **19**, 383–407 (1986)
38. M. Shishikura, The Hausdorff dimension of the boundary of the Mandelbrot set and Julia sets. Ann. Math. **147**, 225–267 (1998)
39. J.H. Silverman, *The Arithmetic of Dynamical Systems*. Graduate Texts in Mathematics, vol. 241 (Springer, New York, 2007)
40. J.H. Silverman, *Moduli Spaces and Arithmetic Dynamics.*, vol. 30 (American Mathematical Society, Providence, 2012)
41. X. Yuan, S.-W. Zhang, The arithmetic hodge index theorem for adelic line bundles II (2013). arXiv:1304.3539
42. A. Zdunik, Parabolic orbifolds and the dimension of the maximal measure for rational maps. Invent. Math. **99**, 627–649 (1990)

Part C
Shimura Varieties

Chapter X: Arakelov Theory on Shimura Varieties

José Ignacio Burgos Gil

1 Introduction

Arakelov theory is at the crossroad between number theory, algebraic geometry and complex analysis. The starting point of this theory is the analogy between $S = \mathrm{Spec}(\mathcal{O}_K)$, the spectrum of the ring of integer $\mathcal{O}_K$ of a number field K, and a smooth affine curve C defined over an arbitrary field k. A smooth affine curve can always be completed to a smooth projective curve $\overline{C}$ by adding a finite number of points. Similarly, the closed points of S correspond to non-Archimedean places of K and S can be "compactified" by adding the Archimedean places. Of course this two compactifications are very different, the curve $\overline{C}$ has a global structure of a smooth projective curve while $\overline{S}$ is just a patch between Archimedean and non-Archimedean places. Nevertheless there are striking formal analogies between these two pictures. For simplicity of the exposition assume that k is algebraically closed.

The first analogy is the product formula. If $0 \neq f \in k(C)$ is a non zero rational function, then

$$\sum_{p \in \overline{C}(k)} \mathrm{ord}_p(f) = 0,$$

while, if we choose an absolute value in each place of K suitably normalized, then, for $0 \neq f \in K$,

$$\sum_{v \in \mathfrak{M}_K} -\log |f|_v = 0,$$

J. I. Burgos Gil (✉)
Instituto de Ciencias Matemáticas, Madrid, Spain

© The Editor(s) (if applicable) and The Author(s), under exclusive license
to Springer Nature Switzerland AG 2021
E. Peyre, G. Rémond (eds.), *Arakelov Geometry and Diophantine Applications*,
Lecture Notes in Mathematics 2276, https://doi.org/10.1007/978-3-030-57559-5_11

where $\mathfrak{M}_K$ is the set of all places of K, Archimedean and non-Archimedean. A second example of such analogies is that, using the appropriate notation (see [22, Chapter 3]) a variant of the Riemann-Roch theorem for algebraic curves is formally identical to Minkowski theorems on the geometry of numbers. The aim of Arakelov theory is to extend this analogy between algebraic geometry and number theory to higher dimension. We refer the reader to the notes by C. Soulé (Chapter I of this volume) for more details on Arakelov Geometry and we will only recall some points of it.

Consider a projective and flat scheme $\mathscr{X}$ defined over $\mathrm{Spec}(\mathbb{Z})$. For instance $\mathscr{X}$ can be obtained from a projective variety over $\mathbb{Q}$ by clearing denominators of a system of defining equations and taking the irreducible component containing the complex points. In other words, $\mathscr{X}$ is obtained from a variety $X \subset \mathbb{P}_{\mathbb{Q}}^N$ by taking its Zariski closure in $\mathbb{P}_{\mathbb{Z}}^N$. Since $\mathrm{Spec}\,\mathbb{Z}$ is affine and not complete, we can not expect $\mathscr{X}$ to behave globally as a complete variety. In the same way that we "compactify" $\mathrm{Spec}\,\mathbb{Z}$ by adding one Archimedian point, we can "compactify" $\mathscr{X}$ by adding one Archimedean fibre $X_{\mathbb{R}} = \mathscr{X} \times \mathrm{Spec}(\mathbb{R})$. The *compound object* $\mathscr{X} \amalg X_{\mathbb{R}}$ should behave formally as a complete variety over a field. This means that many theorems in algebraic geometry should have a number-theoretical analogue in this setting. This analogue will typically involve algebraic geometry in $\mathscr{X}$ and complex analysis in $X_{\mathbb{R}}$. Successful examples of such strategy are the Arithmetic Riemann-Roch Theorem [12] and the Lefschetz fixed point theorem [16].

The main objects considered in Arakelov theory are arithmetic cycles $\overline{Z} = (Z, g_Z)$, where Z is an algebraic cycle on $\mathscr{X}$ and g_Z is a *Green current* for Z in $\mathscr{X}(\mathbb{C})$, and hermitian vector bundles $\overline{E} = (E, h)$, where E is a vector bundle on $\mathscr{X}$ and h is a smooth hermitian metric on the complex vector bundle $E_{\mathbb{C}}$ over $\mathscr{X}(\mathbb{C})$.

There is an arithmetic intersection theory of arithmetic cycles [10] and a theory of arithmetic characteristic classes [11] for hermitian vector bundles. This theory of arithmetic characteristic classes was first developed for smooth hermitian metrics.

Modular and Shimura varieties have many interesting vector bundles that have modular or group theoretical interpretations, like the line bundle of modular forms on a modular curve, or an automorphic vector bundle on a more general Shimura variety. In many instances, these vector bundles come equipped with hermitian metrics that also have modular or group theoretical interpretations. An example of such metrics is the Petersson metric on the line bundle of modular forms.

The geometric and arithmetic invariants of Shimura varieties have a very rich structure. Usually, geometric invariants are related to special values of L-functions, while arithmetic invariants are related to logarithmic derivatives of L functions. Thus, arithmetic invariants give information about the second term in the Taylor expansion of the L function. Kudla's program aims to make precise this idea. See for instance [17] and the references therein.

Many Shimura varieties and automorphic vector bundles have models over arithmetic rings (for instance a ring of integers with some primes inverted). Therefore we have at our disposal almost all the ingredients needed to define and compute arithmetic invariants, at least for projective Shimura varieties. But many modular and Shimura varieties are only quasi-projective and not projective, and, in order to be able to define arithmetic invariants for them, it is useful to find suitable compactifications. Under some conditions (see [21] for precise details) automorphic vector bundles can be extended to such compactifications, but the *interesting* hermitian metrics do not extend to smooth hermitian metrics on the completed vector bundle, but only to logarithmically singular hermitian bundles.

Arakelov theory developed in [10] and [11] only deals with smooth hermitian metrics. So in order to apply Arakelov theory to automorphic vector bundles we need to extend it to logarithmically singular hermitian vector bundles. This extension was made in [8] and [7].

The aim of this note is to introduce the basic ingredients of the extension of Arakelov theory to compactifications of quasi-projective Shimura varieties. In Sect. 2 we will recall some basic facts of Hermitian symmetric spaces. In Sect. 3 we will discuss connected Shimura varieties. Section 4 is devoted to equivariant vector bundles and their invariant metrics. In Sect. 5 we will study log-singular metrics and log–log forms and in Sect. 6 we will put everything together to construct an arithmetic intersection theory suitable to study non-compact Shimura varieties.

2 Hermitian Symmetric Spaces

In this section we will use the modular curve as an illustration of the general theory of hermitian symmetric spaces. We start with the complex upper half plane

$$\mathbb{H} = \{x + iy \in \mathbb{C} \mid y > 0\}$$

and the group $SL_2(\mathbb{R})$. This group acts on $\mathbb{H}$ by Moebius transforms

$$\begin{pmatrix} a & b \\ c & d \end{pmatrix} \cdot \tau = \frac{a\tau + b}{c\tau + d}.$$

The action is transitive and the stabilizer of the point $i \in \mathbb{H}$ is

$$SO_2(\mathbb{R}) = \left\{ \begin{pmatrix} c & -s \\ s & c \end{pmatrix} \middle| c^2 + s^2 = 1 \right\}.$$

Therefore we can write

$$\mathbb{H} = SL_2(\mathbb{R}) / SO_2(\mathbb{R}). \tag{2.1}$$

There is something odd with this identity: the left hand side is a complex manifold, while the right hand side is a real differentiable manifold. Thus this identity has to be seen as an identity of differentiable manifolds. Then, where does the complex structure come from? The next exercise shows that the complex structure of $\mathbb{H}$ is determined (up to complex conjugation) by the groups $SL_2(\mathbb{R})$ and $SO_2(\mathbb{R})$.

Exercise 1 Consider the element

$$J = \frac{1}{\sqrt{2}} \begin{pmatrix} 1 & 1 \\ -1 & 1 \end{pmatrix} \in SO_2(\mathbb{R})$$

1. Show that J has order 8 in $SL_2(\mathbb{R})$ but as an operator on $\mathbb{H}$ has only order 4.
2. Show that the point $i \in \mathbb{H}$ is a fixed point of J and that the action of J on the tangent space $T_i \mathbb{H}$ is multiplication by i.
3. For $g \in SL_2(\mathbb{R})$ and $\tau = g \cdot i$, show that gJg^{-1} stabilizes τ and the action of gJg^{-1} on $T_\tau \mathbb{H}$ is multiplication by i.

Conclude that J defines an equivariant integrable almost complex structure (see [23, Chapter I, Section 3] for the notion of almost complex structure) on the quotient $SL_2(\mathbb{R})/SO_2(\mathbb{R})$ such that Eq. (2.1) becomes an identity of complex manifolds.

The only other choice to cook a complex structure on $SL_2(\mathbb{R})/SO_2(\mathbb{R})$ using elements from $SL_2(\mathbb{R})$ would have been to use the transpose of J. This choice would have produced the complex conjugate space of $\mathbb{H}$, that is, the complex lower half plane.

As we explain below, the method of Exercise 1 can be generalized from $SL_2(\mathbb{R})$ to other groups.

Definition 2.1

1. Let D be a complex manifold with almost complex structure J and g a Riemannian structure on D. We say that g is a *hermitian structure* if for every point $p \in D$ and tangent vectors $v, w \in T_p D$, the condition

$$g(Jv, Jw) = g(v, w).$$

holds, A complex manifold with a hermitian structure is called a *hermitian space*. If D is a hermitian space, the group of holomorphic isometries of D will be denoted by $\text{Is}(D)$. The connected component of the identity in this group will be denoted $\text{Is}^+(D)$.
2. Let D be a connected hermitian space. Then D is called a *hermitian symmetric space* if every point $p \in D$ is an isolated fixed point of an involutive holomorphic isometry of D.

Exercise 2 On the space $\mathbb{H}$ with coordinates x, y we put the Riemannian structure

$$ds^2 = \frac{dx^2 + dy^2}{y^2}. \tag{2.2}$$

1. Show that the Riemannian structure (2.2) is invariant under the action of the group $SL_2(\mathbb{R})$ and it is a hermitian structure.
2. Show that $\mathbb{H}$ is a hermitian symmetric space. *Hint:* J^2 gives us the involution that fixes i and the other needed involutions are obtained by conjugation with elements of $SL_2(\mathbb{R})$.

In fact

$$\mathrm{Is}^+(\mathbb{H}) = SL_2(\mathbb{R})/\{\pm\, \mathrm{Id}\} = PSL_2(\mathbb{R}).$$

For proofs of the next results and more information about symmetric spaces and hermitian symmetric spaces the reader is refered to [1, III, §2] and [14]. In particular, see [14, Chapter VIII] for the definition of irreducible hermitian symmetric spaces and the ones of compact and non-compact type.

Theorem 2.2 ([1, Ch. III §2.1]) *Let D be a hermitian symmetric space. Then there is a decomposition*

$$D = D_0 \times D_1 \times \cdots \times D_n$$

where

1. D_0 *is the quotient of a complex vector space, with a translation invariant hermitian structure, by a discrete group of translations. This factor is called of Euclidean type.*
2. *Each D_i, $i \neq 0$ is an irreducible hermitian symmetric space that is not of Euclidean type.*

In the previous theorem, the factors D_i, $i \neq 0$, that are compact are called of compact type, while the non-compact ones are called of non-compact type.

Theorem 2.3 ([14, Ch. VIII Theorem 6.1])

1. *The irreducible hermitian symmetric spaces of non-compact type are the varieties of the form G/K where G is a connected non-compact simple Lie group with center $\{e\}$ and K is a maximal compact subgroup with non-discrete center.*
2. *The irreducible hermitian symmetric spaces of compact type are the varieties of the form G/K where G is a connected compact simple Lie group with center $\{e\}$ and K is a maximal proper connected subgroup with non-discrete center.*

Let D be a quotient of the form G/K as in the theorem. Let e be the neutral element of G and $o = [e] \in G/K$ its class in D. In both cases of the theorem, to prove that G/K is a hermitian symmetric space we have to construct an almost complex structure and, for each point, an involution as we did for the upper half plane in Exercise 1. To this end one uses the fact that K has non discrete center to find an element J in the center of K whose action in $T_o D$ has order 4. Then the complex structure on the tangent space at o is induced by J and the involution that fixes o is induced by $s = J^2$. Both operations are translated to the whole space by conjugation. See the proof of [14, Ch. VIII Theorem 6.1] for more details.

Example 2.4

1. The group $G = \mathrm{PSL}_2(\mathbb{R}) = \mathrm{SL}_2(\mathbb{R})/\{\pm \mathrm{Id}\}$ is a connected non-compact simple Lie group with center $\{e\}$. Moreover, $K = \mathrm{SO}_2(\mathbb{R})/\{\pm \mathrm{Id}\}$ is a maximal connected compact subgroup isomorphic to U(1). It is abelian and non-discrete. Thus its center is non-discrete. The quotient G/K is the hermitian symmetric space of non-compact type $\mathbb{H}$.

2. The group $G = \mathrm{SU}(2)/\{\pm \mathrm{Id}\}$ is a connected compact simple Lie group with center $\{e\}$ and K as before is a maximal connected proper subgroup isomorphic to U(1). Again it is abelian and non-discrete. Thus its center is non-discrete. The quotient G/K is a hermitian symmetric space of compact type isomorphic to $\mathbb{P}^1(\mathbb{C})$.

The spaces $\mathbb{H}$ and $\mathbb{P}^1(\mathbb{C})$ of Example 2.4 are very close to each other. In fact they are in duality. This is a general procedure that we describe next.

Let D be an irreducible hermitian symmetric space of non-compact type and $o \in D$. Put

$$G = \mathrm{Is}^+(D), \text{ a simple connected Lie group,}$$

$$K = \mathrm{Stab}(o), \text{ the maximal compact subgroup,}$$

$$s_o \in K \text{ the involution fixing } o.$$

Then s_o induces an inner automorphism σ of G. The subgroup G^σ of elements fixed by σ agrees with K. Set

$$\mathfrak{g} = \mathrm{Lie}(G)$$

$$\mathfrak{k} = \mathrm{Lie}(K) = \text{ subspace of } \mathfrak{g} \text{ where } \sigma \text{ acts as } +1,$$

$$\mathfrak{p} = \text{ subspace where } \sigma \text{ acts as } -1,$$

The group G is the identity component of the set of real points of an algebraic group $\mathcal{G}$. Indeed, consider the adjoint action

$$G \longrightarrow \mathrm{GL}(\mathfrak{g})$$

and let $\mathcal{G}$ be the Zariski closure of the image of G. It is a closed subgroup of $\mathrm{GL}(\mathfrak{g})$, hence algebraic and G is the identity component of the group of real points $\mathcal{G}(\mathbb{R})^+$. Write $G_{\mathbb{C}} = \mathcal{G}(\mathbb{C})$.

Inside $\mathfrak{g}_\mathbb{C} = \mathfrak{g} \otimes \mathbb{C} = \mathrm{Lie}(\mathscr{G}(\mathbb{C}))$ we write

$$\mathfrak{k}_c = \mathfrak{k},$$

$$\mathfrak{p}_c = i\mathfrak{p},$$

$$\mathfrak{g}_c = \mathfrak{k}_c \oplus \mathfrak{p}_c.$$

Then $\mathfrak{g}_c$ is a real form of $\mathfrak{g}_\mathbb{C}$ that determines a compact group G_c. The compact dual of D is the symmetric space of compact type

$$\check{D} = G_c/K.$$

This construction can be reversed and the irreducible non Euclidean hermitian symmetric spaces come in pairs, one compact and one non compact (see [1, Ch. III §2.1]).

Definition 2.5 Let D be an irreducible hermitian symmetric space. If D is of non-compact type, its *compact dual* is the previously constructed hermitian symmetric space, while if D is of compact type, its *compact dual* is D itself. If D is a hermitian symmetric space with no Euclidean factors, its *compact dual* is the product of the compact duals of all its factors.

Exercise 3 Consider the case $D = \mathbb{H}$.

1. Show that, in this case $\mathfrak{g}$ is the Lie algebra of 2 by 2 real matrices of trace zero, $\mathfrak{k}$ the subalgebra of skew-symmetric matrices and $\mathfrak{p}$ the subspace of symmetric trace zero matrices.
2. Show that $\mathfrak{g}_c$ is the Lie algebra of 2 by 2 skew-hermitian complex matrices of zero trace. Thus $G_c = \mathrm{SU}(2)$ and the compact dual of $\mathbb{H}$ is isomorphic to $\mathbb{P}^1(\mathbb{C})$.

When working with a general hermitian symmetric space G/K it may be difficult to visualize the complex structure, unless we find a nice representation of it as in the case of the upper half plane. The big advantage of the compact dual is that its complex structure is easier to visualize. Moreover, a non-compact hermitian symmetric space can always be embedded in its compact dual making also apparent its complex structure.

Let $D = G/K$ be a connected hermitian symmetric space without Euclidean factor, and $o \in D$ the image of $e \in G$. Assume that $G = \mathrm{Is}^+(D)$. Then there is a map $u_o: \mathrm{U}(1) \to G$ such that $u_o(z)$ acts as multiplication by z in $T_o D$. Then $J = u_o(i)$ defines the complex structure of $T_o D$ and $s_o = u_o(-1)$ is the involution that fixes o. The subgroup K is the centralizer of $u_o(\mathrm{U}(1))$ and, if D is irreducible, $u_o(\mathrm{U}(1))$ is the center of K. The subspace $\mathfrak{p} \subset \mathfrak{g}$ can be identified with $T_o D$ and the action of K on $T_o D$ agrees with the adjoint action of K on $\mathfrak{p}$. Let

$$\mathfrak{p}_\mathbb{C} := \mathfrak{p} \otimes \mathbb{C} = \mathfrak{p}_+ \oplus \mathfrak{p}_-$$

be the decomposition of $\mathfrak{p}_{\mathbb{C}}$ into $\pm i$ eigenspaces with respect to J (recall that $\mathfrak{p}$ is the -1-eigenspace of $s_o = J^2$). Denote by $P_\pm$ the subgroup of $G_{\mathbb{C}}$ generated by $\exp(\mathfrak{p}_\pm)$. Then $K_{\mathbb{C}}$ normalizes $P_\pm$ and $K_{\mathbb{C}} \cdot P_-$ is a parabolic subgroup of $G_{\mathbb{C}}$ with unipotent radical P_-. Hence $G_{\mathbb{C}}/K_{\mathbb{C}} \cdot P_-$ is a projective algebraic variety that we temporarily denote by X. The following theorem exhibits D as a complex manifold. For a proof of see [14, Ch.8 §7], See also [1, Ch. III Theorem 2.1] for an outline.

Theorem 2.6 (Borel and Harish-Chandra Embedding Theorem)

1. *The map $P_+ \times K_{\mathbb{C}} \times P_- \to G_{\mathbb{C}}$ given by multiplication is injective and its image contains G. Moreover $(K_{\mathbb{C}} \cdot P_-) \cap G = K$.*
2. *We obtain maps*

$$
\begin{array}{ccccc}
G/K & \longrightarrow & P_+ \times K_{\mathbb{C}} \times P_-/K_{\mathbb{C}} \times P_- & \longrightarrow & G_{\mathbb{C}}/K_{\mathbb{C}} \cdot P_- \\
\simeq \uparrow & & \simeq \uparrow & & \simeq \uparrow \\
D & & P_+ & & X \\
& & \simeq \uparrow \mathrm{exp} & & \\
& & \mathfrak{p}_+ & &
\end{array}
$$

They are all open holomorphic immersions. The image of D in $\mathfrak{p}_+$ is a bounded domain and the image of $\mathfrak{p}_+$ in X is a dense open Zariski subset.
3. *The compact form G_c of G acts transitively on X. Moreover $(K_{\mathbb{C}} \cdot P_-) \cap G_c = K$. Therefore $X = G_c/K = \check{D}$ is the compact dual of D.*

Exercise 4 We go back to the example $D = \mathbb{H}$, $G = \mathrm{PSL}_2(\mathbb{R})$, $K = \mathrm{SO}_2(\mathbb{R})/\pm \mathrm{Id}$. It will be easier to work with the double coverings $\widetilde{G} = \mathrm{SL}_2(\mathbb{R})$ and $\widetilde{K} = \mathrm{SO}_2(\mathbb{R})$. The groups P_+ and P_- can be defined as subgroups of $\widetilde{G}_{\mathbb{C}}$, but the map $u_o \colon U(1) \to G$ does not lift to a map to $\widetilde{G}$.

1. Show that $\mathfrak{p}_+$ and $\mathfrak{p}_-$ are the one dimensional subspaces of $\mathfrak{g}_{\mathbb{C}}$ generated respectively by the matrices

$$
\frac{1}{2}\begin{pmatrix} 1 & i \\ i & -1 \end{pmatrix} \quad \text{and} \quad \frac{1}{2}\begin{pmatrix} 1 & -i \\ -i & -1 \end{pmatrix}
$$

Therefore the subgroups $P_\pm$ are given by

$$
P_\pm = \left\{ P_\pm(z) := \begin{pmatrix} 1 + z/2 & \pm iz/2 \\ \pm iz/2 & 1 - z/2 \end{pmatrix} \bigg| z \in \mathbb{C} \right\}.
$$

2. Consider the action of P_+ on $\mathbb{P}^1(\mathbb{C})$ by Moebius transformations and show that

$$
P_+(z) \cdot i = \frac{i(1+z)}{1-z}.
$$

Conclude that the open immersion $\mathbb{H} \to \mathfrak{p}_+$ is given by

$$\tau \mapsto \frac{\tau - i}{\tau + i}$$

and the image of the upper half plane is the interior of the unit disk.

3. The adjoint action of $SO_2(\mathbb{R})$ on $\mathfrak{p}_+$ is given by

$$\mathrm{Ad} \begin{pmatrix} a & -b \\ b & a \end{pmatrix} (M) = \frac{1}{(a+ib)^2} M.$$

4. Let $t \in \mathbb{C}^\times$, the matrix

$$K(t) := \frac{1}{2t} \begin{pmatrix} t^2 + 1 & i(t^2 - 1) \\ i(1 - t^2) & t^2 + 1 \end{pmatrix}$$

belongs to $SO_2(\mathbb{C})$. The map $t \mapsto K(t)$ is a group isomorphism. The adjoint action of $K(t)$ on $\mathfrak{p}_+$ is given by multiplication by t^{-2}.

5. Let

$$\gamma = \begin{pmatrix} a & b \\ c & d \end{pmatrix} \in \widetilde{G},$$

and denote by s_1 the holomorphic map given by the composition

$$\mathbb{H} \to \mathfrak{p}_+ \xrightarrow{\exp} P_+ \to G_{\mathbb{C}}.$$

Then

$$\gamma \cdot s_1(\tau) = s_1(\gamma \cdot \tau) \cdot K(j_1(\gamma, \tau)) \cdot p,$$

where $p \in P_-$ and

$$j_1(\gamma, \tau) = \frac{(a\tau + b) + i(c\tau + d)}{\tau + i}.$$

6. Show that j_1 satisfies the cocycle condition for the action of $\widetilde{G}$ on $\mathbb{H}$. That is

$$j_1(\gamma \cdot \gamma', \tau) = j_1(\gamma, \gamma' \cdot \tau) j_1(\gamma', \tau).$$

7. Let $f : \mathbb{H} \to \mathbb{C}^\times$ be the function $f(\tau) = (\tau + i)$. Show that the cocycle

$$j(\gamma, \tau) = j_1(\gamma, \tau) f(\tau) f(\gamma \cdot \tau)^{-1}.$$

is given by

$$j(\gamma, \tau) = (c\tau + d).$$

Conclude that the holomorphic map $s\colon \mathbb{H} \to \widetilde{G}_{\mathbb{C}}$ given by

$$s(\tau) = s_1(\tau)K(\tau + i) \qquad\qquad (2.3)$$

satisfies

$$\gamma s(\tau) = s(\tau) \cdot K(c\tau + d) \cdot p'$$

for some element $p' \in P_-$.

3 Connected Shimura Varieties

It is time to consider the quotient of a hermitian symmetric space by some interesting discrete subgroups.

Definition 3.1

1. Let G be a group. Two subgroups Γ_1 and Γ_2 are said to be *commensurable* if $\Gamma_1 \cap \Gamma_2$ has finite index in both Γ_1 and Γ_2.
2. Let G be an algebraic group defined over $\mathbb{Q}$ that admits a closed embedding $G \hookrightarrow \mathrm{GL}_n$ also defined over $\mathbb{Q}$. An *arithmetic subgroup* of $G(\mathbb{R})$ is any subgroup of $G(\mathbb{Q})$ that is commensurable with $G(\mathbb{Q}) \cap \mathrm{GL}_n(\mathbb{Z})$.
3. An arithmetic subgroup $\Gamma \subset G(\mathbb{Q}) \subset \mathrm{GL}_n(\mathbb{C})$ is called *neat* if for every $x \in \Gamma$, the subgroup of $\mathbb{C}^\times$ generated by the eigenvalues of x is torsion free.

Remark 3.2

1. The notion of arithmetic subgroup depends on the rational structure of G, but once this is fixed, it is independent of the choice of closed embedding $G \hookrightarrow \mathrm{GL}_n$.
2. Any neat arithmetic subgroup is torsion free. Every arithmetic subgroup has a neat subgroup of finite index.

We are interested in locally symmetric spaces of the form $\Gamma \backslash D$ for $D = G(\mathbb{R})/K$ a hermitian symmetric space, with G an algebraic group over $\mathbb{Q}$ and Γ an arithmetic subgroup of $G(\mathbb{Q})$.

It turns out that, if the image of Γ in $\mathrm{Is}^+(D)$ is torsion free, then it acts freely on D and therefore the quotient $\Gamma \backslash D$ is a smooth complex manifold. Moreover, the fact that Γ is arithmetic implies that $\Gamma \backslash D$ has finite volume. Even more, we not only obtain a complex manifold, but a smooth algebraic variety over $\mathbb{C}$.

Theorem 3.3 (Baily-Borel [2]) *Let $D(\Gamma) = \Gamma \backslash D$ be the quotient of a hermitian symmetric space by a torsion free arithmetic subgroup Γ of $\mathrm{Is}^+(D)$. Then $D(\Gamma)$ has* *a canonical realization as a Zariski-open subset of a projective algebraic variety $D(\Gamma)^*$. In particular, it has a canonical structure of a quasi-projective algebraic variety over $\mathbb{C}$.*

Remark 3.4 In fact, $D(\Gamma)$ has a structure of algebraic variety defined over a number field.

We can now give a definition of Shimura variety. The precise definition of Shimura varieties is rather involved as it describes a family of varieties of a very precise form and requires the language of adéles. We refer the reader to the expository paper by Milne [20] for the precise definition and a different presentation of Shimura varieties. In these notes we just will present the definition of *connected Shimura varieties* using congruence subgroups.

Definition 3.5 Let G be an algebraic reductive group defined over $\mathbb{Q}$. Choose an embedding $G \hookrightarrow \mathrm{GL}_n$ and, for each integer $N \geq 1$, define

$$\Gamma(N) = G(\mathbb{Q}) \cap \{g \in \mathrm{GL}_n(\mathbb{Z}) \mid g \equiv \mathrm{Id} \mod N\}.$$

A *congruence subgroup* of G is a subgroup that contains some $\Gamma(N)$ with finite index.

Every congruence subgroup is an arithmetic subgroup but the converse is not true in general. For instance SL_2 has infinitely many arithmetic subgroups that are not congruence subgroups, see for instance [19, Chs. 6,7].

Definition 3.6 A *connected Shimura datum* is the data of

1. a semisimple algebraic group G defined over $\mathbb{Q}$ of non-compact type,
2. a connected hermitian symmetric space D, and
3. an action of $G^{\mathrm{ad}}(\mathbb{R})^+$ on D defined by a surjective homomorphism $G^{\mathrm{ad}}(\mathbb{R})^+ \to \mathrm{Is}^+(D)$.

Example 3.7 The data of $G = \mathrm{SL}_2$, $D = \mathbb{H}$ and the action of G on $\mathbb{H}$ by Moebius transforms is a connected Shimura datum.

Definition 3.8 The *connected Shimura variety* $\mathrm{Sh}^\circ(G, D)$ is the inverse system of locally symmetric varieties $(\Gamma \backslash D)_\Gamma$, where Γ runs over the set of torsion-free arithmetic subgroups of $G^{\mathrm{ad}}(\mathbb{R})$ whose preimage in $G(\mathbb{R})^+$ is a congruence subgroup.

This definition has some disadvantages. First, for some number theoretic applications it is better to start with a reductive group and not a semisimple group. Second, each space on the tower $(\Gamma \backslash D)_\Gamma$ has a model over a number field, but the number field may change with the subgroup. Therefore, the inductive limit may only be defined over an infinite extension of $\mathbb{Q}$. For instance the modular curve $\Gamma(N) \backslash \mathbb{H}$ has a canonical model over $\mathbb{Q}[\xi_N]$, where ξ_N is a primitive N-th root of 1. Therefore, the connected Shimura variety $\mathrm{Sh}^\circ(\mathrm{SL}_2, \mathbb{H})$ is defined over the whole cyclotomic extension of $\mathbb{Q}$.

The definition of Shimura varieties (as opposed to connected Shimura varieties) solves these disadvantages. In particular Shimura varieties are defined over number fields. See [20, §5] for more details.

In many cases, the algebraic variety $X = D(\Gamma) = \Gamma \backslash D$ is non-compact and it is useful to compactify it. Part of the content of Theorem 3.3 is that there exists a canonical compactification $D(\Gamma)^*$ that is a projective variety. This compactification is canonical but is highly singular. Of course we can appeal to the general theorem of resolution of singularities to find a non-singular compactification. But then we lose control on the compactification. The theory of toroidal compactifications [1] gives us a controlled family of smooth compactifications.

Theorem 3.9 *Let* $D(\Gamma) = \Gamma \backslash D$ *be the quotient of a hermitian symmetric space by an arithmetic subgroup* Γ *of* $\mathrm{Is}^+(D)$. *Then* $D(\Gamma)$ *admits a family of compactifications that can be described combinatorially in terms of the groups* G, K *and* Γ *and are called toroidal compactifications. If* Γ *is neat, there are smooth projective toroidal compactifications* $\overline{X}$ *of* X *with* $\overline{X} \backslash X$ *a simple normal crossings divisor.*

Example 3.10 We recall how the general theory of compactifications of locally symmetric spaces applies to the modular curve. Let $\Gamma \subset \mathrm{SL}_2(\mathbb{Z})$ be a subgroup of finite index such that $-\mathrm{Id} \notin \Gamma$ and write $Y(\Gamma) = \Gamma \backslash \mathbb{H}$. The group Γ acts on $\mathbb{P}^1(\mathbb{Q})$. The curve $Y(\Gamma)$ can be compactified by adding one point for each equivalence class for the action of Γ on $\mathbb{P}^1(\mathbb{Q})$. These points are called cusps. The compactified curve is denoted $X(\Gamma)$.

$$X(\Gamma) = Y(\Gamma) \cup \{\text{cusps}\}$$

In order to understand the algebraic structure of $X(\Gamma)$, we give a local coordinate on $X(\Gamma)$ around each cusp. Any cusp can be sent to the point ∞ by an element $\gamma \in \mathrm{SL}_2(\mathbb{Z})$. Replacing Γ by $\gamma \Gamma \gamma^{-1}$, it is enough to understand what happens at the cusp $\infty = (1:0) \in \mathbb{P}^1(\mathbb{Q})$.

The stabilizer of the point ∞ in Γ is an infinite cyclic group

$$\Gamma_\infty = \left\{ \begin{pmatrix} 1 & nw \\ 0 & 1 \end{pmatrix} \middle| n \in \mathbb{Z} \right\}$$

The integer $w > 0$ is called the width of the cusp ∞. Let Δ be the interior of the unit disk and Δ^* the interior of the unit disk with the point 0 removed. The map

$$\begin{aligned} \mathbb{H} &\longrightarrow \Delta^* \\ \tau &\longmapsto \exp(2\pi i \tau / w) \end{aligned}$$

induces an isomorphism $\Gamma_\infty \backslash \mathbb{H} \to \Delta^*$. The map

$$\Gamma_\infty \backslash \mathbb{H} \to \Gamma \backslash \mathbb{H}$$

sends a neighborhood of $0 \in \Delta$ to a neighborhood of the cusp ∞ in $X(\Gamma)$. Thus we can use $q = \exp(2\pi i \tau / w)$ as a local coordinate around the cusp ∞.

4 Equivariant Vector Bundles and Invariant Metrics

We next move our attention to the construction of holomorphic equivariant vector bundles on a hermitian symmetric space. Being equivariant they will descend to vector bundles on any quotient by an arithmetic subgroup.

Let $D = G/K$ be a hermitian symmetric space, V a complex vector space of finite dimension and $\rho \colon K \to \mathrm{GL}(V)$ a representation of K. The group K acts on G on the right and the map $G \to D$ is a principal K-bundle. Through the representation ρ, K also acts on V. This time on the left. Thus we can form the space

$$\mathscr{V} = G \underset{K}{\times} V := G \times V / \sim$$

where $\sim$ is the equivalence relation

$$(g \cdot k, v) \sim (g, \rho(k)(v)), \text{ for all } g \in G, v \in V, k \in K.$$

The map $\mathscr{V} \to D$ given by $[(g, v)] \mapsto g \cdot o$ is well defined and $\mathscr{V}$ is a differentiable vector bundle over D with fiber V. Moreover $\mathscr{V}$ is a G-equivariant vector bundle with the G-action given by

$$g \cdot [(g', v)] = [(gg', v)].$$

But, is it possible to give to $\mathscr{V}$ the structure of an equivariant *holomorphic* vector bundle? The answer is yes, but not in a unique way.

To this end, we can complexify ρ to a representation of $\rho \colon K_{\mathbb{C}} \to \mathrm{GL}(V)$ and we can extend it to a representation of $K_{\mathbb{C}} \cdot P_-$. One possible way to do this is to declare that the action of P_- is trivial.

We now repeat the previous process to obtain a holomorphic vector bundle

$$\mathscr{V}_h := G_{\mathbb{C}} \underset{K_{\mathbb{C}} \cdot P_-}{\times} V$$

on the compact dual $\check{D}$ of D. The restriction of $\mathscr{V}_h$ to D is a holomorphic vector bundle that has the same underlying differentiable structure as $\mathscr{V}$. We will identify both vector bundles and think of $\mathscr{V}$ as an equivariant holomorphic vector bundle.

Remark 4.1 The condition that the action of P_- is trivial gives a particular choice of holomorphic structure. The holomorphic vector bundles produced using this condition are called *fully decomposable vector bundles*. One has to be careful that some interesting equivariant holomorphic vector bundles on D are not fully decomposable.

Denote by $\pi: G_{\mathbb{C}} \to \check{D}$ the projection deduced from Theorem 2.6 and let $s: D \to G_{\mathbb{C}}$ be a holomorphic map such that the composition $\pi \circ s: D \to \check{D}$ agrees with the embedding of D in $\check{D}$. For instance we can take s as the composition $D \hookrightarrow \mathfrak{p}_+ \xrightarrow{\exp} P_+ \hookrightarrow G_{\mathbb{C}}$.

The holomorphic vector bundle $\pi^* \mathcal{V}_h$ over $G_{\mathbb{C}}$ has a canonical trivialization $\pi^* \mathcal{V}_h = G_{\mathbb{C}} \times V$. Thus $\mathcal{V} \simeq s^* G_{\mathbb{C}} \times V \simeq D \times V$ is a trivialized vector bundle. Although the trivialization depends on the map s.

Let $g \in G$. Since $\pi(g \cdot s(x)) = \pi(s(g \cdot x))$, we deduce that

$$gs(x) = s(gx)j(g,x)p_-(g,x)$$

for well defined elements $j(g,x) \in K_{\mathbb{C}}$ and $p_-(g,x) \in P_-$. The map j is a cocycle in the sense that

$$j(gg',x) = j(g, g' \cdot x)j(g',x).$$

This cocycle depends on the choice of s. The transformation rule for p_- involves conjugating by k, but we will not need it.

Exercise 5 Prove that, in the trivialization determined by s, the action of g on $\mathcal{V}$ is given by

$$g \cdot (x,v) = (g \cdot x, \rho(j(g,x))(v))$$

Example 4.2 We go back to the example $D = \mathbb{H}$. Both groups K and $\widetilde{K}$ are isomorphic to U(1). The irreducible representations of $U(1)$ are one dimensional and classified by its weight. The representation of weight $k \in \mathbb{Z}$ is given, for $z \in U(1) \subset \mathbb{C}^\times$ by

$$\rho_k(z)(v) = z^k v.$$

Since the map $\widetilde{K} \to K$ is a covering of order 2, a representation of K of weight n determines a representation of $\widetilde{K}$ of weight $2n$.

Let k be an even integer. The one-dimensional representation of K of weight $k/2$ induces the representation of $\widetilde{K} = \mathrm{SO}_2(\mathbb{R})$ of weight k. Let $\mathcal{L}_k$ be the corresponding G-equivariant line bundle on $\mathbb{H}$. We use the map (2.3) to trivialize $\mathcal{L}_k$. By Exercise 4 (7) we deduce that the action of G on $\mathcal{L}_k$ is given, in this trivialization, by

$$g \cdot (\tau, v) = (g \cdot \tau, (c\tau + d)^k v).$$

Exercise 6 In this exercise we want to recover the classical definition of modular forms. Let k be an even integer and let $\rho_k \colon SO_2(\mathbb{R}) \to GL_1(\mathbb{C})$ be the representation of weight k. Let $\Gamma \subset SL_2(\mathbb{Z})$ be a neat subgroup of finite index, and write $Y(\Gamma) = \Gamma \backslash \mathbb{H}$ for the corresponding open modular curve. Since Γ is neat, $Y(\Gamma)$ is a smooth quasi-projective curve. Let $\mathscr{L}_k$ be the equivariant line bundle on $\mathbb{H}$ determined by ρ_k and let $\mathscr{M}_k(\Gamma)$ the induced line bundle in $Y(\Gamma)$. Prove that a section s of $\mathscr{L}_k$ descends to a section of $\mathscr{M}_k(\Gamma)$ if and only if it is invariant under Γ. That is, for any $\gamma \in \Gamma$,

$$\gamma s(\tau) = s(\gamma \cdot \tau).$$

Using the trivialization of Exercise 5 we can identify sections of $\mathscr{L}_k$ with holomorphic functions on $\mathbb{H}$. Prove that a holomorphic function f defines a section of $\mathscr{M}_k(\Gamma)$ if and only if, for any $\gamma \in \Gamma$,

$$f(\gamma \cdot \tau) = (c\tau + d)^k f(\tau). \tag{4.1}$$

Remark 4.3 Of course, the condition (4.1) makes sense for non-necessarily neat arithmetic subgroups and for odd weight k. But not always such functions can be interpreted as sections of a line bundle on $Y(\Gamma)$.

The next task is to put metrics on the vector bundles we have constructed. Since the group K is compact, given any complex representation $\rho \colon K \to GL(V)$ there is a (non-unique) hermitian metric on V that is invariant under the action of K. Let $\langle \cdot, \cdot \rangle$ be one such K-invariant metric on V. Then there is a unique way to define a hermitian metric on the vector bundle $\mathscr{V} := G \underset{K}{\times} V \to D$ such that:

1. the restriction to the fiber $\mathscr{V}_o = V$ over the point o agrees with the given metric;
2. it is invariant under the action of G. That is, if $g \in G$, $p \in D$, $\mathscr{V}_p$ the fiber of $\mathscr{V}$ over p and $v, w \in \mathscr{V}_p$, then

$$\langle u, v \rangle_p = \langle g \cdot u, g \cdot v \rangle_{g \cdot p}.$$

In fact, given $p \in D$ and $v, w \in \mathscr{V}_p$, we can choose a $g \in G$ with $g \cdot p = o$ and we necessarily have

$$\langle u, v \rangle_p = \langle g \cdot u, g \cdot v \rangle_o. \tag{4.2}$$

So such invariant metric is unique once we have fixed the metric on V. If g' is another element such that $g' \cdot p = o$, then $g' = kg$ with $k \in K$, and

$$\langle g' \cdot u, g' \cdot v \rangle_o = \langle kg \cdot u, kg \cdot v \rangle_o = \langle g \cdot u, g \cdot v \rangle_o.$$

Thus the product (4.2) does not depend on the choice of g, and the invariant hermitian metric is well defined.

Exercise 7

1. Let $\tau \in \mathbb{H}$ and $g \in SL_2(\mathbb{R})$. Show that

$$\Im(g \cdot \tau) = \frac{\Im\tau}{|c\tau + d|^2},$$

where $\Im\tau$ denotes the imaginary part of τ.

2. Let $\rho_k : SO_2(\mathbb{R})$ be the representation of $SO_2(\mathbb{R})$ of weight k and let $\mathscr{L}_k$ the associated equivariant line bundle over $\mathbb{H}$ and let $C > 0$ be a real positive constant. On the trivialization of Example 4.2, show that the norm

$$\|(\tau, v)\|^2 = |v|^2 (4\pi \Im\tau)^k \tag{4.3}$$

defines an invariant metric on $\mathscr{L}_k$. Show that every invariant metric on $\mathscr{L}_k$ is a constant multiple of this one. The metric on $\mathscr{L}_k$ given by (4.3) is called the *Petersson metric*.

5 Log-Singular Metrics and Log–log Forms

Let $D = G/K$ be a hermitian symmetric space, where G is the connected component of the identity of the set of real points of a semisimple algebraic group defined over $\mathbb{Q}$, and let $\Gamma \subset G$ be an arithmetic subgroup of G. Put $X = \Gamma \backslash D$ for the quotient of the symmetric space by the arithmetic group.

Let $\rho : K \to GL(V)$ be a complex representation of K. Since the associated fully decomposable vector bundle $\mathscr{V}$ over D is G-equivariant, it is also Γ equivariant. Thus it descends to a vector bundle E on X. If V has a K-invariant metric, the vector bundle $\mathscr{V}$ has an induced invariant metric, that descends to a metric on E. Let $\overline{X}$ be a toroidal compactification of X as in Theorem 3.9. Once we have compactified X, it is a natural question to extend to $\overline{X}$ the hermitian vector bundle we have constructed. Let us first examine the case of the modular curve.

Example 5.1 Let s be a section of the line bundle $\mathscr{M}_k(\Gamma)$ on $Y(\Gamma)$ and let f be the corresponding holomorphic function given in Exercise 6. By the transformation rule (4.1), we see that f is invariant under the action of Γ_∞. By Fourier analysis we have an expansion

$$f = \sum_{n \in \mathbb{Z}} a_n q^n \tag{5.1}$$

with $q = \exp(2\pi i \tau / w)$ as in Example 3.10. The section s is said to be meromorphic at the cusp ∞ if the expansion (5.1) has only a finite number of negative terms and is said to be holomorphic if this expansion has only non-negative terms. In fact we define

$$\text{ord}_\infty(s) = \min\{n \mid a_n \neq 0\}.$$

This procedure specifies an extension of $\mathscr{M}_k(\Gamma)$ to $X(\Gamma)$.

Exercise 8 Let s be the section of $\mathscr{M}_k(\Gamma)$ of Example 5.1 and f the corresponding function. Let $\| \cdot \|$ denote the Petersson metric (4.3).

1. Show that, in a neighborhood of the cusp ∞

$$\|s\|^2 = |f|^2 (-w \log |q|^2)^k = \varphi(q)|q|^{2\,\text{ord}_\infty s} (\log|q|^2)^k,$$

where φ is a continuous function with $\varphi(0) \neq 0$ and w is the width of the cusp.

2. The first Chern form of the line bundle $\mathscr{M}_k$ with the metric $\| \cdot \|$ is given by

$$c_1(\mathscr{M}_k, \| \cdot \|) = \frac{i}{2\pi} \partial \bar{\partial}(-\log \|s\|^2).$$

Show that, locally around the cusp, this first Chern form can be written as

$$c_1(\mathscr{M}_k, \| \cdot \|) = \frac{ik}{2\pi} \frac{dq \wedge d\bar{q}}{q\bar{q}(\log q\bar{q})^2}. \tag{5.2}$$

The $(1,1)$-form given in Eq. (5.2) determines a metric on $Y(\Gamma)$ called the *hyperbolic metric* or also the *Poincaré metric*.

As Exercise 8 shows, we can not extend the line bundle $\mathscr{M}_k(\Gamma)$ in such a way that the Petersson metric extends to a smooth (or even continuous) metric on the whole $X(\Gamma)$. The problematic term being $(\log|q|^2)^k$. The best we can hope for is a log-singular metric.

Log–log growth forms is a class of singular differential forms. Among them one finds functions like $\log(-\log|q|)$ and the Poincaré metric (5.2). Log–log growth forms form an algebra and the complex of log–log growth forms has nice cohomological properties. To study the properties of log–log growth forms, until the end of the section, we forget about symmetric spaces and work with arbitrary complex manifolds and normal crossings divisors. Thus we change the notation accordingly.

Let now X be a complex manifold, we will denote by $\mathscr{E}_X^*$ the sheaf of smooth complex valued differential forms and by $\mathscr{E}_{X,\mathbb{R}}^*$ the subsheaf of real valued forms. We will use roman typography to denote the space of global sections. Namely

$$E_X^* = \Gamma(X, \mathscr{E}_X^*).$$

Recall that E_X^* has a bigrading,

$$E_X^n = \bigoplus_{p+q=n} E_X^{p,q},$$

where $\eta \in E_X^{p,q}$ if it can be locally written as

$$\eta = \sum_{I,J} f_{I,J}(z_1, \ldots, z_d) dz_{i_1} \wedge \cdots \wedge dz_{i_p} \wedge d\bar{z}_{j_1} \wedge \cdots \wedge d\bar{z}_{j_q},$$

where the sum runs over subsets $I = \{i_1, \ldots, i_p\}$ and $J = \{j_1, \ldots, j_q\}$ of $\{1, \ldots, d\}$, and $f_{I,J}$ are $\mathscr{C}^\infty$-functions.

There is also a Hodge filtration F given by

$$F^p E_X^n = \bigoplus_{p' \geq p} E_X^{p', n-p'},$$

and a decomposition $d = \partial + \bar{\partial}$, where

$$\partial \colon E_X^{p,q} \to E_X^{p+1,q} \quad \text{and} \quad \bar{\partial} \colon E_X^{p,q} \to E_X^{p,q+1}.$$

Let $D \subset X$ a normal crossings divisor. This means that locally, D is given by the equation

$$z_1 \cdots z_k = 0. \tag{5.3}$$

That is, the divisor D is locally like a collection of coordinate hyperplanes. We will say that an open coordinate neighbourhood $U \subset X$ with coordinates $(z_1, \ldots, z_k)$ is a *small coordinate neighbourhood adapted* to D if D has an equation of the form (5.3) for some $k \leq n$ and each point $p \in U$ has coordinates $(z_1, \ldots, z_i)$ satisfying $|z_i| \leq 1/(2e)$.

In the next definition we will use multi-index notation. That is, for any multi-index $\alpha = (\alpha_1, \ldots, \alpha_d) \in \mathbb{Z}_{\geq 0}^d$, we write

$$|\alpha| = \sum_{i=1}^d \alpha_i, \qquad z^\alpha = \prod_{i=1}^d z_i^{\alpha_i}, \qquad \bar{z}^\alpha = \prod_{i=1}^d \bar{z}_i^{\alpha_i},$$

$$r^\alpha = \prod_{i=1}^d r_i^{\alpha_i}, \qquad (\log(1/r))^\alpha = \prod_{i=1}^d (\log(1/r_i))^{\alpha_i},$$

$$\frac{\partial^{|\alpha|}}{\partial z^\alpha} f = \frac{\partial^{|\alpha|}}{\prod_{i=1}^d \partial z_i^{\alpha_i}} f, \qquad \frac{\partial^{|\alpha|}}{\partial \bar{z}^\alpha} f = \frac{\partial^{|\alpha|}}{\prod_{i=1}^d \partial \bar{z}_i^{\alpha_i}} f.$$

If α and β are multi-indices, we denote by $\alpha + \beta$ the multi-index with components $\alpha_i + \beta_i$. If α is a multi-index and $k \geq 1$ is an integer, we will denote by $\alpha^{\leq k}$ the multi-index

$$\alpha_i^{\leq k} = \begin{cases} \alpha_i, & i \leq k, \\ 0, & i > k. \end{cases}$$

The following definition is taken from [7, Definition 2.17].

Definition 5.2 Let X be a complex manifold and $D \subset X$ a normal crossings divisor. Let U be a small open coordinate neighborhood adapted to D. For every integer $K \geq 0$, we say that a smooth complex function f on $V \backslash D$ has *log–log growth along D of order K*, if there exists an integer N_K such that, for every pair of multi-indices $\alpha, \beta \in \mathbb{Z}_{\geq 0}^d$ with $|\alpha + \beta| \leq K$, it holds the inequality

$$\left| \frac{\partial^{|\alpha|}}{\partial z^\alpha} \frac{\partial^{|\beta|}}{\partial \bar{z}^\beta} f(z_1, \ldots, z_d) \right| \prec \frac{\left| \prod_{i=1}^k \log(\log(1/r_i)) \right|^{N_K}}{|z^{\alpha^{\leq k}} \bar{z}^{\beta^{\leq k}}|}. \tag{5.4}$$

We say that f has *log–log growth along D of infinite order*, if it has log–log growth along D of order K for all $K \geq 0$. The *sheaf of differential forms on X with log–log growth along D of infinite order* is the subalgebra of $\iota_* \mathcal{E}_U^*$ generated, in each small coordinate neighborhood U adapted to D, by the functions with log–log growth along D and the differentials

$$\frac{dz_i}{z_i \log(1/r_i)}, \quad \frac{d\bar{z}_i}{\bar{z}_i \log(1/r_i)}, \qquad \text{for } i = 1, \ldots, k,$$

$$dz_i, d\bar{z}_i, \qquad \text{for } i = k+1, \ldots, d.$$

A differential form with log–log growth along D of infinite order will be called a *log–log growth form*.

A differential form ω on $X \backslash D$ is called *log–log* if ω, $\partial\omega$, $\bar{\partial}\omega$ and $\partial\bar{\partial}\omega$ are log–log growth forms. The sheaf of log–log forms is denoted $\mathcal{E}^*\langle\langle D \rangle\rangle$ and the space of global sections as $E^*\langle\langle D \rangle\rangle$.

Log–log forms have very nice properties. In fact, they are almost as good as smooth forms for many purposes. First $\mathcal{E}^*\langle\langle D \rangle\rangle$ is an algebra and has a bigrading

$$\mathcal{E}^*\langle\langle D \rangle\rangle = \bigoplus_{p+q=n} \mathcal{E}^{p,q}\langle\langle D \rangle\rangle$$

and an associated Hodge filtration

$$F^p \mathcal{E}^*\langle\langle D \rangle\rangle = \bigoplus_{p' \geq p} \mathcal{E}^{p', n-p'}\langle\langle D \rangle\rangle.$$

We will denote by $\mathscr{D}_X^*$ the sheaf of currents on X. It has also a bigrading and a Hodge filtration. Currents are a fundamental tool in Arakelov theory. For more details about currents, the reader is referred to de Rham book [9], also the fourth chapter of [13] or, for a quick introduction the first section of [10].

Recall that a morphisms of complexes (in any abelian category) $f : A^* \to B^*$ is called a quasi-isomorphism if it induces an isomorphism in cohomology objects. A morphism of filtered complexes

$$f : (A^*, F) \to (B^*, F),$$

is called a filtered quasi-isomorphism if all the induced morphisms

$$F^p f : F^p A^* \to F^p B^*$$

are quasi-isomorphisms.

The next result summarizes the main properties of the complex of log–log forms.

Theorem 5.3

1. *Every log–log form is locally integrable. Moreover, they have no residue, this means that the map $\mathscr{E}_X^*\langle\langle D\rangle\rangle \to \mathscr{D}_X^*$ is a morphism of complexes.*
2. *The maps*

$$(\mathscr{E}_X^*, F) \longrightarrow (\mathscr{E}_{X*}\langle\langle D\rangle\rangle, F) \longrightarrow (\mathscr{D}_X^*, F)$$

are filtered quasi-isomorphisms. Therefore, if X is a smooth projective variety, we can use the complex $E_X^*\langle\langle D\rangle\rangle$ to compute the complex cohomology of X with its real structure and its Hodge filtration.

Proof The first statement is proved in [21] for good forms, that are very close to log–log forms. The proof in *loc. cit* can easily be extended to log–log forms. The second statement is [7, Theorem 2.23]. □

Recall that, to any holomorphic hermitian vector bundle we can associate a collection of characteristic forms that represent the characteristic classes of the bundle. Let E be a holomorphic vector bundle on a complex manifold X. Choose a local holomorphic frame for E. The hermitian metric is represented in this frame by a matrix of smooth functions h. The curvature matrix of the metric is the matrix of $(1, 1)$-forms

$$K = \bar{\partial}(\partial h \cdot h^{-1}).$$

The coefficients of the characteristic polynomial of K do not depend on the frame and define global differential forms that represent the characteristic classes of E. For more details, see for instance [13, Chapter 3, §3] or [23, III §3].

Definition 5.4 Let X be a complex manifold, D a normal crossings divisor and E a holomorphic vector bundle on X with a smooth hermitian metric on $X \backslash D$. The metric is said to be *log-singular* along D if, for each small coordinate chart of X adapted to D, and holomorphic frame for E, the following estimates hold.

1. The functions $h_{i,j}$ and $(\det h)^{-1}$ grow at most logarithmically along D.
2. The 1-forms $(\partial h \cdot h^{-1})_{i,j}$ are log–log forms along D.

The interest for us of log-singular metrics and log–log forms comes from the following two theorems by Mumford [21, Theorem 1.4] and [21, Theorem 3.1]. See also [7].

Theorem 5.5 *Let X be a complex manifold and D a normal crossing divisor. Let E be a vector bundle on X with a smooth metric $\| \cdot \|$ on $X \backslash D$ that is log singular along D. Then the characteristic forms of $(E, \| \cdot \|)$ are log–log forms. In particular they are locally integrable and the associated currents represent the Chern classes of E.*

Theorem 5.6 *Let $D = G/K$ be a hermitian symmetric space, with G a semisimple algebraic group defined over $\mathbb{Q}$, Γ a neat arithmetic subgroup of G, V a complex vector space with a hermitian metric and ρ a unitary representation of K on V. To these data we have associated a smooth quasi-projective complex variety $X = \Gamma \backslash G/K$ and a hermitian vector bundle E on X with hermitian metric h. Let $\overline{X}$ be a toroidal compactification of X with $D = \overline{X} \backslash X$ a simple normal crossing divisor. Then E can be extended uniquely to a vector bundle $\overline{E}$ over $\overline{X}$ such that h is log singular along D.*

6   Arakelov Geometry with Log–log Forms

Following the ideas of the previous section, in order to extend Arakelov theory to Shimura varieties, it is enough to replace smooth forms by log–log forms. The theory is almost identical to the classical theory by Gillet and Soulé [10]. For more details the reader can follow Chapter I by Soulé of this volume. We summarize the results in the case of varieties over $\mathrm{Spec}(\mathbb{Z})$. All the changes occur at the generic fibre. Thus there is no difficulty to extend everything to arbitrary arithmetic rings. A variant of the theory presented here has been developed in [7] and [8].

Let $\mathscr{X}$ be a regular flat projective variety over $\mathrm{Spec}\,\mathbb{Z}$. Let $X_{\mathbb{C}}$ be the associated smooth projective complex variety and let $D \subset X_{\mathbb{C}}$ be a normal crossing divisor defined over $\mathbb{R}$. There is an antilinear involution $F_{\infty} \colon X_{\mathbb{C}} \to X_{\mathbb{C}}$ corresponding to complex conjugation of the coordinates. Since the divisor D is defined over $\mathbb{R}$ it is invariant under this involution.

Definition 6.1 Let Z be a codimension p cycle on $\mathscr{X}$. A *log–log Green current* for Z is a current $g_Z \in D_X^{p-1,p-1}$ satisfying a symmetry condition with respect complex conjugation

$$F_\infty^* g_Z = (-1)^{p-1} g_Z$$

and the equation

$$\omega_Z := dd^c g_Z + \delta_Z \in E_X^{p,p} \langle\langle D \rangle\rangle.$$

A *codimension p log–log arithmetic cycle* is a pair (Z, g_Z) with Z a codimension p cycle and g_Z a log–log Green current for Z. We denote by $\widehat{Z}^p(\mathscr{X}, \langle\langle D \rangle\rangle)$ the group of codimension p log–log arithmetic cycles.

The group of rational cycles does not change with respect to the classical Gillet and Soulé theory:

$$\widehat{\mathrm{Rat}}^p(\mathscr{X}, \langle\langle D \rangle\rangle) = \widehat{\mathrm{Rat}}^p(\mathscr{X}) = \mathrm{Span}\{(\mathrm{div}(f), [-\log|f|]) + (0, \partial u + \bar{\partial} v)\}$$

Definition 6.2 The *log–log arithmetic Chow groups* are defined as

$$\widehat{\mathrm{CH}}^p(\mathscr{X}, \langle\langle D \rangle\rangle) = \widehat{Z}^p(\mathscr{X}, \langle\langle D \rangle\rangle)/\widehat{\mathrm{Rat}}^p(\mathscr{X}).$$

A consequence of Theorem 5.3 is that any log–log Green current for a cycle Z, that meets D properly, can be represented by a differential forms with logarithmic singularities along Z and that is log–log along D. This allow one to prove that most of the properties of the arithmetic Chow rings carry over to the log–log arithmetic Chow groups.

For instance, one can define the intersection product as in the classical case, but there is the caveat that we have to move our cycles to meet properly D as well.

Let Z and W be cycles of codimension p and q respectively in $\mathscr{X}$. Assume that $Z_\mathbb{C}$ and $W_\mathbb{C}$ intersect properly and that both intersect properly with D. Let g_Z and g_W be log–log Green currents for Z and W respectively and represent them with differential forms with log and log–log singularities. Then the $*$-product

$$g_Z * g_W = g_Z \wedge \delta_W + \omega_Z \wedge g_W$$

is well defined. As in the classical case, we face the technical problem that Z and W intersect properly on the generic fibre but they do not need to intersect properly globally. In fact the moving lemma is only known for varieties over a field. To remedy this problem following [10], one introduce the Chow groups of $\mathscr{X}$ of cycles that do not meet $\mathscr{X}_\mathbb{Q}$. Denote these groups as $\mathrm{CH}^{p+q}(\mathscr{X})_{\mathrm{fin},\mathbb{Q}}$. Then there is a well defined intersection product

$$(Z, g_Z) \cdot (W, g_W) \in \mathrm{CH}^{p+q}(\mathscr{X})_{\mathrm{fin},\mathbb{Q}} \oplus \widehat{Z}^{p+q}(\mathscr{X}_\mathbb{Q}, \langle\langle D \rangle\rangle).$$

By the moving lemma, this induces an algebra structure on

$$\widehat{\mathrm{CH}}^*(\mathscr{X}, \langle\langle D\rangle\rangle)_{\mathbb{Q}} = \bigoplus_p \widehat{\mathrm{CH}}^p(\mathscr{X}, \langle\langle D\rangle\rangle) \otimes \mathbb{Q}.$$

The inverse image also needs some compatibility with the divisor D. Let $f: \mathscr{Y} \to \mathscr{X}$ be a morphism of varieties over $\mathbb{Z}$. Write $Y = \mathscr{Y}_{\mathbb{C}}$ and $X = \mathscr{X}_{\mathbb{C}}$ for the associated complex varieties, and let $f_{\mathbb{C}}: Y \to X$ be the induced morphism of complex varieties. Clearly, if the image by $f_{\mathbb{C}}$ of any component of Y is contained in $D \subset X$, then there is no hope to define the inverse image. If we add the hypothesis that $E = f_{\mathbb{C}}^{-1}(D)$ is a normal crossing divisor of Y, then there is a well defined inverse image map

$$f^*: \widehat{\mathrm{CH}}^*(\mathscr{X}, \langle\langle D\rangle\rangle) \longrightarrow \widehat{\mathrm{CH}}^*(\mathscr{Y}, \langle\langle E\rangle\rangle).$$

The direct image can be more complicated and we discuss only the case of morphisms to points. Assume that $\mathscr{X}$ is equidimensional of dimension $d+1$ and let $\pi: \mathscr{X} \to \operatorname{Spec} \mathbb{Z}$ be the structural map. Then there are well defined direct images

$$\pi_*: \widehat{\mathrm{CH}}^p(\mathscr{X}, \langle\langle D\rangle\rangle) \longrightarrow \widehat{\mathrm{CH}}^{p-d}(\operatorname{Spec} \mathbb{Z}).$$

Since

$$\widehat{\mathrm{CH}}^p(\operatorname{Spec} \mathbb{Z}) = \begin{cases} \mathbb{Z}, & \text{if } p = 0, \\ \mathbb{R}, & \text{if } p = 1, \\ 0, & \text{otherwise.} \end{cases}$$

only the groups $\widehat{\mathrm{CH}}^{d+1}(\mathscr{X}, \langle\langle D\rangle\rangle)$ and $\widehat{\mathrm{CH}}^d(\mathscr{X}, \langle\langle D\rangle\rangle)$ have a non-zero direct image to $\operatorname{Spec} \mathbb{Z}$.

In the paper [7] it is developed a theory of arithmetic characteristic classes for log-singular hermitian vector bundles extending the theory of [11]. The essential ingredients for this extension are Mumford's Theorems 5.6 and 5.5.

Theorem 6.3 ([7]) *Let $\mathscr{X}$ be a regular projective arithmetic variety over $\operatorname{Spec} \mathbb{Z}$ such that $X = \mathscr{X}_{\mathbb{C}}$ is a finite union of smooth toroidal compactifications of locally symmetric hermitian spaces. Let D be the boundary divisor that we assume to be a normal crossing divisor. Let $\overline{E}$ be a vector bundle on $\mathscr{X}$ with a singular hermitian metric such that, on each component of X, it is the fully decomposable holomorphic vector bundle associated to a unitary representation of the compact subgroup as described in Sect. 3. Then the theory of arithmetic characteristic classes can be extended to define log–log arithmetic characteristic classes*

$$\widehat{c}_i(\overline{E}) \in \widehat{\mathrm{CH}}^i(\mathscr{X}, \langle\langle D\rangle\rangle).$$

Corollary 6.4 *Let $\mathscr{X}$ be an integral model of a Shimura variety of dimension d, Y a codimension p cycle on $\mathscr{X}$ and $\overline{L}_0, \ldots, \overline{L}_{d-p}$ be automorphic line bundles with their Petersson metric. Then the height*

$$h_{\overline{L}_0,\ldots,\overline{L}_{d-p}}(Y)$$

is defined.

Example 6.5 This example is taken from [18]. Let $\Gamma = \mathrm{SL}_2(\mathbb{Z})$. This is an arithmetic group, but it is not neat and the quotient $Y(1) = \Gamma\backslash\mathbb{H}$ has elliptic fixed points. We can ignore the orbifold structure coming from the elliptic fixed points and pretend that $Y(1) \simeq \mathbb{A}^1$. Or, more preciselly, we can choose a neat subgroup $\Gamma' \subset \Gamma$ and work on a covering of $Y(1)$ but it is easier to work directly with $Y(1)$. We compactify $Y(1)$ by adding one cusp to obtain a complete curve $X(1)$. Let k be a positive integer divisible by 12. Then $\mathscr{M}_k(1)$ is a line bundle on $X(1)$ with a log singular hermitian metric on the cusp: the so called the Petersson metric. The fact that we need to go to weight 12 is related with the orbifold structure of $Y(1)$. We can choose a model $\mathscr{X}(1) \simeq \mathbb{P}^1_{\mathrm{Spec}\,\mathbb{Z}}$ of $X(1)$ and $\mathscr{M}_k(1)$ can be extended to a line bundle on $\mathscr{X}(1)$. See [18] for details. We denote by $\overline{\mathscr{M}_k(1)}$ the line bundle on the model with the log singular hermitian metric. Then

$$\widehat{c}_1(\overline{\mathscr{M}_k(1)}) \cdot \widehat{c}_1(\overline{\mathscr{M}_k(1)}) = k^2 \left(\frac{1}{2}\zeta_\mathbb{Q}(-1) + \zeta'_\mathbb{Q}(-1) \right)$$

where $\zeta_\mathbb{Q}$ is Riemann zeta function. As we see in this example, not only the value of the Riemann zeta function appears, but also the value of its derivative.

For more examples of arithmetic intersection numbers on Shimura varieties and its relation with Kudla's program, see [17], [5], [6], [3], [4], [15].

Acknowledgements The author acknowledges partial support from MINECO research projects MTM2016-79400-P and ICMAT Severo Ochoa project SEV-2015-0554.

References

1. A. Ash, D. Mumford, M. Rapoport, Y. Tai, *Smooth Compactification of Locally Symmetric Varieties*. Cambridge Mathematical Library, 2nd edn. (Cambridge University Press, Cambridge, 2010)
2. W.L. Baily, A. Borel, Compactification of arithmetic quotients of bounded symmetric domains. Ann. Math. **84**, 442–528 (1966)
3. J. Bruinier, K. Ono, Heegner divisors, *L*-functions and harmonic weak Maass forms. Ann. Math. **172**, 2135–2181 (2010). MR 2726107
4. J.H. Bruinier, T. Yang, Faltings heights of CM cycles and derivatives of *L*-functions. Invent. Math. **177**, 631–681 (2009). MR 2534103

5. J.H. Bruinier, J.I. Burgos Gil, U. Kühn, Borcherds products and arithmetic intersection theory on Hilbert modular surfaces. Duke Math. J. **139**, 1–88 (2007). MR MR2322676 (2008h:11059)
6. J.H. Bruinier, B. Howard, T. Yang, Heights of Kudla-Rapoport divisors and derivatives of L-functions. Invent. Math. **201**, 1–95 (2015). MR 3359049
7. J.I. Burgos Gil, J. Kramer, U. Kühn, Arithmetic characteristic classes of automorphic vector bundles. Documenta Math. **10**, 619–716 (2005)
8. J.I. Burgos Gil, J. Kramer, U. Kühn, Cohomological arithmetic Chow rings. J. Inst. Math. Jussieu **6**, 1–172 (2007)
9. G. de Rham, *Variétés Différentiables. Formes, Courants, Formes Harmoniques, Hermann, Paris, 1973, Troisième édition revue et augmentée*. Publications de l'Institut de Mathématique de l'Université de Nancago, III, Actualités Scientifiques et Industrielles, No. 1222b (1973)
10. H. Gillet, C. Soulé, Arithmetic intersection theory. Publ. Math. I.H.E.S. **72**, 94–174 (1990)
11. H. Gillet, C. Soulé, Characteristic classes for algebraic vector bundles with hermitian metric I, II. Ann. Math. **131**, 163–203, 205–238 (1990)
12. H. Gillet, C. Soulé, An arithmetic Riemann-Roch theorem. Invent. Math. **110**, 473–543 (1992)
13. P. Griffiths, J. Harris, *Principles of Algebraic Geometry* (Wiley, Hoboken, 1994)
14. S. Helgason, *Differential Geometry, Lie Groups and Symmetric Spaces*. Pure and Applied Mathematics (Academic, Cambridge, 1978)
15. B. Howard, Complex multiplication cycles and Kudla-Rapoport divisors, II. Amer. J. Math. **137**, 639–698 (2015). MR 3357118
16. K. Köhler, D. Rössler, A fixed point formula of Lefschetz type in Arakelov geometry. I. Statement and proof. Invent. Math. **145**, 333–396 (2001)
17. S. Kudla, M. Rappoport, T. Yang, *Modular Forms and Special Cycles on Shimura Curves*. Annals of Mathematics Studies, vol. 161 (Princeton University Press, Princeton, 2006)
18. U. Kühn, Generalized arithmetic intersection numbers. J. Reine Angew. Math. **534**, 209–236 (2001)
19. A. Lubotzky, D. Segal, *Subgroup Growth*. Progress in Mathematics, vol. 212 (Birkhäuser Verlag, Basel, 2003)
20. J.S. Milne, An introduction to Shimura varieties (2004). http://www.jmilne.org/math/xnotes/svi.pdf
21. D. Mumford, Hirzebruch's proportionality theorem in the non-compact case. Invent. Math. **42**, 239–272 (1977)
22. J. Neukirch, *Algebraic Number Theory*. Grundlehren der Mathematischen Wissenschaften, vol. 322 (Springer, Berlin, 1999)
23. R.O. Wells, *Differential Analysis on Complex Manifolds*. Graduate Texts in Mathematics, vol. 65 (Springer, Berlin, 1980)

Chapter XI: The Arithmetic Riemann–Roch Theorem and the Jacquet–Langlands Correspondence

Gerard Freixas i Montplet

1 Introduction

The Riemann–Roch formula in Arakelov geometry [17] is a local-to-global state-ment, that translates some arithmetic intersection numbers of hermitian vector bundles into cohomological invariants. In cases of arithmetic relevance, such as Shimura varieties, it is natural to apply the theorem to automorphic vector bundles. According to general conjectures (e.g. the Maillot-Rössler conjecture [24] and the vast Kudla program), one expects the arithmetic intersection numbers to be related with logarithmic derivatives of L-functions. One may hope that the arithmetic Riemann–Roch theorem provides a cohomological approach to settle cases of this principle. In this geometric setting, the cohomological side of the formula affords an automorphic translation, so that the theory of automorphic representations may be of some use. Unfortunately this approach has currently not been fruitful. In a parallel with the trace formula, arithmetic intersection numbers seem to be analogue to the geometric side, supposed to be easier to deal with than the spectral side, itself analogous to the cohomological part of Riemann–Roch. However, in the theory of automorphic forms, a fruitful idea has been to compare trace formulae, in order to relate automorphic representations for different groups. It is then tempting to combine these relations, when they exist, to relate as well arithmetic intersection numbers for different Shimura varieties. While this does not provide the evaluation of these numerical invariants, it indicates some structural phenomenon that goes beyond the conjectural predictions alluded to above, and that has not been much explored. In these notes, we exemplify this "philosophy" in the case of modular and Shimura curves. This is an excuse to review the arithmetic Riemann–Roch theorem

G. Freixas i Montplet (✉)
CNRS—Institut de Mathématiques de Jussieu-Paris Rive Gauche, Paris, France
e-mail: gerard.freixas@imj-prg.fr

E. Peyre, G. Rémond (eds.), *Arakelov Geometry and Diophantine Applications*,
Lecture Notes in Mathematics 2276, https://doi.org/10.1007/978-3-030-57559-5_12

of Gillet–Soulé, as well as a variant for modular curves due to the author and Anna von Pippich [11, 14] (more generally, arithmetic surfaces with "cusps" and "elliptic fixed points"). Also, we explain in the classical language of modular forms the content of the Jacquet–Langlands correspondence. All these wonderfully combine to provide a relation between arithmetic self-intersection numbers of sheaves of modular forms on modular and Shimura curves [13].

Although we tried to make this survey self-contained, it is nevertheless related to other contributions in this volume. Soulé's paper provides a detailed account on arithmetic intersection theory in arbitrary dimensions. In particular, he expounds the fundamental concept of arithmetic self-intersections or heights. While the general form of the arithmetic Grothendieck–Riemann–Roch theorem is not presented in *loc. cit.*, in Sect. 2 we discuss the special case of hermitian line bundles on arithmetic surfaces. Hopefully, Soulé's survey combined with ours are a useful starting point for a further study of the general statement. Besides, Burgos' contribution is a good complement to our Sect. 3. While Burgos motivates and explains the framework of arithmetic intersections on non-compact Shimura varieties, we focus on the simplest examples of such: modular curves. Actually, our research in this area has been motivated from the beginning by the lack of an arithmetic Grothendieck–Riemann–Roch theorem in the theory developed by Burgos–Kramer–Kühn [7, 6].

2   Riemann–Roch Theorem for Arithmetic Surfaces and Hermitian Line Bundles

2.1   Riemann–Roch Formulae in Low Dimensions

As a matter of motivation, we recall the statement of the Hirzebruch–Riemann–Roch theorem for compact complex manifolds of dimensions 1 and 2 and line bundles on them. Probably one of the most complete reference for the theory of characteristic classes and the general Grothendieck–Riemann–Roch theorem, in the context of algebraic geometry (the one of most interest to us), is Fulton's book [15]. It is also a good source of examples, references and historical remarks.

Let X be a compact Riemann surface and L a holomorphic line bundle on X. To the line bundle L we can associate two integer valued invariants. The first and easiest one is the degree $\deg L$. It is known that L affords non-trivial meromorphic sections, and for such a section s

$$\deg L = \deg(\operatorname{div} s) = \sum_{p \in X} \operatorname{ord}_p(s).$$

The notation $\mathrm{ord}_p(s)$ stands for the order of vanishing or pole of s at the point $p \in X$. This sum is of course finite, and does not depend on the particular choice of s by the residue theorem: the divisor of any meromorphic function has degree zero. The second topological invariant is the holomorphic Euler–Poincaré characteristic:

$$\chi(X, L) = \dim H^0(X, L) - \dim H^1(X, L).$$

The coherent cohomology groups $H^0(X, L)$ and $H^1(X, L)$ are actually finite dimensional $\mathbb{C}$-vector spaces. They can be defined as Čech cohomology. For later motivation, let us provide a geometric differential interpretation of these spaces. The Dolbeault complex of the holomorphic line bundle L is given by the $\bar{\partial}$ operator defining the holomorphic structure of L:

$$A^{0,0}(X, L) \xrightarrow{\bar{\partial}} A^{0,1}(X, L).$$

If $s \in A^{0,0}(X, L)$ is a smooth section of L, and e is a holomorphic trivialization of L on some analytic open subset $U \subset X$, then $s = fe$ for some smooth function f on U and

$$\bar{\partial} s_{|U} = (\bar{\partial} f) \otimes e \in A^{0,1}(U, L).$$

The coherent cohomology of L may then be canonically identified with Dolbeault cohomology (cf. [32, Chap. 5]):

$$H^0(X, L) \simeq \ker \bar{\partial}, \quad H^1(X, L) \simeq \frac{A^{0,1}(X, L)}{\mathrm{Im}\,\bar{\partial}}.$$

The Riemann–Roch formula in this setting relates the numerical invariants we attached to L:

Theorem 2.1 (Riemann–Roch) *The degree and Euler–Poincaré characteristic of L are related by*

$$\chi(L) = \deg L + \frac{1}{2} \deg T_X,$$

$$= \deg L + 1 - g$$

where T_X is the holomorphic tangent bundle and g is the topological genus of X.

The extension of the Riemann–Roch formula to higher compact complex manifolds requires the theory of characteristic classes of holomorphic vector bundles. We will not do this here, and instead we refer to [15] instead for a complete treatment. For the sake of motivation, we state the formula and explain its meaning in complex dimension 2. Hence, let X be a compact complex manifold of dimension 2. Let

L be a holomorphic line bundle on X. Now the holomorphic Euler–Poincaré characteristic is

$$\chi(X, L) = \dim H^0(X, L) - \dim H^1(X, L) + \dim H^2(X, L).$$

In the usual formulation, this invariant is computed by the *Hirzebruch–Riemann–Roch* formula, in terms of the characteristic classes as follows:

$$\chi(X, L) = \int_X (\mathrm{ch}(L)\, \mathrm{td}(T_X))^{(2)}$$

Here ch and td refer to the Chern character and Todd genus, respectively. The index 2 indicates that we only take the codimension 2 contribution of this product of characteristic classes. This can be expanded in terms of Chern classes as

$$\chi(X, L) = \int_X \left\{ c_1(L)^2 + \frac{1}{2} c_1(T_X)\, c_1(L) \right.$$
$$\left. + \frac{1}{12}(c_1(T_X)^2 + c_2(T_X)) \right\}. \tag{1}$$

Let us assume for the sake of simplicity that X is projective. If L and M are line bundles on X, we can find respective meromorphic sections s and t whose divisors $\mathrm{div}\, s = \sum_i m_i D_i$ and $\mathrm{div}\, t = \sum_j n_j E_j$ have smooth components and pairwise transversal intersections (one implicitly invokes Bertini's theorem for that). Then

$$\int_X c_1(L)\, c_1(M) = \sum_{i,j} m_i n_j \#(D_i \cap E_j).$$

This explains the meaning of the first three terms in the Hirzebruch–Riemann–Roch formula (1), once we recall that for a vector bundle E we have $c_1(E) = c_1(\det E)$. The last term involving $c_2(T_X)$ is the Euler number of X, related to the topological Euler-Poincaré characteristic through

$$\chi_{\mathrm{top}}(X, \mathbb{C}) = \sum_{p=0}^{2}(-1)^p \dim H^p(X, \mathbb{C}) = \int_X c_2(T_X).$$

It is a sort of obstruction to the existence of non-vanishing global holomorphic vector fields.

Let us give an example of use of the Hirzebruch–Riemann–Roch formula. Let X be a $K3$ surface. This is a simply connected compact Kähler surface, whose canonical bundle $\omega_X = \wedge^2 \Omega_X^1$ is trivial (i.e. X has a nowhere vanishing

holomorphic 2-form). On the one hand, the Hirzebruch–Riemann–Roch theorem
(1) applied to the trivial line bundle $\mathcal{O}_X$ gives

$$\chi(X, \mathcal{O}_X) = \frac{1}{12} \int_X (c_1(T_X)^2 + c_2(T_X)) = \frac{1}{12} \int_X c_2(T_X),$$

since $c_1(T_X) = -c_1(\omega_X) = 0$. On the other hand, $\chi(X, \mathcal{O}_X) = 2$. Indeed,
$H^0(X, \mathcal{O}_X) = \mathbb{C}$, $H^1(X, \mathcal{O}_X) = 0$ by the Hodge decomposition and simply
connectedness (i.e. $H^1(X, \mathbb{C}) = 0$), and by Serre duality and $K3$ assumption

$$H^2(X, \mathcal{O}_X) \simeq H^0(X, \omega_X)^\vee \simeq H^0(X, \mathcal{O}_X)^\vee = \mathbb{C}.$$

Hence we derive $\int_X c_2(T_X) = 24$, and therefore

$$\chi_{\text{top}}(X, \mathbb{C}) = 24.$$

Again using the simply connectedness and the Hodge decomposition, we infer from
this equality

$$h^{1,1} = \dim H^1(X, \Omega_X^1) = 19.$$

Let us end this section with a question. Assume now that X is the set of complex
points of a proper flat scheme $\mathcal{X}$ over $\mathbb{Z}$. Assume as well that $\mathcal{L}$ is an invertible sheaf
on $\mathcal{X}$. The coherent cohomology groups $H^p(\mathcal{X}, \mathcal{L})$ are $\mathbb{Z}$-modules of finite type. As
such, they have a well-defined rank. By flat base change, the rank is computed after
base changing to $\mathbb{C}$, and hence

$$\sum_p (-1)^p \operatorname{rank} H^p(\mathcal{X}, \mathcal{L}) = \sum_p (-1)^p \dim H^p(\mathcal{X}_{\mathbb{C}}, \mathcal{L}_{\mathbb{C}})$$

can be obtained from the Hirzebruch–Riemann–Roch theorem. One may wonder
what additional information we can catch due to the integral structure. The
arithmetic Riemann–Roch theorem takes this structure into account.

2.2 Arithmetic Intersections on Arithmetic Surfaces

We introduce arithmetic intersections on arithmetic surfaces. A good survey on the
subject is Soulé's Bourbaki's talk [30], where the theory of the Quillen metric and
the arithmetic Riemann–Roch formula is also expounded. We also refer the reader
to Soulé's survey in this volume, where the more general arithmetic intersection
theory in higher dimensions is explained.

Let $\pi : \mathcal{X} \to \operatorname{Spec} \mathbb{Z}$ be an arithmetic surface, i.e. $\mathcal{X}$ is a regular scheme, flat
and projective over $\mathbb{Z}$, of Krull dimension 2. The set of complex points $\mathcal{X}(\mathbb{C})$

has the structure of a possibly non-connected Riemann surface, equipped with an anti-holomorphic involution $F_\infty \colon X(\mathbb{C}) \to X(\mathbb{C})$ induced by complex conjugation. Recall that a (smooth) hermitian line bundle $\overline{\mathcal{L}}$ over X consists in giving a line bundle $\mathcal{L}$ together with a smooth hermitian metric on the associated holomorphic line bundle $\mathcal{L}_{\mathbb{C}}$ on $X(\mathbb{C})$. We require the metric to be invariant under the natural action of F_∞ on the underlying $\mathcal{C}^\infty$ complex line bundle. For a pair of hermitian line bundles $\overline{\mathcal{L}}$ and $\overline{\mathcal{M}}$, we proceed to recall the construction of their *arithmetic intersection number*

$$(\overline{\mathcal{L}} \cdot \overline{\mathcal{M}}) \in \mathbb{R},$$

to be compared with the geometric intersection number

$$\int_X c_1(L) \, c_1(M) \in \mathbb{Z}$$

of two line bundles L, M on a projective complex surface X.

Let ℓ and m be non-trivial rational sections of $\mathcal{L}$ and $\mathcal{M}$ respectively, such that $\operatorname{div} \ell$ and $\operatorname{div} m$ are disjoint on the generic fiber $X_{\mathbb{Q}}$. It is always possible to find such sections, by the projectivity assumption on X. Let us write

$$\operatorname{div} \ell = \sum m_i D_i, \quad \operatorname{div} m = \sum n_j E_j.$$

The D_i and E_j are pairwise generically disjoint. We define finite arithmetic intersection numbers $(D_i \cdot E_j)_{\mathrm{fin}}$ as follows. If D_i is a vertical divisor, hence a variety over $\mathbb{F}_p$ for some prime number p, then we put

$$(D_i \cdot E_j)_{\mathrm{fin}} = (\deg c_1(\mathcal{O}(E_j)) \cap [D_i]) \cdot \log p.$$

Hence we compute the degree of $\mathcal{O}(E_j)$ restricted to the projective curve D_i over $\mathbb{F}_p$, weighted by $\log p$. Assume now that D_i is horizontal, hence flat over $\mathbb{Z}$. Let $x \in D_i \cap E_j$ be an intersection point. We denote by $k(x)$ its (finite) residue field. If $f, g \in \mathcal{O}_{X,x}$ are local equations for D_i and E_j, then the local ring $\mathcal{O}_{X,x}/(f, g)$ has finite length, and we put

$$(D_i \cdot E_j)_{\mathrm{fin}, x} = \operatorname{length} \frac{\mathcal{O}_{X,x}}{(f, g)} \cdot \log(\#k(x)).$$

We define

$$(D_i \cdot E_j)_{\mathrm{fin}} = \sum_{x \in D_i \cap E_j} (D_i \cdot E_j)_{\mathrm{fin}, x}.$$

Finally, we put

$$(\ell, m)_{\text{fin}} = \sum_{i,j} m_i n_j (D_i \cdot E_j)_{\text{fin}} \in \mathbb{R}.$$

Next we introduce the archimedean contribution to the arithmetic intersection pairing:

$$(\ell, m)_\infty = \int_{\mathcal{X}(\mathbb{C})} \left(-\log \|m_\mathbb{C}\| \frac{i}{2\pi} \bar\partial\partial \log \|\ell_\mathbb{C}\|^2 - \log \|\ell_\mathbb{C}\| \delta_{\operatorname{div} m_\mathbb{C}} \right)$$

$$= \int_{\mathcal{X}(\mathbb{C})} \left(-\log \|m_\mathbb{C}\| c_1(\overline{\mathcal{L}}_\mathbb{C}) - \log \|\ell_\mathbb{C}\| \delta_{\operatorname{div} m_\mathbb{C}} \right) \in \mathbb{R},$$

where $c_1(\overline{\mathcal{L}}_\mathbb{C})$ is the first Chern form of the hermitian line bundle $\overline{\mathcal{L}}_\mathbb{C}$. The arithmetic intersection number of $\overline{\mathcal{L}}$ and $\overline{\mathcal{M}}$ is obtained by adding the finite and archimedean intersection pairings above:

$$(\overline{\mathcal{L}} \cdot \overline{\mathcal{M}}) = (\ell, m)_{\text{fin}} + (\ell, m)_\infty \in \mathbb{R}.$$

It can be easily checked, applying Weil's reciprocity and the product formula, that the construction does not depend on the choice of sections ℓ, m. Furthermore, by Stokes' theorem, the arithmetic intersection number is symmetric. It also behaves bilinearly with respect to the tensor product of hermitian line bundles. In the particular case $\overline{\mathcal{L}} = \overline{\mathcal{M}}$, the quantity $(\overline{\mathcal{L}} \cdot \overline{\mathcal{L}})$ is also written $(\overline{\mathcal{L}}^2)$, and is called the *arithmetic self-intersection number* of $\overline{\mathcal{L}}$. It actually equals, by definition, the *height of $\mathcal{X}$ with respect to $\overline{\mathcal{L}}$*, which is also denoted $h_{\overline{\mathcal{L}}}(\mathcal{X})$.

2.3 The Determinant of Cohomology and the Quillen Metric

Let X be a compact Riemann surface and L a line bundle on X. We define the determinant of the cohomology of L as

$$\det H^\bullet(X, L) = \bigwedge^{\max} H^0(X, L) \otimes \bigwedge^{\max} H^1(X, L)^\vee.$$

Recall that the cohomology $H^\bullet(X, L)$ can be computed as the cohomology of the Dolbeault complex

$$A^{0,0}(X, L) \xrightarrow{\bar\partial} A^{0,1}(X, L).$$

In particular, $H^0(X, L)$ can be realized inside $A^{0,0}(X, L)$. If the line bundles T_X and L are equipped with smooth hermitian metrics, one can realize $H^1(X, L)$ inside $A^{0,1}(X, L)$ as well. For if $\overline{\partial}^*$ denotes the formal adjoint of $\overline{\partial}$ with respect to the functional L^2 hermitian products on $A^{0,p}(X, L)$, induced from the choices of metrics (i.e. $\langle \overline{\partial} s, t \rangle = \langle s, \overline{\partial}^* t \rangle$), then

$$H^1(X, L) \simeq \ker \overline{\partial}^*.$$

Through these realizations $H^p(X, L) \subset A^{0,p}(X, L)$, the cohomology spaces inherit L^2 hermitian products. The determinant of cohomology $\det H^\bullet(X, L)$ carries an induced hermitian norm, called the L^2 metric and written h_{L^2} or $\| \cdot \|_{L^2}$. For the sake of completeness, let us just say that the volume form μ needed for the L^2 pairings is normalized to be, locally in holomorphic coordinates z,

$$\mu = \frac{i}{2\pi} \frac{dz \wedge d\overline{z}}{\|dz\|^2}.$$

Define Laplace type operators by $\Delta_{\overline{\partial}, L}^{0,0} = \overline{\partial}^* \overline{\partial}$ and $\Delta_{\overline{\partial}, L}^{0,1} = \overline{\partial} \overline{\partial}^*$ [32, Chap. 5]. These are elliptic differential operators of second order, positive and essentially self-adjoint, with discrete spectrum accumulating only to ∞. The construction of the L^2 metric involves only the 0 eigenspaces of these operators. The Quillen metric includes the rest of the spectrum. Write

$$0 < \lambda_1 \leq \lambda_2 \leq \dots$$

for the strictly positive eigenvalues of $\Delta_{\overline{\partial}, L}^{0,1}$ (or equivalently $\Delta_{\overline{\partial}, L}^{0,0}$), repeated according to multiplicities. Following Ray–Singer [28], we define the *spectral zeta function*

$$\zeta_{\overline{\partial}, L}^{0,1}(s) = \sum_n \frac{1}{\lambda_n^s},$$

which can be shown to be absolutely convergent, hence holomorphic, for $\mathrm{Re}(s) > 1$. It can be meromorphically continued to the whole complex plane, and $s = 0$ is a regular point. Actually, there are asymptotic expansions for the spectral theta function

$$\theta(t) := \mathrm{tr}(e^{-t\Delta_{\overline{\partial}, L}^{0,1}}) - \dim \ker \Delta_{\overline{\partial}, L}^{0,1} = \sum_n e^{-t\lambda_n}, \quad t > 0,$$

as $t \to +\infty$ and $t \to 0^+$, that justify the Mellin transform identity

$$\zeta_{\overline{\partial}, L}^{0,1}(s) = \frac{1}{\Gamma(s)} \int_0^\infty \theta(t) t^{s-1} dt$$

and lead to the meromorphic continuation properties above. The zeta regularized determinant of $\Delta_{\bar{\partial},L}^{0,1}$ is then defined to be

$$\det \Delta_{\bar{\partial},L}^{0,1} = \exp\left(-\frac{d}{ds}\Big|_{s=0} \zeta_{\bar{\partial},L}^{0,1}(s)\right) \text{``} = \prod_n \lambda_n \text{''}.$$

The Quillen metric is obtained by rescaling the L^2 metric:

$$h_Q := \left(\det \Delta_{\bar{\partial},L}^{0,1}\right)^{-1} h_{L_2}.$$

For the purpose of this article, we refer to the work of Bismut–Gillet–Soulé [1, 2, 3] on the theory of the Quillen metric in the context of Arakelov geometry.

Let now $X \to \operatorname{Spec}\mathbb{Z}$ be an arithmetic surface and $\overline{\mathcal{L}}$ a hermitian line bundle. Assume that $T_{X(\mathbb{C})}$ is endowed with a hermitian metric, with the usual invariance property under the action of F_∞. The groups $H^p(X, \mathcal{L})$ are $\mathbb{Z}$-modules of finite rank, and by means of 2 term free resolutions, one can define their determinants $\det H^p(X, \mathbb{Z})$. These are free $\mathbb{Z}$-modules of rank 1. We put

$$\det H^\bullet(X, \mathcal{L}) = \det H^0(X, \mathcal{L}) \otimes \det H^1(X, \mathcal{L})^\vee.$$

This construction commutes with base change, and in particular we can endow $\det H^\bullet(X, \mathcal{L})$ with the Quillen metric after base changing to $\mathbb{C}$. We usually denote $\det H^\bullet(X, \mathcal{L})_Q$ to indicate the resulting hermitian line bundle over $\mathbb{Z}$. Then we can attach to this object a numerical invariant called the *arithmetic degree*: if e is any basis of the $\mathbb{Z}$-module $\det H^\bullet(X, \mathcal{L})_Q$, then the arithmetic degree is

$$\widehat{\deg} \det H^\bullet(X, \mathcal{L})_Q = -\log \|e\|_Q \in \mathbb{R}.$$

Observe this is well-defined, since e is unique up to sign. The arithmetic degree of the determinant of cohomology is the arithmetic counterpart of the Euler–Poincaré characteristic in the complex geometric setting. The arithmetic degree of a hermitian $\mathbb{Z}$-module is defined analogously, and is sometimes called *Arakelov degree*. This is the terminology employed by Bost, Chen and Gaudron in their contributions in this volume, to which the reader is referred for a more systematic treatment.

2.4 The Arithmetic Riemann–Roch Theorem of Gillet–Soulé

Let $\pi: X \to \operatorname{Spec}\mathbb{Z}$ be an arithmetic surface, $\overline{\mathcal{L}}$ a hermitian line bundle and fix a F_∞ invariant hermitian metric on $T_{X(\mathbb{C})}$. The arithmetic Riemann–Roch formula of Gillet–Soulé [17] computes $\widehat{\deg} \det H^\bullet(X, \mathcal{L})_Q$ in terms of arithmetic intersections, in a formally analogous expression to the Hirzebruch–Riemann–Roch theorem in dimension 2. To state the theorem, we briefly need to discuss the relative dualizing sheaf and the analogue of the Euler class.

Because the morphism $\pi : \mathcal{X} \to \operatorname{Spec}\mathbb{Z}$ is projective and $\mathcal{X}$ and $\operatorname{Spec}\mathbb{Z}$ are regular schemes, there is a relative dualizing line bundle $\omega_{\mathcal{X}/\mathbb{Z}}$. Its complexification is dual to $T_{\mathcal{X}(\mathbb{C})}$ and hence we can endow it with the dual hermitian metric. The line bundle can be explicitly constructed from any factorization

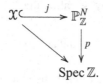

The immersion j is regular, and its conormal sheaf is thus a vector bundle N_j on $\mathcal{X}$. The relative cotangent bundle Ω_p is a locally free rank N vector bundle on $\mathbb{P}_{\mathbb{Z}}^N$. One can then prove that

$$\omega_{\mathcal{X}/\mathbb{Z}} = \det j^*\Omega_p \otimes \det N_j$$

is a dualizing sheaf. Using the exact sequence

$$0 \longrightarrow N_j^\vee \longrightarrow j^*\Omega_p \longrightarrow \Omega_{\mathcal{X}/\mathbb{Z}} \longrightarrow 0$$

and the theory of Bott–Chern secondary classes, one can define an arithmetic second Chern class $\widehat{c}_2(\Omega_{\mathcal{X}/\mathbb{Z}})$. It is actually defined independently of any metrized datum. Its arithmetic degree can be expressed in terms of *localized Chern classes*

$$\delta_\pi = \widehat{\deg}\,\widehat{c}_2(\Omega_{\mathcal{X}/\mathbb{Z}}) = \sum_p \deg c_2^{\mathcal{X}_{\mathbb{F}_p}}(\Omega_{\mathcal{X}/\mathbb{Z}_p}) \log p.$$

The localized Chern classes $c_2^{\mathcal{X}_{\mathbb{F}_p}}(\Omega_{\mathcal{X}/\mathbb{Z}_p})$ belong the *Chow groups with supports* $\mathrm{CH}^2_{\mathcal{X}_{\mathbb{F}_p}}(\mathcal{X})$. They measure the bad reduction of π at p. If π is semi-stable, then its degree is the number of singular points in the geometric fiber of π at p. We don't elaborate further here, since we will later discard the term δ_π. The curious reader will refer to Fulton [15, Chap. 17] for the theory of Chow groups with support and localized characteristic classes. A discussion adapted to (and inspired by) Arakelov geometry is also provided by Bloch–Gillet–Soulé [4].

We can now state the arithmetic Riemann–Roch theorem of Gillet–Soulé for hermitian line bundles on arithmetic surfaces.

Theorem 2.2 (Gillet–Soulé) *Let $\pi : \mathcal{X} \to \operatorname{Spec}\mathbb{Z}$ be an arithmetic surface, $\overline{\mathcal{L}}$ a hermitian line bundle, and fix a F_∞ invariant metric h on $T_{\mathcal{X}(\mathbb{C})}$. Endow the relative* *dualizing sheaf $\omega_{X/\mathbb{Z}}$ with the metric dual to h. Then there is an equality of real numbers*

$$12 \widehat{\deg} \det H^\bullet(X, \mathcal{L})_Q - \delta_\pi = (\overline{\omega}^2_{X/\mathbb{Z}}) + 6(\overline{\mathcal{L}} \cdot \overline{\mathcal{L}} \otimes \overline{\omega}^{-1}_{X/\mathbb{Z}})$$

$$- (2g - 2)\#\pi_0(X(\mathbb{C})) \left(\frac{\zeta'(-1)}{\zeta(-1)} + \frac{1}{2} \right),$$

where g is the genus of any connected component of $X(\mathbb{C})$ and ζ is the Riemann zeta function.

A particular relevant case of the theorem is the *arithmetic Noether formula*, beautifully presented by Moret–Bailly in [25], obtained by specializing to $\overline{\mathcal{L}} = \overline{\mathcal{O}}_X$ the trivial hermitian line bundle:

$$12 \widehat{\deg} \det H^\bullet(X, \mathcal{O})_Q - \delta_\pi = (\overline{\omega}^2_{X/\mathbb{Z}})$$

$$- (2g - 2)\#\pi_0(X(\mathbb{C})) \left(\frac{\zeta'(-1)}{\zeta(-1)} + \frac{1}{2} \right).$$

3 An Arithmetic Riemann–Roch Formula for Modular Curves

3.1 The Setting

A natural geometric situation of arithmetic interest to which we would like to apply the arithmetic Riemann–Roch theorem is the case of integral models of compactified modular curves. Let $X \to \operatorname{Spec} \mathbb{Z}$ be an arithmetic surface such that

$$X(\mathbb{C}) = \bigsqcup_j (\Gamma_i \backslash \mathbb{H} \cup \{\text{cusps}\}),$$

where the $\Gamma_i \subset \operatorname{PSL}_2(\mathbb{R})$ are congruence subgroups (e.g. $\Gamma_0(N)$, $\Gamma_1(N)$ or $\Gamma(N)$). In the arithmetic Riemann–Roch theorem, we need to fix a hermitian metric on the holomorphic tangent bundle. In the present setting, it could be tempting to choose a Poincaré type metric induced by the uniformization by $\mathbb{H}$: if $\tau = x + iy$ is the usual complex coordinate on the upper half plane, then the tensor

$$\frac{|d\tau|^2}{(\operatorname{Im} \tau)^2}$$

defines a $\operatorname{PSL}_2(\mathbb{R})$ invariant metric, and it is unique with this property, up to a constant (fixing the constant is tantamount to prescribing the constant negative gaussian curvature). However, the quotient metric on each factor $\Gamma_i \backslash \mathbb{H} \cup \{\text{cusps}\}$ has singularities. The obvious ones happen at the cusps, where the metric is not even defined. Also, the groups Γ_i may have fixed points on $\mathbb{H}$, responsible for conical type metric singularities on the quotient. Observe these features are not specific of congruence subgroups, but this is a general fact for Fuchsian groups of the first kind, i.e. discrete subgroups Γ of $\mathsf{PSL}_2(\mathbb{R})$ such that $\Gamma\backslash\mathbb{H}$ is a complex algebraic curve. An introduction to the theory of Fuchsian groups, and the spectral theory we will later invoke (including the trace formula), is Iwaniec's book [20].

Let Γ be a Fuchsian group of the first kind. The serious difficulty we have to face happens at the level of spectral theory. Let us work with the Poincaré metric, and the trivial hermitian line bundle on $\Gamma\backslash\mathbb{H}$. The corresponding laplace operator on $A^{0,0}(\Gamma\backslash\mathbb{H}) := A^{0,0}(\mathbb{H})^\Gamma$ is, up to a constant, induced by the $\mathsf{PSL}_2(\mathbb{R})$ invariant operator

$$-y^2\left(\frac{\partial}{\partial x^2} + \frac{\partial}{\partial y^2}\right).$$

This is an elliptic positive differential operator of order 2, essentially self-adjoint (with respect to the natural L^2 structure on $A^{0,0}(\Gamma\backslash\mathbb{H})$), but it has discrete as well as continuous spectrum. Hence the definitions of the spectral zeta function and the regularized determinant do not make sense. And even if we can find a sensible definition, Theorem 2.2 does not automatically adapt to this case.

In this section we discuss a version of the arithmetic Riemann–Roch theorem that applies to the previous setting and the trivial hermitian line bundle (so it is actually a version of the arithmetic Noether formula). We consider an arithmetic surface $\pi\colon \mathcal{X} \to \operatorname{Spec}\mathcal{O}_K$ over the ring of integers of a number field K, together with generically disjoint sections

$$\sigma_1,\ldots,\sigma_n\colon \operatorname{Spec}\mathcal{O}_K \to \mathcal{X},$$

such that

$$\mathcal{X}(\mathbb{C}) = \bigsqcup_{\tau\colon K\hookrightarrow\mathbb{C}} (\Gamma_\tau\backslash\mathbb{H} \cup \{\sigma_1(\tau),\ldots,\sigma_r(\tau)\}),$$

and $\sigma_{r+1}(\tau),\ldots,\sigma_n(\tau)$ correspond to the (elliptic) fixed points of Γ_τ with stabilizers of orders $e_{r+1},\ldots,e_n \geq 2$, respectively. The Poincaré hermitian metric induces a log-singular metric on the $\mathbb{Q}$-line bundle

$$\omega_{\mathcal{X}/\mathcal{O}_K}\left(\sum_i (1 - e_i^{-1})\sigma_i\right),$$

where we put $e_i = \infty$ for $i = 1,\ldots,r$ (i.e. we declare that the *cusps* $\sigma_1,\ldots,\sigma_r$ have infinite order). We will indicate the choice of this metric by an index "hyp". Hence there is still a well-defined arithmetic intersection number

$$\left(\omega_{\mathcal{X}/\mathcal{O}_K}\left(\sum_i (1 - e_i^{-1})\sigma_i\right)^2_{\mathrm{hyp}}\right) \in \mathbb{R},$$

according to the formalism developed by Bost and Kühn [5, 23], and later generalized by Burgos–Kramer–Kühn [7, 6] to any dimension, allowing log-singular metrics. See also Burgos' article in this volume for a general treatment of arithmetic intersections for such singular metrics. This will be the main numerical invariant on the right hand side of the arithmetic Riemann–Roch formula. The need of twisting by the sections σ_i will be compensated by a suitable "boundary" contribution.

3.2 Renormalized Metrics (Wolpert Metrics)

Let Γ be a Fuchsian group of the first kind, and endow the quotient $\Gamma \backslash \mathbb{H}$ with the metric induced by $|d\tau|^2/(\operatorname{Im} \tau)^2$. We discuss the existence of canonical coordinates at cusps and elliptic fixed points, that serve to renormalize the singularities of the quotient metric.

Recall that a cusp of Γ corresponds to a point in $\mathbb{P}^1(\mathbb{R})$ with non-trivial stabilizer under Γ. This stabilizer is conjugated in $\mathrm{PSL}_2(\mathbb{R})$ to the group generated by the translation $\tau \mapsto \tau + 1$. It follows that for a cusp p of $\Gamma \backslash \mathbb{H} \cup \{\mathrm{cusps}\}$, there exists a holomorphic coordinate z such that the hyperbolic metric tensor becomes

$$\frac{|dz|^2}{(|z| \log |z|)^2}.$$

The coordinate z is unique up to a constant of modulus one, and hence the assignment

$$\|dz\|_{w,p} = 1$$

is a well-defined hermitian metric on the holomorphic cotangent space of X at p, $\omega_{X,p}$. In the theory of modular forms, this variable z is usually denoted q, and appears in the so-called q-expansions (Fourier series expansions).

Elliptic fixed points correspond to points in $\mathbb{H}$ whose stabilizer is non-trivial. The stabilizer is then conjugated in $\mathrm{PSL}_2(\mathbb{R})$ to a finite group of rotations centered at $i \in \mathbb{H}$. A neighborhood of an elliptic fixed point q in $\Gamma \backslash \mathbb{H}$ is thus isometric to a quotient $D(0, \varepsilon)/\mu_k$, where the disk is endowed with the hyperbolic metric $|dw|^2/(4(1 - |w|^2)^2)$ and $\mu_k = \langle e^{2\pi i/k} \rangle$ acts by multiplication. This quotient can again be identified to $D(0, \varepsilon)$, via the map $w \mapsto z = w^k$. On the quotient, the hyperbolic metric tensor becomes

$$\frac{|dz|^2}{4|z|^{2-2/k}(1 - |z|^{2/k})^2}.$$

Again, such a coordinate is unique up to a factor of modulus one, and the assignment

$$\|dz\|_{w,q} = 1$$

defines a hermitian metric on $\omega_{X,q}$.

The renormalized hyperbolic metrics defined above for cusps and elliptic fixed points were first introduced by Wolpert [33, Def. 1] (in the case of cusps). We adopt the terminology introduced in [11, 12], and we call them *Wolpert metrics*.

Let now $\pi : \mathcal{X} \to \operatorname{Spec} \mathcal{O}_K$ be an arithmetic surface and $\sigma_1, \ldots, \sigma_n$ be sections corresponding to elliptic fixed points of orders e_i or cusps, as before. For every section σ_i, the line bundle $\psi_i := \sigma_i^*(\omega_{\mathcal{X}/\mathcal{O}_K})$ can be endowed (after base change through $K \hookrightarrow \mathbb{C}$) with the corresponding Wolpert metric. We indicate this choice of hermitian metric by an index W. We define a $\mathbb{Q}$-hermitian line bundle

$$\psi_W = \sum_i (1 - e_i^{-2}) \psi_{i,W}.$$

The arithmetic degree of ψ_W is a measure of how far the transcendental canonical coordinates just discussed are from being formal algebraic. Actually, this can be used to construct heights on moduli spaces of curves with marked points (see [12]).

3.3 A Quillen Type Metric

Let $X = \Gamma \backslash \mathbb{H} \cup \{\text{cusps}\}$ be a compact Riemann surface, where Γ is a Fuchsian group of the first kind. We endow $\mathbb{H}$ with the hyperbolic metric, and we would like to define a Quillen type metric on the determinant of the cohomology of the trivial line bundle $\mathcal{O}_X$.

The L^2 metric poses no problem. Indeed, there is a well-defined L^2 metric on $H^0(X, \mathcal{O}_X) = \mathbb{C}$, given by taking $\|1\|^2$ as the volume, which is finite. We normalize the volume form so that

$$\|1\|^2 = 2g - 2 + \sum_i (1 - e_i^{-1}).$$

For $H^1(X, \mathcal{O}_X)$, we invoke the Serre duality isomorphism

$$H^1(X, \mathcal{O}_X) \simeq H^0(X, \omega_X)^\vee$$

and introduce the L^2 scalar product on the latter given by

$$\langle \alpha, \beta \rangle = \frac{i}{2\pi} \int_X \alpha \wedge \overline{\beta}.$$

Let Δ_{hyp} be the hyperbolic Laplacian, acting on $\mathcal{C}^\infty(\mathbb{H})^\Gamma$ as

$$\Delta_{\text{hyp}} = -y^2 \left(\frac{\partial}{\partial x^2} + \frac{\partial}{\partial y^2} \right).$$

The points of the spectrum of Δ_{hyp} can be classified into three types:

- *Cuspidal spectrum.* It consists of eigenvalues $\lambda > 0$ whose eigenvectors are L^2 functions (with respect to the hyperbolic volume form) with vanishing Fourier coefficients at cusps. The cuspidal spectrum constitutes discrete set. It is even conjectured to be finite for generic Fuchsian groups, by Phillips–Sarnak [26].
- *Continuous spectrum.* It arises from scattering theory. Let $\Gamma_0 \subset \Gamma$ be the stabilizer of a cusp. We define a corresponding Eisenstein series

$$E_0(\tau, s) = \sum_{\gamma \in \Gamma_0 \backslash \Gamma} \operatorname{Im}(\gamma \tau)^s,$$

which absolutely convergens for $\operatorname{Re}(s) > 1$. It is not L^2 with respect to the hyperbolic measure, but satisfies

$$\Delta_{\text{hyp}} E_0(\tau, s) = s(1 - s) E_0(\tau, s).$$

It can be shown that $E_0(\tau, s)$ has a meromorphic continuation in $s \in \mathbb{C}$. Its residues are contained in the real interval $(1/2, 1]$, and $s = 1$ is a simple pole with constant residue. The points on $\operatorname{Re}(s) = 1/2$ are regular. Finally, if we put all the Eisenstein series for all cusps in a vector $\mathcal{E}(\tau, s)$, then there is a functional equation

$$\mathcal{E}(\tau, s) = \Phi(s)\mathcal{E}(\tau, 1 - s),$$

where $\Phi(s)$ is a square matrix with meromorphic function entries, satisfying $\Phi(s)\Phi(1 - s) = \operatorname{id}$ and such that $\Phi(s)$ is unitary for $\operatorname{Re}(s) = 1/2$. The matrix $\Phi(s)$ is called the scattering matrix, and can be computed from the Fourier expansions of the Eisenstein series at cusps. More precisely, if $E_1, \ldots, E_r$ are all the Eisenstein series, then the Fourier expansion of E_i at the j-th cusp has the form

$$y^s + \varphi_{ij}(s) y^{1-s} + \rho_{ij}(\tau, s),$$

where ρ_{ij} is L^2 for the hyperbolic measure. Then $\Phi(s) = (\varphi_{ij}(s))$. This picture is analogous to the phenomena of incoming and outcoming waves in quantum mechanics. Finally, the continuous spectrum is formed by $1/4 + t^2$, where $t \in \mathbb{R}$.

- *residual spectrum.* It arises from residues of Eisenstein series. If $E_0(\tau, s)$ is an Eisenstein series with a pole at $s_0 \in (1/2, 1]$, then

$$u(\tau) = \operatorname{res}_{s=s_0} E_0(\tau, s)$$

is an L^2, non-cuspidal, eigenfunction of Δ_{hyp}, with eigenvalue $s_0(1 - s_0)$. These constitute a finite set of eigenvalues.

The cuspidal spectrum and the residual spectrum together form the discrete spectrum of Δ_{hyp}. The spectral theorem [20, Thm. 7.3] asserts that for any L^2 function f on $\Gamma\backslash\mathbb{H}$, there is an expansion (valid in L^2)

$$f(\tau) = \sum_i \langle f, \varphi_j \rangle_{L^2} \varphi_j(\tau)$$

$$+ \sum_j \frac{1}{4\pi} \int_{-\infty}^{\infty} \langle f, E_j(\tau, \frac{1}{2} + it) \rangle_{L^2} E_j(\tau, \frac{1}{2} + it) dt,$$

where φ_j are orthonormal L^2 eigenfunctions for the discrete spectrum, and the $E_j(\tau, s)$ constitute the finite set of Eisenstein series.

We introduce a spectral zeta function that takes into account both the discrete spectrum $\{\lambda_n\}_n$ and the continuous spectrum. Let $\varphi(s) = \det \Phi(s)$ be the determinant of the scattering matrix.

Definition 3.1 We define the *hyperbolic spectral zeta function* by

$$\zeta_{\mathrm{hyp}}(\Gamma, s) = \sum_{\lambda_n > 0} \frac{1}{\lambda_n^s}$$

$$- \frac{1}{4\pi} \int_{-\infty}^{\infty} \frac{\varphi'(1/2 + it)}{\varphi(1/2 + it)} (1/4 + t^2)^{-s} dt + 4^{s-1} \mathrm{tr}(\Phi(\frac{1}{2})).$$

This expression is inspired by the spectral side of the Selberg trace formula [20, Chap. 10] applied to a suitable test function. Actually, by means of the Selberg trace formula, one can show that $\zeta_{\mathrm{hyp}}(s)$ is holomorphic for $\mathrm{Re}(s) > 1$ and has a meromorphic continuation to $\mathbb{C}$. Moreover it is holomorphic at $s = 0$. See [14, Sec. 8] for the proof. We then define the *zeta regularized determinant of the hyperbolic Laplacian*

$$\det \Delta_{\mathrm{hyp}, \Gamma} = \exp\left(-\frac{d}{ds}\Big|_{s=0} \zeta_{\mathrm{hyp}}(\Gamma, s) \right).$$

Finally, we define a Quillen type metric by mimicking the usual definition:

$$h_{Q, \mathrm{hyp}} = (\det \Delta_{\mathrm{hyp}, \Gamma})^{-1} h_{L^2}.$$

Remark 3.2 For a torsion free and co-compact Fucshian group, the Quillen metric thus defined agrees with the Quillen metric of Sect. 2.3 up to an *explicit* constant, depending only on the topological type of Γ. This is because the Dolbeault Laplacian on functions differs by a constant from the scalar hyperbolic Laplacian. For the sake of a cleaner presentation, we prefer to skip this normalization issue.

3.4   An Arithmetic Riemann–Roch Formula

Let $\pi : \mathfrak{X} \to \operatorname{Spec} \mathcal{O}_K$ be an arithmetic surface, with geometrically connected fibers. We suppose given sections $\sigma_1, \ldots, \sigma_n$, such that

$$\mathfrak{X}(\mathbb{C}) = \bigsqcup_{\tau : K \hookrightarrow \mathbb{C}} (\Gamma_\tau \backslash \mathbb{H} \cup \{\sigma_1(\tau), \ldots, \sigma_r(\tau)\}),$$

for some Fuchsian groups of the first kind Γ_τ having $\sigma_1(\tau), \ldots, \sigma_r(\tau)$ as cusps and $\sigma_{r+1}(\tau), \ldots, \sigma_n(\tau)$ as elliptic fixed points of orders e_i. For the cusps we put $e_i = \infty$. We endow $\mathfrak{X}(\mathbb{C})$ with the singular hyperbolic metric.

Theorem 3.3 (Freixas–von Pippich [14], Thm. 10.1) *With the notations and assumptions as above, there is an equality of real numbers*

$$12 \,\widehat{\deg} \det H^\bullet(\mathfrak{X}, \mathcal{O}_\mathfrak{X})_Q - \delta_\pi + \widehat{\deg} \, \psi_W = (\omega_{\mathfrak{X}/\mathcal{O}_K} (\sum_i (1 - e_i^{-1}) \sigma_i)_{\mathrm{hyp}}^2)$$

$$- [K : \mathbb{Q}] C(g, \{e_i\}),$$

for some explicit constant $C(g, \{e_i\})$ depending only on the genus g and the orders e_i (i.e. the type of the groups Γ_τ).

The proof of the theorem is long and technical, and combines basic facts of arithmetic intersection theory, glueing properties of determinants of Laplacians, and explicit computations of determinants of Laplacians on "model" hyperbolic cusps and cones. The latter are inspired by results drawn from theoretical physics (the theory of branes). The explicit value of the constant is actually relevant in some applications, but in these lectures we prefer to focus on the rest of the terms of the arithmetic Riemann–Roch formula.

Remark 3.4 In later computations, we will appeal to a weak version of the theorem, where instead of working with an arithmetic surface over $\mathcal{O}_K$, we directly work with a smooth projective curve over K. The consequence for the numerical invariants will be that they are then only well defined modulo the $\mathbb{Q}$ vector space spanned by the real numbers $\log p$, for p prime. Indeed, one may choose an auxiliary regular model over $\mathcal{O}_K$ and apply the theorem. The numerical invariants for two different models differ by an element in $\mathbb{Q} \otimes_{\mathbb{Z}} \log |\mathbb{Q}^\times| \hookrightarrow \mathbb{R}$.

4   Modular and Shimura Curves

4.1   Modular Curves

Modular curves are moduli spaces of elliptic curves, possibly with some extra structure. The point of departure is the mapping

$$\mathbb{H} \longrightarrow \{\text{complex tori}\}$$
$$\tau \longmapsto \mathbb{C}/(\mathbb{Z} + \tau\mathbb{Z}).$$

It is known that every elliptic curve over $\mathbb{C}$ is isomorphic to a torus as above. The isomorphism relation on such corresponds to the action of $\mathsf{PSL}_2(\mathbb{Z})$ on $\mathbb{H}$. The quotient $\mathbb{H}/\mathsf{PSL}_2(\mathbb{Z})$ is the open modular curve, whose points are thus in bijection with isomorphism classes of elliptic curves over $\mathbb{C}$. The j-invariant of elliptic curves defines a biholomorphic map

$$\mathbb{H}/\mathsf{PSL}_2(\mathbb{Z}) \xrightarrow{\sim} \mathbb{C}.$$

The cusp compactification of $\mathbb{H}/\mathsf{PSL}_2(\mathbb{Z})$ corresponds to $\mathbb{C} \cup \{\infty\}$, hence to $\mathbb{P}^1_{\mathbb{C}}$. A holomorphic neighborhood of the cusp has holomorphic coordinate $q = e^{2\pi i \tau}$, and it is best understood as parametrizing elliptic curves uniformized in the form

$$\mathbb{C}^{\times}/q^{\mathbb{Z}}.$$

It is thus reasonable to declare that the cusp $q = 0$ corresponds to the torus $\mathbb{C}^{\times}$. Or equivalently, to a so-called *generalized elliptic curve*: the singular nodal genus one curve

$$\mathbb{P}^1_{\mathbb{C}}/\{0 \sim \infty\},$$

together with its multiplicative algebraic group structure when deprived from the cannot singular point $0 \sim \infty$. This complex geometric picture can be extended to $\mathrm{Spec}\,\mathbb{Z}$, and gives rise first to a Deligne-Mumford stack $\overline{\mathcal{M}}_1$, then to the *coarse moduli scheme* of generalized elliptic curves $\mathbb{P}^1_{\mathbb{Z}} \to \mathrm{Spec}\,\mathbb{Z}$, with the cusp at ∞ as a section. One can more generally introduce additional *level structures* to elliptic curves, and build corresponding integral moduli stacks or schemes. All this is formalized in the work of Deligne–Rapoport [9] and Katz–Mazur [22], which provide a reference for the discussion below.

In these lectures we will be mostly interested in the moduli of elliptic curves with a torsion point of exact order N, and slight variants. Fix $N \geq 1$ an integer. Over a general base scheme S, we consider elliptic curves $E \to S$ (i.e. smooth proper schemes over S, with geometrically connected fibers of dimension 1 and a relative group scheme structure) together with a section $P \colon S \to E$, generating a finite flat subgroup scheme of order N. The coarse moduli of elliptic curves with a point of order N "classifies" such couples $(E/S, P)$ up to isomorphism. One can prove that it defines a proper flat normal scheme $Y_1(N)$ over $\operatorname{Spec}\mathbb{Z}$, with geometrically connected fibers of pure dimension 1. It is smooth over $\mathbb{Z}[1/N]$. Moreover, forgetting the point of order N defines a finite flat morphism

$$j\colon Y_1(N) \longrightarrow \mathbb{A}^1_{\mathbb{Z}} \subset \mathbb{P}^1_{\mathbb{Z}}.$$

A compactification of $Y_1(N)$ is obtained by taking the integral closure of $\mathbb{P}^1_{\mathbb{Z}}$ with respect to the morphism j. The compactification actually affords a moduli interpretation as a coarse moduli scheme of generalized elliptic curves. We will not need this description. For our purposes, it will be enough to know that $X_1(N) \backslash Y_1(N)$ is a relative Cartier divisor over $\mathbb{Z}$, that becomes rational (i.e. given by sections) over $\operatorname{Spec}[\zeta_N]$. Finally, if $N \geq 5$, then $X_1(N)$ is actually a fine moduli scheme. In applications, we will stick to the restriction $N \geq 5$, and we will actually consider $X_1(N)$ and its variants as defined over $\mathbb{Q}$.

The Riemann surface $Y_1(N)(\mathbb{C})$ can be uniformized as

$$Y_1(N)(\mathbb{C}) = \mathbb{H}/\Gamma_1(N),$$

where

$$\Gamma_1(N) = \left\{ \begin{pmatrix} a & b \\ c & d \end{pmatrix} \in \mathsf{SL}_2(\mathbb{Z}) \mid a - 1 \equiv c \equiv 0 \mod N \right\},$$

or rather its image in $\mathsf{PSL}_2(\mathbb{Z})$. When $N \geq 5$, $\Gamma_1(N)$ has no torsion and is actually identified with a Fuchsian group.

For the purpose of defining Hecke operators later on, and in some intermediary steps, we need a variant of the geometric objects above. Instead of moduli of elliptic curves with a point of exact order N, we add to the data a cyclic subgroup of order M prime to N. Namely, we classify triples $(E/S, P, C)$ where P is an S-point of order N of E and $C \subset E[M]$ is a cyclic finite flat subgroup of order M. The outcome is a proper, normal and flat scheme $X_1(N, M)$ over $\operatorname{Spec}\mathbb{Z}$, with geometrically connected fibers and smooth over $\operatorname{Spec}\mathbb{Z}[1/NM]$. Again, in applications we will actually consider it to be defined over $\mathbb{Q}$. Over the complex numbers, it can be presented as

$$(\mathbb{H} \cup \mathbb{P}^1(\mathbb{Q}))/\Gamma_1(N) \cap \Gamma_0(M),$$

where now

$$\Gamma_0(M) = \left\{ \begin{pmatrix} a & b \\ c & d \end{pmatrix} \in \mathsf{SL}_2(\mathbb{Z}) \mid c \equiv 0 \mod M \right\}.$$

We will write $\Gamma_1(N, M) = \Gamma_1(N) \cap \Gamma_0(M)$.

4.2 Modular Forms

Classically, a modular form of weight k for $\Gamma_1(N)$ (or more generally $\Gamma_1(N, M)$) is a holomorphic map

$$f: \mathbb{H} \longrightarrow \mathbb{C},$$

such that

$$f\left(\frac{a\tau + b}{c\tau + d}\right) = (c\tau + d)^k f(\tau) \quad \text{for} \begin{pmatrix} a & b \\ c & d \end{pmatrix} \in \Gamma_1(N),$$

that extends holomorphically to the cusps. For the cusp at ∞, this condition is stated as follows. Since $f(\tau + 1) = f(\tau)$ by the equivariance property above, f has a Fourier series expansion in $q = e^{2\pi i \tau}$

$$f(\tau) = \sum_{n \in \mathbb{Z}} a_n q^n.$$

We require that $a_n = 0$ for $n < 0$. We moreover say that f is a cusp form if $a_0 = 0$. We will only need weight 2 cusp forms, so that in the following we restrict to modular forms of even weight. These can be best understood as differential forms on $X_1(N)(\mathbb{C})$. First observe that the equivariance property of a modular form of weight $2k$, say f, is equivalent to the invariance of the tensor $f(\tau)d\tau^{\otimes k}$. In the coordinate q, this tensor becomes

$$\sum_{n \geq 0} a_n q^n \left(\frac{1}{2\pi i}\frac{dq}{q}\right)^{\otimes k}.$$

If we denote by **cusps** the divisor consituted by the cusps, we thus see that modular forms are global holomorphic sections of the sheaf

$$(\omega_{X_1(N)/\mathbb{C}}(\textbf{cusps}))^{\otimes k},$$

and cusp forms are global holomorphic sections of the sheaf

$$\omega_{X_1(N)/\mathbb{C}}^{\otimes k}((k-1)\,\textbf{cusps}).$$

Typical notations are

$$M_{2k}(\Gamma_1(N)) = H^0(X_1(N)(\mathbb{C}), (\omega_{X_1(N)/\mathbb{C}}(\textbf{cusps}))^{\otimes k})$$

for the space of modular forms of weight $2k$ and

$$S_{2k}(\Gamma_1(N)) = H^0(X_1(N)(\mathbb{C}), \omega^{\otimes k}_{X_1(N)/\mathbb{C}}((k-1) \mathbf{cusps}))$$

for the subspace of cusp forms.

The spaces of modular and cusp forms have rational and integral structures provided by the rational and integral models of $X_1(N)$. In particular, in weight 2, the space of cusp forms has a rational structure

$$S_{2k}(\Gamma_1(N), \mathbb{Q}) := H^0(X_1(N)_{\mathbb{Q}}, \omega_{X_1(N)/\mathbb{Q}}).$$

This is exactly the $\mathbb{Q}$-vector space of cusp forms with rational Fourier coefficients (at all cusps). The same would be true for any subfield of $\mathbb{C}$, and this is known as the *q-expansion principle*. Due to its relation to the arithmetic Riemann–Roch theorem for modular curves, we will focus on this space from now on.

The space $S_2(\Gamma_1(N), \mathbb{Q})$ has an action of the algebra of Hecke correspondences and diamond operators. This is an algebra of endomorphisms of the Jacobian $J_1(N)/\mathbb{Q}$ of $X_1(N, M)$, constructed as follows. First, for the Hecke operators, let M be an integer prime to N. We introduce the auxiliary curve $X_1(N, M)$. There are two natural morphisms from $X_1(N, M)$ to $X_1(N)$. The first one, that we call α, is just forgetting the cyclic subgroup of order M. The second one is given by

$$\beta: (E/S, P, C) \mapsto (E/C/S, P \mod C),$$

where E/C is the well-defined quotient of the (possibly generalized) elliptic curve E by the finite flat subgroup C, and $P \mod C$ is the induced S-point of order N. In the language of correspondences, the M-th Hecke operator T_M is $\alpha_* \circ \beta^*$. For the diamond operators, let $d \in (\mathbb{Z}/N\mathbb{Z})^\times$. There is an induced automorphism

$$\langle d \rangle: X_1(N) \longrightarrow X_1(N)$$

$$(E/S, P) \longmapsto (E/S, dP).$$

The Hecke operators together with the diamond operators span a free $\mathbb{Z}$-subalgebra of finite type of $\mathrm{End}_{\mathbb{Q}}(J_1(N))$, formed by pairwise commuting endomorphisms. It is called the Hecke algebra, and denoted $\mathbb{T}_N$. By functoriality, the Hecke algebra acts on

$$H^0(J_1(N), \Omega^1_{J_1(N)/\mathbb{Q}}) \simeq H^0(X_1(N), \omega_{X_1(N)/\mathbb{Q}}).$$

An analogous construction carries over to modular curves $X_1(N, M)$. An important property that we cannot skip is that after extension scalars to $\overline{\mathbb{Q}}$, spaces of cusp forms have bases of simultaneous eigenfunctions for the Hecke algebra.

To conclude this section, we discuss the notion of d-new forms. From a modular curve $X_1(N, d)$, there are several morphisms down to $X_1(N, d')$, for $d' \mid d$, through which cusp forms can be pulled-back. The resulting cusp forms are called *d-old*. Let us fix d' such a divisor. For any divisor of d/d', say m, we have a morphism

$$X_1(N, d) \longrightarrow X_1(N, d/m)$$

$$(E/S, C) \longmapsto (E/C[m]/S, C/C[m]),$$

where as above $C[m]$ denotes the m-torsion part of C. Then, since d' divides d/m, we have a forgetful map

$$X_1(N, d/m) \longrightarrow X_1(N, d')$$

$$(E/S, C) \longmapsto (E/S, C[d']).$$

By composing these two arrows, we obtain a so-called *degeneracy* morphism depending both on d' and m, and denoted $\gamma_{d',m}$. In terms of the degeneracy maps, the d-old subspace of $S_2(\Gamma_1(N, d), \mathbb{Q})$ is

$$S_2(\Gamma_1(N, d), \mathbb{Q})^{d-old} := \sum_{d'|d} \sum_{m|(d/d')} \gamma_{d',m}^* S_2(\Gamma_1(N, d'), \mathbb{Q}).$$

Actually, the sum can be seen to be direct. The d-new quotient of $S_2(\Gamma_1(N, d), \mathbb{Q})$ is defined to be

$$S_2(\Gamma_1(N, d), \mathbb{Q})^{d-new} := S_2(\Gamma_1(N, d), \mathbb{Q}) / S_2(\Gamma_1(N, d), \mathbb{Q})^{d-old}.$$

Actually, the $\mathbb{Q}$-vector space of d-new forms can be realized inside $S_2(\Gamma_1(N, d), \mathbb{Q})$ in such a way that, with respect to the natural hermitian structure on $H^0(X_1(N, d), \omega_{X_1(N,d)/\mathbb{Q}})$, we have

$$S_2(\Gamma_1(N, d), \mathbb{Q}) = S_2(\Gamma_1(N, d), \mathbb{Q})^{d-new} \overset{\perp}{\oplus} S_2(\Gamma_1(N, d), \mathbb{Q})^{d-old}.$$

4.3 Shimura Curves and Quaternionic Modular Curves

We consider "compact" counterparts of modular curves, arising from arithmetic quaternionic groups. Moduli theoretically, they classify abelian surfaces with a faithful action of a maximal order in an indefinite quaternion algebra over $\mathbb{Q}$. They share many features with modular curves (integral models, Hecke operators, etc). They have the advantage of being automatically "compact", and at the same time the disadvantage of not having q-expansions, precisely due to the lack of cusps. We refer to Buzzard [8] for a complete treatment parallel to Deligne–Rapoport. A suitable reference on quaternion algebras is Vignéras' book [31].

Let B be an indefinite quaternion division algebra over $\mathbb{Q}$, so that $B \otimes_{\mathbb{Q}} \mathbb{R}$ is isomorphic to the matrix algebra $M_2(\mathbb{R})$. For every finite prime p, the quaternion algebra $B \otimes_{\mathbb{Q}} \mathbb{Q}_p$ is either a matrix algebra or a division algebra. In the latter case, we say that B is ramified at p. The set of primes where B is ramified is finite and of even cardinality. The discriminant of B is the product of all such primes, and is denoted $\mathrm{disc}(B)$. Let $N \geq 1$ be an integer prime to $\mathrm{disc}(B)$. We can choose an order $\mathcal{O}$ in B (hence both a subring and lattice in B), with the following properties:

- for every $p \nmid \mathrm{disc}(B)N$,

$$\mathcal{O} \otimes \mathbb{Z}_p \xrightarrow{\sim} M_2(\mathbb{Z}_p).$$

- for every $p \mid N$,

$$\mathcal{O} \otimes \mathbb{Z}_p \xrightarrow{\sim} \left\{ \begin{pmatrix} a & b \\ c & d \end{pmatrix} \in M_2(\mathbb{Z}_p) \mid a - 1 \equiv c \equiv 0 \mod N \right\}.$$

- for every prime $p \mid \mathrm{disc}(B)$, $\mathcal{O} \otimes \mathbb{Z}_p$ is a maximal order in $B \otimes \mathbb{Q}_p$.

The analogue of the group $\Gamma_1(N)$ in this setting will be $\Gamma_1^B(N) := \mathcal{O}^{\times,1}$, namely the subgroup of the units in $\mathcal{O}$ of reduced norm 1. Actually the notation $\Gamma_1^B(N)$ is ambiguous in that it does not render the choice of $\mathcal{O}$ explicit. The group $\Gamma_1^B(N)$ can be realized into $\mathsf{SL}_2(\mathbb{R})$: take its image under a fixed algebra isomorphism $B \otimes \mathbb{R} \simeq M_2(\mathbb{R})$. As a group of fractional linear transformations of $\mathbb{H}$, it is cocompact. Moreover, if $N \geq 5$ it is torsion free as well. The quotient

$$X_1^B(N) := \mathbb{H}/\Gamma_1^B(N)$$

is thus a compact Riemann surface, and is called a Shimura curve. Its points are in bijective correspondence with complex abelian surfaces A with a faithful action of a maximal order in B, together with a level structure of type $\Gamma_1(N)$. We will not make these structures more explicit. It will suffice to say that the Shimura curve $X_1^B(N)$ has a smooth projective model over $\mathrm{Spec}\,\mathbb{Z}[1/N\,\mathrm{disc}(B)]$, and in particular over $\mathbb{Q}$. For $N \geq 5$, we have again a fine moduli space.

The spaces of quaternionic modular forms are defined in analogy to the classical modular forms. We are particularly interested in weight 2 quaternionic forms and their rational structures:

$$S_2(\Gamma_1^B(N), \mathbb{Q}) = H^0(X_1^B(N), \omega_{X_1^B(N)/\mathbb{Q}}).$$

In contrast with modular curves, these rational structures cannot be read in Fourier expansions, since there are no cusps at our disposal. Finally, an algebra of Hecke operators can as well be defined in the quaternionic setting, in a similar way as for modular curves. We will not give further details.

5 The Jacquet–Langlands Correspondence and the Arithmetic Riemann–Roch Theorem

The cohomological side of the arithmetic Riemann–Roch theorem for modular (or Shimura) curves can be interpreted in terms of automorphic forms. As we saw in the previous chapter, the global sections of the canonical sheaf on a compactified modular curve correspond to cusp forms of weight 2. Similarly, the regularized determinant of the hyperbolic Laplacian is the contribution of the non-holomorphic modular forms, commonly known as Maass forms. Both holomorphic and non-holomorphic modular forms can be seen as vectors in spaces of automorphic representations. One can thus expect that general principles in the theory of (global) automorphic representations, combined with the arithmetic Riemann–Roch theorem, can be useful for a better understanding of some arithmetic intersection numbers. In this chapter we explain the relationship between the arithmetic Riemann–Roch theorem and the Jacquet–Langlands correspondence relating automorphic representations of $\mathsf{GL}_{2/\mathbb{Q}}$ to those of $B^\times$, for a division quaternion algebra B over $\mathbb{Q}$. We will however not enter into the details of automorphic representations in order to keep the size of this course reasonable, and only state the consequences we need in the classical language of modular forms. For an introduction to the theory of automorphic forms for GL_2, and the Jacquet–Langlands correspondence, the reader can consult Gelbart [16].

5.1 On the Jacquet–Langlands Correspondence for Weight 2 Forms

We fix an integer $N \geq 1$, and B an indefinite division quaternion algebra over $\mathbb{Q}$, whose discriminant $d = \mathrm{disc}(B)$ is prime to N. We deal with rational weight 2 cusp forms for $\Gamma_1(N, d)$ and rational quaternionic modular forms for $\Gamma_1^B(N)$. Recall that the notation for the latter hides a choice of order in B. Recall as well that the spaces of such classical or quaternionic modular forms

$$S_2(\Gamma_1(N, d), \mathbb{Q}) = H^0(X_1(N, d), \omega_{X_1(N,d)/\mathbb{Q}}),$$

$$S_2(\Gamma_1^B(N), \mathbb{Q}) = H^0(X_1^B(N), \omega_{X_1^B(N)/\mathbb{Q}}),$$

come equipped with the action of the respective Hecke algebras. The following statement follows from the Jacquet–Langlands correspondence [21], together with work of Ribet on Hecke algebras [29] and Faltings' isogeny theorem [10]. For a detailed proof we refer to Helm [19, Cor. 2.4].

Theorem 5.1 *There is a $\mathbb{Q}$-linear and Hecke equivariant isomorphism*

$$H^0(X_1(N,d), \omega_{X_1(N,d)/\mathbb{Q}})^{d-new} \xrightarrow{\sim} H^0(X_1^B(N), \omega_{X_1^B(N)/\mathbb{Q}}),$$

compatible with the natural hermitian structures up to $\mathbb{Q}^\times$.

An immediate corollary of the theorem is the corresponding relationship between arithmetic degrees of spaces of modular forms, endowed with the L^2 norm:

Corollary 5.2 *We have an identity in $\mathbb{R}/\log|\mathbb{Q}^\times|$*

$$\widehat{\deg}\, H^0(X_1(N,d), \omega_{X_1(N,d)/\mathbb{Q}})^{d-new}_{L^2} = \widehat{\deg}\, H^0(X_1^B(N), \omega_{X_1^B(N)/\mathbb{Q}})_{L^2}.$$

Some comments are in order. First of all, by the very definition of the spaces of newforms, the equality of the statement is equivalent to

$$\widehat{\deg}\, H^0(X_1^B(N), \omega_{X_1^B(N)/\mathbb{Q}})_{L^2}$$
$$= \sum_{d'|d} \mu(d')\sigma(d/d')\,\widehat{\deg}\, H^0(X_1(N,d'), \omega_{X_1(N,d')/\mathbb{Q}})_{L^2},$$

where μ is the Möbius function and σ is the divisor counting function. Second, if one works over $\operatorname{Spec}\mathbb{Z}[1/Nd]$, thanks to the relation of the quantities above to Faltings heights of Jacobians, together with work by Prasanna [27], one can refine the identity to an equality of real numbers modulo $\log p's$, for p in a controlled finite set of primes with a precise arithmetic meaning (apart from primes $p \mid Nd$, one has to take into account the so-called Eisenstein primes). This integral refinement is carefully dealt with in [13].

Finally, it is immediate from the corollary and the volume computations for modular and Shimura curves that:

Corollary 5.3

$$12\,\widehat{\deg}\det H^\bullet(X_1^B(N), \mathcal{O}_{X_1^B(N)})_{L^2}$$
$$= 12\sum_{d'|d} \mu(d')\sigma(d/d')\,\widehat{\deg}\det H^\bullet(X_1(N,d'), \mathcal{O}_{X_1(N,d')})_{L^2}$$

in $\mathbb{R}/\log|\mathbb{Q}^\times|$.

5.2 The Jacquet–Langlands Correspondence for Maass Forms

We are now concerned with eigenspaces of the hyperbolic Laplacian acting on functions of modular and Shimura curves. Let $\lambda > 0$ be a positive real number, and define

$$V_\lambda(\Gamma_1(N, d)) = \text{cuspidal eigenspace of } \Delta_{\text{hyp}} \text{ of eigenvalue } \lambda.$$

Hence, this is a finite dimensional complex vector space spanned by non-holomorphic cusp forms, proper under the action of the hyperbolic Laplacian, with eigenvalue λ. In an analogous way to classical modular forms, one can define an action of the Hecke algebra on $V_\lambda(\Gamma_1(N, d))$. The notion of d-new and d-old forms makes sense as well. The space $V_\lambda(\Gamma_1(N, d))$ is contained in the L^2 functional space with respect to the hyperbolic measure. There is an orthogonal decomposition

$$V_\lambda(\Gamma_1(N, d)) = V_\lambda(\Gamma_1(N, d))^{d-new} \overset{\perp}{\oplus} V_\lambda(\Gamma_1(N, d))^{d-old}.$$

Similarly we define spaces $V_\lambda(\Gamma_1^B(N))$ for the quaternionic modular group $\Gamma_1^B(N)$, with no need of any cuspidality condition.

The theory of automorphic representations works as well to decompose the spaces V_λ in irreducible modules for the action of the Hecke algebra. The Jacquet–Langlands correspondence applies too: it actually makes no distinction between automorphic representations arising from holomorphic modular forms or non-holomorphic ones.

Theorem 5.4 (Jacquet–Langlands [21]) *There is a Hecke equivariant isomorphism of finite dimensional complex vector spaces*

$$V_\lambda(\Gamma_1(N, d))^{d-new} \overset{\sim}{\longrightarrow} V_\lambda(\Gamma_1^B(N)).$$

In particular, we have the relation

$$\dim V_\lambda(\Gamma_1^B(N)) = \sum_{d'|d} \mu(d')\sigma(d/d') \dim V_\lambda(\Gamma_1(N, d')).$$

To relate the hyperbolic spectral zeta functions for $\Gamma_1^B(N)$ and the several $\Gamma_1(N, d')$ (Definition 3.1), we only need the multiplicities of eigenspaces provided by the theorem. The compatibility with the action of Hecke algebras does not actually play any role. There are two additional subtleties we need to address, related to the non-compactness of the modular curves. The spectral zeta function in the modular curve case involves the residual spectrum and the scattering matrices of the vector of Eisenstein series (i.e. the continuous spectrum), while for the Shimura curve there are no such contributions. It turns out that the residual spectrum for congruence subgroups is actually trivial: the only poles of the Eisenstein series happen at $s = 1$. And the spectral zeta function requires only strictly positive eigenvalues! For the scattering matrices, we have the following fact:

Lemma 5.5 *Denote by $\Phi(\Gamma, s)$ the scattering matrix for a Fuchsian group Γ, and $\varphi(\Gamma, s)$ its determinant. Then we have*

$$\sum_{d'|d} \mu(d')\sigma(d/d')\,\mathrm{tr}(\Phi(\Gamma_1(N, d'), s) = 0,$$

$$\prod_{d'|d} \varphi(\Gamma_1(N, d'), s)^{\mu(d')\sigma(d/d')} = 1.$$

The lemma follows from explicit computations of scattering matrices, or simply by relating the cusps (and their stabilizers) for a group $\Gamma_1(X, d')$ with those coming from smaller levels through the degeneracy mappings. We refer to Hejhal [18] for such computations. By taking logarithmic derivatives on the second relation, we find

$$\sum_{d'|d} \mu(d')\sigma(d/d') \int_{-\infty}^{\infty} \frac{\varphi'(\Gamma_1(N, d'), 1/2 + it)}{\varphi(\Gamma_1(N, d'), 1/2 + it)} (1/4 + t^2)^{-s}\,dt = 0.$$

Taking into account the very definition of the spectral zeta function, the theorem and the lemma readily imply:

Theorem 5.6 *Let $\zeta_{\mathrm{hyp}}(\Gamma, s)$ be the hyperbolic spectral zeta function for a Fuchsian subgroup Γ. Then*

$$\zeta_{\mathrm{hyp}}(\Gamma_1^B(N), s) = \sum_{d'|d} \mu(d')\sigma(d/d')\zeta_{\mathrm{hyp}}(\Gamma_1(N, d'), s).$$

In particular, for the regularized zeta determinants

$$\det \Delta_{\mathrm{hyp}, \Gamma_1^B(N)} = \prod_{d'|d} (\det \Delta_{\mathrm{hyp}, \Gamma_1(N, d')})^{\mu(d')\sigma(d/d')}.$$

Together with Corollary 5.3 we obtain:

Corollary 5.7 *There is an equality in $\mathbb{R}/\log |\mathbb{Q}^\times|$*

$$12\,\widehat{\deg} \det H^\bullet(X_1^B(N), \mathcal{O}_{X_1^B(N)})_Q$$

$$= 12 \sum_{d'|d} \mu(d')\sigma(d/d')\,\widehat{\deg} \det H^\bullet(X_1(N, d'), \mathcal{O}_{X_1(N, d')})_Q.$$

5.3 Relating Arithmetic Intersection Numbers

The discussion of the previous paragraphs easily leads to the following conclusion.

Theorem 5.8 (Freixas i Montplet [13]) *Let $N \geq 5$ and B an indefinite division quaternion algebra of discriminant d coprime to B. There is a relation between arithmetic intersection numbers*

$$\frac{(\omega_{X_1(N)/\mathbb{Q}}(\textbf{\textit{cusps}})^2_{\mathrm{hyp}})}{\deg \omega_{X_1(N)/\mathbb{Q}}(\textbf{\textit{cusps}})} = \frac{(\omega^2_{X_1^B(N)/\mathbb{Q},\mathrm{hyp}})}{\deg \omega_{X_1^B(N)/\mathbb{Q}}} \quad in \; \mathbb{R}/\mathbb{Q} \log |\mathbb{Q}^\times|,$$

where **cusps** *is the reduced boundary divisor with support $X_1(N) \backslash Y_1(N)$.*

In the statement we wrote $\mathbb{Q} \log |\mathbb{Q}^\times|$ for the $\mathbb{Q}$ vector space spanned by $\log |\mathbb{Q}^\times|$, namely $\mathbb{Q} \otimes_{\mathbb{Z}} \log |\mathbb{Q}^\times|$. As for arithmetic degrees of spaces of cusp forms, the theorem can be refined to an equality of real numbers up to some $\log p$'s, for p running over a controlled finite set of primes. The assumption $N \geq 5$ is made to avoid the presence of elliptic fixed points and simplify the discussion.

Let us say some words about the proof. First of all, by the functoriality properties of arithmetic intersection numbers, the quotients appearing in the statement of the theorem are independent of the level. In particular, we have

$$\frac{(\omega_{X_1(N)/\mathbb{Q}}(\textbf{cusps})^2_{\mathrm{hyp}})}{\deg \omega_{X_1(N)/\mathbb{Q}}(\textbf{cusps})} = \frac{(\omega_{X_1(N,d')/\mathbb{Q}}(\textbf{cusps})^2_{\mathrm{hyp}})}{\deg \omega_{X_1(N,d')/\mathbb{Q}}(\textbf{cusps})}$$

for any $d' \mid d$. These arithmetic intersection numbers appear in the arithmetic Riemann-Roch theorem for the modular curves $X_1(N, d')$, up to a small detail: the cusp divisor **cusps** becomes rational only after base changing to $\mathbb{Q}(\zeta_N)$. Nevertheless, again by functoriality properties of arithmetic intersection numbers, the quotients above are also invariant under extension of the base field. Therefore, we can work over $\mathbb{Q}(\zeta_N)$ instead of $\mathbb{Q}$ and assume that **cusps** is formed by rational points. Then, after Corollary 5.7 and the arithmetic Riemann–Roch theorem for modular curves, we are reduced to showing that for every cusp σ (written as a section for coherence of notations) of $X_1(N, d')$ we have

$$\widehat{\deg}\, \sigma^*(\omega_{X_1^B(N,d')/\mathbb{Q}(\zeta_N)})_W = 0 \quad \text{in } \mathbb{R}/\mathbb{Q} \log |\mathbb{Q}^\times|.$$

To simplify the exposition, we will proceed for the cusp at ∞. Observe we are allowed to increase the level N, by the previous remarks. It is then known that for N big enough, the canonical sheaf of $X_1(N, d')$ is ample. We can then find a global section s of $\omega^{\otimes k}_{X_1^B(N,d')/\mathbb{Q}(\zeta_N)}$ that does not vanish at the cusp ∞. Fix an embedding $\mathbb{Q}(\zeta_N) \subset \mathbb{C}$. The Fourier expansion of s at the cusp ∞ then looks like

$$s = (\sum_{n \geq 0} a_n q^n) dq^{\otimes k},$$

with $a_n \in \mathbb{Q}(\zeta_N)$ and $a_0 \neq 0$. This is a consequence of the q-expansion principle, if we think of s as a modular form. Moreover, for any automorphism τ of $\mathbb{Q}(\zeta_N)$, the conjugate section s^τ has q-expansion

$$s^\tau = (\sum_{n \geq 0} \tau(a_n) q^n) dq^{\otimes k}.$$

By construction, the Wolpert norm of $\sigma^*(dq)$ is 1. We thus see that

$$\widehat{\deg}\, \sigma^*(\omega_{X_1^B(N,d')/\mathbb{Q}(\zeta_N)})_W^{\otimes k} = -\log|N_{\mathbb{Q}(\zeta_N)/\mathbb{Q}}(a_0)| = 0 \quad \text{in } \mathbb{R}/\log|\mathbb{Q}^\times|,$$

and hence the claim. For the proof of the theorem to be complete, one actually needs the precise value of the topological constant $C(g, \{e_i\})$ in the arithmetic Riemann–Roch theorem, and check the needed relations between the constants for the groups $\Gamma_1(N, d')$ and $\Gamma_1^B(N)$. We don't provide the details since this is deprived of any conceptual interest.

Acknowledgements I am indebted to Huayi Chen, Emmanuel Peyre and Gaël Rémond for giving me the opportunity to participate in the summer school of the "Institut Fourier" in Grenoble, and their warm hospitality that made a kidney stone attack much less painful. Thanks as well to the students and other colleagues for attending the lectures and making encouraging comments on this circle of ideas.

References

1. J.-M. Bismut, H. Gillet, C. Soulé, Analytic torsion and holomorphic determinant bundles. I. Bott-Chern forms and analytic torsion. Comm. Math. Phys. **115**, 49–78 (1988). MR 929146
2. J.-M. Bismut, H. Gillet, C. Soulé, Analytic torsion and holomorphic determinant bundles. II. Direct images and Bott-Chern forms. Comm. Math. Phys. **115**, 79–126 (1988). MR 929147
3. J.-M. Bismut, H. Gillet, C. Soulé, Analytic torsion and holomorphic determinant bundles. III. Quillen metrics on holomorphic determinants. Comm. Math. Phys. **115**, 301–351 (1988). MR 931666
4. S. Bloch, H. Gillet, C. Soulé, Non-Archimedean Arakelov theory. J. Algebraic Geom. **4**, 427–485 (1995). MR 1325788
5. J.-B. Bost, Potential theory and Lefschetz theorems for arithmetic surfaces. Ann. Sci. École Norm. Sup. (4) **32**, 241–312 (1999)
6. J.I. Burgos Gil, J. Kramer, U. Kühn, Arithmetic characteristic classes of automorphic vector bundles. Doc. Math. **10**, 619–716 (2005)
7. J.I. Burgos Gil, J. Kramer, U. Kühn, Cohomological arithmetic Chow rings. J. Inst. Math. Jussieu **6**, 1–172 (2007)
8. K. Buzzard, Integral models of certain Shimura curves. Duke Math. J. **87**, 591–612 (1997)
9. P. Deligne, M. Rapoport, *Les Schémas de Modules de courbes Elliptiques*. Lecture Notes in Mathematical, vol. 349 , 143–316 (1973)
10. G. Faltings, Finiteness theorems for abelian varieties over number fields, in *Arithmetic Geometry* (Storrs, Conn., 1984) (Springer, New York, 1986). Translated from the German original [Invent. Math. **7**, 349–366; ibid. **75**, 381, pp. 9–27 (1984)
11. G. Freixas i Montplet, An arithmetic Riemann-Roch theorem for pointed stable curves. Ann. Sci. Éc. Norm. Supér. (4) **42**, 335–369 (2009)

12. G. Freixas i Montplet, An arithmetic Hilbert-Samuel theorem for pointed stable curves. J. Eur. Math. Soc. (JEMS) **14**, 321–351 (2012)
13. G. Freixas i Montplet, The Jacquet-Langlands correspondence and the arithmetic Riemann-Roch theorem for pointed curves. Int. J. Number Theory **8**, 1–29 (2012)
14. G. Freixas i Montplet, A. von Pippich, Riemann-Roch isometries in the non-compact orbifold setting. J. Eur. Math. Soc. **22**, 3491–3564 (2020)
15. W. Fulton, *Intersection Theory*, 2nd edn., Ergebnisse der Mathematik und ihrer Grenzgebiete. 3. Folge. A Series of Modern Surveys in Mathematics [Results in Mathematics and Related Areas. 3rd Series. A Series of Modern Surveys in Mathematics], vol. 2 (Springer, Berlin, 1998)
16. S.S. Gelbart, *Automorphic Forms on adèle Groups* (Princeton University/University of Tokyo, Princeton/Tokyo, 1975). Annals of Mathematics Studies, No. 83
17. H. Gillet, C. Soulé, An arithmetic Riemann-Roch theorem. Invent. Math. **110**, 473–543 (1992)
18. D.A. Hejhal, *The Selberg Trace Formula for* PSL(2, **R**), vol. 2. Lecture Notes in Mathematics, vol. 1001 (Springer, Berlin, 1983)
19. D. Helm, On maps between modular Jacobians and Jacobians of Shimura curves. Israel J. Math. **160**, 61–117 (2007). MR 2342491
20. H. Iwaniec, *Spectral Methods of Automorphic Forms*, 2nd edn., in *Graduate Studies in Mathematics*, vol. 53 (American Mathematical Society/Revista Matemática Iberoamericana, Providence/Madrid, 2002)
21. H. Jacquet, R.P. Langlands, in *Automorphic forms on* GL(2). Lecture Notes in Mathematics, vol. 114 (Springer, Berlin, 1970). MR 0401654
22. N.M. Katz, B. Mazur, Arithmetic moduli of elliptic curves, in *Annals of Mathematics Studies*, vol. 108 (Princeton University, Princeton, 1985). MR 772569
23. U. Kühn, Generalized arithmetic intersection numbers. J. Reine Angew. Math. **534**, 209–236 (2001)
24. V. Maillot, D. Roessler, Conjectures sur les dérivées logarithmiques des fonctions L d'Artin aux entiers négatifs. Math. Res. Lett. **9**, 715–724 (2002)
25. L. Moret-Bailly, La formule de Noether pour les surfaces arithmétiques. Invent. Math. **98**, 491–498 (1989)
26. R.S. Phillips, P. Sarnak, On cusp forms for co-finite subgroups of PSL(2, **R**). Invent. Math. **80**, 339–364 (1985). MR 788414
27. K. Prasanna, Integrality of a ratio of Petersson norms and level-lowering congruences. Ann. Math. (2) **163**, 901–967 (2006)
28. D.B. Ray, I.M. Singer, Analytic torsion for complex manifolds. Ann. Math. (2) **98**, 154–177 (1973). MR 383463
29. K.A. Ribet, On modular representations of Gal($\overline{\mathbf{Q}}/\mathbf{Q}$) arising from modular forms. Invent. Math. **100**, 431–476 (1990). MR 1047143
30. C. Soulé, *Géométrie d'Arakelov des surfaces arithmétiques*, Astérisque (177–178), Exp. No. 713 (1989), pp. 327–343. Séminaire, Bourbaki, vol. 1988/89
31. M.-F. Vignéras, *Arithmétique des algèbres de Quaternions*. Lecture Notes in Mathematics, vol. 800 (Springer, Berlin, 1980). MR 580949
32. C. Voisin, Hodge theory and complex algebraic geometry. I, english ed., *Cambridge Studies in Advanced Mathematics*, vol. 76 (Cambridge University, Cambridge, 2007), Translated from the French by Leila Schneps. MR 2451566
33. S.A. Wolpert, Cusps and the family hyperbolic metric. Duke Math. J. **138**, 423–443 (2007)

Chapter XII: The Height of CM Points on Orthogonal Shimura Varieties and Colmez's Conjecture

Fabrizio Andreatta

1 Introduction

These are the notes for a 6 h course given by the author during the summer school "Géométrie d'Arakelov", organized by the Institut Fourier (Grenoble) in the summer of 2017. They are based on the paper "Faltings' Heights of Abelian Varieties with Complex Multiplication" [2] by myself, Eyal Goren, Ben Howard and Keerthi Madapusi Pera and on notes by myself and Eyal Goren. No new results are presented. The goal is to describe the strategy to reduce the proof of an averaged version of Colmez's conjecture to a conjecture of Bruinier, Kudla and Yang, an instance of what is known as the *Kudla's program*. Infact a weak version of this conjecture, which is proven in [2] and is briefly discussed in the present text, implies this weaker version of Colmez's conjecture. Note that the latter has been used by Tsimerman [17] to provide an unconditional proof of the André-Oort conjecture for abelian varieties of Hodge type. Around the same time as [2] also X. Yuan and S.-W. Zhang [20] proved, using different techniques, the averaged form of Colmez's conjecture.

2 The Averaged Colmez's Conjecture

Let $E \subset \mathbb{C}$ be a CM field of degree $2d$ with a totally real subfield F of degree d. Let A be an abelian variety over $\mathbb{C}$ of dimension d with an action of the ring of integers $\mathcal{O}_E$ of E. A theorem of Shimura and Taniyama, see [7, Thm. 1.7.2.1], implies that

F. Andreatta (✉)
Dipartimento di Matematica "F. Enriques", Università degli Studi di Milano, Milano, Italy
e-mail: fabrizio.andreatta@unimi.it

© The Editor(s) (if applicable) and The Author(s), under exclusive license to Springer Nature Switzerland AG 2021
E. Peyre, G. Rémond (eds.), *Arakelov Geometry and Diophantine Applications*,
Lecture Notes in Mathematics 2276, https://doi.org/10.1007/978-3-030-57559-5_13

A can be defined over a number field K. Furthermore A has potentially everywhere good reduction thanks to [16, Thm. 6]. Thus, possibly after enlarging K, we may further assume that A extends to an abelian scheme $\mathscr{A}$ over the ring of integers $\mathscr{O}_K$ of K.

2.1 Faltings' Height

We denote by $\omega_{\mathscr{A}}$ the $\wedge^d$-power of the relative invariant differentials $\mathrm{H}^0\left(\mathscr{A}, \Omega^1_{\mathscr{A}/\mathscr{O}_K}\right)$ of $\mathscr{A}$. It is a projective $\mathscr{O}_K$-module of rank 1. Given a generator s of $\omega_{\mathscr{A}} \otimes_{\mathscr{O}_K} K$ define

$$h^{\mathrm{Falt}}_\infty(A, s) = \frac{-1}{2[K : \mathbb{Q}]} \sum_{\sigma : K \to \mathbb{C}} \log \left| \int_{A^\sigma(\mathbb{C})} s^\sigma \wedge \overline{s^\sigma} \right|,$$

where for every embedding $\sigma : K \to \mathbb{C}$ we let A^σ and s^σ be the base change of A and the section s to $\mathbb{C}$. When we write $A^\sigma(\mathbb{C})$ we consider the underlying complex analytic structure so that the integral makes sense. Define

$$h^{\mathrm{Falt}}_f(A, s) = \frac{1}{[K : \mathbb{Q}]} \sum_{\mathfrak{p} \subset \mathscr{O}_K} \mathrm{ord}_{\mathfrak{p}}(s) \cdot \log \mathrm{N}(\mathfrak{p}),$$

where $\mathrm{ord}_{\mathfrak{p}}(s)$ is defined as the order with respect to a generator of $\omega_{\mathscr{A}} \otimes_{\mathscr{O}_K} \mathscr{O}_{K,\mathfrak{p}}$ as $\mathscr{O}_{K,\mathfrak{p}}$-module. Finally define

$$h^{\mathrm{Falt}}(A) = h^{\mathrm{Falt}}_f(A, s) + h^{\mathrm{Falt}}_\infty(A, s),$$

the *Faltings' height* or *modular height* of A.

Remark 2.1 As the notation suggests the quantity $h^{\mathrm{Falt}}(A)$ is independent of the choice of the section s thanks to the product formula. Thanks to the normalization factor $\frac{1}{[K:\mathbb{Q}]}$ it is also invariant under field extension.

The Faltings' height can also be defined as the arithmetic degree of the metrized line bundle $\widehat{\omega}_{\mathscr{A}}$ over $\mathrm{Spec}(\mathscr{O}_K)$ as in Example 2.8, Chap. 8, where $\omega_{\mathscr{A}}$ is the underlying $\mathscr{O}_K$-module and the metrics at infinity are defined using integration as above.

2.2 Colmez's Theorem

The fact that A has an action of $\mathscr{O}_E$ singles out a subset $\Phi \subset \mathrm{Hom}(E, \mathbb{C})$: the action of E on $\mathrm{H}^0(A, \Omega^1_{A/\mathbb{C}})$ decomposes into a sum of d one dimensional eigenspaces for the action of E and on each of them E acts via an embedding $E \to \mathbb{C}$. We let

Φ be the subset of embeddings $E \subset \mathbb{C}$ appearing in this way. It is called the *CM type* of A. It is a subset of cardinality d and $\mathrm{Hom}(E, \mathbb{C}) = \Phi \sqcup \overline{\Phi}$ (here $\overline{\Phi}$ stands for the image of Φ under complex conjugation on $\mathbb{C}$). We then have the following Theorem of Colmez, [6, Théorème 0.3]:

Theorem 2.2 *Under the assumption that A is an abelian variety of dimension d with action of the ring of integers $\mathscr{O}_E$ of E and with CM type Φ, the Faltings' height $h^{\mathrm{Falt}}(A)$ depends only on the pair (E, Φ), and not on the choice of the abelian variety A.*

We write $h^{\mathrm{Falt}}_{(E,\Phi)}$ for the quantity $h^{\mathrm{Falt}}(A)$ as it is independent of the choice of A.

In the same paper Colmez provided a conjectural formula, see [6, Conj. 0.4], that computes $h^{\mathrm{Falt}}_{(E,\Phi)}$ in terms special values of L-functions of Artin characters. When $d = 1$, so that E is a quadratic imaginary field, Colmez's conjecture is a form of the famous *Chowla–Selberg formula*:

$$h^{\mathrm{Falt}}_{(E,\Phi)} = -\frac{1}{2}\frac{L'(\varepsilon, 0)}{L(\varepsilon, 0)} - \frac{1}{2}\log(2\pi) - \frac{1}{4}\log|D_E|,$$

where D_E is the discriminant of E and $L(\varepsilon, s)$ is the Hecke L-function associated to the quadratic character ε defined by the quadratic extension $\mathbb{Q} \subset E$.

In these lectures we'll be interested in an averaged form of Colmez's conjecture where we fix the CM field E but we sum over the set $\mathrm{CM}(E)$ of all CM types Φ (i.e., subsets Φ of $\mathrm{Hom}(E, \mathbb{C})$ of cardinality d such that $\mathrm{Hom}(E, \mathbb{C}) = \Phi \sqcup \overline{\Phi}$). His conjectural formula amounts to the following:

Colmez's Averaged Conjecture

$$\frac{1}{2^d}\sum_{\Phi \in \mathrm{CM}(E)} h^{\mathrm{Falt}}_{(E,\Phi)} = -\frac{1}{2} \cdot \frac{L'(\chi, 0)}{L(\chi, 0)} - \frac{1}{4} \cdot \log\left|\frac{D_E}{D_F}\right| - \frac{d}{2} \cdot \log(2\pi)$$

$$= -\frac{1}{2} \cdot \frac{\Lambda'(\chi, 0)}{\Lambda(\chi, 0)} - \frac{d}{4}\log(16\pi^3 e^\gamma),$$

where $\chi : \mathbb{A}_F^\times \to \{\pm 1\}$ is the quadratic Hecke character determined by the extension E/F, $L(\chi, s)$ is the associated L-function, $\Lambda(\chi, s) = |\frac{D_E}{D_F}|^{s/2}\Gamma_{\mathbb{R}}(s+1)^d L(\chi, s)$ is the completed L-function (so that $\Lambda(\chi, 1-s) = \Lambda(\chi, s)$), D_E and D_F are the discriminants of E and F respectively and $\gamma = -\Gamma'(1)$ is the Euler–Mascheroni constant.

2.3 Colmez's Conjecture

Let us reformulate Colmez's conjecture more precisely. Let $\mathbb{Q}^{\mathrm{CM}} \subset \mathbb{C}$ be the composite of all algebraic CM extensions of $\mathbb{Q}$. It is a Galois extension of $\mathbb{Q}$, containing the composite of all algebraic abelian extensions of $\mathbb{Q}$. Denote by $\mathscr{G} = \mathrm{Gal}(\mathbb{Q}^{\mathrm{CM}}/\mathbb{Q})$ its Galois group; write $c \in \mathscr{G}$ for the complex conjugation. Let $\mathscr{CM}^0$ be the $\mathbb{Q}$-vector space of locally constant, *central* functions (i.e., functions constant on conjugacy classes) $a \colon \mathscr{G} \to \mathbb{Q}$ such that the function $\mathscr{G} \ni g \mapsto a(g) + a(cg)$ is constant. Any such a is a $\mathbb{C}$-linear combination $a = \sum_\eta a(\eta)\eta$ of Artin characters. Since $c \in \mathscr{G}$ is a central element any such character satisfies $\eta(c) = \pm\eta(\mathrm{Id})$. The assumption that $a(g) + a(cg)$ is constant implies that for all non-trivial η for which $a(\eta) \neq 0$ we have $\eta(c) = -\eta(\mathrm{Id})$ so that $L(\eta, 0) \neq 0$. In particular to $a \in \mathscr{CM}^0$ we can associate the complex number

$$Z(a) = -\sum_\eta a(\eta) \left(\frac{L'(\eta, 0)}{L(\eta, 0)} + \frac{\log(f_\eta)}{2} \right),$$

where f_η is the Artin conductor of η.

Now start with a CM field $E \subset \mathbb{Q}^{\mathrm{CM}}$ and a CM type Φ. Define the locally constant function on $\mathscr{G}$:

$$a_{(E,\Phi)}(\sigma) = |\Phi \cap \sigma \circ \Phi| \quad \forall \sigma \in \mathscr{G}.$$

The average

$$a^0_{(E,\Phi)} = \frac{1}{[\mathscr{G} : \mathrm{Stab}(\Phi)]} \sum_{\tau \in \mathscr{G}/\mathrm{Stab}(\Phi)} a_{(E,\tau\Phi)}$$

lies in $\mathscr{CM}^0$. In fact

$$a_{(E,\Phi)}(\sigma) + a_{(E,\Phi)}(c\sigma) = |\Phi|$$

is independent of $\sigma \in \mathscr{G}$ and hence also $a^0_{(E,\Phi)}(\sigma) + a^0_{(E,\Phi)}(c\sigma)$. Here $\mathrm{Stab}(\Phi) \subset \mathscr{G}$ is the subgroup stabilizing the CM type Φ. We then have the following conjecture.

Colmez's Conjecture If A is an abelian variety with CM by the ring of integers $\mathscr{O}_E$ of E and with CM type Φ, we have

$$h^{\mathrm{Falt}}_{(E,\Phi)} = Z\big(a^0_{(E,\Phi)}\big).$$

Colmez verified the correctness of his conjecture, up to rational multiples of $\log(2)$, for E an abelian extension of $\mathbb{Q}$. Obus [15] removed this error term for abelian extensions. When $d = 2$, Yang [19] was the first to prove the formula for non-abelian extensions.

2.4 Some Consequences and Reduction Steps

We will use this (conjectural) combinatorial expression in two ways. First of all one can compute the following equality of virtual representations

$$\frac{1}{[E : \mathbb{Q}]} \sum_{\Phi} a^0_{(E, \Phi)} = 2^{d-2} \left(1 + \frac{1}{d} \mathrm{Ind}^{\mathscr{G}}_{\mathscr{G}_F}(\chi)\right),$$

which provides upon taking $Z(\_)$ the expression appearing on the right hand side of Colmez's averaged conjecture. Here $\mathscr{G}_F = \mathrm{Gal}(\mathbb{Q}^{\mathrm{CM}}/F)$. In fact

- $\log(2\pi) = \frac{\zeta'(0)}{\zeta(0)}$ is the value at $s = 0$ of the logarithmic derivative of the Riemann zeta function which is the L-function associated to the trivial character;
- the L-function of the induced representation $\mathrm{Ind}^{\mathscr{G}}_{\mathscr{G}_F}(\chi)$ is the L-function of χ (over the field F);
- the Artin conductor of $\mathrm{Ind}^{\mathscr{G}}_{\mathscr{G}_F}(\chi)$ is $|D_E/D_F|$.

Secondly, we relate the averaged sum over the CM types to the height of the total reflex algebra $E^\sharp$ and the reflex CM type $\Phi^\sharp$. This will play a crucial role in the sequel. We let $E^\sharp$ be the étale $\mathbb{Q}$-algebra defined, via Grothendieck's formalism of Galois theory, by the $\mathrm{Gal}(\overline{\mathbb{Q}}/\mathbb{Q})$-set $\mathrm{CM}(E)$ of all CM types on E, i.e., $E^\sharp$ is characterized by the fact that $\mathrm{Hom}(E^\sharp, \overline{\mathbb{Q}}) \cong \mathrm{CM}(E)$ as $\mathrm{Gal}(\overline{\mathbb{Q}}/\mathbb{Q})$-sets. We let $\Phi^\sharp \subset \mathrm{Hom}(E^\sharp, \overline{\mathbb{Q}})$ consist of all $\Phi \in \mathrm{CM}(E)$ such that the given embedding $\iota_0 \colon E \to \mathbb{C}$ lies in Φ. In fact one can prove that $E^\sharp = \prod_i E'_i$ is a product of CM fields (as many as the orbits of $\mathrm{Gal}(\overline{\mathbb{Q}}/\mathbb{Q})$ on $\mathrm{CM}(E)$) and $\Phi^\sharp \subset \mathrm{Hom}(E^\sharp, \overline{\mathbb{Q}}) = \amalg_i \mathrm{Hom}(E'_i, \overline{\mathbb{Q}})$ is the disjoint union $\Phi^\sharp = \amalg_i \Phi'_i$ of CM types for the E'_i. Furthermore if E'_i corresponds to the orbit of $\Phi_i \in \mathrm{CM}(E)$ then (E, Φ_i) and (E'_i, Φ'_i) is a reflex pair. Then

$$a^0_{(E^\sharp, \Phi^\sharp)} = \frac{1}{[E : \mathbb{Q}]} \sum_{\Phi \in \mathrm{CM}(E)} a^0_{(E, \Phi)}. \tag{1}$$

Colmez further proves, see [6, Théorème 0.3], the following result:

Proposition 2.3 *There exists a unique $\mathbb{Q}$-linear map* $\mathrm{ht} \colon \mathscr{CM}^0 \to \mathbb{R}$ *such that* $h^{\mathrm{Falt}}_{(E, \Phi)} = \mathrm{ht}(a^0_{(E, \Phi)})$.

We deduce that $h^{\mathrm{Falt}}_{(E^\sharp, \Phi^\sharp)} = \frac{1}{[E : \mathbb{Q}]} \sum_{\Phi \in \mathrm{CM}(E)} h^{\mathrm{Falt}}_{(E, \Phi)}$. Colmez's averaged conjecture amounts then to prove the following:

Colmez's Averaged Conjecture Revisited

$$[E:\mathbb{Q}]h^{\text{Falt}}_{(E^{\sharp},\Phi^{\sharp})} = -\frac{1}{2}\cdot\frac{\Lambda'(\chi,0)}{\Lambda(\chi,0)} - \frac{d}{4}\log(16\pi^3 e^{\gamma}).$$

Our Strategy The Chowla–Selberg formula implies Colmez's averaged conjecture for $d = 1$. Therefore we may and will assume through this text that $d \geq 2$. We will define a certain normal scheme $\mathcal{Y}_0$, finite over $\text{Spec}(\mathcal{O}_E)$, carrying an abelian scheme $\mathcal{A}^{\sharp}$ with action of $\mathcal{O}_{E^{\sharp}}$ and CM type $\Phi^{\sharp}$. In particular $\mathcal{Y}_0$ will carry the metrized line bundle $\widehat{\omega}_{\mathcal{A}^{\sharp}}$ whose degree, divided by the degree of $\mathcal{Y}_0$, will compute $h^{\text{Falt}}_{(E^{\sharp},\Phi^{\sharp})}$ by definition.

On the other hand, for $L \subset E$ a suitable lattice, we will also define auxiliary morphisms $\mathcal{Y}_L \to \mathcal{M}_L$ where $\mathcal{Y}_L \to \mathcal{Y}_0$ is a finite morphism and $\mathcal{M}_L$ are certain models of Shimura varieties of orthogonal type. The key point is that the metrized line bundle $\widehat{\omega}_{\mathcal{A}^{\sharp}}$ over $\mathcal{Y}_L$ is the pull–back of a combination of *arithmetic special divisors* on $\mathcal{M}_L$, at least away from a finite set of primes depending on L that we will denote D_L. Using this and work of Bruinier et al. [5] we will get a way to compute $h^{\text{Falt}}_{(E^{\sharp},\Phi^{\sharp})}$ as the arithmetic intersection between these special divisors and $\mathcal{Y}_L$. This provides the RHS in the formula for Colmez's averaged conjecture (revisited), up to rational linear combination of $\log p$ for $p \in D_L$. As we may choose lattices so that $\cap_L D_L = \emptyset$ and logarithms of primes are linearly independent, Colmez's averaged conjecture follows.

3 Shimura Varieties of Orthogonal Type and CM Cycles

3.1 GSpin Shimura Varieties

Let V be a $\mathbb{Q}$-vector space of dimension $n+2$ with $n \geq 0$, equipped with a quadratic form

$$Q: V \to \mathbb{Q}$$

which is non degenerate, of signature $(n, 2)$. Consider the associated bilinear form

$$[\_,\_]: V \times V \to \mathbb{Q}, \quad [x, y] = Q(x + y) - Q(x) - Q(y).$$

We have the associated Clifford algebra $C(V) = C(V, Q)$. It is a $\mathbb{Q}$-algebra, with an inclusion

$$V \hookrightarrow C(V)$$

satisfying the following universal property: for any $\mathbb{Q}$ algebra R with a $\mathbb{Q}$-linear map $j : V \to R$ such that

$$j(v)j(v) = Q(v)$$

there exists a unique homomorphism of $\mathbb{Q}$-algebras

$$C(V) \to R$$

such that the composite with the inclusion $V \subset C(V)$ is j. In particular for any v and $w \in V$, we have

$$v \cdot w + w \cdot v = [v, w] \in C(V),$$

where $v \cdot w$ (and $w \cdot v$) is the product in $C(V)$.

The construction of the Clifford algebra is quite straightforward. In fact,

$$C(V) := (\bigoplus_{n=0}^{\infty} V^{\otimes m})/(v \otimes v - Q(v)|v \in V),$$

the quotient of the tensor algebra of V by the two sided ideal generated by the elements $v \otimes v - Q(v)$ for $v \in V$. As such an ideal is generated by elements lying in even degree (in the tensor algebra considered with its natural grading), the $\mathbb{Z}/2\mathbb{Z}$-grading on the tensor algebra (into even and odd tensors) passes to the Clifford algebra that correspondingly splits into a direct sum

$$C(V) = C^+(V) \oplus C^-(V).$$

Note that $C^+(V)$ is a subalgebra while $C^-(V)$ is just a two-sided module for $C^+(V)$. Furthermore we have the following formulas:

$$\dim_{\mathbb{Q}}(C(V)) = 2^{n+2}, \quad \dim_{\mathbb{Q}}^+(C(V)) = \dim_{\mathbb{Q}}(C^-(V)) = 2^{n+1}.$$

(Recall that V has dimension $n + 2$.)

Next we construct the algebraic group $\mathrm{GSpin}(V, Q)$. Given a commutative $\mathbb{Q}$-algebra R, its R-valued points are

$$\mathrm{GSpin}(V)(R) := \{ x \in (C^+(V) \otimes_{\mathbb{Q}} R)^* : x(V \otimes_{\mathbb{Q}} R)x^{-1} \subset V \otimes_{\mathbb{Q}} R \}.$$

In particular any $x \in \mathrm{GSpin}(V)(R)$ acts on $V \otimes_{\mathbb{Q}} R$ and, since for any $y \in V \otimes_{\mathbb{Q}} R$

$$Q(y) = Q(xyx^{-1}),$$

this action factors through the orthogonal group $O(V, Q)(R)$. Notice that the units of R form a subgroup of the center of $\text{GSpin}(V)(R)$ and in particular they act trivially on $V \otimes_{\mathbb{Q}} R$. Furthermore, one can prove that the action of $\text{GSpin}(V, Q)$ on V factors through the special orthogonal group $\text{SO}(V, Q)$ and one gets an exact sequence of algebraic groups:

$$0 \longrightarrow \mathbb{G}_m \longrightarrow \text{GSpin}(V) \longrightarrow \text{SO}(V, Q) \longrightarrow 0 \qquad (2)$$

3.2   Examples of GSpin Groups

Example I: The Case $n = 0$

In this case

$$V = \mathbb{Q}e_1 \oplus \mathbb{Q}e_2,$$

So $C^-(V) = V$ and $C^+(V)$ is an algebra of dimension 2:

$$C^+(V) = \mathbb{Q} \oplus \mathbb{Q}e_1 \cdot e_2.$$

Denote $x = e_1 \cdot e_2$, let $a_1 = Q(e_1)$ and $a_2 = Q(e_2)$. They are negative rational numbers and if we write $b = [e_1, e_2] \in \mathbb{Q}$, we have

$$x^2 = e_1 e_2 e_1 e_2 = -e_1^2 e_2^2 + [e_1, e_2]e_1 e_2 = -a_1 a_2 + bx,$$

so we have

$$x^2 - bx + a_1 a_2 = 0.$$

As an algebra, this gives

$$C^+(V) = \mathbb{Q}[x]/(x^2 - bx + a_1 a_2),$$

which is an imaginary quadratic field K, and we see that

$$\text{GSpin}(V) = C^+(V)^\times = \text{Res}_{K/\mathbb{Q}} \mathbb{G}_m.$$

In particular its base change to $\mathbb{R}$ is the so-called Deligne torus $\text{Res}_{\mathbb{C}/\mathbb{R}} \mathbb{G}_{m,\mathbb{R}}$.

Example II: The Case $n = 1$

Consider the $\mathbb{Q}$-vector space

$$V \subset M_{2 \times 2}(\mathbb{Q}) = \{x \in M_{2 \times 2} : Tr(x) = 0\}.$$

Fix some $N \in \mathbb{N}$ such that $N \geq 1$. Let Q_N be the quadratic form

$$A \mapsto N \cdot \det A.$$

Then

$$\mathrm{GSpin}(V) \cong \mathrm{GL}_2,$$

where GL_2 acts on V by conjugation.

3.3 Hermitian Symmetric Spaces

Fix the algebraic group $G := \mathrm{GSpin}(V, Q)$. As a first step in order to construct a Shimura variety, we need to construct a Hermitian symmetric space; see §2 of Ch. X for this notion. It admits several realizations:

1. as a complex manifold $D_{\mathbb{C}} = \{z \in V_{\mathbb{C}} \backslash \{0\} : Q(z) = 0, [z, \bar{z}] < 0\}/\mathbb{C}^* \subset \mathbb{P}(V_{\mathbb{C}})$
2. as a Riemannian manifold $D_{\mathbb{R}} = \{$ Negative definite oriented planes $H \subset V_{\mathbb{R}}\}$
3. using Deligne torus $\mathbb{S} = \mathrm{Res}_{\mathbb{C}/\mathbb{R}}\mathbb{G}_{m,\mathbb{R}}$, and realizing

$$D = G(\mathbb{R}) \text{ conjugacy class of } h \colon \mathbb{S} \to G_{\mathbb{R}}.$$

In fact, there are natural bijections between $D_{\mathbb{C}}$, $D_{\mathbb{R}}$ and D. Let us explain how we can go back and forth between these incarnations.

Given $H = \mathbb{R}e_1 \oplus \mathbb{R}e_2$ as in (2), with $\{e_1, e_2\}$ an orthogonal basis with $Q(e_1) = Q(e_2)$, we let $z = e_1 + ie_2$, then take the line $[z]$ to get the realization (1). To get realization (3), we simply take $\mathbb{S} \cong \mathrm{GSpin}H \hookrightarrow G_{\mathbb{R}}$ using that $H \subset V_{\mathbb{R}} := V \otimes_{\mathbb{Q}} \mathbb{R}$.

Example 3.1 In the case I discussed in Sect. 3.2 the Hermitian space consists of two points

$$D_{\mathbb{R}} = \{ \text{ two possible orientations on } V_{\mathbb{R}} = \mathbb{R}e_1 \oplus \mathbb{R}e_2\}.$$

Example 3.2 In the case II of Sect. 3.2 we have

$$D_{\mathbb{R}} \cong \mathbb{H}^+ \sqcup \mathbb{H}^- \subset \mathbb{C}$$

which are the Poincaré upper and lower half planes. The inverse of the map is given as

$$\mathbb{R} \cdot Re \begin{pmatrix} z & -z^2 \\ 1 & -z \end{pmatrix} \oplus \mathbb{R} \cdot \text{Im} \begin{pmatrix} z & -z^2 \\ 1 & -z \end{pmatrix} \leftarrow z = x + iy.$$

Pick $[z] \in D_{\mathbb{C}}$, then

$$V_{\mathbb{C}} = \mathbb{C}z \oplus (\mathbb{C}z \oplus \mathbb{C}\overline{z})^{\perp} \oplus \mathbb{C}\overline{z}$$

and the tangent space of Q in $\mathbb{P}(V_{\mathbb{C}})$ at $[z]$ can be computed as the Zariski tangent space at $[z]$, namely the set of lines $[z + \delta\epsilon + \gamma\epsilon\overline{z}]$, with $\delta \in (\mathbb{C}z \oplus \mathbb{C}\overline{z})^{\perp}$ and ϵ a formal variable with square $\epsilon^2 = 0$, such that $Q([z + \delta\epsilon + \gamma\epsilon\overline{z}]) = 0$, i.e, if and only if $\gamma = 0$. Thus the tangent space of $D_{\mathbb{C}}$ at $[z]$ is isomorphic to $(\mathbb{C}z \oplus \mathbb{C}\overline{z})^{\perp}$ and $\dim D_{\mathbb{C}} = n$.

3.4 GSpin-*Shimura Varieties*

Given V and D and the algebraic group $G := \text{GSpin}(V, Q)$ as in the previous section, define the complex manifold:

$$M_K(\mathbb{C}) = G(\mathbb{Q})\backslash D \times G(\mathbb{A}_f)/K = \amalg_{g \in G(\mathbb{Q})\backslash G(\mathbb{A}_f)/K} \Gamma_g \backslash D$$

for some compact open subgroup K of the adelic points $G(\mathbb{A}_f)$ of G. Here $\Gamma_g := G(\mathbb{Q}) \cap (gKg^{-1})$ is an arithmetic subgroup of $G(\mathbb{Q})$. It has a natural structure of algebraic variety over $\mathbb{C}$.

Given a quadratic lattice $L \subset V$, i.e., a lattice on which Q is integral valued, one can construct a compact open subgroup K_L by taking

$$K_L = G(\mathbb{A}_f) \cap C^+(\widehat{L})^{\times} \subset C^+(V)^{\times}(\mathbb{A}_f),$$

where $\widehat{L} := L \otimes_{\mathbb{Z}} \widehat{\mathbb{Z}}$. We will be especially interested in the case that L is maximal among the integral lattices. This will guarantee the existence of good integral models over $\mathbb{Z}$ for the Shimura variety $M_{K_L}(\mathbb{C})$. If the compact open subgroup is of the type K_L for some lattice L, we simply write $M_L(\mathbb{C})$.

Now let us look at the examples again.

Example 3.3 In the I case of Sect. 3.2 $M(\mathbb{C})$ consists of finitely many points.

Example 3.4 In the II case of Sect. 3.2 we have

$$V = M_{2\times 2}(\mathbb{Q})^{Tr=0}$$

and using Q_N we get $\mathrm{GSpin}(V) \cong \mathrm{GL}_2$. Take

$$L := \left\{ \begin{pmatrix} a & \frac{-b}{N} \\ c & -a \end{pmatrix} : a, b, c \in \mathbb{Z} \right\},$$

one can check that

$$K_L \cong \prod_p \widetilde{K}_p,$$

where

$$\widetilde{K}_p = \left\{ \begin{pmatrix} \alpha & \beta \\ \gamma & \delta \end{pmatrix} \in \mathrm{GL}_2(\mathbb{Z}_p) : \gamma \in N\mathbb{Z}_p \right\}.$$

In this case one sees that

$$M_2(\mathbb{C}) \cong Y_0(N)(\mathbb{C})$$

which is the modular curve of level $\Gamma_0(N)$, classifying cyclic isogenies $\rho : E \rightarrow E'$ of degree N of elliptic curves.

Warning The case of elliptic curves is misleading as it might appear that $M_L(\mathbb{C})$ has a natural moduli interpretation. This is not the case if the dimension $n + 2$ of V is large. In this case $M_L(\mathbb{C})$ does *not* have, in general, a natural interpretation as a PEL type moduli problem, that classifies abelian varieties with given polarization, endomorphisms and level structures. As we will see, this is the source of complications when one attempts to provide integral models for $M_L(\mathbb{C})$.

4 Extra Structures on GSpin-Shimura Varieties

Recall the notation from the previous sections: we fixed a vector space V over $\mathbb{Q}$ of dimension $n + 2$, and a quadratic form $Q : V \rightarrow \mathbb{Q}$ of signature $(n, 2)$, with a maximal quadratic lattice $L \subset V$. We let G be the algebraic group $\mathrm{GSpin}(V)$ and D the associated Hermitian symmetric space. Then we defined

$$M_L(\mathbb{C}) = G(\mathbb{Q}) \backslash D \times G(\mathbb{A}_f)/K_L$$

for a particular choice of compact open subgroup K_L associated to the lattice L.

We have a natural functor from the category of algebraic representations of G to the category of local systems of $\mathbb{Q}$-vector spaces on $M_L(\mathbb{C})$, given by

$$\left(G \rightarrow GL(W) \right) \mapsto \left(W_{\mathrm{Betti}, \mathbb{Q}} \longrightarrow M_L(\mathbb{C}) \right),$$

where

$$W_{\text{Betti},\mathbb{Q}} := G(\mathbb{Q})\backslash(W \times D) \times G(\mathbb{A}_f)/K_L.$$

Note that this gives us a unique pair (W_{dR}, ∇) of a locally free $\mathcal{O}_{M_L(\mathbb{C})}$-module with integral connection. Namely $W_{\text{dR}} := W_{\text{Betti},\mathbb{Q}} \otimes_{\mathbb{Q}} \mathcal{O}_{M_L(\mathbb{C})}$ with the connection $\nabla = 1 \otimes d$. Then (W_{dR}, ∇) is characterized as the vector bundle with integrable connection such that

$$W_{\text{dR}}^{\nabla=0} = W_{\text{Betti},\mathbb{Q}} \otimes_{\mathbb{Q}} \mathbb{C}.$$

We further have the following extra properties:

(a) For any z in the symmetric space D, the map

$$h_z: \mathbb{S} \to G_{\mathbb{R}} \to GL(W_{\mathbb{R}}),$$

where $G_{\mathbb{R}} \to GL(W_{\mathbb{R}})$ is defined by the representation $G \to GL(W)$, induces a map

$$\mathbb{S}(\mathbb{C}) = \mathbb{C}^* \times \mathbb{C}^* \to GL(W_{\mathbb{R}} \otimes_{\mathbb{R}} \mathbb{C}) = GL(W_{\mathbb{C}})$$

and the fiber $W_{\text{dR},z}$ at z has a bigraduation $\oplus_{p,q} W_{\text{dR},z}^{p,q}$ obtained by the decomposition of $W_{\mathbb{C}}$ according to the action of $\mathbb{C}^* \times \mathbb{C}^*$.

(b) W_{dR} is endowed with a decreasing filtration $\text{Fil}^J(W_{\text{dR}}) \subset W_{\text{dR}}$ by holomorphic sub-bundles of W_{dR}, defined pointwise by

$$\text{Fil}^J(W_{\text{dR},z}) := \oplus_{p \geq J} W_{\text{dR},z}^{p,q}.$$

A $\mathbb{Q}$-local system on $M_L(\mathbb{C})$ with these properties is called *a variation of $\mathbb{Q}$-Hodge structures*. In particular, $W_{\text{Betti},\mathbb{Q}}$ is a variation of $\mathbb{Q}$-Hodge structures.

4.1 Example A: The Representation V

Consider in particular the homomorphism

$$G \to SO(V),$$

given by

$$x \mapsto \{y \mapsto xyx^{-1}\}$$

as a map $V \to V$ inside $SO(V)$. Now we get as before a variation of $\mathbb{Q}$-Hodge structures $\mathbb{V}_{\mathrm{Betti}}$ and even a variation of $\mathbb{Z}$-Hodge structures $\mathbb{V}_{\mathrm{Betti}}$. In particular we get a vector bundle with connection $\mathbb{V}_{\mathrm{dR}}$. They are all endowed with a quadratic form Q_{Betti} and Q_{dR} respectively.

For any $(z, g) \in M_L(\mathbb{C})$ with $z \in D_{\mathbb{C}}$ and $g \in G(\mathbb{A}_f)$, where $D_{\mathbb{C}}$ is the incarnation of the symmetric space as the isotropic lines in $V_{\mathbb{C}}$, the morphism h_z defines a decomposition

$$V_{\mathbb{C}} = \mathbb{C}_z \oplus (\mathbb{C}_z \oplus \mathbb{C}_{\bar{z}})^{\perp} \oplus \mathbb{C}_{\bar{z}} \subset \mathrm{End}\big(H_{1,\mathrm{dR}}(A_z)\big).$$

The filtration is given by

$$\mathrm{Fil}^1(\mathbb{V}_{\mathrm{dR},z}) = \mathbb{C}z$$

$$\mathrm{Fil}^0(\mathbb{V}_{\mathrm{dR},z}) = \mathbb{C}z \oplus (\mathbb{C}z \oplus \mathbb{C}\bar{z})^{\perp}$$

$$\mathrm{Fil}^{-1}(\mathbb{V}_{\mathrm{dR},z}) = \mathbb{C}z \oplus (\mathbb{C}z \oplus \mathbb{C}\bar{z})^{\perp} \oplus \mathbb{C}\bar{z}.$$

4.2 Example B: The Kuga–Satake Abelian Scheme

Consider the representation $C(V)$ of $G = \mathrm{GSpin}(V)$ via the inclusion $G \subset C^+(V)^*$ and the map $C^+(V)^* \subset \mathrm{GL}\big(C(V)\big)$ provided by left multiplication. It provides a variation of Hodge structures: for any $z \in D$, we have

$$C(V)_z = C(V)_z^{-1,0} \oplus C(V)_z^{0,-1};$$

this is equivalent to giving a complex structure on $C(V_{\mathbb{R}})$ and we obtain a complex abelian variety

$$A_z := C(V_{\mathbb{R}})/C(L),$$

called the *Kuga–Satake abelian variety*. In this way one gets an abelian scheme

$$\pi : A \longrightarrow M_L(\mathbb{C}).$$

The associated vector bundle with connection $C(V)_{\mathrm{dR}}$ coincides with the relative de Rham homology

$$C(V)_{\mathrm{dR}} = \mathcal{H}_{1,\mathrm{dR}}(A),$$

the connection is the so called *Gauss–Manin connection* and the filtration is given by the Hodge filtration

$$0 \longrightarrow \mathrm{R}_1\pi_*(\mathscr{O}_A)^\vee \to \mathscr{H}_{1,\mathrm{dR}}(A) \longrightarrow \pi_*(\Omega^1_A)^\vee \longrightarrow 0.$$

Example 4.1 Recall that in this case $n = 0$ and $C^+(L) \subset C^+(V) = K$ is an order in the quadratic imaginary field K. Then

$$A_z = A_z^+ \times A_z^-,$$

where A_z^+ is an elliptic curve with complex multiplication by $C^+(L)$ and

$$A_z^- = A_z^+ \otimes_{C^+(L)} L.$$

Example 4.2 In this case II of Sect. 3.2 $V = M_{2\times2}(\mathbb{Q})^{Tr=0}$, we have

$$M(\mathbb{C}) \cong Y_0(N)(\mathbb{C})$$

by $z \mapsto [E_z \to E'_z]$ and

$$A_z = A_z^+ \times A_z^-$$

and

$$A_z^+ = A_z^- = E_z \times E'_z.$$

Consider $\ell\colon V \hookrightarrow \mathrm{End}\big(C(V)\big)$ given by

$$v \mapsto \ell_v := \{ \text{ left multiplication by } v \text{ on } C(V)\}.$$

It is a morphism in the category of representation of G. Indeed, recall that G acts on $C(V)$ via the inclusion into $C^+(V)^*$ and the left multiplication action of $C^+(V)^*$ on $C(V)$. The induced action of $C^+(V)^*$ on $\mathrm{End}\big(C(V)\big)$ takes $\alpha \in C^+(V)^*$ and $g \in \mathrm{End}\big(C(V)\big)$ to $\alpha g(\alpha^{-1}\_) \in \mathrm{End}\big(C(V)\big)$ which restricted to ℓ_v gives $\alpha\ell_v(\alpha^{-1}(\_)) = \ell_{\alpha v \alpha^{-1}}(\_)$. Hence, we get a morphism of Hodge structures

$$\ell_{\mathrm{Betti}}\colon V_{\mathrm{Betti}} \hookrightarrow \mathrm{End}\big(\mathscr{H}_1(A,\mathbb{Q})\big) \tag{3}$$

and a morphism of vector bundles with connections over $M_L(\mathbb{C})$

$$\ell_{\mathrm{dR}}\colon V_{\mathrm{dR}} \hookrightarrow \mathrm{End}\big(\mathscr{H}_{1,\mathrm{dR}}(A)\big). \tag{4}$$

5 The Big CM Points

We follow here [5]. Let E be a CM field of degree $2d$ with a totally real subfield F (of degree d). We label the embeddings $\{\sigma_0, \ldots, \sigma_{d-1}\} = \mathrm{Hom}(F, \mathbb{R})$. For every integer $0 \leq i \leq d-1$ label by σ_i and $\overline{\sigma}_i : E \to \mathbb{C}$ the two conjugate embeddings of E extending $\sigma_i : F \to \mathbb{R}$.

Let $\lambda \in F$ be an element such that $\sigma_0(\lambda) < 0$ and $\sigma_i(\lambda) > 0$ for $1 \leq i \leq d-1$. Consider the quadratic space $W = E$ of dimension 2 over F with bilinear pairing

$$B_W : W \times W \longrightarrow F, \quad (x, y) \mapsto B_W(x, y) = \mathrm{Tr}_{E/F}(\lambda x \overline{y}).$$

It is negative definite at one place and positive definite at the others. Namely, for all embedding $\sigma_i : F \to \mathbb{R}$ with $i \neq 0$, writing $W_{\sigma_i} := W \otimes_F^{\sigma_i} \mathbb{R}$, the induced bilinear form $B_{\sigma_i} : W_{\sigma_i} \times W_{\sigma_i} \longrightarrow \mathbb{R}$ is positive definite for $i = 1, \ldots, d-1$, and it is negative definite at the remaining place σ_0. We let

$$Q_W(x) = \frac{1}{2} B_W(x, x)$$

be the associated quadratic form. The even Clifford algebra $C_F^+(W)$ over F is identified with E.

One then lets

$$V = \mathrm{Res}_{F/\mathbb{Q}}(W).$$

That is, V is simply E, as $\mathbb{Q}$-vector space of dimension $2d$ so that

$$V \otimes_{\mathbb{Q}} \mathbb{R} = W \otimes_F \mathbb{R} = \bigoplus_j W_{\sigma_j}$$

equipped with the form

$$B_V(x, y) = \mathrm{Tr}_{E/\mathbb{Q}}(\lambda x \overline{y}).$$

Example Let $L = \mathfrak{a}$ be a fractional ideal of E. Let $\overline{L} := \overline{\mathfrak{a}}$ be the image of $\mathfrak{a}$ under complex conjugation. Then B_V is integrally valued on L if and only if $\lambda \mathfrak{a} \overline{\mathfrak{a}} \subset \mathfrak{D}_{E/\mathbb{Q}}^{-1}$. In this case $L^\vee = \left(\lambda \mathfrak{D}_{E/\mathbb{Q}} \overline{\mathfrak{a}}\right)^{-1}$ and

$$L^\vee / L \cong \mathscr{O}_E / \lambda \mathfrak{D}_{E/\mathbb{Q}} \mathrm{Norm}_{E/F}(\mathfrak{a}) \tag{5}$$

Definition 5.1 Given a quadratic lattice $L \subset V$ we say that the prime p is *good for* L if the following conditions hold. For every prime $\mathfrak{p}$ of F over p set

$$L_{\mathfrak{p}} = (L \otimes \mathbb{Z}_p) \cap V_{\mathfrak{p}}.$$

We demand that

- For every $\mathfrak{p} \mid p$ unramified in E, the $\mathbb{Z}_p$-lattice $L_{\mathfrak{p}}$ is $\mathscr{O}_{E,\mathfrak{p}}$-stable and self-dual for the induced $\mathbb{Z}_p$-valued quadratic form.
- For every $\mathfrak{p} \mid p$ ramified in E, the $\mathbb{Z}_p$-lattice $L_{\mathfrak{p}}$ is maximal for the induced $\mathbb{Z}_p$-valued quadratic form, and there exists an $\mathscr{O}_{E,\mathfrak{q}}$-stable lattice $\Lambda_{\mathfrak{p}} \subset V_{\mathfrak{p}}$ such that

$$\Lambda_{\mathfrak{p}} \subset L_{\mathfrak{p}} \subsetneqq \mathfrak{d}_{E_{\mathfrak{q}}/F_{\mathfrak{p}}}^{-1} \Lambda_{\mathfrak{p}}.$$

where $\mathfrak{q} \subset \mathscr{O}_E$ is the unique prime above $\mathfrak{p}$.

All but finitely many primes are good: Choose any $\mathscr{O}_E$-stable lattice $\Lambda \subset L$. Then, for all but finitely many primes p, $\Lambda_{\mathbb{Z}_p} = L_{\mathbb{Z}_p}$ will be self-dual and hence good. We let $D_{L,\mathrm{bad}}$ be the product of all primes that are *not* good for L.

For every $j = 0, \ldots, d - 1$ we have $C^+(W_{\sigma_j}) \cong C_F^+(W) \otimes_F^{\sigma_j} \mathbb{R}$. There are maps

$$\prod_j C^+(W_{\sigma_j}) \longrightarrow \otimes_j C^+(W_{\sigma_j}) \longrightarrow C^+\left(\bigoplus_j W_{\sigma_j}\right) = C^+(V \otimes_{\mathbb{Q}} \mathbb{R}).$$

(The tensor product is over $\mathbb{R}$ or more generally any field over which F splits completely.) The first map is multiplicative and multi-linear and the second map is a ring homomorphism. The ring $C_F^+(W)$ has a basis over F given by $\{1, e_1 e_2\}$. A general element has the form $a + b e_1 e_2$; via the isomorphism $C_F^+(W) \otimes_{\mathbb{Q}} \mathbb{R} \cong \prod_j C^+(W_{\sigma_j})$ and the maps above, the image of this element in $C^+(\bigoplus_j W_{\sigma_j})$ is $\prod_j (\sigma_j(a) + \sigma_j(b) e_1^j e_2^j)$ (product in the Clifford algebra). In particular, $a \in F$ is mapped to $\mathrm{Norm}_{F/\mathbb{Q}}(a)$ and $e_1 e_2$ to $\prod_j e_1^j e_2^j$.

Passing to GSpin, making use of $\mathrm{GSpin}_F(W) = C_F^+(W)^*$, we get a homomorphism of groups over $\mathbb{R}$ (or any field splitting F):

$$\prod_j \mathrm{GSpin}_{\mathbb{R}}(W_{\sigma_j}) \longrightarrow \mathrm{GSpin}_{\mathbb{R}}\left(\bigoplus_j W_{\sigma_j}\right). \tag{6}$$

The homomorphism (6) descends to a homomorphism of algebraic groups over $\mathbb{Q}$

$$g : \mathrm{Res}_{F/\mathbb{Q}}(\mathrm{GSpin}_F(W)) \longrightarrow \mathrm{GSpin}(V) \tag{7}$$

(cf. [5, §2]). Let T be the image in $\mathrm{GSpin}(V)$ of this homomorphism; it is a torus and there is an exact sequence (loc. cit.)

$$1 \longrightarrow T_F^{\mathrm{Nm}=1} \longrightarrow T_E \xrightarrow{g} T \longrightarrow 1.$$

Here $T_E = \mathrm{Res}_{E/\mathbb{Q}}(\mathbb{G}_{m,E}) = \mathrm{Res}_{F/\mathbb{Q}}(\mathrm{GSpin}_F(W))$ and $T_F^{\mathrm{Nm}=1}$ is the subgroup of norm 1 elements of $T_F = \mathrm{Res}_{F/\mathbb{Q}}(\mathbb{G}_{m,F})$.

Note that there is a unique, up to isomorphism, rank 2 non-split torus over $\mathbb{R}$ and, thus, there is an isomorphism

$$h: \mathbb{S} \cong \mathrm{GSpin}(W_{\sigma_0}) \subset \mathrm{Res}_{F/\mathbb{Q}}(\mathrm{GSpin}_F(W))(\mathbb{R}). \qquad (8)$$

This gives a point $z_0 \in D_{\mathbb{R}}$. Fixing a quadratic lattice $L \subset V$ we get the Shimura variety $M_L(\mathbb{C})$ as explained in Sect. 3.4. Taking a compact open subgroup $K_T \subset K_L \cap T(\mathbb{A}_f)$ we can define the Shimura variety

$$Y_L(\mathbb{C}) = T(\mathbb{Q})\backslash \{z_0\} \times T(\mathbb{A}_f)/K_T,$$

which consists of finitely many points, and a homomorphism

$$Y_L(\mathbb{C}) \to M_L(\mathbb{C}) = G(\mathbb{Q})\backslash D \times G(\mathbb{A}_f)/K_L,$$

whose image is called the *big CM cycle* associated to (E, σ_0, λ) (the specific subgroup K_T we will consider depends only on L). It is defined over its reflex field E and we let $\mathcal{Y}_L$ be the normalization of $\mathrm{Spec}(\mathcal{O}_E)$ in Y_L.

Remark 5.2 The world "big" suggests the existence of *small CM cycles* as well. This is the case and they are constructed starting from quadratic imaginary extensions of $\mathbb{Q}$ instead of a CM field extension of degree $n+2$. There are interesting conjectures in this setting as well in the spirit of Kudla's program, elaborated by Bruinier and Yang in [4]. These conjectures have been proven under some mild assumptions in [1].

5.1 The Total Reflex Algebra

Write

$$V_{\mathbb{C}} = E \otimes_{\mathbb{Q}} \mathbb{C} = \bigoplus_{\rho \in \mathrm{Hom}(E,\mathbb{C})} \mathbb{C} \cdot e_\rho. \qquad (9)$$

where the e_ρ's are orthogonal idempotents of the algebra $E \otimes_{\mathbb{Q}} \mathbb{C}$. We recall the following results from [9].

Lemma 5.3 *Define for $\rho \in \mathrm{Hom}(E, \mathbb{C})$ an element of $C^+(V_{\mathbb{C}})$,*

$$\delta_\rho = \frac{1}{\rho(\lambda)} e_\rho e_{\bar\rho}.$$

There are 2d such elements δ_ρ. They all commute. Furthermore $\delta_\rho^2 = \delta_\rho$ and $\delta_\rho \delta_{\bar\rho} = 0$.

Denote by CM(E) the set of CM types on E.

Lemma 5.4 *For every $\phi \in \mathrm{CM}(E)$ define the following element of $C^+(V_{\mathbb{C}})$:*

$$\Delta_\phi := \prod_{\rho \in \mathrm{CM}(E)} \delta_\rho$$

1. *There are 2^d such elements Δ_ϕ. They all commute.*
2. *$\Delta_\phi^2 = \Delta_\phi$.*
3. *$\Delta_{\phi_1} \Delta_{\phi_2} = 0$ if $\phi_1 \neq \phi_2$.*
4. *for $\alpha \in \mathrm{Aut}(\mathbb{C}/\mathbb{Q})$, $\alpha(\Delta_\phi) = \Delta_{\alpha\phi}$.*
5. *$\sum_{\phi \in \Phi} \Delta_\phi = 1$.*

In particular the $\overline{\mathbb{Q}}$-span of the Δ_Φ's in $C^+(V) \otimes \overline{\mathbb{Q}}$ is an étale subalgebra, corresponding to the $\mathrm{Gal}(\overline{\mathbb{Q}}/\mathbb{Q})$-set $\mathrm{CM}(E)$. Taking $\mathrm{Gal}(\overline{\mathbb{Q}}/\mathbb{Q})$-invariants one realizes $E^\sharp$ as a $\mathbb{Q}$-subalgebra of $C^+(V)$.

Using (6), we also obtain a multiplicative homomorphism, called the *complete reflex norm*, that factors through T:

$$T_E \longrightarrow T_{E^\sharp} \hookrightarrow \mathrm{GSpin}(V), \qquad \sum_\rho x_\rho e_\rho \mapsto \prod_j (x_{\rho_j} \delta_{\rho_j} + x_{\bar\rho_j} \delta_{\bar\rho_j}). \qquad (10)$$

In particular the inclusion $T \subset G$ factors via $T_{E^\sharp}$. We denote by $V^\sharp$ the $\mathbb{Q}$-vector space $E^\sharp$, viewed as a $\mathbb{Q}$-representation of T. The induced map $T_E \to \mathrm{GL}(V^\sharp)$ is given by sending $\alpha \in E^*$ to the automorphism $V^\sharp \to V^\sharp$ sending Δ_Φ to $\prod_{\rho \in \Phi} \rho(\alpha) \Delta_\Phi$.

We further get a homomorphism:

$$\ell: V \to \mathrm{End}(V^\sharp) \qquad (11)$$

defined by sending $e_\rho \in V_{\mathbb{C}} = E_{\mathbb{C}}$ to the $\mathbb{C}$-linear endomorphism of $V_{\mathbb{C}}^\sharp = \oplus_{\Phi \in \mathrm{CM}(E)} \mathbb{C} \Delta_\Phi$ given by $e_\rho \mapsto \sum_{\rho \in \Phi \in \mathrm{CM}(E)} \Delta_\Phi \otimes \Delta_{\Phi \cup \{\bar\rho\} \setminus \{\rho\}}^{\vee}$.

Corollary 5.5 *Let $h: \mathbb{S} \xrightarrow{\cong} \mathrm{GSpin}(W_{\sigma_0}) \subset \mathrm{Res}_{F/\mathbb{Q}}(\mathrm{GSpin}_F(W))(\mathbb{R}) = (E \otimes_{\mathbb{Q}} \mathbb{R})^\times$ be as in (8). There is a unique way to choose h so that $g \circ h(i)$, where g is the homomorphism (7), is the element*

$$\varphi(ie_{\rho_0} - ie_{\bar\rho_0} + \sum_{\rho \notin \{\rho_0, \bar\rho_0\}} e_\rho).$$

That is,

$$g \circ h(i) = \sum_{\phi \in \Phi} \epsilon(\phi) \cdot i \cdot \Delta_\phi, \qquad \epsilon(\phi) = \begin{cases} 1 & \rho_0 \in \phi \\ -1 & \rho_0 \notin \phi. \end{cases}$$

In the following, we shall denote $g \circ h \colon \mathbb{S} \longrightarrow \mathrm{GSpin}(V_{\mathbb{R}})$ *simply by* h.

5.2   Extra Structure on the Big CM Cycle

We start with

Lemma 5.6 *The action of* $E = C^+(W)$ *on* $W = C^-(W)$ *gives an action of* E *on* V, *as* V *is equal to* W *only viewed as a rational vector space. This action will be denoted* $\beta \cdot v$ *for* $\beta \in E$ *and* $v \in V$. *On the other hand, through the homomorphisms*

$$E^\times \xrightarrow{g} \mathrm{GSpin}(V) \xrightarrow{\pi} \mathrm{SO}(V),$$

we get another action of $E^\times$ *on* V *that we call* $\rho(\beta)v$, $\beta \in E$, $v \in V$ ($\rho = \pi \circ g$). *The actions are related as follows:*

$$\rho(\beta)v = (\beta\bar{\beta}^{-1}) \cdot v.$$

In particular the homomorphism ℓ *of (11) is a homomorphism as representations of* T.

Proof The two actions of E on W are related by the formula $\rho(\beta)w = (\beta\bar{\beta}^{-1}) \cdot w$; indeed, $\rho(\beta)w = \beta w \beta^{-1} = (\beta\bar{\beta}^{-1})w = (\beta\bar{\beta}^{-1}) \cdot w$. The lemma now follows applying restriction of scalars and the commutativity of the following diagram:

$$
\begin{array}{ccccc}
C^+(W) \otimes_{\mathbb{Q}} \mathbb{R} & \xrightarrow{\cong} & \prod_j C^+(W_j) & \longrightarrow & \mathrm{GSpin}(V) \\
& & \downarrow & & \downarrow \\
& & \prod_j \mathrm{SO}(W_j) & \longrightarrow & \mathrm{SO}(V).
\end{array}
$$

$\square$

Proposition 5.7 *The endomorphism ring of* V *as a rational Hodge structure is precisely* E *with the "dot action". Moreover,* E, *viewed in* $\mathrm{End}(V)$ *via the "dot action" is the* $\mathbb{Q}$-*linear span of the image of* $T(\mathbb{Q})$.

Proof Let T_E be the $\mathbb{Q}$-torus associated to E and let $\gamma \colon T_E \longrightarrow T_E$ be the morphism given by $\alpha \mapsto \alpha/\bar{\alpha}$ on $\mathbb{C}$-points. This morphism is defined over $\mathbb{Q}$ and its image, by Lemma 5.6, is nothing but the image of $T(\mathbb{Q})$ in $GL(V)$. One can check that E is the $\mathbb{Q}$-span of the elements of $T(\mathbb{Q})$, namely $\{\alpha/\bar{\alpha} : \alpha \in E^{\times}\}$.   □

Notice that V has the extra structure of E-vector space that defines endomorphisms of $V_{\text{Betti}}|_{Y(\mathbb{C})}$. In fact, for any $z \in Y(\mathbb{C})$ the E action on $V_{\text{Betti},\mathbb{Q},z}$ induces a decomposition

$$V_{\text{Betti},\mathbb{Q},z} \otimes_{\mathbb{Q}} \mathbb{C} = V_{\text{dR},z} = \oplus_{i=0}^{d} V_{\text{dR},z}(\sigma_i) \oplus V_{\text{dR},z}(\bar{\sigma}_i),$$

where $V_{\text{dR},z}(\sigma)$ is the 1-dimensional $\mathbb{C}$-vector space on which E acts via $\sigma : E \to \mathbb{C}$. Then

$$\text{Fil}^1 V_{\text{dR},z} = V_{\text{dR},z}(\sigma_0), \, \text{Gr}^{-1} V_{\text{dR},z} = V_{\text{dR},z}(\bar{\sigma}_0)$$

and

$$\text{Gr}^0 V_{\text{dR},z} = \oplus_{i=1}^{d} V_{\text{dR},z}(\sigma_i) \oplus V_{\text{dR},z}(\bar{\sigma}_i).$$

We summarize our findings. Write $V := E$, considered as $\mathbb{Q}$-vector space. Then:

- there is a morphism $T_E \to GSpin(V)$, factoring through T, where $\alpha \in T_E(\mathbb{Q}) = E^{\times}$ acts on $V = E$ through multiplication by $\alpha\bar{\alpha}^{-1}$;
- we can realize $E^{\sharp}$ as a $\mathbb{Q}$-subalgebra of $C^+(V)$ such that the homomorphism $T \to GSpin(V)$ factors via the subtorus $T_{E^{\sharp}} \subset GSpin(V)$.
- there is a morphism $h : \mathbb{S} \to T$ inducing a Hodge structure on V where

$$\text{Fil}^1 V_{\mathbb{C}} = E \otimes_E^{\sigma_0} \mathbb{C}, \, \text{Gr}^{-1} V_{\mathbb{C}} = E \otimes_E^{\bar{\sigma}_0} \mathbb{C}, \, \text{Gr}^0 V_{\mathbb{C}} = \oplus_{i=1}^{d-1} \left(E \otimes_E^{\sigma_i} \mathbb{C} \oplus E \otimes_E^{\bar{\sigma}_i} \mathbb{C} \right)$$

and a Hodge structure of type $(-1, 0)$ and $(0, -1)$ on $V^{\sharp}$ where, writing $V^{\sharp}_{\mathbb{C}} = E^{\sharp} \otimes_{\mathbb{Q}} \mathbb{C} = \oplus_{\Phi \in CM(E)} \mathbb{C}\Delta_{\Phi}$ as a sum of idempotents,

$$\left(V^{\sharp}_{\mathbb{C}} \right)^{(-1,0)} = \oplus_{\sigma_0 \in \Phi \in CM(E)} \mathbb{C}\Delta_{\Phi}, \quad \left(V^{\sharp}_{\mathbb{C}} \right)^{(0,-1)} = \oplus_{\bar{\sigma}_0 \in \Phi \in CM(E)} \mathbb{C}\Delta_{\Phi}.$$

- there is an embedding $j : V \to C^+(V)$ such that multiplication on $C^+(V)$ induces a T-equivariant morphism $\ell : V \to End(V^{\sharp})$.

We now consider a second level structure $K_0 \subset T(\mathbb{A}_f)$ defining a CM cycle

$$Y_0(\mathbb{C}) = T(\mathbb{Q})\backslash\{z_0\} \times T(\mathbb{A}_f)/K_0.$$

We define $K_0 = \prod_p K_{0,p}$ where $K_{0,p} \subset T(\mathbb{Q}_p)$ sits in an exact sequence

$$1 \to \mathbb{Z}_p^{\times} \to K_{0,p} \to \frac{(\mathbb{Z}_p \otimes_{\mathbb{Z}} \mathcal{O}_E)^{\times}}{(\mathbb{Z}_p \otimes_{\mathbb{Z}} \mathcal{O}_F)^{\times}} \to 1$$

via the exact sequence

$$1 \to \mathbb{G}_m \to T \to T_E/T_F \to 1$$

defined by the natural projection $T = T_E/T_F^{\mathrm{Nm}=1} \to T_E/T_F$. Then Y_0 is defined over E and we let $\mathscr{Y}_0$ be the normalization of $\mathrm{Spec}(\mathscr{O}_E)$ in Y_0.

Consider the integral structures $L_0 := \mathscr{O}_E \subset E = V$ and $L_0^\sharp := \mathscr{O}_{E^\sharp} \subset E^\sharp = V^\sharp$. Then K_0 preserves their profinite completions and the Hodge structure $V^\sharp$ of type $(-1, 0)$ and $(0, -1)$ defines an abelian variety $A^\sharp$ over Y_0 as follows: given $z := (z_0, g) \in Y_0(\mathbb{C}) := T(\mathbb{Q})\backslash\{z_0\} \times T(\mathbb{A}_f)/K_0$ then $A_z^\sharp := V_\mathbb{R}^\sharp/(V^\sharp \cap g\hat{L}_0^\sharp g^{-1})$. It extends to an abelian scheme $\mathscr{A}^\sharp$ over $\mathscr{Y}_0$. We let $\mathbb{H}^\sharp$ be its first de Rham homology group. It is a filtered $\mathscr{O}_{\mathscr{Y}_0}$-module. Thanks to the choice of the compact open subgroup K_0 we have

Proposition 5.8 $\mathscr{Y}_0$ *is étale over* $\mathrm{Spec}(\mathscr{O}_E)$. *Furthermore the inclusion* $\ell \colon V \to \mathrm{End}(V^\sharp)$ *defines a strict morphism of filtered* $\mathscr{O}_{\mathscr{Y}_0}$-*modules*

$$\ell \colon \mathbb{V}_{0,\mathrm{dR}} \to \mathrm{End}(\mathbb{H}^\sharp)$$

and the $\mathscr{O}_E$-*action on* L_0 *extends to a* $\mathscr{O}_E$-*action on* $\mathbb{V}_{0,\mathrm{dR}}$.

Proof The first claim follows from Shimura's reciprocity laws. The second claim follows from Kisin's theory. See [2, Prop. 3.5.5] □

In particular we define the invertible $\mathscr{O}_{\mathscr{Y}_0}$-module $\omega_0 := \mathrm{Fil}^1\mathbb{V}_{0,\mathrm{dR}}$. We endow it with a metric at infinity induced by the standard Hermitian metric Q_0 on $V_\mathbb{C} = E \otimes_\mathbb{Q} \mathbb{C}$, $x \mapsto Q_0(x) := \mathrm{Tr}_{F/\mathbb{Q}}(x\bar{x})$.

6 Integral Models

In order to proceed with the computation of the intersections numbers we want, we need models for $M_L(\mathbb{C})$ and the CM cycle $Y_L(\mathbb{C})$ over $\mathbb{Z}$.

6.1 Integral Models of GSpin-Shimura Varieties

We know from work of Shimura and Deligne [8] that $M_L(\mathbb{C})$ is the complex analytic space associated to a quasi-projective variety M_L over a number field K called the *reflex field*. In the case of GSpin, for $n \geq 1$, the reflex field is $\mathbb{Q}$. For $n = 0$ it is the quadratic imaginary field $K = C^+(V)$.

We assume next that $n \geq 1$. Let Δ_L be the discriminant $\Delta_L := [L^\vee : L]$ where $L^\vee$ is the $\mathbb{Z}$-dual of L and the inclusion $L \subset L^\vee$ is defined using the bilinear form

[_, _]. By work of Vasiu [18] and Kisin (see in particular [11]) the scheme M_L has a canonical integral model $\mathcal{M}_{L,\mathbb{Z}\left[2^{-1}|\Delta_L|^{-1}\right]}$, smooth over $\mathbb{Z}\left[2^{-1}|\Delta_L|^{-1}\right]$. Also V_{dR} has a model $\mathbb{V}_{\mathrm{dR}}$ which is a locally free $\mathcal{O}_{\mathcal{M}_{L,\mathbb{Z}\left[2^{-1}|\Delta_L|^{-1}\right]}}$-module, endowed with a descending filtration $\mathrm{Fil}^\bullet V_{\mathrm{dR}}$ by locally free submodules, and an integrable connection satisfying Griffiths' transversality. For the purpose of computing some arithmetic intersection we wish to have a model

$$\mathcal{M} \longrightarrow \mathrm{Spec}(\mathbb{Z})$$

to which some of the extra structures described above extend as well.

If L is maximal among the quadratic lattices of V and is self dual at 2, Madapusi Pera [14] and Kim-Madapusi Pera [10] constructed such a canonical integral model which has singular fibers at the primes dividing Δ_L. Unfortunately at primes whose square divides Δ_L this model is not well behaved: for our purposes the CM cycle Y_L will have a model $\mathcal{Y}_L$, finite over $\mathrm{Spec}(\mathcal{O}_E)$ but the morphism $Y_L \to M_L$ on the generic fiber does *not* in general extend to a morphism from $\mathcal{Y}$ to the Madapusi Pera model. In [2, §4.4] we proceeded differently. Let p a prime dividing Δ_L and let $L \subset L^\diamond$ be an isometric embedding of quadratic lattices with signature $(n, 2)$ and $(n^\diamond, 2)$ respectively. We take $L^\diamond$ that is self-dual at p. By functoriality we will have a morphism of Shimura varieties $M_L \to M_{L^\diamond}$ and $M_{L^\diamond}$ will admit an integral model $\mathcal{M}_{L^\diamond,\mathbb{Z}[\Delta_{L^\diamond}^{-1}]}$ smooth over $\mathrm{Spec}(\mathbb{Z}[\Delta_{L^\diamond}^{-1}])$. We define $\mathcal{M}_{L,\mathbb{Z}[\Delta_{L^\diamond}^{-1}]}$ to be the normalization of $\mathcal{M}_{L^\diamond,\mathbb{Z}[\Delta_{L^\diamond}^{-1}]}$ in M_L. The restriction of the tautological bundle on $\mathcal{M}_{L^\diamond,\mathbb{Z}[\Delta_{L^\diamond}^{-1}]}$ defines a line bundle on $\omega_{L^\diamond}$ on $\mathcal{M}_{L,\mathbb{Z}[\Delta_{L^\diamond}^{-1}]}$. We have the following proposition proven in [2, Prop. 4.4.1].

Proposition 6.1 *The models $\mathcal{M}_{L,\mathbb{Z}[\Delta_{L^\diamond}^{-1}]}$ and the line bundles $\omega_{L^\diamond}$ glue for varying embeddings $L \subset L^\diamond$ to a normal model $\mathcal{M}_L$ over $\mathrm{Spec}(\mathbb{Z})$ and a line bundle ω that agree with the construction on $\mathcal{M}_{L,\mathbb{Z}\left[2^{-1}|\Delta_L|^{-1}\right]}$ provided by Kisin and Vasiu.*

Via the uniformization map

$$\mathcal{D} \to M_L(\mathbb{C}), \quad z \mapsto (z, g)$$

the fiber of ω at $z \in \mathcal{D}$ is the isotropic line $\mathbb{C}z = V_{\mathbb{C}}^{(1,-1)} \subset V_{\mathbb{C}}$. We then endow ω with the metric $||z||^2 := -[z, \bar{z}]$, called the Petersson metric, and we obtain a metrized line bundle in the sense of Chap. 8, Sect. 2.3,

$$\hat{\omega} \in \widehat{\mathrm{Pic}}(\mathcal{M}_L).$$

6.2 Integral Models of Big CM Cycles

Consider now the big CM cycles $\mathscr{Y}_L$ (associated to the compact open subgroup $K_T := K_L \cap K_0$) and $\mathscr{Y}_0$ constructed in the previous sections. The definition $\mathscr{M}_L$ is made so that the morphism $Y_L(\mathbb{C}) \to M_L(\mathbb{C})$ extends to a morphism $\mathscr{Y}_L \to \mathscr{M}_L$. As $K_T \subset K_0$ we get the following diagram

$$
\begin{array}{ccc}
\mathscr{Y}_L & \longrightarrow & \mathscr{M}_L \\
\downarrow & & \\
\mathscr{Y}_0. & &
\end{array}
$$

In particular we have two metrized line bundles on $\mathscr{Y}_L$: the pull back $\widehat{\omega}_{\mathscr{Y}_L}$ of the metrized line bundle $\widehat{\omega}$ over $\mathscr{M}_L$ and the pull back $\widehat{\omega}_{0,\mathscr{Y}_L}$ of the metrized line bundle $\widehat{\omega}_0$ over $\mathscr{Y}_0$.

For the first we will be able to compute the arithmetic degree using work of Bruinier, Kudla and Yang. The second is related to the Faltings' height of the abelian variety $A^{\sharp}$ which has an action of $\mathscr{O}_{E^{\sharp}}$. Let us start with this connection. Recall from Proposition 5.8 that we have a strict morphism of filtered $\mathscr{O}_{\mathscr{Y}_0}$-module

$$
\ell : \mathbb{V}_{0,\mathrm{dR}} \to \mathrm{End}(\mathbb{H}^{\sharp}),
$$

where $\mathbb{H}^{\sharp}$ is the de Rham homology of $\mathscr{A}^{\sharp}$ (the extension of $A^{\sharp}$ to an abelian scheme over $\mathscr{Y}_0$). Then $\omega_0 = \mathrm{Fil}^1 \mathbb{V}_{0,\mathrm{dR}}$ maps to the endomorphisms of $\mathbb{H}^{\sharp}$ sending $\mathrm{Gr}^{-1}\mathbb{H}^{\sharp} \to \mathrm{Fil}^0\mathbb{H}^{\sharp}$, i.e., upon taking determinants we have a map of $\mathscr{O}_{\mathscr{Y}_0}$-modules

$$
\omega_0^{2^{d-1}} \otimes_{\mathscr{O}_{\mathscr{Y}_0}} \det \mathrm{Gr}^{-1}\mathbb{H}^{\sharp} \longrightarrow \det \mathrm{Fil}^0\mathbb{H}^{\sharp}.
$$

By definition of the Hodge filtration of $\mathbb{H}^{\sharp}$ we have $\left(\det \mathrm{Gr}^{-1}\mathbb{H}^{\sharp}\right)^{-1} = \omega^{\sharp}$, the Hodge bundle of $\mathscr{A}^{\sharp}$. With this notation we have the following result proven in [2, Thm 9.4.2].

Theorem 6.2 *The following holds*

$$
\frac{1}{2^d} \sum_{\Phi} h_{(E,\Phi)}^{\mathrm{Falt}} = \frac{1}{4} \frac{\widehat{\deg}(\widehat{\omega}_0)}{\deg_{\mathbb{C}}(Y_0)} + \frac{1}{4} \log |D_F| - \frac{1}{2}d \cdot \log(2\pi).
$$

Proof We first use (1) to relate $\sum_{\Phi} h_{(E,\Phi)}^{\mathrm{Falt}}$ with $h_{(E^{\sharp},\Phi^{\sharp})}^{\mathrm{Falt}}$ via the formula $2d h_{(E^{\sharp},\Phi^{\sharp})}^{\mathrm{Falt}} = \sum_{\Phi} h_{(E,\Phi)}^{\mathrm{Falt}}$. Second, by definition the degree of the metrized Hodge bundle $\widehat{\Omega}^{\sharp}$ of $\mathscr{A}^{\sharp}$ satisfies

$$
\frac{\deg \widehat{\Omega}^{\sharp}}{\deg Y_0} = 2d h_{(E^{\sharp},\Phi^{\sharp})}^{\mathrm{Falt}}.
$$

Finally we study the inclusion of invertible sheaves:

$$\omega_0^{2^{d-1}} \subset \left(\det \mathrm{Gr}^{-1}\mathbb{H}^\sharp\right)^{-2} \otimes_{\mathscr{O}_{\mathscr{Y}_0}} \det\mathbb{H}^\sharp \cong (\omega^\sharp)^2 \otimes_{\mathscr{O}_{\mathscr{Y}_0}} \det\mathbb{H}^\sharp.$$

One proves that the metrics on the two sides coincide; here $\det\mathbb{H}^\sharp$ is endowed with the standard metric given by integration over $A^\sharp(\mathbb{C})$ of top degree C^∞ de Rham classes. Its arithmetic degree is computed in [2, Lemma 9.4.3] and coincides with

$$\widehat{\deg\det}\mathbb{H}^\sharp = 2^d d \cdot \deg Y_0 \cdot \log(2\pi).$$

We are left to study the cokernel of the displayed inclusion. A delicate algebra computation, see [2, Prop. 9.4.1], shows that the difference of these line bundles has degree $2^{d-1}\deg Y_0 \log|D_F|$ and the claim follows.   □

Next we compare the metrized line bundles $\widehat{\omega}_{\mathscr{Y}_L}$ and $\widehat{\omega}_{0,\mathscr{Y}_L}$.

Proposition 6.3 *We have* $\deg(\widehat{\omega}_{\mathscr{Y}_L}) \sim_L \deg(\widehat{\omega}_{0,\mathscr{Y}_L}) + \log|D_F|$ *where* $\sim_L$ *means equal up to rational linear combinations of log of primes dividing* $D_{L,\mathrm{bad}}$; *see Definition 5.1.*

Proof Over $\mathbb{Q}$ the two sheaves coincide as they are associated to the same Hodge structure, namely V. Via this identification the metric on ω is the metric on ω_0 times $\sigma_0(\lambda)$ (recall that the quadratic form on V is defined by $x \mapsto \mathrm{Tr}_{F/\mathbb{Q}}(\lambda x \bar{x})$). We have only to check that over $\mathscr{Y}_L[D_{L,\mathrm{bad}}^{-1}]$ we have $\omega = \lambda\mathfrak{D}_{F/\mathbb{Q}}^{-1} \otimes_{\mathscr{O}_F} \omega_0$. Using Kisin's correspondence one is reduced to prove this statement for the lattices L and L_0 of V and the claim follows by local calculations (see the proof of [2, Prop. 9.5.1] using Definition 5.1).   □

7 The Bruinier, Kudla, Yang Conjecture

Thanks to Theorem 6.2 and Proposition 6.3 and the $\mathbb{Q}$-linear independence of log of primes, in order to conclude the proof of the averaged version of Colmez's conjecture we need to prove the following

Theorem 7.1 *The degree of* $\widehat{\omega}_{\mathscr{Y}_L}$ *satisfies*

$$\frac{\deg\widehat{\omega}_{\mathscr{Y}_L}}{\deg Y_L} \sim_L -\frac{2\Lambda'(\chi,0)}{\Lambda(\chi,0)} - d\log(4\pi e^\gamma).$$

Moreover, $\cap_L D_{L,\mathrm{bad}} = \emptyset$.

The assertion that $\cap_L D_{L,\mathrm{bad}} = \emptyset$ can be turned via (5) into a class field theory question that we do no discuss here; we refer to [2, Prop. 9.5.2] for details.

The main idea is then to realize $\widehat{\omega}$ as a combination of arithmetic divisors using Borcherds theory and results of Bruinier, Kudla, Yang that compute the contribution at infinity of the intersection of these arithmetic divisors with $\mathscr{Y}_L$. Bruinier, Kudla, Yang provided also a conjecture for the contribution at finite places that we verify in [2] in sufficiently many cases to get the result.

7.1 Special Divisors

In this section we will show how, given an element $\lambda \in V$ with $Q(\lambda) > 0$, we can construct a divisor in $M_L(\mathbb{C})$ as a Shimura subvariety. These will give the special divisors mentioned in the introduction. The fact that we have such a large supply of easily constructed divisors, and in general of cycles of higher codimension obtained by intersecting such divisors, makes the theory of GSpin-Shimura varieties extremely rich.

Given λ as above, set $V_\lambda := \lambda^\perp \subset V$. This is a dimension $(n-1)+2$ subspace of V and $Q_\lambda := Q|_{V_\lambda}$ is a quadratic form of signature $(n-1, 2)$. Then we get a subgroup

$$G_\lambda = \mathrm{GSpin}(V_\lambda, Q_\lambda) \subset \mathrm{GSpin}(V) = G.$$

The symmetric space D_λ for G_λ is identified with

$$D_\lambda = \left\{ [z] \in D_{\mathbb{C}} \subset V_{\mathbb{C}} \backslash \{0\} : z \in V_{\lambda, \mathbb{C}} = \lambda^\perp \right\} / \mathbb{C}^*.$$

Let $L_\lambda := L \cap V_\lambda$ so we have $K_\lambda \subset G_\lambda(\mathbb{A}_f)$ and we get a GSpin-Shimura variety

$$M_\lambda(\mathbb{C}) = G_\lambda(\mathbb{Q}) \backslash D_\lambda \times G_\lambda(\mathbb{A}_f) / K_\lambda$$

together with a finite analytic map

$$M_\lambda(\mathbb{C}) \to M_L(\mathbb{C}) = G(\mathbb{Q}) \backslash D \times G(\mathbb{A}_f) / K_L.$$

Notice that such map is in general not an injection but the image of this map consists of divisors of $M_L(\mathbb{C})$. The discrepancy between $M_\lambda(\mathbb{C})$ and its image in $M_L(\mathbb{C})$ makes the intersection theory of special divisors more involved. We will ignore this issue here for sake of simplicity and pretend that we can identify $M_\lambda(\mathbb{C})$ and its image. We refer to [2, Def. 4.5.6] for the correct treatment using stacks. Also, in order to have a good theory of the integral models of such divisors, following Borcherds we group together all the $M_\lambda(\mathbb{C})$'s considering the λ's with equal norm $Q(\lambda) = m$ and equal class $\mu \in L^\vee / L$:

Definition 7.2 For any $m \in \mathbb{N}_{>0}$ and every $\mu \in L^\vee / L$, let

$$Z(m, \mu)(\mathbb{C}) := \amalg_{g \in G(\mathbb{Q}) \backslash G(\mathbb{A}_f) / K_L} \Gamma_g \backslash \left(\amalg_{\substack{\lambda \in \mu_g + L_g \\ Q(\lambda) = m}} D_\lambda \right).$$

where $\Gamma_g = G(\mathbb{Q}) \cap g K_L g^{-1}$, $L_g \subset V$ is the lattice $V \cap (g\widehat{L}g^{-1})$, and $\mu_g \in L_g^\vee/L_g$ is the class of $g\mu g^{-1}$.

Recall that $M_L(\mathbb{C}) = \amalg_{g \in G(\mathbb{Q})\backslash G(\mathbb{A}_f)/K_L} \Gamma_g \backslash D_\mathbb{C}$ so that we have a natural morphism

$$Z(m, \mu)(\mathbb{C}) \to M_L(\mathbb{C})$$

whose image is the union of the images of various $M_\lambda(\mathbb{C})$.

The image of $Z(m, \mu)(\mathbb{C})$ singles out points z of $M_L(\mathbb{C})$ where the $\mathbb{Z}$-Hodge structure $\mathbb{V}^\vee_{\mathrm{Betti},z}$ acquires a Hodge $(0,0)$-class λ_z of norm $Q_{\mathrm{Betti}}(\lambda_z) = m$ and whose class in $\mathbb{V}^\vee_{\mathrm{Betti},z}/\mathbb{V}_{\mathrm{Betti},z}$, which can be canonically identified with $L^\vee/L$, is $\mu \in L^\vee/L$.

We can give an intrinsic characterization of the image of $Z(m, \mu)(\mathbb{C})$ in $M_L(\mathbb{C})$ as follows; take $(z, g) \in M_L(\mathbb{C})$ and consider the corresponding element $\ell_{\mathrm{Betti}}(\lambda) \in \mathrm{End}(H_{1,\mathrm{Betti}}(A_{(z,g)}))$ (using (3)). Then,

Proposition 7.3 *The point (z, g) is in the image of $\mathscr{D}_\lambda$ if and only if $\ell_{\mathrm{Betti}}(\lambda)$ arises as the Betti realization of a rational endomorphism $\ell_\lambda \in \mathrm{End}(A_{(z,g)})$.*

Proof The point (z, g) lies in D_λ if and only if

$$\lambda \in z^\perp = \mathrm{Fil}^0 \mathbb{V}_{\mathrm{dR},z} \subset \mathbb{V}_{\mathrm{dR},z}$$

if and only if the element $\ell_{\mathrm{dR}}(\lambda) \in \mathrm{End}(H_{1,\mathrm{dR}}(A_{(z,g)}))$ of (4) lies in $\mathrm{Fil}^0\mathrm{End}(H_{1,\mathrm{dR}}(A_{(z,g)}))$, i.e., $\ell_{\mathrm{dR}}(\lambda)$ preserves the Hodge filtration of $H_{1,\mathrm{dR}}(A_{(z,g)})$. This is equivalent to require that the element $\ell_{\mathrm{Betti}}(\lambda)$ defines an endomorphism of $A_{(z,g)}$. $\square$

As before we take $\lambda \in V$ be an element with $Q(\lambda) > 0$. The next lemma shows that the images of $Y_L(\mathbb{C})$ and $M_\lambda(\mathbb{C})$ in $M_L(\mathbb{C})$ do not intersect. This will imply that for the associated arithmetic objects, i.e., the associated objects over $\mathbb{Z}$, we have *proper intersection*. This is not at all the case for the small CM points of Remark 5.2 where we have *improper intersection*; see [1] for a discussion.

Lemma 7.4 *The intersection of the images $Y_L(\mathbb{C})$ and $M_\lambda(\mathbb{C})$ in $M_L(\mathbb{C})$ is empty.*

Proof Assume that we have an element $z \in D_\lambda$ whose image on $(z, g) \in M_L(\mathbb{C})$ lies in the image of $Y_L(\mathbb{C})$. This is equivalent to saying that $\lambda \in V_{\mathrm{Betti},\mathbb{Q},z}$ is such that $\lambda \in (\mathbb{C}z \oplus \mathbb{C}\bar{z})^\perp$. But $V_{\mathrm{Betti},\mathbb{Q},z} = V = E$ as a $\mathbb{Q}$-vector space and we get that

$$V = E \cdot \lambda \subset (\mathbb{C}z \oplus \mathbb{C}\bar{z})^\perp$$

as the pairing on V is E-hermitian. This is clearly a contradiction as $(\mathbb{C}z \oplus \mathbb{C}\bar{z})^\perp \oplus (\mathbb{C}z \oplus \mathbb{C}\bar{z}) = V_\mathbb{C}$. $\square$

7.2   Integral Models of Special Divisors

We need to define models of the special divisors $Z(m, \mu)$ over $\mathbb{Q}$ and even integral models $\mathscr{Z}(m, \mu)$ over $\mathbb{Z}$ in order to compute arithmetic intersection theory. Clearly we can not use complex uniformization as we did in the previous section. Proposition 7.3 leads us to an intrinsic definition that makes perfect sense over $\mathbb{Z}$ and provides us with the definition we want. We set $\mathscr{Z}(m, \mu)$ to be the scheme representing the functor of pairs (ρ, f) where $\rho : S \to \mathscr{M}_L$ and $f \in \operatorname{End}^0(\mathscr{A} \times_{\mathscr{M}_L} S)$ is a rational endomorphism of the Kuga–Satake abelian scheme $\mathscr{A}_S := \mathscr{A} \times_{\mathscr{M}_L} S$ such that

i. $f \circ f = [m]$ (the multiplication by m map on $\mathscr{A}_S$);
ii. the endomorphism defined by f on the de Rham homology of $\mathscr{A}_S$, on the Tate module of $\mathscr{A}_S$ and on the crystal defined by $\mathscr{A}_S$ is in the image of the de Rham realization, étale or crystalline realization of $L^\vee$ of class $\mu \in L^\vee / L$.

We explain what me mean in (ii). Over $\mathbb{Z}[\Delta_L^{-1}]$ the lattice $L \subset V$ defines a *motive* $\mathbb{V}$ over $\mathscr{M}_L$, namely a $\mathbb{Z}$-variation of Hodge structures over $M_L(\mathbb{C})$, a filtered vector bundle with connection over $\mathscr{M}_L$, a lisse ℓ-adic étale sheaf over $\mathscr{M}_L[\ell^{-1}]$, a crystal over $\mathscr{M}_L \otimes \mathbb{F}_p$. The map ℓ defines an embedding of motives $\mathbb{V} \subset \operatorname{End}(\mathbb{H})$ where $\mathbb{H}$ is the motive associated to the abelian scheme $\mathscr{A}$: its de Rham homology, its ℓ-adic Tate module, its Dieudonné module. In (ii) we ask that the realization of f lies in the image of $\mathbb{V}$.

To extend this notion to the whole of $\mathbb{Z}$ one works with auxiliary lattices $L \subset L^\diamond$, that are self dual at a given prime p and demands this condition working on $\mathscr{M}_{L^\diamond}$. A result analogous to Proposition 6.1 implies that we get a well-posed definition, independent of the auxiliary choice of $L^\diamond$.

It is proven in [2, Prop. 4.5.8] that the models $\mathscr{Z}(m, \mu)$ have good properties, namely they do not have vertical components that would create troubles in computing arithmetic intersections:

Proposition 7.5 *If V has dimension ≥ 5 then the $\mathscr{Z}(m, \mu)$'s are flat over $\mathbb{Z}[1/2]$ and even over $\mathbb{Z}$ if L is self dual at 2.*

7.3   Special Divisors and $\widehat{\omega}$

We now come to the main result expressing $\widehat{\omega}$ as a combination of arithmetic special divisors. Thanks to Borcherds' theory in fact the special divisors $\mathscr{Z}(m, \mu)$ are endowed with natural Green functions $\Phi_{m,\mu}$ (see [4, Thm. 4.2] using [3]). And we can consider the pair $\widehat{\mathscr{Z}}(m, \mu) := (\mathscr{Z}(m, \mu), \Phi_{m,\mu}) \in \widehat{CH}^1(\mathscr{M}_L)$. In the following we will freely use the equivalence $\widehat{\operatorname{Pic}}(\mathscr{M}_L) \cong \widehat{CH}^1(\mathscr{M}_L)$, between metrized line bundles and arithmetic divisors (modulo principal arithmetic divisors). We have the following fundamental result, see [2, Thm. 4.8.1]:

Theorem 7.6 *Suppose that $n \geq 3$. There are finitely many integers $c(-m, \mu)$ for $m \geq 0$ and $\mu \in L^{\vee}/L$ with $c(0,0) \neq 0$ and there exists a rational section Ψ of $\omega^{\otimes c(0,0)}$, defined over $\mathbb{Q}$, such that*

$$\widehat{\omega}^{\otimes c(0,0)} = \widehat{\mathrm{div}}(\Psi)$$

$$= \sum_{m,\mu} c(-m, \mu) \widehat{\mathcal{Z}}(m, \mu) - c_f(0,0) \cdot \left(0, \log(4\pi e^{\gamma})\right) + \widehat{\mathcal{E}},$$

where $(0, \log(4\pi e^{\gamma}))$ denotes the trivial divisor endowed with the constant Green function $\log(4\pi e^{\gamma})$ and $\widehat{\mathcal{E}} = (\mathcal{E}, 0)$ is a vertical divisor with the trivial Green function that has a decomposition

$$\mathcal{E} = \sum_{p \mid D_L} \mathcal{E}_p,$$

such that $\mathcal{E}_p$ is supported on the special fiber $\mathcal{M}_{L, \mathbb{F}_p}$, and:

- *If p is odd and $p^2 \nmid D_L$ then $\mathcal{E}_p = 0$;*
- *If $n \geq 5$ then $\mathcal{E} = \mathcal{E}_2$ is supported on $\mathcal{M}_{L, \mathbb{F}_2}$.*
- *If $n \geq 5$ and $L_{(2)}$ is self-dual, then $\mathcal{E} = 0$.*

We apply this result first to the lattice L, coming from the CM field E, and then also for some auxiliary lattice $L \subset L^{\diamond}$, self dual at a given bad prime p. In this case some care is needed to assure that the special divisors on $\mathcal{M}_{L^{\diamond}}$ do *not* contain the big CM cycle, i.e., that we have proper intersection. In particular $\mathcal{E}_p = 0$ and we will compute the contribution at p of $\deg(\widehat{\omega}_{\mathcal{Y}_L})$ using the expression of $\widehat{\omega}$ in terms of the divisors $\widehat{\mathcal{Z}}(m, \mu)$ provided by the theorem above.

7.4 Arithmetic Intersection and Special Values

Consider the divisor $\widehat{\mathcal{Z}} := \sum_{m,\mu} c(-m, \mu) \widehat{\mathcal{Z}}(m, \mu)$ of Theorem 7.6. Write $[\mathcal{Y}_L : \widehat{\mathcal{Z}}]$ for the arithmetic degree of the base change of $\widehat{\mathcal{Z}}$ to $\mathcal{Y}_L$. The main result that finishes the proof of the averaged version of Colmez's conjecture is the following:

Theorem 7.7 *We have*

$$\frac{[\mathcal{Y}_L : \widehat{\mathcal{Z}}]}{\deg(Y)} \sim_L -2 \frac{\Lambda'(0, \chi) \cdot c(0,0)}{\Lambda(0, \chi)}$$

(recall that $\sim_L$ means equality up to a $\mathbb{Q}$-linear combinations of $\{\log(p) : p \mid D_{L,\mathrm{bad}}\}$).

The first ingredient for the proof is the computation of the contribution at infinity of the degree, provided by the following theorem which is the main result of [5].

Thanks to Lemma 7.4 this is computed as the value of the Green function for the arithmetic divisor $\widehat{\mathscr{X}}$ at the $\mathbb{C}$-points of $\mathscr{Y}_L$:

Theorem 7.8 *Let Φ be the Green function associated to the arithmetic divisor $\widehat{\mathscr{X}}$ of the previous theorem.*

$$\frac{\Phi(\mathscr{Y}_L^\infty)}{2\deg(Y_L)} = \sum_{\substack{\mu \in L^\vee/L \\ m \geq 0}} \frac{a(m,\mu) \cdot c(-m,\mu)}{\Lambda(0,\chi)},$$

where $\mathscr{Y}^\infty = Y_L \times_{\mathbb{Q}} \mathbb{C}$ and $\Phi(\mathscr{Y}^\infty)$ is the weighted sum of the values of Φ

$$\Phi(\mathscr{Y}^\infty) = \sum_{y \in \mathscr{Y}^\infty(\mathbb{C})} \frac{\Phi(y)}{|\mathrm{Aut}(y)|}.$$

Here the $a(m,\mu)/\Lambda(0,\chi)$'s are the coefficients of the formal q-expansion of the restriction of the derivative of a suitable weight 1 Hilbert modular Eisenstein series introduced by Kudla [12]. In order to deduce Theorem 7.7 from the result of [5] above one needs to:

1. compute the coefficients $a(m,\mu)$. Typically they are expressed as orbital integrals and one wants to get explicit quantities;
2. prove that the *finite* part of the intersection is $\mathscr{X}(m,\mu)$ along $\mathscr{Y}_L$ is $a(m,\mu)/\Lambda(0,\chi)$.

Both calculations need to be done up to $\mathbb{Q}$-linear combinations of $\log(p)$'s for $p \mid D_{L,\mathrm{bad}}$. Computation (1) is due to Kudla and Yang [13] (and [2, §6] for the contribution at the prime $p = 2$). Claim (2) is proven in §7 of [2]. It is what Bruinier, Kudla and Yang conjecture to be true in [5], even in a more general setting than what is needed and proven in Theorem 7.7. It is a remarkable instance of Kudla's program relating generating series of (arithmetic) intersection numbers and automorphic forms.

References

1. F. Andreatta, E. Goren, B. Howard, K. Madapusi Pera, Height pairings on orthogonal Shimura varieties. Compos. Math. **153**, 474–534 (2017)
2. F. Andreatta, E. Goren, B. Howard, K. Madapusi Pera, Faltings' heights of abelian varieties with complex multiplication. Ann. Math. **187**, 391–531 (2018)
3. J.H. Bruinier, J. Funke, On two geometric theta lifts. Duke Math. J. **125**, 45–90 (2004)
4. J.H. Bruinier, T. Yang, Faltings heights of CM cycles and derivatives of L-functions. Invent. Math. **177**, 631–681 (2009)
5. J.H. Bruinier, S.S. Kudla, T. Yang, Special values of Green functions at big CM points. Int. Math. Res. Not. **9**, 1917–1967 (2012)

6. P. Colmez, Périodes des variétés abéliennes à multiplication complexe. Ann. Math. **138**, 625–683 (1993)
7. C.-L. Chai, B. Conrad, F. Oort, *Complex Multiplication and Lifting Problems.* Mathematical Surveys and Monographs, vol. 195 (American Mathematical Society, Providence, 2014), x+387pp. ISBN: 978-1-4704-1014-8
8. P. Deligne, Varietés de Shimura: interprétation modulaire, et techniques de construction de modèles canoniques, in *Automorphic Forms, Representations and L Functions (Proceedings of Symposia in Pure Mathematics, Oregon State University, Corvallis, Oregon, 1977)*, Part 2, vol. XXXIII, pp. 247–289 (American Mathematical Society, Providence, 1977)
9. A. Fiori, Characterization of special points of orthogonal symmetric spaces. J. Algebra **372**, 397–419 (2012)
10. W. Kim, K. Madapusi Pera, 2-Adic integral canonical models. Forum Math. Sigma **4**, e28 (2016). https://doi.org/10.1017/fms.2016.23
11. M. Kisin, Integral models for Shimura varieties of abelian type. J. Am. Math. Soc. 23, 967–1012 (2010)
12. S.S. Kudla, Central derivatives of Eisenstein series and height pairings. Ann. Math. **146**, 545–646 (1997)
13. S.S. Kudla, T. Yang, Eisenstein series for SL(2). Sci. China Math. **53**, 2275–2316 (2010)
14. K. Madapusi Pera, Integral canonical models for Spin Shimura varieties. Compos. Math. **152**, 769–824 (2016)
15. A. Obus, On Colmez's product formula for periods of CM-abelian varieties. Math. Ann. **356**, 401–418 (2013)
16. J.-P. Serre, J. Tate, Good reduction of abelian varieties. Ann. Math. **88**, 492–517 (1968)
17. J. Tsimerman, A proof of the André-Oort conjecture for A_g. Ann. Math. **187**, 379–390 (2018)
18. A. Vasiu, Good reductions of Shimura varieties of Hodge type in arbitrary unramified mixed characteristic I, II (2007). arXiv:0707.1668, arXiv:0712.1572
19. T. Yang, Arithmetic intersection on a Hilbert modular surface and the Faltings height. Asian J. Math. **17**, 335–381 (2013)
20. X. Yuan, S.-W. Zhang, On the averaged Colmez conjecture. Ann. Math. **187**, 533–638 (2018)

Glossary

Index

LECTURE NOTES IN MATHEMATICS

Editors in Chief: J.-M. Morel, B. Teissier;

Editorial Policy

1. Lecture Notes aim to report new developments in all areas of mathematics and their applications – quickly, informally and at a high level. Mathematical texts analysing new developments in modelling and numerical simulation are welcome.

   Manuscripts should be reasonably self-contained and rounded off. Thus they may, and often will, present not only results of the author but also related work by other people. They may be based on specialised lecture courses. Furthermore, the manuscripts should provide sufficient motivation, examples and applications. This clearly distinguishes Lecture Notes from journal articles or technical reports which normally are very concise. Articles intended for a journal but too long to be accepted by most journals, usually do not have this "lecture notes" character. For similar reasons it is unusual for doctoral theses to be accepted for the Lecture Notes series, though habilitation theses may be appropriate.

2. Besides monographs, multi-author manuscripts resulting from SUMMER SCHOOLS or similar INTENSIVE COURSES are welcome, provided their objective was held to present an active mathematical topic to an audience at the beginning or intermediate graduate level (a list of participants should be provided).

   The resulting manuscript should not be just a collection of course notes, but should require advance planning and coordination among the main lecturers. The subject matter should dictate the structure of the book. This structure should be motivated and explained in a scientific introduction, and the notation, references, index and formulation of results should be, if possible, unified by the editors. Each contribution should have an abstract and an introduction referring to the other contributions. In other words, more preparatory work must go into a multi-authored volume than simply assembling a disparate collection of papers, communicated at the event.

3. Manuscripts should be submitted either online at www.editorialmanager.com/lnm to Springer's mathematics editorial in Heidelberg, or electronically to one of the series editors. Authors should be aware that incomplete or insufficiently close-to-final manuscripts almost always result in longer refereeing times and nevertheless unclear referees' recommendations, making further refereeing of a final draft necessary. The strict minimum amount of material that will be considered should include a detailed outline describing the planned contents of each chapter, a bibliography and several sample chapters. Parallel submission of a manuscript to another publisher while under consideration for LNM is not acceptable and can lead to rejection.

4. In general, **monographs** will be sent out to at least 2 external referees for evaluation.

   A final decision to publish can be made only on the basis of the complete manuscript, however a refereeing process leading to a preliminary decision can be based on a pre-final or incomplete manuscript.

   Volume Editors of **multi-author works** are expected to arrange for the refereeing, to the usual scientific standards, of the individual contributions. If the resulting reports can be forwarded to the LNM Editorial Board, this is very helpful. If no reports are forwarded or if other questions remain unclear in respect of homogeneity etc, the series editors may wish to consult external referees for an overall evaluation of the volume.

5. Manuscripts should in general be submitted in English. Final manuscripts should contain at least 100 pages of mathematical text and should always include

   – a table of contents;
   – an informative introduction, with adequate motivation and perhaps some historical remarks: it should be accessible to a reader not intimately familiar with the topic treated;
   – a subject index: as a rule this is genuinely helpful for the reader.
   – For evaluation purposes, manuscripts should be submitted as pdf files.

6. Careful preparation of the manuscripts will help keep production time short besides ensuring satisfactory appearance of the finished book in print and online. After acceptance of the manuscript authors will be asked to prepare the final LaTeX source files (see LaTeX templates online: https://www.springer.com/gb/authors-editors/book-authors-editors/manuscriptpreparation/5636) plus the corresponding pdf- or zipped ps-file. The LaTeX source files are essential for producing the full-text online version of the book, see http://link.springer.com/bookseries/304 for the existing online volumes of LNM). The technical production of a Lecture Notes volume takes approximately 12 weeks. Additional instructions, if necessary, are available on request from lnm@springer.com.

7. Authors receive a total of 30 free copies of their volume and free access to their book on SpringerLink, but no royalties. They are entitled to a discount of 33.3 % on the price of Springer books purchased for their personal use, if ordering directly from Springer.

8. Commitment to publish is made by a *Publishing Agreement*; contributing authors of multiauthor books are requested to sign a *Consent to Publish form*. Springer-Verlag registers the copyright for each volume. Authors are free to reuse material contained in their LNM volumes in later publications: a brief written (or e-mail) request for formal permission is sufficient.

Addresses:
Professor Jean-Michel Morel, CMLA, École Normale Supérieure de Cachan, France
E-mail: moreljeanmichel@gmail.com

Professor Bernard Teissier, Equipe Géométrie et Dynamique,
Institut de Mathématiques de Jussieu – Paris Rive Gauche, Paris, France
E-mail: bernard.teissier@imj-prg.fr

Springer: Ute McCrory, Mathematics, Heidelberg, Germany,
E-mail: lnm@springer.com

Printed in the United States
by Baker & Taylor Publisher Services